Advances in the Chemistry and Physics of Materials

Overview of Selected Topics

Advances in the Chemistry and Physics of Materials

Overview of Selected Topics

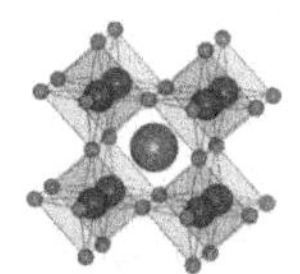

Editors

Subi J George

Chandrabhas Narayana

C N R Rao

Jawaharlal Nehru Centre for Advanced Scientific Research (JNCASR), India

NEW JERSEY · LONDON · SINGAPORE · BEIJING · SHANGHAI · HONG KONG · TAIPEI · CHENNAI · TOKYO

Published by

World Scientific Publishing Co. Pte. Ltd.

5 Toh Tuck Link, Singapore 596224

USA office: 27 Warren Street, Suite 401-402, Hackensack, NJ 07601

UK office: 57 Shelton Street, Covent Garden, London WC2H 9HE

Library of Congress Control Number: 2019039115

British Library Cataloguing-in-Publication Data
A catalogue record for this book is available from the British Library.

ADVANCES IN THE CHEMISTRY AND PHYSICS OF MATERIALS
Overview of Selected Topics

ISBN 978-981-121-132-4

For any available supplementary material, please visit
https://www.worldscientific.com/worldscibooks/10.1142/11580#t=suppl

Preface

Materials Science has emerged as an important cross-disciplinary area of research in the last three to four decades. Chemists and physicists have contributed to this area in a big way and have created many new directions. The subject is taught in most educational institutions in the world at both undergraduate and post-graduate levels. In view of the great importance of materials science and the ever-expanding scope of the subject, we considered it important to bring out a book on the *Advances in the Chemistry and Physics of Materials*. Such a book would be useful to students and teachers as well as for practitioners of the subject. Such a book is possible with the help of experts in the School of Advanced Materials at the Jawaharlal Nehru Centre for Advanced Scientific Research. The School has faculty with expertise in most areas of chemistry and physics of materials and they have helped us in bringing out this volume.

The topics in materials research that we have covered in the book include synthesis, phenomena, properties, various types of materials, specially nanomaterials, and aspects of energy. Besides experimental findings, we have included theoretical efforts in materials research. We do hope that the book will be found useful by all concerned.

Editors

Contents

C. Energy Materials

D. Theoretical and Computational Materials Science

Chapter 1

Nanotubes

Leela Srinivas Panchakarla[a] and C. N. R. Rao[b,*]

[a]*Department of Chemistry, Indian Institute of Technology Bombay, Powai, Mumbai 400076, Maharashtra, India.*
[b]*Chemistry and Physics of Materials Unit, International Centre for Materials Science, Sheikh Saqr Laboratory and School of Advanced Materials, Jawaharlal Nehru Centre for Advanced Scientific Research, Jakkur, Bangalore 560064, Karnataka, India*
[*]*cnrrao@jncasr.ac.in*

Nanotubes are unique quasi-one-dimensional nanomaterials. After carbon nanotubes (CNTs), there are different layered and non-layered materials, which are made into nanotubular structures. Nanotubes show significant electrical, mechanical and thermal properties. A variety of high- and low-temperature strategies are employed for the synthesis of carbon nanotubes. Single-walled CNTs are either metallic or semiconducting depending on the rolling. The separation of these two types in the post-synthesis mixture is an important issue for using them in further applications. Different techniques have been utilised to separate them. Doping and surface functionalization of carbon nanotubes have been developed to fine-tune the physical and chemical properties. Synthesis and characterization of inorganic nanotubes are developed in parallel with CNTs. Different electrical, mechanical properties of CNTs and inorganic nanotubes are measured on individual nanotube levels. Synthesis of nanotubes from complex misfit-layered compounds has also been achieved recently. This chapter comprises an up-to-date overview of carbon and inorganic nanotubes covering different significant aspects.

    L.S. Panchakarla & C.N.R. Rao

1. Introduction

Inherent crystal asymmetry was considered the only way to roll layered structures until fullerenes were made for the first time.[1] It was realized later that any layered material can be rolled into nanotubular structures under appropriate conditions.[2-4] Free energy for catalytically-driven growth of nanotubes is minimum when the low surface free energy (002) basal plane is exposed outside and the high energy dangling bonds are sealed.[5]

Nanotubes are one-dimensional nanostructures. Both carbon and inorganic materials are known to form nanotubular structures.[3,5,6] Naturally occurring nanotubular structures based on asbestos minerals like kaolinite have been known for a long time. It is the discovery of carbon nanotubes that brought real attention to this field as the nanotubes show extraordinary electrical, thermal and mechanical properties. Carbon nanotubes (CNTs) are crystalline allotropes of carbon and can be considered as the rolled forms of graphene sheets with the ends sealed.[3] The sp^2 bonded graphene sheets can be rolled in different ways to form zigzag, arm-chair or chiral nanotubes depending on the orientation of the tube axis with respect to the hexagonal lattice (see Figure 1). Properties of single-walled carbon nanotubes are dependent on the diameter and chirality defined by the chiral vector (how the nanotube is rolled), $C_h = a_1 n + a_2 m$, where a_1, a_2 are unit vectors and n,m are chiral indices.[7-11] The type of carbon nanotube is defined by the n,m values such as zigzag (one of the indices is zero, the chiral angle is $0°$), armchair ($n=m$ and chiral angle is $30°$) or chiral (other than zigzag and armchair). Typical transmission electron microscope (TEM) images of single-walled (SWNTs), double-walled (DWNTs) or multi-walled (MWNTs) carbon nanotubes are shown in Figure 2.

Soon after the discovery of carbon nanotubes, inorganic nanotubes formed by WS_2 and MoS_2 were realized in the laboratory.[2,12] Now there are so many materials with layered structures which form nanotubular structures.[5] Nanotubes from misfit layered compounds (MLCs) are the recent additions to the nanotube family.[13,14]

In this chapter, we discuss the different synthesis and characterization methods of carbon and inorganic nanotubes. As carbon nanotube properties change with chirality, we have outlined important synthetic advances to selectively make one type of nanotube over the other. Some of the important properties of both carbon nanotubes and inorganic nanotubes will be presented. We have also described different synthesis and characterization methods of MLCs in this chapter.

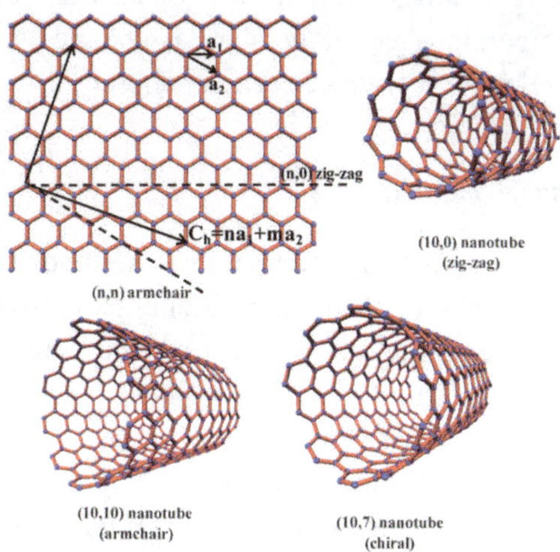

Fig. 1. Relationship between graphene and SWNTs is schematically shown. Unit vectors (a_1 and a_2) and chiral vector C_h are shown. Schematic of zigzag, armchair and chiral nanotubes is given. Adapted with permission from Ref. 15.

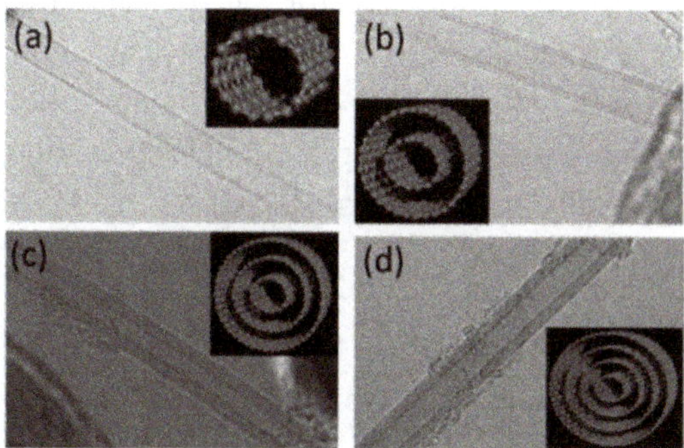

Fig. 2. Typical TEM images of single-walled (a) double-walled (b) triple-walled (c) and multi-walled (d) carbon nanotubes. Insets in all figures show corresponding schematic nanotubes. Adapted with permission from Ref. 16.

2. Carbon nanotubes

2.1. *Synthesis*

We shall first discuss carbon nanotubes. As the bending of material demands high energy, most of the synthesis techniques developed earlier are high energy techniques such as arc discharge, laser ablation. A relatively low-temperature technique such as chemical vapor deposition has also been developed to produce carbon nanotubes.

2.1.1. *Arc discharge method*

Arc discharge is one of the high-temperature methods to produce high-quality carbon nanotubes.[3,17] DC arc discharge between two closely brought (1-2 mm) graphite electrodes produces carbon nanotubes.[18] Atmosphere, pressure, current and voltage play a role in the morphology and quality of nanotubes in arc discharge technique. Typical voltages and currents used in the arc-discharge method are 15-30 V and 50-120 A, respectively.[19] Lower voltages (~10 V) were found to produce fullerenes whereas high voltage favors CNT formation and its quality. Electrical breakdown of the gas by arc discharge generates plasma with very high temperatures (above 3000°C) that evaporate the carbon and produces carbon nanotubes, fullerenes and other carbonaceous products in the product. Iijima used the arc process for the first time to generate carbon nanotubes.[3] Single-walled carbon nanotubes could be produced when one of the electrodes was replaced with transition metal-filled graphite.[20] Fe, Co and Ni are usually used as catalysts along with catalytic promoters S, Y and Cr in the arc-discharge process.[18]

2.1.2. *Laser ablation*

Smalley and co-workers applied laser ablation to make the carbon nanotubes in 1995.[21] In the present case, the laser is used to evaporate the graphite or metal filled graphite electrode. Usually, a graphite target is placed in a quartz tube kept at a high temperature (~1000 °C) tube furnace with continuous purging of an inert gas.[4] A pulsed laser (YAG type, KrF excimer or CO_2) is used to evaporate the target and the nanotubes thus formed are collected from the cooler surfaces of the reactor. Metal filled (Fe or Ni) graphite electrodes are vital for SWNTs' synthesis.[19] Laser ablation technique produces superior quality carbon nanotubes.

2.1.3. *Chemical vapor deposition (CVD)*

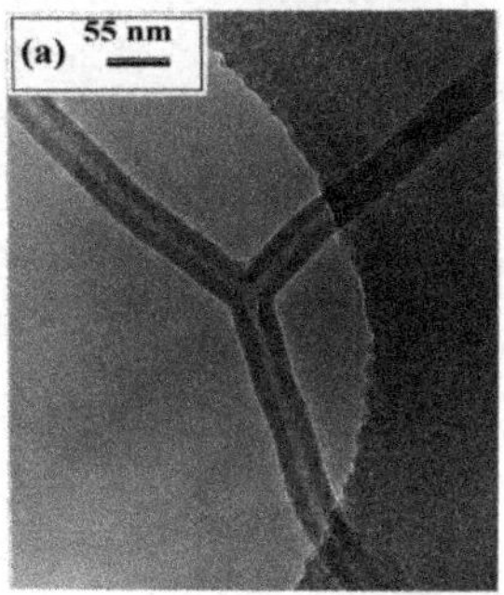

Fig. 3. TEM images of Y-junction carbon nanotube. Adapted with permission from Ref. 28.

CVD to produce carbon nanotubes is as old as 1890.[22] Today, CVD has become the most popular and versatile technique to produce carbon nanotubes due to its low cost and high production yield. The CVD process generally consists of a tubular chamber where cracking of hydrocarbon vapors on metallic particles (e.g., Fe, Co, Ni) are achieved above hydrocarbon decomposition temperatures (450-1200°C).[23,24] Along with a carbon source, a carrier gas such as argon, nitrogen, etc. is used in the process. Choice of the metal, catalyst–substrate interaction, carbon source and temperature affect the quality and yield of the carbon nanotubes in the CVD process.[23] Ethanol turns out to be the popular choice as it is found to produce clean nanotubes and removes amorphous carbon in the synthesis.[25,26] Water vapor along with carbon source has also been used to produce clean and super-growth carbon nanotubes.[27] Introducing thiophene along with organometallic precursors such as nickelocene was found to yield Y-junction carbon nanotubes.[28] A typical TEM image of Y-junction carbon nanotube is shown in Figure 3.

2.2. *Selective generation of CNTs*

Selectively making CNTs with semiconducting or metallic or known chirality is highly demanding. Several synthesis and post-synthesis techniques are used to separate semiconducting and metallic nanotubes.[15,29] High currents passing through bundle of SWNTs, selectively eliminate metallic nanotubes.[30] The difference in chemical reactivity is being used to selectively etch metallic nanotubes to the corresponding hydrocarbon by methane plasma at 400°C,[31] or by H_2O vapor.[32] Dielectrophoresis technique is also be used to separate metallic vs. semiconducting nanotubes from highly dispersed nanotube solutions.[33] Selective functionalization of SWNTs with surfactants or DNA has been used to separate metallic and semiconducting nanotubes.[34,35] Simple molecular charge transfer between potassium salt of coronene tetracarboxylic acid with SWNTs was found to separate metallic vs. semiconducting nanotubes.[36]

Synthesis methods to selectively make one of the types are highly desirable. There is some progress in this regard.[29,37] Rao and co-workers synthesized selective metallic SWNTs (94% yield) using arc discharge between graphite electrodes filled with $Ni+Y_2O_3$ in the presence of $Fe(CO)_5$.[38] 65% enriched metallic SWNTs (m-SWNTs) are produced on Fe–Co/MgO catalyst using long chains of monohydroxy aliphatic alcohols.[39] On the other hand, 90% pure semiconducting SWNTs (s-SWNTs) are synthesized by plasma-enhanced chemical vapor deposition (PECVD) at 600°C.[40,41] UV-Vis absorption, Raman spectroscopy and conductivity measurements are used to characterize the samples.

Interacting SWNTs with electron donating molecules (tetrathiafulvalene, aniline) convert s-SWNTs to m-SWNTs while electron withdrawing molecules (tetracyanoethylene, nitrobenzene) convert m-SWNTs to s-SWNTs.[42] Interestingly, gold and platinum decoration on SWNTs converts s-SWNTs to m-SWNTs.[43]

Each semiconducting nanotube shows a different band gap depending on its diameter and chirality. To use SWNTs in electronic applications, it is very important to control the chirality. Chirality selective nanotubes are produced by taking fullerene as precursors.[44] Post-synthesis separation of surfactant-coated SWNTs with respect to diameter and chirality have been achieved by density gradient ultracentrifugation.[45] DNA assisted ion-exchange chromatography has also been employed to separate SWNTs based on chirality.[46] We will not discuss the progress of the process in this chapter, rather, we shall recommend a review on recent developments.[47]

2.3. *Characterization*

Raman spectroscopy, transmission electron microscope (TEM), scanning electron microscope (SEM) and scanning tunneling microscope (STM) are suitable tools to characterize carbon nanotubes. SEM can be used to identify yield, purity, length, etc., while TEM is useful to find out diameter, the number of walls and structure (through electron diffraction or by aberration-corrected TEM). STM is a powerful tool to characterize CNTs, where direct imaging is possible on conducting substrates (Figure 4(a)).[9,10] This technique provides atomic resolution images, thus atomic structure, chirality and diameter, and simultaneously density of states can be observed by scanning tunneling spectroscopy (Figure 4(b)).[48]

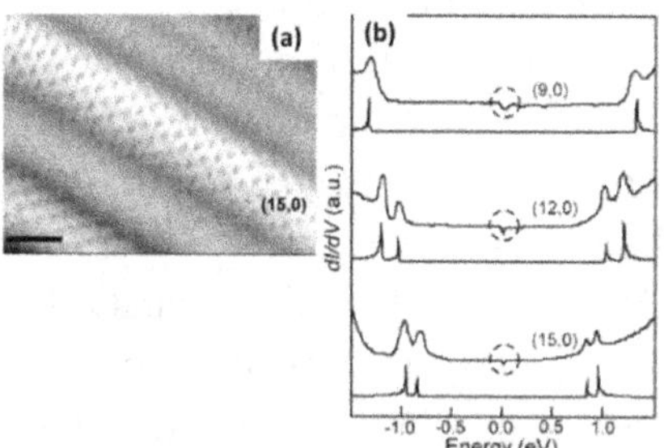

Fig. 4. (a) Typical STM image of single-walled carbon nanotubes of type (15, 0). (b) Experimental tunneling conductance data (dI/dV) along with calculated density of states of SWNTs of type (9,0); (12,0) and (15,0). Adapted with permission from Ref. 48.

Raman spectroscopy is one of the most prevalent and non-destructive tools to characterize carbon nanotubes.[49] In the case of carbon nanotubes, Raman spectroscopy provides information of phononic, electronic structures as well as defects.

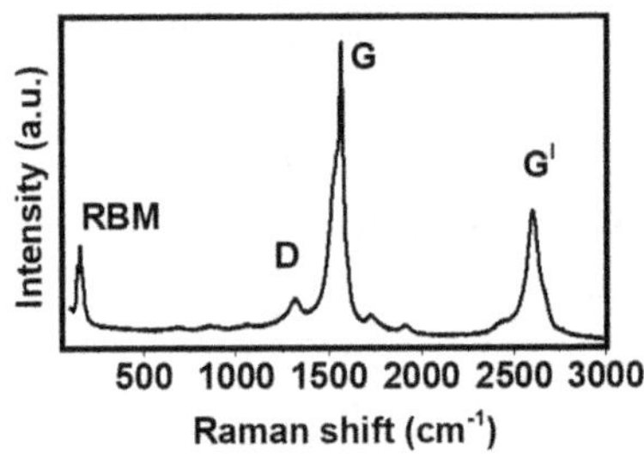

Fig. 5. Raman spectrum of SWNTs.

Figure 5 shows the typical Raman spectrum of single-walled carbon nanotubes. Both radial breathing mode (RBM), and G-band are first-order Raman modes.[49] RBM phonon modes (frequency observed in 100-500 cm^{-1} region) is coherent out-of-plane stretching of carbon atoms in the radial direction. One can estimate SWNTs diameter, d, by Raman frequency using the following formula:

$$\omega_{RBM} = (A/d)+B \tag{1}$$

where ω_{RBM} is the RBM frequency and A, B are constants. B is dependent on tube-tube interaction and A is dependent on nanotube-substrate interaction.

The G-band, observed around 1580 cm^{-1}, is due to stretching of the neighboring sp^2 carbon atoms in the opposite direction along the surface of a nanotube. The line shape of the G-band readily shows whether it is semiconducting (Lorentzian) or metallic (Breit–Wigner–Fano).[49] D and G′ (note: in modern literature this band is named as 2D) bands respectively at 1350 and 2700 cm^{-1} originate from a second-order process, where two scatterings are involved. In the D-band, one of the two scattering processes is elastic originating due to any symmetry breaking defect in graphene. The intensity ratio between the D-band and the G-band can be used to find out the quality of nanotubes or the defects in nanotubes.

2.4. *Purification*

As synthesized carbon nanotubes contain metallic nanoparticles and amorphous carbon along with it. Purification by acid treatment (HNO_3, HCl, etc.) is generally used to remove the metallic nanoparticles.[50] Treatment with ozone or oxygen or H_2O_2 or reducing by H_2 gas at high temperature is used to remove the amorphous carbon.[50,51] As the graphitic carbon nanotubes are less reactive compared to amorphous carbon, either oxidation or reduction selectively remove amorphous carbon. In the oxidation/acid treatment process, some of the closed-end tubes open up and get functionalized with hydroxyl or carboxylic groups.[50]

2.5. *Functionalization*

Functionalization can be broadly divided into two types: (1) Doping (chemical doping, electrochemical and charge transfer doping, etc.) and (2) surface modification (by covalent and non-covalent means). Doping is one of the effective ways of tuning the electric properties and has been explored well in the literature.[52] Nitrogen and boron are primarily dopants for carbon nanotubes, making CNTs n-type and p-type, respectively. NH_3 or pyridine is a commonly used precursor to dope nitrogen whereas B_2H_6 has been used to dope boron in CNTs.[52] Molecular charge transfer is another effective way of doping CNTs and carbon materials in general.[38] Strong electron-withdrawing and electron-donating molecules such as tetracyanoethylene (TCNE), tetrathiafulvalene (TTF) are used to make adducts with SWNTs. TCNE withdraws electrons from SWNTs and makes them more semiconducting while TTF donates electrons to SWNTs and makes them metallic.[42]

For most of the applications, carbon nanotubes in their pristine form are not desirable as they bundle up due to strong van der Waals interaction. Suitable functionalization makes carbon nanotubes to disperse in different solvents and also allow to bind to different polymers to form good composites.[53,54]

Covalent functionalization of CNTs is achieved by treating CNTs by reflexing or sonication in an acid or a mixture of acids such as HNO_3, $H_2SO_4+HNO_3$ or $KMnO_4+H_2SO_4$.[23] Acids treatment functionalize CNTs with carboxylic acid and other groups. The carboxylic acid group can be further converted to an amide by reacting with $SOCl_2$ followed by long-chain amine or amine-terminated polymers.[54] These functionalized nanotubes can be dispersed in different organic or aqueous solvents.

Reaction of CNTs with $F_2/HF/IF_5$ at room-temperature or elevated temperatures are used to fluorinate CNTs. Fluorinated nanotubes are found to have good solubility. Different fluorination techniques for carbon materials have been recently summarized by Adamska & Narkiewicz.[55] Different other functionalizations such as reacting with dichlorocarbenes to sidewall C = C to make cyclopropane ring, the cycloaddition of nitrenes by reacting with alkyl azides, 1,3-dipolar cycloaddition of azomethine ylides have been achieved.[56] These products are found to have good solubility in DMSO. 1,3-dipolar cycloaddition on CNTs has been achieved by ozonolysis at low-temperatures.[53,57] Silylation of CNTs is achieved by treating acid-functionalized CNTs with different organosilanes.[58] Radical reactions via thermal, photochemical or electrochemical means have also been successfully tried on CNTs.[58,59]

Non-covalent functionalization by interacting CNTs with surfactants, conjugated polymers or aromatic compounds such as pyrenyl group is another important step in solubilizing CNTs in different solvents.[23,60,61] Non-covalent functionalization doesn't alter the electronic properties of nanotubes as much as covalent functionalization.[62] Wrapping of CNTs with polyvinyl pyrrolidone (PVP), substituted poly (m-phenylenevinylene), and dendrimers have been achieved by sonication.[23,63] Functionalization with pyrene-containing molecules, crosslinked and amphiphilic copolymers have also been achieved.[23,64] CNTs have also been functionalized with different biomolecules including DNA, enzymes, nucleic acids and proteins, etc. for making sensor and other applications.[65,66] Functionalized CNTs are used as catalytic supports and atomic force microscope tips, and have also found applications in optical and biomedical fields.[58,66]

2.6. *Properties of carbon nanotubes*

Carbon nanotubes exhibit excellent electrical, thermal, mechanical and field emission properties. Several studies on individual nanotubes and nanotube ropes (bundle of nanotubes) have been conducted.[67] Ballistic electron transport is predicted in these quantum tubes. Mean free paths of electrons in CNTs are predicted to be several microns and experimentally found to be one micron.[68,69] Conductivity in MWNTs was found to be a non-uniform distribution of current among different walls within the nanotubes. SWNTs bundle resistivity has been measured and found to be as low as $\sim 10^{-4}$ ohm-cm at room-temperature, with current densities reaching up to 10^7 A/cm^2.[70] Field effect transistors have been made with semiconducting nanotubes. Diodes and transistors have also been made with CNT interconnects.[67] It is found that the speed of the electrical signal in SWNTs is high (up to 10 GHz).[71] Nanoelectromechanical devices have been made with individual carbon nanotubes. Oscillation in the conductivity has been observed in single nanotubes by twisting the nanotubes.[72] These oscillations are attributed to the changing of chirality by mechanical torsion, which changes the corners of the first Brillouin zone of graphene.

Due to the strong bonds between the carbons in the graphene network, the elastic modulus is very high. Carbon nanotubes have been demonstrated to be tips in atomic force microscopes due to their high strength. Young's modulus of an SWNT is reported to be 1.8 TPa.[73] Due to its remarkable strength, carbon nanotubes are used as strengthening agents. Mixing CNTs with other materials to make composites with superior properties are well explored. We suggest other reviews for further reading.[74,75] CNTs have been tested to be good field emitters.[76] Sharp tips of CNTs, develop large electric fields even for small applied voltages. These are good candidates for flat-panel displays.

3. Inorganic Nanotubes

Apart from carbon, other compounds with layered structures are also known to make nanotubes. Inorganic nanotubes based on layered transition metal dichalcogenides such as WS_2 and MoS_2 have been synthesized as early as 1992 by Tenne et al.[2,12] Unlike carbon nanotubes, where each layer is one atom thick, the WS_2 layer consists of three atoms in one layer and the transition metal is sandwiched between the sulfur layers (see Figure 6). Nanotubes from other layered materials have been synthesized including other transition metal dichalcogenides (TiS_2, SnS_2, etc.), boron nitride, metal hydroxides, etc.[5,6,14,77] The basic driving force for the formation of inorganic nanotubes could be similar to that in the formation mechanism of carbon nanotubes that is to minimize high energy dangling bonds when they are in low-dimensions.

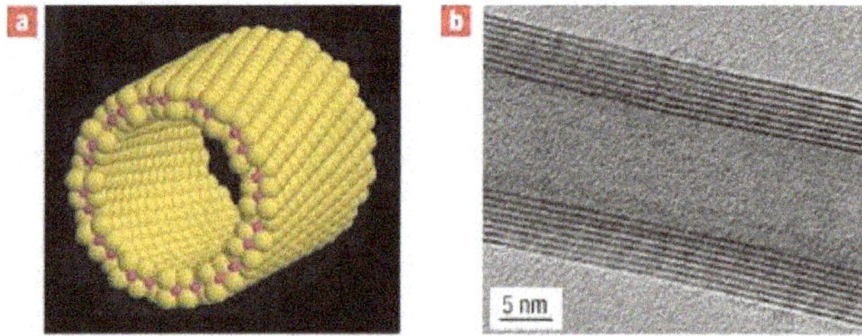

Fig. 6. (a) Schematic three-atom thick single-wall WS_2 nanotube, where pink and yellow represent W and S atoms respectively. (b) A typical TEM image of a multiwall WS_2 nanotube. Adapted with permission from Ref. 78.

3.1. *Synthesis*

The general strategy to make WS_2 nanotubes is by high-temperature reduction of WO_{3-x} nanorods in the presence of H_2S in a tube furnace.[79] As oxide to sulfide conversion is a diffusion-controlled process, the procedure is not applicable for other materials where reactivity is less as in TiO_2 and NbO_2. Heating metal halides or carbonyls in the presence of H_2S or H_2Se, generates the respective metal chalcogenide nanotubes. This process became the general technique to generate a variety of rolled structures including NbS_2, TiS_2, WS_2, etc.[5] Rao et al. found that heating of trisulfides in hydrogen yields nanotubes of various metal sulfides including MoS_2, WS_2, TaS_2, NbS_2.[80,81] Recently, Pb has been added to stabilize MoO_{3-x} rods at high temperatures and convert to MoS_2 at high temperatures under concentrated solar energy.[82] This procedure also yields $MoSe_2$, WS_2 and WSe_2 under appropriate starting materials. High yield synthesis of MoS_2 is also achieved by carbon nanotubes as the template.[83] MoS_2 nanotubes with uniform length and number of layers are obtained by electrospinning a mixture of polyacrylonitrile and molybdenum thiosalt precursors (ammonium tetrathiomolybdate, potassium tetrathiomolybdate) to make nanofibers followed by controlled nanofiber calcination in the presence of O_2.[84] Green synthesis of WS_2 nanotubes has been achieved by laser ablation of WS_2 bulk powder admixed with Pb/PbO.[85]

Transition metal dichalcogenides with three atom-thick layers suffer from high elastic energy making it difficult to be rolled into single-walled nanotubes. However, Remskar et al. reported sub-nanometer MoS_2 nanotubes by chemical vapor transport using C_{60} as the growth promoter.[86] Recently, inductively coupled radiofrequency plasma treatment on WS_2 nanotubes were found to yield small diameter WS_2 nanotubes with one to three layers.[87] Ni et al. report the formation of 1-3 wall nanotubes of WS_2 by sulphurization of ultrathin sputtered tungsten films.[88] Apart from MoS_2 and WS_2, hydrothermal synthesis has been optimized to get varieties of single-wall inorganic nanotubes including indium sulfide, $Co(OH)_2$, nickel phosphate.[89] TiO_2 is another well-explored material as it is used for different applications including photocatalysis and electron transport in solar cells. Hydrothermal and template replication methods are well explored to get nanotubular structures.[90] Layered nanotubular structures are either amorphous or titanate, which is further annealed around 450 °C to get the TiO_2 (anatase). Ultra-long TiO_2 nanotubes (1-2 μm) are achieved by pre-sonication of TiO_2 powder in basic solution followed by hydrothermal at 150°C.[91] Electrochemical anodization is used to grow aligned TiO_2 nanotubes, but yields large diameter (~50 nm) nanotubes. Wang et al. report the two-step anodization process to obtain small diameters (< 30 nm) and ultra-long lengths up to 30 microns.[92] In this process, a protective layer is created in the first step by high voltages (50 V) followed by a low-voltage (15 V) in the second-step for a prolonged time under stirring. These ultra-long nanotubes have been found to improve dye-sensitized solar cells efficiency.

3.2. *Properties of inorganic nanotubes*

Inorganic nanotubes show a wide range of properties, from metallic (TiS_2, TaS_2, NbS_2) to semiconducting (MoS_2, WS_2) and from insulating (BN) to superconducting ($NbSe_2$) properties. Unlike quantum dots where the band gap increases with decreasing size, in inorganic fullerenes or nanotubes of MoS_2/WS_2 the band gap narrows with decreasing diameter.[5,93-95] Due to the high strain, bending three-atom thick layers, modifies the chemical bond between the atoms leading to the observed trend.

Recent field-effect transistors studies on individual WS_2 nanotubes found to have a field-effect mobilities of 50 cm^2 V^{-1} s^{-1}.[96,97] with current carrying capacity (~630 µA). MoS_2 nanotube-based transistors reaching on/off current ratios more than 10^3 have been reported.[98] Superconductivity has been very recently observed in WS_2 nanotubes by electrochemical doping.[99] The torsional electro-mechanical properties of individual suspended WS_2 nanotubes showed excellent deformation-induced electrical properties that would be useful in sensors for nanometric motors.[100]

Due to excellent mechanical properties, inorganic nanotubes have been commercialized as an additive in solid-state lubricants.[101] Inorganic nanotubes have also been used as additives to improve the mechanical properties of polymers.[14,102] Although inorganic nanotubes are inferior to CNTs in mechanical properties, they disperse better in polymers, show less agglomeration tendency and less entanglement compared to CNTs. Mechanical properties of individual WS_2 nanotubes have been studied and it was found that they are strong as well as elastic, and their tensile strength (~16 GPa) reaches the theoretical ideal strength due to the absence of defects. In this respect, inorganic nanotubes are potential candidates for future composites. BN is another important material that shows high mechanical strength. *In situ* TEM measurements on the individual BN nanotubes show Young's modulus and tensile strength respectively of 924 and 18.8 GPa.[103] Small diameter BN (~10 nm) shows excellent bending/compression strength of more than 1200 MPa.[104]

4. Misfit Layered Nanotubes

Misfit layered compounds (MLCs) can be considered as two-dimensional intercalation compounds.[105,106] The general formula of chalcogenide-based misfit compounds are $(MX)_{1+x}TX_2$ (or simply O-T) where M=lanthanide, Pb, Bi, Sn, etc., T=Sn, Ti, V, Cr, etc. and X=S or Se (see Figure 7(a) for schematic structure). Alternative stacking of MX and TX_2 layers can be more complex than simple O-T-O-T… stacking. O-T-O-T-T…, O-T-T-O-T-T… etc., also known in the literature. Properties of each type would differ.[107] The lattice parameter of MX does not match with TX_2 in at least one direction and this leads to mutual structural modulation and incommensurability. Misfit compounds are thermodynamically stable compared to their individual components. Some of the interesting magnetic, superconducting and charge density waves have been found in these materials.[105] Recently, they have been found to have better thermoelectric properties in bulk or thin films.[108-110] One-dimensional analogs could further improve their properties.[111] Misfit stress between the adjacent layers in MLCs helps to roll into a nanotubular structure.[13] Several synthesis techniques have been developed to make misfit nanotubes. Laser ablation of an SnS_2 target, evaporated some of the sulfur for the lattice and produced SnS-SnS_2 nanotubular misfit superstructure with less yield.[112] Chemical vapor transport (CVT) of Sn and S in vacuum-sealed tubes were subjected to the thermal gradient (800 ^{o}C-150 ^{o}C) to yield good quantities of SnS-SnS_2 misfit nanotubes.[113] After this, several compounds were found to make misfit nanotubes by this technique including PbS-NbS_2, PbS, SbS, BiS intercalated TaS_2, LnS intercalated TaS_2 (Ln= La, Ce, Nd, Ho, Er, Pr, Sm, Gd, Yb).[114-119] The selection of temperature and temperature gradient is crucial to get a good yield of nanotubes. SEM images of some of the misfit compounds synthesized by CVT process are shown in Figure. 7(b).

High-temperature annealing of respective hydroxide mixture in the presence of H_2S was found to yield large quantities of misfit compounds. A variety of nanotubes of formula LnS-TS_2 (Ln= La, Ce, Nd, Gd, Tb and T=Cr and V), are synthesized by this method.[120,121]

One of the important reasons for the stability of this compound is attributed to the charge transfer between the layers in addition to van der Walls forces and electrostatic interactions.[105] In the case of lanthanide-based compounds, LnS donates charge to TS_2 layer. Contrary to this, the theoretical calculation on SnS-SnS_2 was found to show a small charge transfer from SnS_2 to SnS. A significant change in the vibrational properties probed by Raman spectroscopy also conform to the interlayer charge transfer.[113,114]

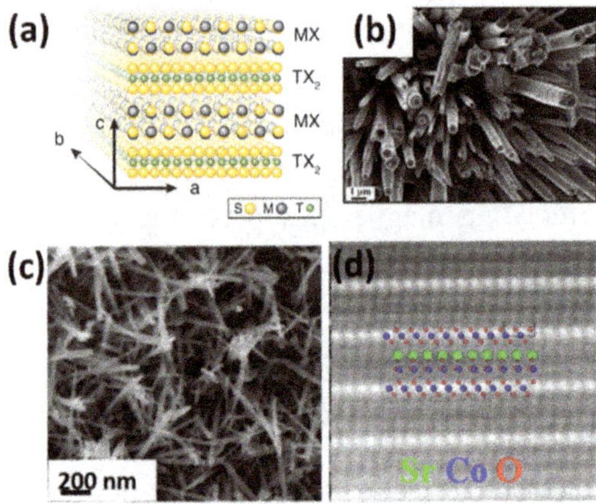

Fig. 7. (a) Schematic presentation of a typical chalcogenide-based misfit layered structure. Adapted with permission from Ref. 115. (b) Typical SEM image of misfit SbS–TaS_2 nanotubes. Adapted with permission from Ref. 13. (c) SEM image of misfit SrCoO-CoO_2 nanotubes and (d) HR-STEM ADF micrograph of a SrCoO-CoO_2 nanotube superposed with $SrCoO_2$–CoO_2 relaxed atomic structure. Adapted with permission from Ref. 122.

Apart from sulfide- and selenide-based misfit compounds, oxide-based misfit compounds are also known in the literature. $Ca_3Co_4O_{9-x}$ (Ca_2CoO_3-CoO_2) is an interesting class of misfit compounds where a CoO_2 layer is sandwiched between Ca_2CoO_3 blocking layers.[123] Neither CoO_2 nor Ca_2CoO_3 are stable by themselves. The Seebeck coefficient of these materials is high in spite of being metallic in nature.[124] One-dimensional analogs of these compounds have been synthesized recently.[122,125] Hydrothermal synthesis of $Ca_3Co_4O_9$ or $Sr_3Co_4O_9$ in basic condition was found to yield calcium/strontium deficient phase of calcium/strontium cobalt oxide nanotubes.[122,125] These structures are found to be $CaCoO_2$ or $SrCoO_2$ intercalated CoO_2. SEM image of the misfit $SrCoO$-CoO_2 nanotubes is shown in Figure 7(c). Figure 7(d) shows high resolution scanning transmission electron microscope (HR-STEM) annular dark field (ADF) micrograph of a $SrCoO$-CoO_2 nanotube superposed with $SrCoO_2$−CoO_2 relaxed atomic structure. Interestingly, both experimental and theoretical calculations of calcium-deficient phases were found to be semiconducting in nature, in contrast to their bulk metallic nature and are predicted to yield better thermoelectric properties.[126,127]

5. Outlook

There has been a significant amount of research work in the last two to three decades on one-dimensional nanotubes. Their unique structure as well as mechanical, thermal and electrical properties, makes them outstanding materials for applications. Synthesis conditions can be optimized to meet industrial needs (several tons/year). For example, some of the companies are presently selling CNTs and CNT- based products including CNT conductive coating, CNT epoxy composites and CNT thermal radiation coatings. MoS_2 and WS_2 fullerenes and nanotubular structures have also been commercialized as solid-state lubricant additives. However, there are certain drawbacks in this field to use them to their fullest potential. The next immediate application would be in flat-panel displays (thanks to the field emission properties). Prototype displays have been demonstrated and CNTs can be potential candidates for electronics as interconnectors. However, major challenges are to be addressed before realizing the above applications. SWNTs are either metallic or semiconducting and each semiconducting nanotube shows different properties depending on chirality. Mass production of metallic nanotubes, as well as semiconducting nanotubes, separately are crucial. An equally essential factor is to synthesize SWNTs with particular chirality in larger quantities. Another important consideration is to pattern these nanotubes on desired substrates. MWNTs are produced in large quantities whereas developing industry scale SWNTs still requires improvement.

Realizing applications of inorganic nanotubes is in need of further attention. Except for a few inorganic nanotubes (WS_2, BN nanotubes), synthesis of a large variety of them in bulk quantities still needs to be developed. MoS_2 or WS_2 single layers have been proven to have great potential for some useful applications. Making single-walled nanotubes from inorganic nanotubes remains a challenge. Misfit layered nanotubes are still in its infancy. Developing more synthetic strategies and studying properties of individual nanotubes and bulk materials require research to further understand their properties and possible applications. The scope of theoretical study of these nanotubes is limited and needs further attention.

Each new development and product leaves a footprint that impacts the environment. To take these products to the next stage for safe utilization requires a study involving long-term environmental effects and related aspects.

References

1. H. W. Kroto, J. R. Heath, S. C. O'Brien, R. F. Curl and R. E. Smalley, C_{60}: Buckminsterfullerene. *Nature*, **318**(6042), 162-163, (1985).
2. R. Tenne, L. Margulis, M. Genut and G. Hodes, Polyhedral and cylindrical structures of tungsten disulphide. *Nature*, **360**(6403), 444-446, (1992).
3. S. Iijima, Helical microtubules of graphitic carbon. *Nature*, **354**(6348), 56-58, (1991).
4. C. N. R. Rao and A. Govindaraj, (2006) Nanotubes and Nanowires, RSC Series on Nano*Science*: Royal Society of Chemistry, London.
5. R. Tenne, Inorganic nanotubes and fullerene-like nanoparticles. *Nat. Nanotechnol.*, **1**(2), 103-111, (2006).
6. C. N. R. Rao, B. C. Satishkumar, A. Govindaraj and M. Nath, Nanotubes. *ChemPhysChem*, **2**(2), 78-105, (2001).
7. N. Hamada, S.-i. Sawada and A. Oshiyama, New one-dimensional conductors: Graphitic microtubules. *Phys. Rev. Lett.*, **68**(10), 1579-1581, (1992).
8. M. S. Dresselhaus, G. Dresselhaus and R. Saito, Carbon fibers based on C_{60} and their symmetry. *Phy. Rev. B*, **45**(11), 6234-6242, (1992).
9. J. W. G. Wilder, L. C. Venema, A. G. Rinzler, R. E. Smalley and C. Dekker, Electronic structure of atomically resolved carbon nanotubes. *Nature*, **391**(6662), 59-62, (1998).
10. T. W. Odom, J.-L. Huang, P. Kim and C. M. Lieber, Atomic structure and electronic properties of single-walled carbon nanotubes. *Nature*, **391**(6662), 62-64, (1998).

11. A. Saito, G. Dresselhaus and M. S. Dresselhaus, Physical Properties of Carbon Nanotubes. London : Imperial College Press, (1998).

12. L. Margulis, G. Salitra, R. Tenne and M. Talianker, Nested fullerene-like structures. *Nature*, **365**(6442), 113-114, (1993).

13. L. S. Panchakarla, G. Radovsky, L. Houben, R. Popovitz-Biro, R. E. Dunin-Borkowski and R. Tenne, Nanotubes from misfit layered compounds: a new family of materials with low dimensionality. *J. Phys. Chem. Lett.*, **5**(21), 3724-3736, (2014).

14. B. Višić, L. S. Panchakarla and R. Tenne, Inorganic Nanotubes and Fullerene-like Nanoparticles at the Crossroads between Solid-State Chemistry and Nanotechnology. *J. Am. Chem. Soc.*, **139**(37), 12865-12878, (2017).

15. C. N. R. Rao, R. Voggu and A. Govindaraj, Selective generation of single-walled carbon nanotubes with metallic, semiconducting and other unique electronic properties. *Nanoscale*, **1**, 96-105, (2009).

16. Y. A. Kim, H. Muramatsu, T. Hayashi, M. Endo, M. Terrones and M. S. Dresselhaus, Fabrication of high-purity, double-walled carbon nanotube buckypaper. *Chem. Vap. Deposition*, 2006, **12**(6), 327-330.

17. T. W. Ebbesen and P. M. Ajayan, Large-scale synthesis of carbon nanotubes. *Nature*, **358**(6383), 220-222, (1992).

18. N. Arora and N. N. Sharma, Arc discharge synthesis of carbon nanotubes: Comprehensive review. *Diam. Relat. Mater.*, **50**, 135-150, (2014).

19. J. Prasek, J. Drbohlavova, J. Chomoucka, J. Hubalek, O. Jasek, V. Adam and R. Kizek, Methods for carbon nanotubes synthesis—review. *J. Mater. Chem.*, **21**, 15872-15884, (2011).

20. S. Iijima and T. Ichihashi, Single-shell carbon nanotubes of 1-nm diameter. *Nature*, **363**(6430), 603-605, (1993).

21. T. Guo, P. Nikolaev, A. Thess, D. T. Colbert and R. E. Smalley, Catalytic growth of single-walled manotubes by laser vaporization. *Chem. Phys. Lett.*, **243**(1-2), 49-54, (1995).

22. P. S. L. Schützenberger, *C. R. Acad. Sci.*, **111**, 774, (1890).

23. Y. Yan, J. Miao, Z. Yang, F.-X. Xiao, H. B. Yang, B. Liu and Y. Yang, Carbon nanotube catalysts: recent advances in synthesis, characterization and applications. *Chem. Soc. Rev.*, **44**(10), 3295-3346, (2015).

24. Kumar, M. and Y. Ando, Chemical vapor deposition of carbon nanotubes: A review on growth mechanism and mass production. *J. Nanosci. Nanotechnol.* **10**(6), 3739-3758, (2010).

25. S. Maruyama, R. Kojima, Y. Miyauchi, S. Chiashi and M. Kohno, Low-temperature synthesis of high-purity single-walled carbon nanotubes from alcohol. *Chem. Phys. Lett.*, **360**(3-4), 229-234, (2002).

26. O. Shingo, S. Takeshi, S. Shinzo, A. Yohji, T. Kazuhito, A. Yoshinobu and K. Hiromichi, Purification of single-wall carbon nanotubes synthesized from alcohol by catalytic chemical vapor deposition. *Jpn. J. Appl. Phys.*, **43**, L396, (2004).

27. K. Hata, D. N. Futaba, K. Mizuno, T. Namai, M. Yumura and S. Iijima, Water-assisted highly efficient synthesis of impurity-free single-walled carbon nanotubes. *Science*, **306**(5700), 1362-1364, (2004).

28. B. C. Satishkumar, P. J. Thomas, A. Govindaraj and C. N. R. Rao, Y-junction carbon nanotubes. *Appl. Phys. Lett.*, **77**(16), 2530-2532, (2000).

29. I. Ibrahim, T. Gemming, W. M. Weber, T. Mikolajick, Z. Liu and M. H. Rümmeli, Current progress in the chemical vapor deposition of type-selected horizontally aligned single-walled carbon nanotubes. *ACS Nano*, **10**(8), 7248-7266, (2016).

30. P. G. Collins, M. S. Arnold and P. Avouris, Engineering carbon nanotubes and nanotube circuits using electrical breakdown. *Science*, **292**(5517), 706-709, (2001).

31. G. Zhang, P. Qi, X. Wang, Y. Lu, X. Li, R. Tu, S. Bangsaruntip, D. Mann, L. Zhang and H. Dai, Selective etching of metallic carbon nanotubes by gas-phase reaction. *Science*, **314**(5801), 974-977, (2006).

32. F. Yang, X. Wang, J. Si, X. Zhao, K. Qi, C. Jin, Z. Zhang, M. Li, D. Zhang, J. Yang, Z. Zhang, Z. Xu, L.-M. Peng, X. Bai and Y. Li, Water-assisted preparation of high-purity semiconducting (14,4) carbon nanotubes. *ACS Nano*, **11**(1), 186-193, (2017).

33. R. Krupke, F. Hennrich, H. v. Löhneysen and M. M. Kappes, Separation of metallic from semiconducting single-walled carbon nanotubes. *Science*, **301**(5631), 344-347, (2003).

34. M. Zheng, A. Jagota, M. S. Strano, A. P. Santos, P. Barone, S. G. Chou, B. A. Diner, M. S. Dresselhaus, R. S. Mclean, G. B. Onoa, G. G. Samsonidze, E. D. Semke, M. Usrey and D. J. Walls, Structure-based carbon nanotube sorting by sequence-dependent DNA assembly. *Science*, **302**(5650), 1545-1548, (2003).

35. M. Zheng, A. Jagota, E. D. Semke, B. A. Diner, R. S. McLean, S. R. Lustig, R. E. Richardson and N. G. Tassi, DNA-assisted dispersion and separation of carbon nanotubes. *Nat. Mater.*, **2**(5), 338-342, (2003).

36. R. Voggu, K. V. Rao, S. J. George and C. N. R. Rao, A simple method of separating metallic and semiconducting single-walled carbon nanotubes based on molecular charge transfer. *J. Am. Chem. Soc.*, **132**(16), 5560-5561, (2010).

37. J. Li, C.-T. Ke, K. Liu, P. Li, S. Liang, G. Finkelstein, F. Wang and J. Liu, Importance of diameter control on selective synthesis of semiconducting single-walled carbon nanotubes. *ACS Nano*, **8**(8), 8564-8572, (2014).

38. R. Voggu, S. Ghosh, A. Govindaraj and C. N. R. Rao, New strategies for the enrichment of metallic single-walled carbon nanotubes. *J. Nanosci. Nanotechnol.*, **10**(6), 4102-4108, (2010).

39. Y. Wang, Y. Liu, X. Li, L. Cao, D. Wei, H. Zhang, D. Shi, G. Yu, H. Kajiura and Y. Li, Direct enrichment of metallic single-walled carbon nanotubes induced by the different molecular composition of monohydroxy alcohol homologues. *Small*, **3**(9), 1486-1490, (2007).

40. Y. Li, D. Mann, M. Rolandi, W. Kim, A. Ural, S. Hung, A. Javey, J. Cao, D. Wang, E. Yenilmez, Q. Wang, J. F. Gibbons, Y. Nishi and H. Dai, Preferential growth of

semiconducting single-walled carbon nanotubes by a plasma enhanced CVD method. *Nano Lett.,* **4**(2), 317-321, (2004).

41. L. Qu, F. Du and L. Dai, Preferential syntheses of semiconducting vertically aligned single-walled carbon nanotubes for direct use in FETs. *Nano Lett.*, **8**(9), 2682-2687, (2008).

42. R. Voggu, C. S. Rout, A. D. Franklin, T. S. Fisher and C. N. R. Rao, Extraordinary sensitivity of the electronic structure and properties of single-walled carbon nanotubes to molecular charge-transfer. *J. Phys. Chem. C*, **112**(34), 13053-13056, (2008).

43. R. Voggu, P. Shrinwantu, K. P. Swapan and C. N. R. Rao, Semiconductor to metal transition in SWNTs caused by interaction with gold and platinum nanoparticles. *J. Phys. Condens. Matter.*, **20**(21), 215211, (2008).

44. F. Rao, T. Li and Y. Wang, Growth of "all-carbon" single-walled carbon nanotubes from diamonds and fullerenes. *Carbon*, **47**(15), 3580-3584, (2009).

45. A M. S. Arnold, A. A. Green, J. F. Hulvat, S. I. Stupp and M. C. Hersam, Sorting carbon nanotubes by electronic structure using density differentiation. *Nat. Nanotechnol.*, **1**(1), 60-65, (2006).

46. M. Zheng and B. A. Diner, Solution redox chemistry of carbon nanotubes. *J. Am. Chem. Soc.*, **126**(47), 15490-15494, (2004).

47. B. Liu, F. Wu, H. Gui, M. Zheng and C. Zhou, Chirality-controlled synthesis and applications of single-wall carbon nanotubes. *ACS Nano*, **11**(1), 31-53, (2017).

48. M. Ouyang, J.-L. Huang, C. L. Cheung and C. M. Lieber, Energy gaps in "metallic" single-walled carbon nanotubes. *Science*, **292**(5517), 702-705, (2001).

49. M. S. Dresselhaus, G. Dresselhaus, R. Saito and A. Jorio, Raman spectroscopy of carbon nanotubes. *Phys. Rep.*, **409**(2), 47-99, (2005).

50. P.-X. Hou, C. Liu and H.-M. Cheng, Purification of carbon nanotubes. *Carbon*, **46**(15), 2003-2025, (2008).

51. S. R. C. Vivekchand, A. Govindaraj, M. M. Seikh and C. N. R. Rao, New method of purification of carbon nanotubes based on hydrogen treatment. *J. Phys. Chem. B*, **108**(22), 6935-6937, (2004).

52. L. S. Panchakarla, A. Govindaraj and C. N. R. Rao, Boron- and nitrogen-doped carbon nanotubes and graphene. *Inorg. Chim. Acta*, **363**(15), 4163-4174, (2010).

53. D. Tasis, N. Tagmatarchis, A. Bianco and M. Prato, Chemistry of carbon nanotubes. *Chem. Rev.*, **106**(3), 1105-1136, (2006).

54. V. Georgakilas, K. Kordatos, M. Prato, D. M. Guldi, M. Holzinger and A. Hirsch, Organic functionalization of carbon nanotubes. *J. Am. Chem. Soc.*, **124**(5), 760-761, (2002).

55. M. Adamska and U. Narkiewicz, Fluorination of carbon nanotubes – A review. *J. Fluorine Chem.*, **200**, 179-189, (2017).

56. N. Tagmatarchis and M. Prato, Functionalization of carbon nanotubes via 1,3-dipolar cycloadditions. *J. Mater. Chem.*, **14**(4), 437-439, (2004).

57. S. Banerjee and S. S. Wong, Rational sidewall functionalization and purification of single-walled carbon nanotubes by solution-phase ozonolysis. *J. Phys. Chem. B*, **106**(47), 12144-12151, (2002).

58. Y.-P. Sun, K. Fu, Y. Lin and W. Huang, Functionalized carbon nanotubes: properties and applications. *Acc. Chem. Res.*, **35**(12), 1096-1104, (2002).

59. Y. Ying, R. K. Saini, F. Liang, A. K. Sadana and W. E. Billups, Functionalization of carbon nanotubes by free radicals. *Org. Lett.*, **5**(9), 1471-1473, (2003).

60. Y.-L. Zhao and J. F. Stoddart, Noncovalent functionalization of single-walled carbon nanotubes. *Acc. Chem. Res.*, **42**(8),1161-1171, (2009).

61. P. Bilalis, D. Katsigiannopoulos, A. Avgeropoulos and G. Sakellariou, Non-covalent functionalization of carbon nanotubes with polymers. *RSC Adv.*, **4**(6), 2911-2934, (2014).

62. L. S. Panchakarla and A. Govindaraj, Covalent and non-covalent functionalization and solubilization of double-walled carbon nanotubes in nonpolar and aqueous media. *J. Chem. Sci.*, **120**(6), 607-611, (2008).

63. Y.-P. Sun, W. Huang, Y. Lin, K. Fu, A. Kitaygorodskiy, L. A. Riddle, Y. J. Yu and D. L. Carroll, Soluble dendron-functionalized carbon nanotubes: Preparation, characterization, and properties. *Chem. Mater.*, **13**(9), 2864-2869, (2001).

64. R. J. Chen, Y. Zhang, D. Wang and H. Dai, Noncovalent sidewall functionalization of single-walled carbon nanotubes for protein immobilization. *J. Am. Chem. Soc.*, **123**(16), 3838-3839, (2001).

65. C. Hu, Y. Zhang, G. Bao, Y. Zhang, M. Liu and Z. L. Wang, DNA functionalized single-walled carbon nanotubes for electrochemical detection. *J. Phys. Chem. B*, **109**(43), 20072-20076, (2005).

66. L. Santiago-Rodríguez, G. Sánchez-Pomales and C. R. Cabrera, DNA-Functionalized carbon nanotubes: synthesis, self-assembly, and applications. *Isr. J. Chem.*, **50**(3), 277-290, (2010).

67. A. Lekawa-Raus, J. Patmore, L. Kurzepa, J. Bulmer and K. Koziol, Electrical properties of carbon nanotube based fibers and their future use in electrical wiring. *Adv. Func. Mater.*, **24**(24) 3661-3682, (2014).

68. W C. T. White and T. N. Todorov, Carbon nanotubes as long ballistic conductors. *Nature*, **393**(6682), 240-242, (1998).

69. M. S. Purewal, B. H. Hong, A. Ravi, B. Chandra, J. Hone and P. Kim, Scaling of resistance and electron mean free path of single-walled carbon nanotubes. *Phys. Rev. Lett.*, **98**(18), 186808, (2007).

70. B. Wei, R. Vajtai and P. M. Ajayan, Reliability and current carrying capacity of carbon nanotubes. *Appl. Phys. Lett.*, **79**(8), 1172-1174, (2001).

71. Z. Yu and P. J. Burke, Microwave transport in metallic single-walled carbon nanotubes. *Nano Lett.*, **5**(7), 1403-1406, (2005).

72. T. Cohen-Karni, L. Segev, O. Srur-Lavi, S. R. Cohen and E. Joselevich, Torsional electromechanical quantum oscillations in carbon nanotubes. *Nat. Nanotechnol.*, **1**(1), 36-41, (2006).

73. J.-P. Salvetat, J.-M. Bonard, N. H. Thomson, A. J. Kulik, L. Forró, W. Benoit and L. Zuppiroli, Mechanical properties of carbon nanotubes. *Appl. Phys. A*, **69**(3), 255-260, (1999).

74. J. N. Coleman, U. Khan, W. J. Blau and Y. K. Gun'ko, Small but strong: A review of the mechanical properties of carbon nanotube–polymer composites. *Carbon*, **44**(9), 1624-1652, (2006).

75. H. Qian, E. S. Greenhalgh, M. S. P. Shaffer and A. Bismarck, Carbon nanotube-based hierarchical composites: a review. *J. Mater. Chem.*, **20**(23), 4751-4762, (2010).

76. M. H. M. O. Hamanaka, V. P. Mammana and P. J. Tatsch, Review of field emission from carbon nanotubes: Highlighting measuring energy spread, in NanoCarbon 2011: Selected works from the Brazilian Carbon Meeting, C. Avellaneda, Editor. 2013, Springer Berlin Heidelberg: Berlin, Heidelberg. p. 1-32.

77. C. N. R. Rao and M. Nath, Inorganic nanotubes. *Dalton Trans.*, 1, 1-24, (2003).

78. R. Tenne, *Issue 15 cover image. Phys. Chem. Chem. Phys.* (2004), **6**

79. R. Tenne, Advances in the synthesis of inorganic nanotubes and fullerene-like nanoparticles. *Angew. Chem., Int. Ed.*, **42**(42), 5124-5132, (2003).

80. M. Nath and C. N. R. Rao, New metal disulfide nanotubes. *J. Am. Chem. Soc.*, **123**(20), 4841-4842, (2001).

81. M. Nath, A. Govindaraj and C. N. R. Rao, Simple synthesis of MoS_2 and WS_2 nanotubes. *Adv. Mater.*, **13**(4), 283-286, (2001).

82. O. Brontvein, D. G. Stroppa, R. Popovitz-Biro, A. Albu-Yaron, M. Levy, D. Feuerman, L. Houben, R. Tenne and J. M. Gordon, New high-temperature Pb-catalyzed synthesis of inorganic nanotubes. *J. Am. Chem. Soc.*, **134**(39), 16379-16386, (2012).

83. Y. Wang, Z. Ma, Y. Chen, M. Zou, M. Yousaf, Y. Yang, L. Yang, A. Cao and R. P. S. Han, Controlled synthesis of core–shell carbon@MoS_2 nanotube sponges as high-performance battery electrodes. *Adv. Mater.*, **28**(46), 10175-10181, (2016).

84. D.-H. Nam, H.-Y. Kang, J.-H. Jo, B. K. Kim, S. Na, U. Sim, I.-K. Ahn, K.-W. Yi, K. T. Nam and Y.-C. Joo, Controlled molybdenum disulfide assembly inside carbon nanofiber by boudouard reaction inspired selective carbon oxidation. *Adv. Mater.*, **29**(12), 1605327, (2017).

85. K. Savva, B. Višić, R. Popovitz-Biro, E. Stratakis and R. Tenne, Short pulse laser synthesis of transition-metal dichalcogenide nanostructures under ambient conditions. *ACS Omega*, **2**(6), 2649-2656 (2017).

86. M. Remskar, A. Mrzel, Z. Skraba, A. Jesih, M. Ceh, J. Demšar, P. Stadelmann, F. Lévy and D. Mihailovic, Self-Assembly of subnanometer-diameter single-wall MoS_2 nanotubes. *Science*, **292**(5516), 479-481, (2001).

87. V. Brüser, R. Popovitz-Biro, A. Albu-Yaron, T. Lorenz, G. Seifert, R. Tenne and A. Zak, Single- to triple-wall WS_2 nanotubes obtained by high-power plasma ablation of WS_2 multiwall nanotubes. *Inorganics*, **2**(2), 177-190, (2014).

88. E. Hossain, A. A. Rahman, R. D. Bapat, J. B. Parmar, A. P. Shah, A. Arora, R. Bratschitsch and A. Bhattacharya, Facile synthesis of WS_2 nanotubes by

sulfurization of tungsten thin films: formation mechanism, and structural and optical properties. *Nanoscale*, **10**(35), 16683-16691, (2018).

89. B. Ni, H. Liu, P.-p. Wang, J. He and X. Wang, General synthesis of inorganic single-walled nanotubes. *Nat. Commun.*, **6**, 8756, (2015).

90. P. Roy, S. Berger and P. Schmuki, TiO_2 Nanotubes: synthesis and applications. *Angew. Chem., Int. Ed.*, **50**(13), 2904-2939, (2011).

91. N. Viriya-empikul, N. Sano, T. Charinpanitkul, T. Kikuchi and W. Tanthapanichakoon, A step towards length control of titanate nanotubes using hydrothermal reaction with sonication pretreatment. *Nanotechnology*, **19**(3), 035601, (2018).

92. X. Wang, L. Sun, S. Zhang and X. Wang, Ultralong, small-diameter TiO_2 Nanotubes achieved by an optimized two-step anodization for efficient dye-sensitized solar cells. *ACS Appl. Mater. Interfaces*, **6**(3), 1361-1365, (2014).

93. I. Milošević, B. Nikolić, E. Dobardžić, M. Damnjanović, I. Popov and G. Seifert, Electronic properties and optical spectra of MoS_2 and WS_2 nanotubes. *Phys. Rev. B*, **76**(23), 233414, (2007).

94. G. Seifert, H. Terrones, M. Terrones, G. Jungnickel and T. Frauenheim, Structure and electronic properties of MoS_2 nanotubes. *Phys. Rev. Lett.*, **85**(1), 146-149, (2000).

95. G. L. Frey, R. Tenne, M. J. Matthews, M. S. Dresselhaus and G. Dresselhaus, Optical properties of MS_2 (M = Mo, W) inorganic fullerenelike and nanotube material optical absorption and resonance raman measurements. *J. Mater. Res.*, **13**(9), 2412-2417, (1998).

96. M. Remskar, A. Mrzel, M. Virsek, M. Godec, M. Krause, A. Kolitsch, A. Singh and A. Seabaugh, The MoS_2 nanotubes with defect-controlled electric properties. *Nanoscale Res Lett*, **6**(1), 26, (2010).

97. R. Levi, O. Bitton, G. Leitus, R. Tenne and E. Joselevich, Field-effect transistors based on WS_2 nanotubes with high current-carrying capacity. *Nano Lett.*, **13**(8), 3736-3741, (2013).

98. S. Fathipour, M. Remskar, A. Varlec, A. Ajoy, R. Yan, S. Vishwanath, S. Rouvimov, W. Hwang, H. Xing, D. Jena and A. Seabaugh, Synthesized multiwall MoS_2 nanotube and nanoribbon field-effect transistors. *Appl. Phys. Lett.*, **106**(2), 022114, (2015).

99. F. Qin, W. Shi, T. Ideue, M. Yoshida, A. Zak, R. Tenne, T. Kikitsu, D. Inoue, D. Hashizume and Y. Iwasa, Superconductivity in a chiral nanotube. *Nat. Commun.* **8**, 14465 (2017)

100. R. Levi, J. Garel, D. Teich, G. Seifert, R. Tenne and E. Joselevich, Nanotube electromechanics beyond carbon: The case of WS_2. *ACS Nano*, **9**(12),12224-12232, (2015).

101. L. S. Panchakarla and R. Tenne, Inorganic nanotubes and fullerene-like nanoparticles at the crossroad between materials *Science* and nanotechnology and their applications with regards to sustainability, in Nanotechnology for Energy

Sustainability, Raj, Van de Voorde, and Mahajan, Editors. 2017: Wiley-VCH Verlag GmbH & Co. KGaA

102. M. Naffakh, A. M. Díez-Pascual and C. Marco, Polymer blend nanocomposites based on poly(l-lactic acid), polypropylene and WS_2 inorganic nanotubes. *RSC Adv.*, **6**(46), 40033-40044, (2016).

103. D. Golberg, P. M. F. J. Costa, M.-S. Wang, X. Wei, D.-M. Tang, Z. Xu, Y. Huang, U. K. Gautam, B. Liu, H. Zeng, N. Kawamoto, C. Zhi, M. Mitome and Y. Bando, Nanomaterial engineering and property studies in a transmission electron microscope. *Adv. Mater.*, **24**(2), 177-194, (2012).

104. Y. Huang, J. Lin, J. Zou, M.-S. Wang, K. Faerstein, C. Tang, Y. Bando and D. Golberg, Thin boron nitride nanotubes with exceptionally high strength and toughness. *Nanoscale*, **5**(11), 4840-4846, (2013).

105. G. A. Wiegers, Misfit layer compounds: Structures and physical properties. *Prog. Solid St. Chem.*, **24**(1–2), 1-139, (1996).

106. A. Meerschaut, Misfit layer compounds. *Curr. Opin. Solid St. M.*, **1**(2), 250-259, (1996).

107. D. Merrill, D. Moore, S. Bauers, M. Falmbigl and D. Johnson, Misfit layer compounds and ferecrystals: Model systems for thermoelectric nanocomposites. *Materials*, **8**(4), 2000-2029, (2015).

108. P. Jood, M. Ohta, H. Nishiate, A. Yamamoto, O. I. Lebedev, D. Berthebaud, K. Suekuni and M. Kunii, Microstructural control and thermoelectric properties of misfit layered sulfides $(LaS)_{1+m}TS_2$ (T = Cr, Nb): The natural superlattice systems. *Chem. Mater.*, **26**(8), 2684-2692, (2014).

109. Z. Diao, H. N. Lee, M. F. Chisholm and R. Jin, Thermoelectric properties of $Bi_2Sr_2Co_2O_y$ thin films and single crystals. *Physica B: Condensed Matter.*, **511**, 42-46, (2017).

110. A. Ansari, A. Kaushik and G. Rajaraman, Mechanistic insights on the ortho-hydroxylation of aromatic compounds by non-heme iron complex: A computational case study on the comparative oxidative ability of ferric-hydroperoxo and high-valent Fe^{IV}=O and Fe V=O intermediates. *J. Am. Chem. Soc.*, **135**(11), 4235-4249, (2013).

111. L. D. Hicks and M. S. Dresselhaus, Thermoelectric figure of merit of a one-dimensional conductor. *Phys. Rev. B*, **47**(24), 16631-16634, (1993).

112. S. Y. Hong, R. Popovitz-Biro, Y. Prior and R. Tenne, Synthesis of SnS_2/SnS fullerene-like nanoparticles: A superlattice with polyhedral shape. *J. Am. Chem. Soc.*, **125**(34),10470-10474, (2003).

113. G. Radovsky, R. Popovitz-Biro, M. Staiger, K. Gartsman, C. Thomsen, T. Lorenz, G. Seifert and R. Tenne, Synthesis of copious amounts of SnS_2 and SnS_2/SnS nanotubes with ordered superstructures. *Angew. Chem. Inter. Ed.*, **50**(51), 12316-12320, (2011).

114. G. Radovsky, R. Popovitz-Biro, T. Lorenz, J.-O. Joswig, G. Seifert, L. Houben, R. E. Dunin-Borkowski and R. Tenne, Tubular structures from the LnS-TaS$_2$ (Ln =

La, Ce, Nd, Ho, Er) and LaSe-TaSe$_2$ misfit layered compounds. *J. Mater. Chem. C,* **4**(1), 89-98, (2016).

115. G. Radovsky, R. Popovitz-Biro, D. G. Stroppa, L. Houben and R. Tenne, Nanotubes from chalcogenide misfit compounds: Sn–S and Nb–Pb–S. *Acc. Chem. Res,* **47**(2), 406-416, (2004).

116. G. Radovsky, R. Popovitz-Biro and R. Tenne, Nanotubes from the misfit layered compounds MS–TaS$_2$, where M = Pb, Sn, Sb, or Bi: Synthesis and study of their structure. *Chem. Mater.,* **26**(12), 3757-3770, (2014).

117. G. Radovsky, R. Popovitz-Biro and R. Tenne, Study of tubular structures of the misfit layered compound SnS$_2$/SnS. *Chem. Mater.,* **24**(15), 3004-3015, (2012).

118. S M. Serra, D. Stolovas, L. Houben, R. Popovitz-Biro, I. Pinkas, F. Kampmann, J. Maultzsch, E. Joselevich and R. Tenne, Synthesis and characterization of nanotubes from misfit (LnS)$_{1+y}$TaS$_2$ (Ln=Pr, Sm, Gd, Yb) Compounds. *Chem. Eur. J,* **24**(44), 11354-11363, (2018).

119. L. Lajaunie, G. Radovsky, R. Tenne and R. Arenal, Quaternary chalcogenide-based misfit nanotubes LnS(Se)-TaS(Se)$_2$ (Ln = La, Ce, Nd, and Ho): Synthesis and atomic structural studies. *Inorg. Chem.,* **57**(2), 747-753, (2018).

120. L. S. Panchakarla, R. Popovitz-Biro, L. Houben, R. E. Dunin-Borkowski and R. Tenne, Lanthanide-based functional misfit-layered nanotubes. *Angew. Chem. Int. Ed.,* **53**(27), 6920-6924, (2014).

121. L. S. Panchakarla, L. Lajaunie, R. Tenne and R. Arenal, Atomic structural studies on thin single-crystalline misfit-layered nanotubes of TbS-CrS$_2$. *J. Phys. Chem. C,* **120**(29), 15600–15607, (2016).

122. L. S. Panchakarla, L. Lajaunie, A. Ramasubramaniam, R. Arenal and R. Tenne, Strontium cobalt oxide misfit nanotubes. *Chem. Mater.,* **28**(24) 9150-9157 (2016).

123. A. C. Masset, C. Michel, A. Maignan, M. Hervieu, O. Toulemonde, F. Studer, B. Raveau and J. Hejtmanek, Misfit-layered cobaltite with an anisotropic giant magnetoresistance: Ca$_3$Co$_4$O$_9$. *Phys. Rev. B,* **62**(1), 166-175, (2000).

124. Yuzuru Miyazaki, Kazutaka Kudo, Megumi Akoshima, Yasuhiro Ono, Yoji Koike and Tsuyoshi Kajitani, Low-temperature thermoelectric properties of the composite crystal [Ca$_2$CoO$_{3.34}$]$_{0.614}$ [CoO$_2$]. Japanese *J. Appl. Phys.,* **39**(6A), L531 (2000).

125. L. S. Panchakarla, L. Lajaunie, A. Ramasubramaniam, R. Arenal and R. Tenne, Nanotubes from oxide-based misfit family: The case of calcium cobalt oxide. *ACS Nano,* **10**(6), 6248–6256 (2016).

126. A. Ramasubramaniam, First-principles studies of the electronic and thermoelectric properties of misfit layered phases of calcium cobaltite. *Isr. J. Chem.,* **57**(6), 522-528 (2007).

127. L. Lajaunie, A. Ramasubramaniam, L. S. Panchakarla and R. Arenal, Optoelectronic properties of calcium cobalt oxide misfit nanotubes. *Appl. Phys. Lett.,* **113**(3), 031102, (2018).

Chapter 2

Graphene and Other 2D Materials

Uttam Gupta and C.N.R. Rao[*]

International Centre for Materials Science, Sheikh Saqr Laboratory, and School of Advanced Materials, Jawaharlal Nehru Centre for Advanced Scientific Research, Jakkur, Bangalore 560064, Karnataka, India
[*]*cnrrao@jncasr.ac.in*

The discovery of the amazing properties of graphene has evoked great interest in the various classes of two-dimensional (2D) materials. 2D inorganic analogues of graphene such as the transition metal dichalcogenides are investigated widely and these materials, especially MoS_2, exhibit many properties of interest. In particular, they possess properties of direct use in energy devices. In this chapter, we give a brief overview of the synthesis and features of the 2D materials and applications such as transistors, sensors, supercapacitors and batteries as well as in oxygen reduction and hydrogen evolution reactions.

1. Introduction

Advancements in materials science provide a major driving force of technological progress. The discovery of high electron mobility in graphene established that the properties of two-dimensional (2D) materials could be different and sometimes far superior to those of the bulk counterparts. Graphene is a one-atom-thick carbon nanosheet, which was isolated from the parent graphite in 2004 and became the first 2D nanostructure.[1] Later, it captured the interest of researchers from different fields, such as electronics, photonics, materials science, engineering and sensors, by serving as a model two-dimensional system. The unique physical and chemical properties of graphene such as the large specific surface area, superior electrical conductivity, excellent thermal stability and conductivity with oxidation resistance temperatures up to 600°C, remarkable mechanical strength, the excellent optical transmittance of 97.7% has found many applications. For example, graphene and its derivatives are widely studied in the field of electrochemistry as they demonstrate interesting electrochemical properties with a wide-ranging electrochemical potential and a low charge transfer resistance.

Graphene, with a plethora of unique properties, has its drawbacks which limit its applications. For example, graphene lacks an intrinsic band gap. Expediently, after graphene's discovery enormous interest toward other 2D materials has increased, which possess novel properties rivaling graphene. Other elemental materials also give rise to graphene-like nanosheets (Phosphorene, silicene, germanene and so on). Hexagonal boron nitride (hBN), transition metal dichalcogenides (MoS_2, TaS_2, TiS_2, and others), transition metal trichalcogenides ($NbSe_3$, $TaSe_3$), transition metal oxides ($LaVO_3$, $LaMnO_3$) and other layered materials belong to the pool of 2D materials which cover a range of properties (Figure 1).[2] We have discussed graphene, transition metal dichalcogenides and recently emerging material borocarbonitrides (BCN) in this article.

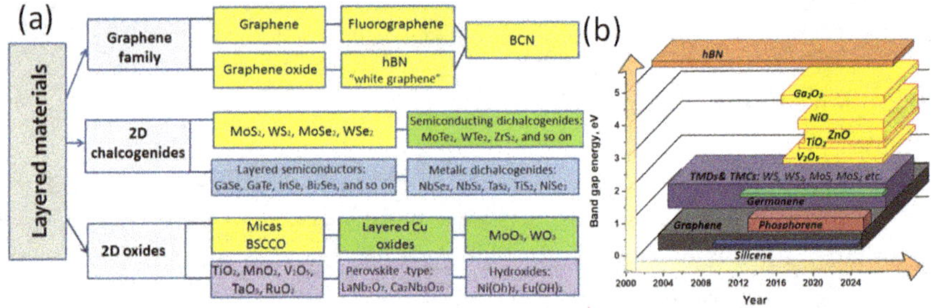

Fig. 1. (a). The chart is elucidating the categorized library of 2D materials. The yellow and green box contains material which is stable and potentially stable in ambient conditions respectively, the blue box lists materials stable only in an inert atmosphere, and the purple box has materials that have been exfoliated into monolayers. (b) Illustration of the evolution of the family of 2D materials as a function of time (horizontal axis) and their respective band gap values (vertical axis). The yellow components represent the expected 2D metal oxides contribution. Adapted from Ref. 2. Copyright 2016 MDPI Journals.

Graphene and transition metal dichalcogenides (TMDs) and such 2D-layered materials exhibit fascinating properties which can be employed for applications in transistors, sensing, energy storage and catalysis. However, due to the gapless band structure of graphene, it cannot be used in certain applications. Strategies to create a band gap in graphene by hetero-atom doping or by surface modification have been pursued. Nitrogen-doped graphene shows good electron mobility with n-type semiconducting behavior whereas the boron doped is ambipolar.[3] Increasing interest in graphene has led to the exploration of other 2D-materials. Borocarbonitrides containing hexagonal BN and graphene domains in varying proportions show interesting electronic and gas adsorption properties. TMDs include materials with a variety of properties like insulating HfS_2, semiconducting MoS_2 metallic TaS_2 and $NbSe_2$. TMDs also exhibit exotic features with new properties arising from the valley polarization from spin-orbit splitting, low-temperature superconductivity, charge-density waves and Mott metal to insulator transition.[4] In addition to the physiochemical properties, the layered materials have efficiently been used as materials in transistors, sensors and in energy devices.[5]

In this chapter, we give an overview of synthetic strategies, properties and applications of the 2D materials especially graphene, borocarbonitrides and transition metal dichalcogenides.

2. Structural Aspects

2.1. *Graphene*

Graphene is a two-dimensional sheet of extended honeycomb network of sp^2-hybridized carbon (Figure 2(a)). It can be considered as a fundamental building block of other important allotropes of carbon where it can be wrapped to form 0D fullerenes, rolled to form 1D nanotubes and can be stacked to form a 3D graphite. Charge transport in graphene is ambipolar.[3,6] This implies that carriers can be tuned continuously between holes and electrons by supplying the requisite gate bias due to the unique band structure of graphene (Figure 2(b)). Under

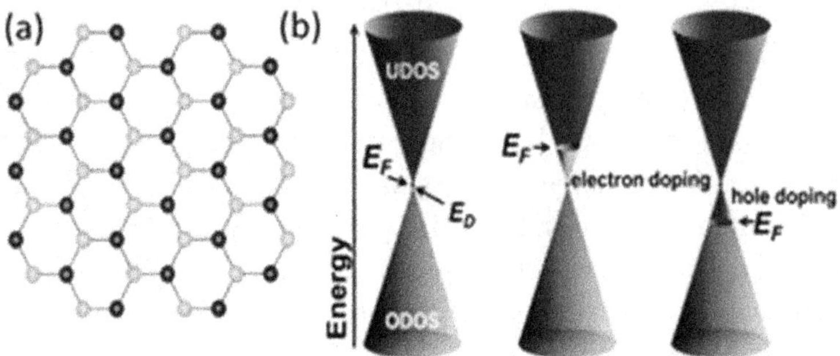

Fig. 2. (a) The hexagonal arrangement of carbon atoms in a graphene sheet depicting two unique positions per unit cell in green and blue. (b) Representation of the Dirac cone (ED) at the K point in (b) depicting the linear dispersion of conduction and valence bands that meet at the Dirac point. Rigid band shifts due to charge transfer to and from graphene are depicted as electron or hole doping, respectively, wherein the Fermi level (EF) is displaced to higher or lower energy with respect to the Dirac point. Adapted from Ref. 6. Copyright 2014 Royal Society of Chemistry.

negative gate bias, the Fermi level drops below the Dirac point, introducing a significant population of holes into the valence band. Under positive gate bias, the Fermi level rises above the Dirac point, promoting a significant population of electrons into the conduction band (Figure 2(b)). The type of carriers can be manipulated by doping with B-atom (holes) or N-atom (electrons) in graphene. The bandgap can be opened up by breaking the symmetry which can be achieved by functionalization or by substituting atoms in the graphene network.[7]

2.2. *Borocarbonitrides*

The properties of borocarbonitrides (BCN) vary over a wide range, with the composition. BCN exists as hexagonal nanosheets containing graphene and BN domains, possibly with BCN rings (Figure 3(a)). The energetic stabilities of the bonds in BCN are in the order B-N > C-C > C-N > C-B > B-B > N-N and therefore the presence of B-B or N-N bonds is unlikely. Further, a single carbon atom in the BN matrix is also energetically unfavorable (Figure 3(a)).[8] In C-rich samples, BN islands are found in a graphene matrix and vice versa. The zig-zag interface of graphene-BN is more favorable as compared to the armchair.[9] Moreover, at the interface of graphene-BN, Stone-Wales (SW) defects, the formation of a pair of pentagon and heptagon (5/-7 dislocations), may be observed due to the 90° rotation of a C-C, C-B, C- or B-N bond. The SW defects also depend on the length of the graphene and BN domains, which may give rise to interesting properties in the BCN.[10]

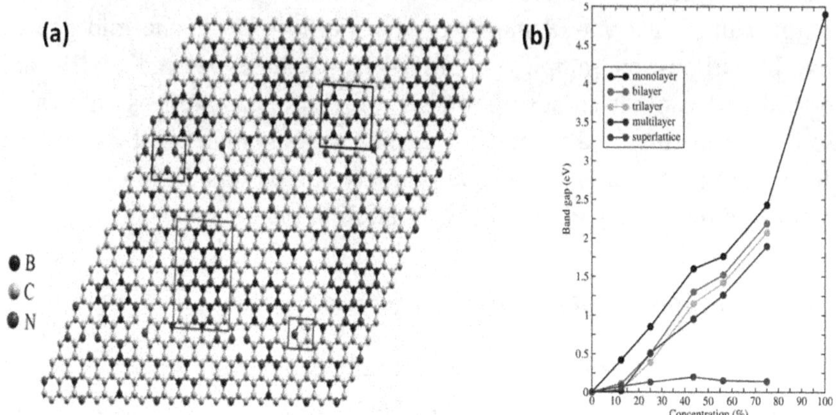

Fig 3. (a) Schematic of borocarbonitride, $B_xC_yN_z$. Adapted from Ref. 10. (b) Bandgap as a function of the substitution of BN for various BN-doped graphene systems for monolayer, bilayer, trilayer and multilayer for concentrations between 0% and 75%. Adapted from Ref. 11. Copyright 2014 Royal Society of Chemistry. Copyright 2014 AIP Publishing.

The bandgap in graphene can be introduced by breaking the sub-lattice symmetry. The magnitude of the bandgap and distribution of mid-gap states in BCN depend upon the number of B and N atoms. If the number of B is greater than N, then the sample is hole-doped, while the N-rich sample is electron-doped. However, if the B and N atoms are equally substituted in graphene, the concentrations of e^- and h^+ are equal, and the density of states above and below Fermi-energy are symmetric.[10] Schwingenschlögl and co-workers calculated the bandgap in BCN with the increasing concentration of BN, with an equal number of B and N atoms for monolayer, bi-layer, tri-layer and multilayer of BCN (Figure 3(b)). The bandgap increases progressively with increase in the concentration of BN for all the layers with increasing number of layers due to interlayer coupling (Figure 3(b)).[11]

Experimentally, absorption in the visible light region is greatly enhanced with the red-shift in the absorption edge with an increase in the C-content of the BCN nanosheets with a reduction optical band gap of-of BN (5.5 eV) to 2.6 eV for ~24% of C in BCN.[12] Additionally, the electronic properties of BCN are influenced by the domain structure.[10] In a BN rich system, the carbon domains exhibit narrow quantum dot type energy bands, just above and below the Fermi-level in the mid-gap of BN. In carbon-rich samples, HOMO-LUMO states related to BN are formed and can act as adsorption sites for gases as well as catalysis.[9] Both experimental and theoretical studies suggest high tunability of the electronic properties of BCN. The property can be tuned from insulating to metallic by varying the BN concentration in the carbon.

2.3.   *Transition metal dichalcogenides*

Group 4–7 and 10 transition metal dichalcogenides are generally found to have layered structures. The single-layer of transition metal dichalcogenides with the general formula of MX_2, depicts M as a transition metal atom and X as a chalcogenide atom. Metal atoms are hexagonally packed and sandwiched between two layers of chalcogen atoms. The intra-layer M–X bonds are covalent in nature, whereas the interlayers are weakly bonded by van der Waals force.[13] They are two co-ordination type, trigonal prismatic (with point group as D_{3h}) or trigonal antiprismatic or octahedral (with point group as D_{3d}) which gives rise to three crystal phases (Figure 4(a)) trigonal, hexagonal and rhombohedral. They are depicted as 1T, 2H and 3R where the letters stand for trigonal, hexagonal and rhombohedral, respectively and the numbers represent the number of MX_2 units in the unit cell. Both 3R and 2H forms have trigonal prismatic coordination, while 1T has an octahedral coordination. This crystal structure can be differentiated by using different techniques like high-resolution transmission electron micrography in the annular dark field mode.[14]

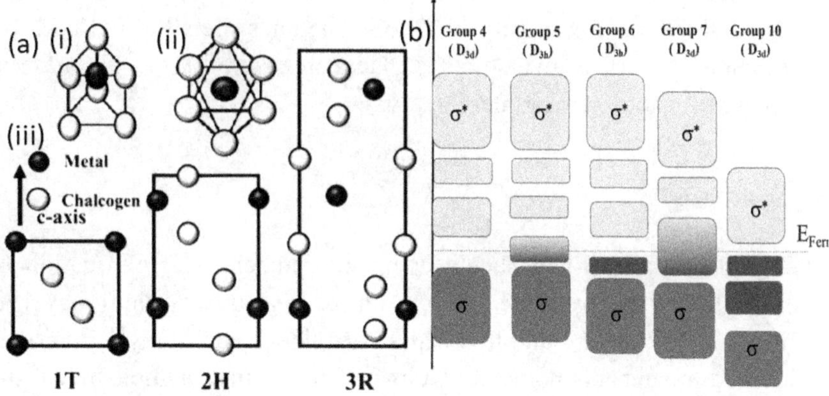

Fig. 4. (a) Schematic of (i) trigonal prismatic, (ii) octahedral coordination and (iii) the stacking sequences in the 1T, 2H and 3R polytype. The structures are shown as projections onto the (110) plane. They are depicted as 1T, 2H and 3R where letters stand for trigonal, hexagonal and rhombohedral, respectively, and the number represents the number of MX_2 units in the unit cell. Both 3R and 2H forms have trigonal prismatic coordination, while 1T has octahedral coordination. (b) Schematic representation showing filling of electrons in d orbitals that are located within the bandgap of bonding (s) and non-binding (s*) bands of transition metal dichalcogenides of Group 4-7, and 10 of their naturally abundant state. The filled and unfilled are shaded as gray and light gray respectively. Adapted from Ref. 13. Copyright 2013 Nature Publishing Group.

The electronic structure of transition metal dichalcogenides depends upon the d-electron count of the metal atom. A progressive increase in the number of d-electrons from Group 4 to 10 gives a different electronic property as illustrated in Figure 4(b). For example, MoS_2 is naturally present in a 2H-form, whereas TaS_2 is present as an 1T-form. In both the forms the non-bonding d-bands of the transition metal dichalcogenides are located in the gap between the bonding (σ) and non-bonding (σ*) bands of metal chalcogen bonds shown in Figure 4(b) with respect to the Fermi-level. Transition metal dichalcogenides of Group 4 (d^0), Group 6 (d^2) and Group 10 (d^{10}) can be present as a semiconductor and the band gaps vary depending on the metal atom and chalcogen atom. Group 5 (d^1) and Group 7 (d^3) are metallic in nature as the orbital is partially filled and the Fermi level lies within the band. Interestingly, one can change the electronic property of the transition metal dichalcogenides by changing the crystal structure. For example, lithium intercalation in

MoS$_2$ followed by exfoliation transforms from 2H to 1T polymorph implying that semiconducting 2H-MoS$_2$ becomes metallic in 1T-MoS$_2$, in principle.[14, 15] The reverse takes places in case of TaS$_2$, 1T-phase to 2H-phase transition, on Li intercalation.[16]

3. Synthesis

The synthesis of 2D-nanosheets using micromechanical exfoliation of bulk single crystals by scotch tape produces high-quality single-layer as well as a few-layer samples which are ideal for studying various condensed matter phenomena. Geim et al. [17] first demonstrated the micromechanical cleavage of graphite crystal using the scotch tape method to synthesize the single-layer graphene. However the yields were very poor and therefore to study their properties liquid exfoliation was preferred. Chemical vapor deposition on different metal surfaces yields a single layer to few-layer graphene with varying sources of carbon (methane, ethylene, acetylene). Scalable synthesis of high quality doped (N and B atoms) and undoped few-layer graphene (2-4 layers) can result in the arc-discharge of graphite in a suitable environment.[18] The thermal treatment of graphene oxide in the presence of ammonia or chemical vapor deposition can give a higher percentage of N-doping.[19] Borocarbonitrides (BCN), which have both graphene and boron nitride domains can be synthesized by heating activated charcoal, urea and boric acid at a high temperature in an inert atmosphere.[9] The composition of BCN can be tuned by varying the contents of urea and boric acid.

Direct sonication in a suitable solvent exfoliates 2D-materials and TMDs like MoS$_2$, MoSe$_2$, WS$_2$, TaS$_2$ into single and multiple layers (Figure 5(a)). The dispersing medium, therefore, must have surface energies comparable to the van der Waals force to overcome the cohesive energy between the adjacent layers (Figure 5(a)).[20] Mechanical exfoliation, however, provides no control over the number of layers as well as the lateral dimension. Chemical exfoliation can be used for mass production (~100%) of exfoliated TMD nanosheets as exemplified by the use of n-butyl lithium dissolved in hexane as the intercalation agent (Figure 5 (b)).[21] Interestingly, Li-intercalation exfoliates sheets from the

3D-bulk crystal but sometimes changes the glass and electronic structures. For example, the semiconducting 2H-form of MoS_2 changes to the metallic 1T-MoS_2 after Li-intercalation and exfoliation. [13, 22] There are various routes for exfoliating 2D materials from their bulk which can be suitably chosen based on the material.[23]

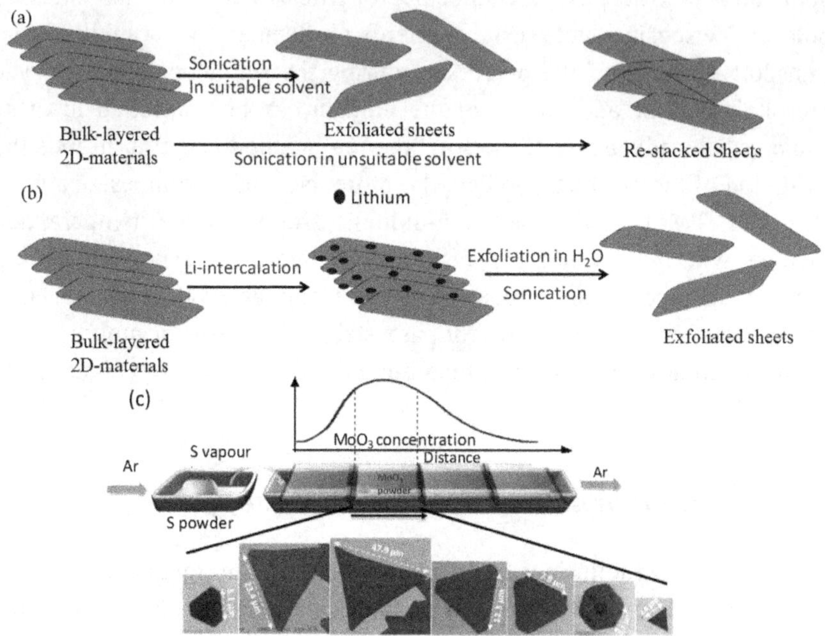

Fig. 5. (a) Exfoliation of bulk 2D-layered materials either by probe or bath sonication in a suitable solvent can give high-quality single- to few-layer sheets. In unsuitable solvents, the exfoliated sheets can be restacked. (b) Chemical exfoliation by Li-intercalation followed by exfoliation in water gives large-scale exfoliated single- to few-layer sheets. (c) Large-area sheets of 2D TMDs under suitable conditions can be synthesized. A schematic representation of the synthesis of MoS_2 nanoflakes with different shapes. The concentration of MoS_2 is the determining factor in the evolution of these shapes during synthesis. The Ar flow rate is 100 sccm and reaction occurs on 300 nm SiO_2 300 nm SiO_2 coated Si-chip. Adapted with permission from Ref. 24. Copyright 2014 American Chemical Society.

Large area films of layered-TMDs can be grown by physical or chemical vapor deposition under suitable conditions. High-quality TMDs are synthesized by vapor-phase transport recrystallization from powders, vapor-phase deposition methods which include sulfurization or selenization of metals or metal oxides or thermal decomposition of thiosalts.[24, 18] The ratio of precursors controls the physical and chemical properties of the films (Figure 5(c)).[24]

Solvothermal synthesis can give a large-scale synthesis of TMDs for bulk applications like catalysis, supercapacitors and other similar applications. The physicochemical properties of the solvents, such as polarity, viscosity and surface energy influence the solubility and transport behavior of the precursors during the synthetic process.[25] The morphology, size and phases of the product can be controlled in these solvents. The advantage of the solvothermal method is that it inhibits the oxidation of the product, and can, therefore, be used to synthesize a wide range of TMDs and other non-oxides. The moderate temperatures together with the advantage of fast reaction kinetics and short processing time render a promising route to prepare similar products on a large scale, economically. In the recent past, stable 1T-phases of molybdenum dichalcogenides have been prepared solvothermally as a bottom-up approach. [26]

4. Giant Magnetoresistance and Superconductivity

Distorted layered tellurides of M and W exhibit giant magnetoresistance (MR) and pressure-driven superconductivity (Figures 15(a) and (b)) and have thus attracted attention. Single crystals of semimetallic orthorhombic Td-WTe$_2$ exhibit giant, unidirectional (along W–W chains) positive MR of 452,700% at 4.5K in a magnetic field of 14.7 T, and 13 million % at 0.53K in a field of 60 T has been reported. [27] Interestingly, at very high applied magnetic fields, MR in Td-WTe$_2$ does not saturate, which is ascribed to a balanced electron-hole resonance. Bulk monoclinic 1T-MoTe$_2$ with Mo–Mo zigzag chains is semimetallic and also exhibits a giant MR of 16 000% in a magnetic field of 14 T at 1.8K.[28] Superconductivity in Td-WTe$_2$ appears at a pressure of 2.5 GPa with the maximum critical temperature (T_c) of 7K at around 16.8 GPa which monotonically decreases with increasing pressure.[29] The superconductivity and MR show a unique behavior with increasing pressure, MR is gradually suppressed and disappears at a critical pressure of 10.5 GPa, where superconductivity begins to emerge.[30] Similarly, pressure-induced superconductivity has also been seen in single crystals of orthorhombic T_d-MoTe$_2$ (below 120 K), a low-temperature polymorph of 1T-MoTe$_2$. A maximum T_c of 8.2K is observed at 11.7 GPa.[31]

5. Transistors

Graphene exhibits an extraordinary mobility of ~25000 cm^2 V^{-1} s^{-1} at room temperature and is most widely studied for transistor properties.[32] Graphene output characteristics can be influenced by gate voltage in a FET device since it is a zero-band-gap semiconductor. However, it shows poor on-off ratios (<10) due to the delocalized electrons, and therefore, it is not possible to switch off the FET. Optimization of graphene, like creating nanoribbons which creates a small band can result in attractive FET properties. Graphene nanoribbons (10 nm), with a small band-gap of 140 meV, show conventional band transport at room temperature.[33] Chemically synthesized RGO has been used for a transistor, but the performance has been poor as compared to the mechanically exfoliated graphene.[34] An optimal balance between conductivity and band gap is essential for functional FETs. Borocarbonitrides (BCN) have exhibited FET properties with mobility ranges from 6.4-10 cm^2/V s for carbon-rich samples.[35] Thus, other 2D-materials like transition metal dichalcogenides (TMDs), silicene and phosphorene have been recently studied. The electronic mobility shows a relationship with electronic structure and is demonstrated for graphene, TMDs and other 2D-materials with state of the art materials and other 3D-materials (Figure 6(a)).[36]

The bandgap of TMDs can increase with a decrease in the number of layers and varies from 1.1 eV to 2.2 eV.[13] Initial uses of TMDs in the FETs were reported in 2004 with WSe$_2$ crystals, where mobility was comparable to the best single crystal Silicon FETs (500 cm^2 V^{-1} s^{-1} for p-type conductivity with on/off ratio 10^4 on/off at 60 K). [37] Soon after, FETs of mechanically exfoliated MoS$_2$ of thickness 8-40 nm with back-gated configuration have shown mobilities in the range of 0.1-10 cm^2 V^{-1} s^{-1}. [17, 38] With regard to the back-voltage-gated FET configuration at a gate voltage of -50 V, n-type transport characteristics were observed with the mobility of up to 50 cm^2 V^{-1} s^{-1} and with the high on-off ratio of 10^5.[38] MoS_2 has attracted attention over other TMDs since it is robust, naturally abundant and can be freely mechanically exfoliated. The top-gated monolayer MoS_2 transistor based fabricated by Kis and co-workers with n-type conductivity and mobility of >200 cm^2 V^{-1} s^{-1} had an on/off ratio of 10^8 at room temperature (Figure 6(b)).[39] The high-k dielectric HfO_2 benefits the mobility of the monolayer of MoS_2, while the top gate geometry can allow a reduction in the voltage, necessary to switch the device that enables integration of multiple devices on a single substrate. Top-gated high dielectric p-type FET with an active channel made of monolayer flake of WSe_2, exhibited a room-temperature performance of ~250 cm^2 V^{-1} s^{-1} hole mobility with an on/off ratio of 10^6.[40] The lithography patterning of multiple sets of the electrode on a single piece of monolayer MoS_2 by Radisavljevic et al. built a functional electronic circuit capable of performing digital logic operations as a logical NOR gate which is considered to be one of the universal gates (Figure 6(c)).[41] The ambipolar transport was demonstrated in a thin (10 nm thick) MoS_2 electric double layer transistor using an ionic liquid as the gate to reach extremely high carrier concentrations of 10^{14} cm^2 with an on/off ratio of 200.[42] Heterojunctions of 2D-materials have shown good capacity as interband or interlayer tunneling transistors operating at low potentials.[43]

In contrast to electronic and spintronics devices which exploit the charge and spin of electrons respectively, valleytronics employs the conduction (valence) bands of the materials which possesses two or more degenerate minima (maxima), separated in the momentum space.[44] A valleytronic device controls the number of carriers during valley-polarization. The crystal possesses two bands being of spin down ($E\downarrow$) and spin up ($E\uparrow$) in character owing to the broken spin degeneracy with time-reversal symmetry which leads to an inherent coupling of the valley and spin of the valence bands resulting in the valley-dependent optical selection rule. Since the circular component of the band edge luminescence is of the same polarization as that of the circularly polarized excitation, the inter-band transitions in the environs of the K- or K valleys couple exclusively to the right or left a circularly polarized light, respectively.[45] However, the inversion and time-reversal symmetry

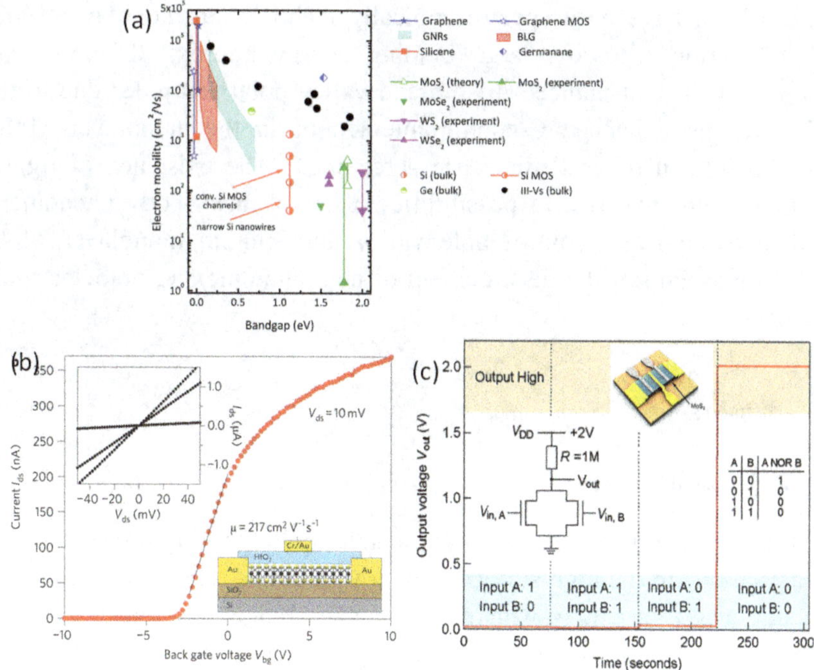

Fig. 6. (a) Comparison of room-temperature electron mobility vs. band gap for different materials with 2D materials like graphene and TMDs. Adapted with permission from Ref. 36. (b) Schematic of a cross-sectional view of the top-gate FET of a monolayer MoS₂ (inset in right-hand corner). Transfer characteristic for the FET with applied bias voltage V_{ds} (10 mV). The back-gate voltage (V_{bg}) is applied to the substrate while the top gate is disconnected. Drain-source current (I_{ds}) –drain-source voltage (V_{ds}) curves acquired for V_{bg} at 0, 1 and 5 V. Adapted from Ref. 39. (c) Schematic of a device for NOR gate based on single-layer MoS₂ (inset middle top). Two monolayer MoS₂ transistors are connected in parallel using an external 1 MΩ resistor as a load (left inset) the logic circuit is formed. The output voltage (V_{out}) is shown in various regimes for four different sets of input states (1,0), (1,1), (0,1) and (0,0) with response only, with both inputs in the low state. The truth table shows that these combinations can be shown as NOR gate operations (inset). Adapted from Ref. 41. Copyright 2016 Elsevier. Copyright 2011 Springer Nature. Copyright 2011 American Chemical Society.

is preserved in bilayer MoS₂, the spin-degeneracy is restored by splitting the fourfold degenerate valence bands into two spin-degenerate valence bands. However, in bilayer MoS₂ valley and spin are decoupled, the selection rule is not allowed which results in negligible circular polarization.[45] Quasi-Particle, negative trions, composed of two electrons and a hole, have been spectroscopically identified in monolayer MoS_2 FET.[46] Trions (possess large binding energy approx. 20 meV), are formed with light-induced in spin and valley polarized holes due to the significantly enhanced Coulomb interactions in the monolayer MoS_2, arising from a reduced dielectric screening.[46] The existence of tightly bound trions give rise to possibilities of novel many-body phenomena which dynamically control hole valley and spin in monolayer MoS_2 which may impact the development of new photonic and optoelectronic devices.

6. Sensors

6.1. *Material sensors*

2D-layered geometry offers a unique solution to sensing properties of a material due to the high surface area, low-single to noise ratio compared to 0D or 1D structures which varies in the range of interactions from weak van der Waals interaction to strong covalent bonds.[1,47] The slightest interaction of the molecules with graphene changes the localized impurity states, which can be manifested by variation in the conductivity of the system (change in the conductivity which is ΔR).[48] The presence of an electron-donating molecule like NH_3 decreases the conductivity by reducing the concentration of p-type carrier and its conductivity while, NO_2, an electron-withdrawing molecule increases the concentration of p-type carriers and its conductivity (Figure 7(a)). Single layer graphene gas sensor has a detection limit of ppb for NO_2 molecule, which was further improved by using Hall geometry to provide a strong response by changing the carrier density near the Dirac point (Figure 7(b)).[49] Single layer graphene has been an attractive material to detect gases and small organic molecules like methanol, THF, acetonitrile, and chloroform.[18] Vapors of different molecules induce noise of non-identical frequency which can be employed as a signal to detect these molecules (Figure 7(c)).[50] However, pristine graphene is highly sensitive to contaminants and this can be overcome on illumination with UV-light. The UV-light illuminated sensor is highly sensitive to N_2O, NH_3, SO_2, and H_2O with detection limits in the range of 39-136 parts per trillion.[51]

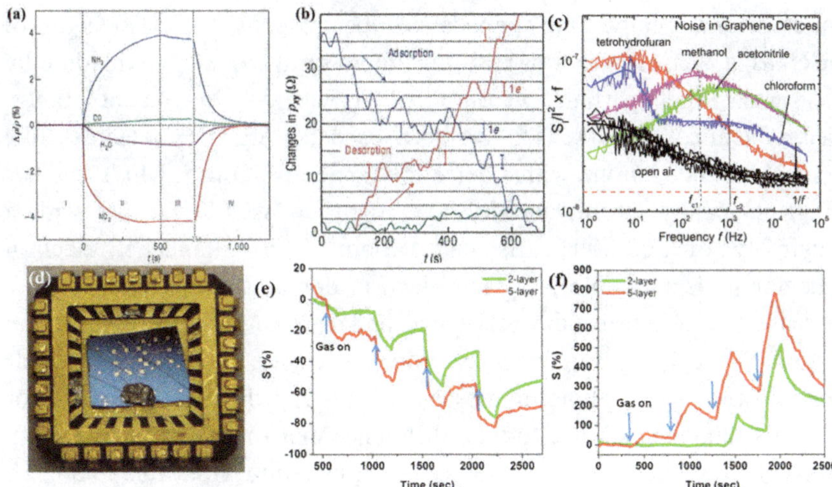

Fig. 7. (a) The graph represents a change in resistivity with respect to time over different stages of gas sensing procedure of various gases (1 ppm). Vacuum (Region I), when gas is introduced (Region II) on exposure to gas, evacuation of the experimental set-up (Region III) and annealing at 150°C to de-adsorb gas (Region IV). The positive sign indicates electron doping while the negative sign indicates hole doping. Adapted from Ref. 49. (b) The change of Hall resistivity ρ_{xy} observed near the neutrality point through adsorption (blue curve) and desorption (red curve) of diluted NO_2. The green curve is when the sensor is exposed to pure He, which is used as a reference. Adapted from Ref. 52. (c) Analysis of product of noise spectral density (SI/I₂) multiplied by frequency (f) vs. f for the sensor upon exposure to different vapour samples along with open air. (d) Photograph of the transistor-based MoS_2 -sensing device mounted on the chip. Sensing behaviour of (e) NH_3 and (f) NO_2 gas with and without applying a back-gate voltage (V_g=15 V) for two-layer MoS_2 (arrows indicate the concentration of gases at 100, 200, 500 and 1000 ppm of each gas). Adapted from Ref. 64. Copyright 2017 Springer Nature. Copyright 2017 American Chemical Society.

Chemically synthesized graphene-like RGO shows good sensing properties and are highly sensitive to H_2O vapor and gases like NO_2 and Cl_2 at low concentrations. The sensitivity is further increased by chemical modification.[18] Chemically synthesized few-layer graphene effectively detects biological molecules like acetaminophen, uric acid or ascorbic acid, neurotransmitters, oxidase / dehydrogenase-related molecules, DNA bases, nicotinamide adenine dinucleotide (NADH), and other biomolecules.[52] FET based patterned RGO sensors can do real-time detection of biomolecules or heavy metals at low concentrations. Electrochemical sensors based on RGO detect biomolecules at very low concentrations employing impedance spectroscopy.[53] Earlier uses of BCN as a sensor was for real-time monitoring of slag oxidation by measuring the partial pressure of oxygen.[54] In recent times, electrochemical BCN sensor has been used for organic molecules like uric acid and dopamine with a detection range of 2-20µM.[55] FET devices of mechanically exfoliated MoS_2 are reported as NO sensors with a sensitivity of 300-800 parts per billion.[56] Like graphene, electron acceptor molecules like NO_2 and electron donating molecule like NH_3 increases or decreases the resistance and can be useful for detection (Figure 7 d-f).[57] Single layer MoS_2 detects DNA and similar biomolecules using photoluminescence and can differentiate single or double-stranded DNA due to their difference in affinity. [58] In conclusion, graphene has a better sensing ability as compared to MoS_2 due to a large change in conductivity (ΔR) on perturbation with the contaminants. The sensing properties of the gases by graphene and other TMDs is highlighted in Table 1.

Table 1. The sensing properties of graphene and TMDs.

Sensor material	Target gas	Sensitivity and limit of detection
Graphene		
[49] Mechanically exfoliated graphene	NO_2	$\Delta R > 2.5\Omega$ for one electron Single molecule detection
[50] Mechanically exfoliated graphene	CH_3OH, C_2H_5OH, THF, CH_3CN	Using frequency vs. resistivity pattern variation of each vapor molecule
[59] CVD grown G	NO_2, O_2, SO_2, NO	$\Delta R = 0.2\text{-}0.5\%\ ppm^{-1}$ 158 ppq-103 ppt
[60] RGO (printable sensor)	NO_2, Cl_2	500 ppb-100 ppm
MoS_2		
Mechanically exfoliated 1L-MoS_2		Selective detection of electron donors like Triethylamine, selectivity better than CNT 10 ppb
Mechanically exfoliated 2&5L MoS_2	NO_2, NH_3, RH	$\Delta R/R = 1.37\%\ ppm^{-1}$

6.2. *Photodetectors*

Graphene has optical absorption over a wide range with strong interband transitions which can be used for ultrafast photodetectors in single layer FET-based detectors.[3,18] Chemically synthesized graphene offers tunable, cheaper, scalable solution processed sensors which have exhibited UV and IR detectors with good photoresponse and an on/off ratio.[61] The detectors are sensitive to signals as weak as the human-hand. However, a suitable band gap is preferred for photodetector and photovoltaic applications.

Layered TMDs being bandgap semiconductors are suitable for photodetector and photovoltaic applications. Single and double-layer MoS_2 based detectors respond to green light while three-layered MoS_2 are responsive to red light. [62] Solution-processed chemically exfoliated MoS_2 and $MoSe_2$ detect near IR absorbance better than graphene with and MoS_2.[63] Phototransistors of single-layer MoS_2 show quick response and good photosensitivity with a fast switching rate which could be tailored by the gate voltage, making them even better than graphene-based devices.[5] Heterostructures of TMDs with other materials can show strong sensing properties towards both light and matter.[64]

7. Supercapacitors

Supercapacitors are electrochemical energy storing devices which find applications because of the high energy and power densities. Carbon-based materials such as carbon nanotubes, carbon nanofibers and graphene have shown excellent supercapacitor properties due to their high surface area and electrical conductivity.[65] Chemical (doping or functionalization) and physical modifications (crumbling, surface area, meso- and micro- pore) in graphene can significantly enhance the supercapacitor properties.[61, 66] Nitrogen plasma processed graphene shows nearly four times higher capacitance as compared to unprocessed graphene with excellent cycling stability and power densities which can be further enhanced by increasing pore volume, nitrogen content and surface area.[67] Pyrrolic and pyridinic nitrogen are electrochemically active which resultantly increase in capacitance whereas the graphitic nitrogen increases the conductivity of the material.[68] In addition to nitrogen-doped graphene, boron-doped graphene also shows an increase in capacitance when compared to the pristine graphene due to the formation of oxygen-containing functional groups such as BC_2O/BCO_2.[69] Porous B-doped graphene synthesized by fried ice method exhibits a high surface area of 622 m^2 g^{-1} and a capacitance of around 281 F g^{-1}.[70] Borocarbonitrides (BCN) show excellent supercapacitor properties with carbon-rich BCN manifesting a high specific capacitance of 178 F/g and 240 F/g in aqueous electrolyte and ionic liquid respectively.[35] Layered transition metal dichalcogenides (TMDs) such as MoS_2 and WS_2 have been explored for supercapacitor applications.[5] Chemically exfoliated 1T MoS_2 nanosheets exhibit high capacitance values in the range of 400 to 700 F cm^{-3} for aqueous electrolytes and the material can be charged up to a high-voltage (3.5 V) in organic electrolytes.[71] The capacitance of MoS_2 can be tuned by making composites with graphene. MoS_2-graphene composites have shown extraordinary supercapacitor properties.[5, 72]

8. Oxygen Reduction Reaction

Oxygen reduction reaction (ORR) plays a vital role in fuel cell technology, and the reaction kinetics are sluggish. Nitrogen-doped graphene has shown phenomenal ORR activity in most of the cases with an electron transfer of 4,[5] for example, nitrogen-doped graphene synthesized via various chemical routes such as chemical vapor deposition, pyrolysis of polyaniline-graphene or GO-melamine turns out to be an excellent robust ORR catalyst in alkaline medium with low overpotential.[61] Theoretical calculations predict that the nitrogen doping in graphene introduces asymmetry spin density and atomic charge density which in turn enhances the electrocatalytic activity.[73] BCN with different proportions of nitrogen and boron have shown excellent oxygen reduction reaction activity with an electron transfer of 4 in alkaline medium.[74] Few-layer MoS_2 particles exhibit improved catalytic activity towards the ORR when compared to bulk.[75] The exposed Mo edges of the MoS_2 and nanostructuring plays an important role in the enhanced activity. Composites of MoS_2 and graphene and other carbon materials have shown significant ORR activity when compared to the pristine MoS_2.[76]

9. Lithium-Ion Batteries

Lithium-ion batteries are energy storage devices with high energy density and power density and play a vital role in mobile electronics, etc. 2D materials like graphene and MoS_2, because of the fast ion-conduction and reversible Li-intercalation and de-intercalation properties, are ideal hosts for the batteries.[77] The relatively low density of lithium in graphite leads to the relatively low specific capacity of graphite. Individual graphene sheets increase the storage capacity since lithium is stored on both sides of the graphene sheet. The low lateral dimension allows Li^+ to diffuse into the interlayer space much easily with high reversibility.[78] Nanostructuring or creating pores (holes) in graphene sheets increases high-rate discharge capability of Li-ion battery anodes which can be functionalized to facilitate selective ion passage to further enhance the charge storage capacity.[78] Borocarbonitrides have shown the exceptional stability of more than 100 cycles and exhibit specific capacities ranging from 710 mAh g^{-1} to 150 mAh g^{-1}. MoS_2 based Li-batteries reported by Haering et al. have shown a charge capacity of 750 mAh g^{-1} after 20 cycles at a current density of 50 mAg^{-1}.[79] The charge capacity has been further increased by making the MoS_2 nanosheets through chemical means where the charge capacity resulted in an increase of ~1000 mAh g^{-1}.[80] Composites of graphene and MoS_2 showed an improved reversible capacity of 1290 mAh/g at a current density of 100 mAg^{-1} up to 50 cycles.[81]

10. Hydrogen Evolution Reaction

Hydrogen evolution reaction (HER), refers to splitting water into hydrogen, usually in the presence of the sacrificial agent to enhance the reaction to overcome the sluggish water oxidation by a facile one. HER can be achieved either by electrochemical, photochemical or photo-electrochemical or thermochemical means, with photochemical being the most preferred means. Graphene and other 2D-materials have shown great potential as HER catalysts with some of them possessing a thermo-neutral onset potential ($\Delta G \sim 0$) along with high current density like Pt. Graphene-based catalysts are stable and durable with desirable electrical and thermal properties beneficial to the diffusion and conduction of the heat generated during reactions, thus making them good candidates as electrocatalysts and their supports. The chemical inertness of graphene offers scope to induce catalytic properties by tuning the electronic properties. Introducing edge, dislocations or doping with non-metals in graphene can assist the catalytic process.[61,82] For example, borocarbonitrides, a non-metal catalyst, especially BC_7N, has shown high potential to replace Pt for hydrogen evolution reaction as electrocatalyst since it has low onset-potential, high current density and can be scaled for commercial use.[83] Functionalizing with oxygen groups like oxygen, hydrogen or halogens can alter electronic properties of graphene.[82] Graphene oxide (GO), possesses a band gap which depends on the degree of functionalization and exhibits hydrogen evolution reaction in pure water, and steady H_2 evolution in methanol solution on irradiation with mercury lamp which is further enhanced by loading Ni or NiO on the surface of GO.[7,18] Graphene acts as an electron transport bridge or redox mediator in HER.

2D- Transition metal dichalcogenides (TMDs) such as MoS_2 have created significant interest and offers tremendous scope to improve the activity through various strategies.[84] It is well established that the edge-sites of MoS_2 are active and not basal planes for HER and strategies to activate basal planes or increase edge sites are being pursued. In a recent development, Rao and co-workers have demonstrated that inorganic fullerenes which have a high density of edge-sites showed high current density. They have doped both n-type (Re atom) and p-type (Nb atom) in the fullerene structure and found that the n-type atoms have better HER activity in low pH and vice versa. The compounds performed well at all pH ranges and even in sea-water.[85] Phase-engineered $1T\text{-}MoS_2$ has both active basal planes and edge sites, and it is found that high in-plane conductivity increases HER.[86]

Strategies for tuning the catalytic property of TMDs are being pursued in photochemical HER similar to electrocatalysis where they act as co-catalysts or HER sites. MoS_2 is generally coupled with a dye or a semiconductor for utilizing its HER potential.[84] Bulk MoS_2 has reduced activity, but it was improved significantly by growing few-layers of MoS_2 via the solvothermal route. The activity was further enhanced by growing MoS_2 on graphene sheets and by synthesizing conducting phase engineered 1T-form.[14] These strategies showed huge enhancement in HER activity. Another route is to stabilize the 1T-phase and use it for HER, and recent results seem to be promising.[87] Interface-based catalysts with controlled chemical composition, morphology and increased active sites, along with efficient charge separation and utilization needs to be improved. Covalently cross-linked catalysts have shown a significant development in this area. For example, cross-linked MoS_2 with C_3N_4 and other 2D-materials have exhibited superior activity as compared to individual compounds. The enhanced activity was attributed to charge transfer across the network through space and bonds because of direct overlapping of the layers.[88] Thus HER activity of 2D-materials shows tremendous scope of development and may be part of commercial catalysts in the near future. Thus, the linking of BCN with MoS_2 offers immense benefits in catalyzing HER.

11. Conclusion and Outlook

The previous sections review the interesting properties exhibited by graphene, transition metal dichalcogenides, and other 2D materials. These materials are being pursued widely and may find practical applications. Single-layer MoS_2, being a direct bandgap material, exhibits unique features such as intense photoluminescence, circular luminescence, and phenomena associated with it such as valleytronics. Photoluminescence of single-layer MoS_2 can be used for sensor applications. Graphene has found applications in supercapacitors and other areas related to the electronic field, such as sensors and energy devices. MoS_2 has shown excellent properties as electrodes in Na-ion batteries with desirable characteristics. HER properties of MoS_2 and BCN-based materials exhibit excellent activity. They are likely to be applied in heterojunctions, covalently cross-linked structures as well as van der Waals structures. The technology for obtaining 2D materials is integrally related to their layered structures and the weak van der Waals bonds that exist between the layers. The scotch-tape approach, sonication in solvents, chemical intercalation and exfoliation of 2D flakes, and direct growth techniques (physical and chemical vapor deposition) are widely used methods to synthesize 2D-materials.

The unusual and multi-functional properties associated with the 2D-materials, will trigger further research and hopefully overcome the constraints related to practical applications. At the present stage, it is desirable to perfect synthetic strategies for manufacturing of cheap single-phase and defect free 2D-materials with desired properties. Theoretical predictions can guide us and might be helpful in designing experiments. Novel heterostructures are required for next-generation materials for fabricating faster, efficient, superior devices and materials. Implantation of 2D-materials in electronic devices is still at the research level. Integration of metallic multilayer 2D-materials can reduce the problem pertaining to contact resistance. The heterostructure design can result in the thinnest possible FETs leading to several opportunities in 2D-crystal electronics. Layered oxides are strong candidates for dielectric and biosensor applications.

References

1. K.S. Novoselov, A. K. Geim, S. V. Morozov, D. Jiang, Y. Zhang, S. V. Dubonos, I. V. Grigorieva and A. A. Firsov, Electric field effect in atomically thin carbon films, *Science* **306**, 666-669 (2004).
2. K. Shavanova, Y. Bakakina, I. Burkova, I. Shtepliuk, R. Viter, A. Ubelis, V. Beni, N. Starodub, R.Yakimova and V. Khranovskyy, Application of 2D non-graphene materials and 2D oxide nanostructures for biosensing technology, *Sensors* **16**(2), 223 (2016).
3. M. J. Allen, V. C. Tung and R. B. Kaner, Honeycomb carbon: a review of graphene, *Chem. Rev.* **110**, 132-145 (2010).
4. M. K. Jana and C. N. R. Rao, Two-dimensional inorganic analogues of graphene: transition metal dichalcogenides, *Philos Trans A Math Phys Eng Sci.* **374**, 20150318 (2016).
5. C. N. R. Rao, K. Gopalakrishnan and U. Maitra, Comparative study of potential applications of graphene, MoS_2, and other two-dimensional materials in energy

devices, sensors, and related areas, *ACS Appl. Mater. Interfaces* **7**, 7809-7832 (2015).

6. B. J. Schultz, R. V. Dennis, V. Lee and S. Banerjee, An electronic structure perspective of graphene interfaces, *Nanoscale* **6**, 3444-3466 (2014).

7. T. F. Yeh, J. M. Syu, C. Cheng, T.H. Chang and H. Teng, Graphite oxide as a photocatalyst for hydrogen production from water, *Adv. Func. Mat.* **20**, 2255-2262 (2010).

8. H. Nozaki and S. Itoh, Structural stability of BC_2N, *J. Phys. Chem. Solids* **57**, 41-49 (1996).

9. C. N. R. Rao and K. Pramoda, Borocarbonitrides, $B_xC_yN_z$, 2D nanocomposites with novel properties, *Bull. Chem. Soc. Jpn* **92**(2), 441-468 (2018).

10. Kumar N., Moses K., Pramoda K., Shirodkar S. N., Mishra A. K., Waghmare U. V., Sundaresan A. and Rao C. N. R., Borocarbonitrides, BxCyNz, *J. Materials Chemistry A* **1**, 5806-5821 (2013).

11. T. P. Kaloni, R. P. Joshi, N. P. Adhikari and U. Schwingenschlögl, Band gap tunning in BN-doped graphene systems with high carrier mobility, *Appl. Phys. Lett.* **104**, 073116 (2014).

12. C. Sun, F. Ma, L. Cai, A. Wang, Y. Wu, M. Zhao, W. Yan and X. Hao, Metal-free ternary BCN nanosheets with synergetic effect of band gap engineering and magnetic properties, *Sci. Rep.* **7**, 6617 (2017).

13. M. Chhowalla, H. S. Shin, G. Eda, L.J. Li, K. P. Loh and H. Zhang, The chemistry of two-dimensional layered transition metal dichalcogenide nanosheets, *Nat. Chem.* **5**, 263-275 (2013).

14. U. Maitra, U. Gupta, M. De, R. Datta, A. Govindaraj and C. N. R. Rao, Highly effective visible-light-induced H_2 generation by single-layer 1T-MoS_2 and a nanocomposite of few-layer 2H-MoS_2 with heavily nitrogenated graphene, *Angew. Chem. Int. Ed.* **52**, 13057-13061 (2013).

15. U. Gupta, B. S. Naidu, U. Maitra, A. Singh, S. N. Shirodkar, U. V. Waghmare and C. N. R. Rao, Characterization of few-layer 1T-$MoSe_2$ and its superior performance in the visible-light induced hydrogen evolution reaction, *APL Materials* **2**, 092802 (2014).

16. P. Ganal, W. Olberding, T. Butz and G. Ouvrard, Soft chemistry induced host metal coordination change from octahedral to trigonal prismatic in 1T-TaS_2, *Solid State Ionics* **59**, 313-319 (1993).

17. K. S. Novoselov, D. Jiang, F. Schedin, T. J. Booth, V. V. Khotkevich, S. V. Morozov and A. K. Geim, Two-dimensional atomic crystals, *Proc. Natl. Acad. Sci. U.S.A* **102**, 10451-10453 (2005).

18. U. Gupta, K. Gopalakrishnan and C. N. R. Rao, Synthesis and properties of graphene and its 2D inorganic analogues with potential applications, *Bull. Mater. Sci* **41**, 129 (2018).

19. D. Wei, Y. Liu, Y. Wang, H. Zhang, L. Huang and G. Yu, Synthesis of N-doped graphene by chemical vapor deposition and its electrical properties, *Nano Lett.* **9**, 1752-1758 (2009).

20. R. J. Smith, P. J. King, M. Lotya, C. Wirtz, U. Khan, S. De, A. O'Neill, G. S. Duesberg, J. C. Grunlan, G. Moriarty, J. Chen, J. Wang, A. I. Minett, V. Nicolosi and J. N. Coleman 2011, Large-scale exfoliation of inorganic layered compounds in aqueous surfactant solutions, *Adv. Mat.* **23**, 3944-3948 (2011).

21. P. Joensen, R. F. Frindt and S. R. Morrison, Single-layer MoS_2, *Mat. Res. Bull.* **21**, 457-461 (1986).

22. S. J. Rowley-Neale, J. M. Fearn, D. A. C. Brownson, G. C. Smith, X. Ji and C. E. Banks, 2D molybdenum disulphide ($2D-MoS_2$) modified electrodes explored towards the oxygen reduction reaction, *Nanoscale* **8**, 14767-14777 (2016).

23. V. Nicolosi, M. Chhowalla, M. G. Kanatzidis, M. S. Strano and J. N. Coleman, Liquid exfoliation of layered materials, *Science* **340**, 1226419 (2013).

24. S. Wang, Y. Rong, Y. Fan, M. Pacios, H. Bhaskaran, K. He and J. H. Warner, Shape evolution of monolayer MoS_2 crystals grown by chemical vapor deposition, *Chem. Mat.* **26**, 6371-6379 (2014).

25. M. R. Gao, Y. F Xu., J. Jiang and S. H.Yu, Nanostructured metal chalcogenides: synthesis, modification, and applications in energy conversion and storage devices, *Chem. Soc. Rev.* **42**, 2986-3017 (2013).

26. Z. Lei, J. Zhan, L. Tang, Y. Zhang and Y. Wang, Recent development of metallic (1T) phase of molybdenum disulfide for energy conversion and storage, *Adv. Energy Mater.* **8**, 1703482 (2018).

27. M. N. Ali, J. Xiong, S. Flynn, J. Tao, Q. D Gibson., L. M. Schoop, T. Liang, N. Haldolaarachchige, M. Hirschberger, N. P. Ong and R. J. Cava, Large, non-saturating magnetoresistance in WTe_2, *Nature* **514**, 205-208 (2014).

28. D. H. Keum, S. Cho, J. H. Kim, D.H. Choe, H.J. Sung, M. Kan, H. Kang, J.Y. Hwang, S. W. Kim, H. Yang, K. J. Chang and Y. H. Lee, Bandgap opening in few-layered monoclinic $MoTe_2$, *Nat. Phys* **11**, 482-486 (2015).

29. X.C. Pan, X. Chen, H. Liu, Y. Feng, Z. Wei, Y. Zhou, Z. Chi, L. Pi, F. Yen, F. Song, X. Wan, Z. Yang, B. Wang, G. Wang and Y. Zhang, Pressure-driven dome-shaped superconductivity and electronic structural evolution in tungsten ditelluride, *Nat Commun.* **6**, 7805 (2015).

30. D. Kang, Y. Zhou, W. Yi, C. Yang, J. Guo, Y. Shi, S. Zhang, Z. Wang, C. Zhang, S. Jiang, A. Li, K. Yang, Q. Wu, G. Zhang, L. Sun and Z. Zhao, Superconductivity emerging from a suppressed large magnetoresistant state in tungsten ditelluride, *Nat Commun.* **6**, 7804 (2015).

31. Y. Qi, P. G. Naumov, M. N Ali., C. R. Rajamathi, W. Schnelle, O. Barkalov, M. Hanfland, S.C. Wu, C. Shekhar, Y. Sun, V. Süß, M. Schmidt, U. Schwarz, E. Pippel, P. Werner, R. Hillebrand, T. Förster, E. Kampert, S. Parkin, R. J. Cava, C. Felser, B. Yan and S. A. Medvedev, Superconductivity in Weyl semimetal candidate $MoTe_2$, *Nat. Commun.* **7**, 11038 (2016).

32. A. H. Castro Neto, F. Guinea, N. M. R. Peres, K. S. Novoselov and A. K. Geim, The electronic properties of graphene, *Rev. Mod. Phys.* **81**, 109-162 (2009).

33. T. Fang, A. Konar, H. Xing and D. Jena, Mobility in semiconducting graphene nanoribbons: Phonon, impurity, and edge roughness scattering, *Phys. Rev. B* **78**, 205403 (2008).

34. J. Daeha, A. Chunder, Z. Lei and I. K. Saiful, High yield fabrication of chemically reduced graphene oxide field effect transistors by dielectrophoresis, *Nanotechnology* **21**, 165202 (2010).

35. M. B. Sreedhara, K. Gopalakrishnan, B. Bharath, R. Kumar, G. U. Kulkarni and C. N. R. Rao, Properties of nanosheets of 2D-borocarbonitrides related to energy devices, transistors and other areas, *Chem. Phys. Lett.* **657**, 124-130 (2016).

36. F. Schwierz, J. Pezoldt and R. Granzner, Two-dimensional materials and their prospects in transistor electronics, *Nanoscale* **7**, 8261-8283 (2015).

37. V. Podzorov, M. E. Gershenson, C. Kloc, R. Zeis and E. Bucher, High-mobility field-effect transistors based on transition metal dichalcogenides, *Appl. Phys. Lett.* **84**, 3301-3303 (2004).

38. A. Ayari, E. Cobas, O. Ogundadegbe and M. S. Fuhrer, Realization and electrical characterization of ultrathin crystals of layered transition-metal dichalcogenides, *J. Appl. Phys.* **101**, 014507 (2007).

39. B. Radisavljevic, A. Radenovic, J. Brivio, V. Giacometti and A. Kis, Single-layer MoS_2 transistors, *Nat. Nanotechnol* **6**, 147-150 (2011).

40. H. Fang, S. Chuang, T. C. Chang, K. Takei, T. Takahashi and A. Javey, High-performance single layered WSe_2 p-FETs with chemically doped contacts, *Nano Lett.* **12**, 3788-3792 (2012).

41. B. Radisavljevic, M. B. Whitwick and A. Kis, Integrated circuits and logic operations based on single-layer MoS_2, *ACS Nano* **5**, 9934-9938 (2011).

42. Y. Zhang, J. Ye, Y. Matsuhashi and Y. Iwasa, Ambipolar MoS_2 thin flake transistors, *Nano Lett.* **12**, 1136-1140 (2012).

43. V. K. Sangwan and M. C. Hersam, Electronic transport in two-dimensional materials, *Annu. Rev. Phys. Chem* **69**, 299-325 (2018).

44. K. F. Mak, K. He, J. Shan and T. F. Heinz, Control of valley polarization in monolayer MoS_2 by optical helicity, *Nat. Nanotechnol* **7**, 494-498 (2012).

45. H. Zeng, J. Dai, W. Yao, D. Xiao and X. Cui, Valley polarization in MoS_2 monolayers by optical pumping, *Nat. Nanotechnol* **7**, 490-494 (2012).

46. K. F. Mak, K. He, C. Lee, G. H. Lee, J. Hone, T. F. Heinz and J. Shan, Tightly bound trions in monolayer MoS_2, *Nat. Mater* **12**, 207-211 (2012).

47. W. Yuan and G. Shi, Graphene-based gas sensors, *J. Mater. Chem. A* **1**, 10078-10091 (2013).

48. T. O. Wehling, M. I. Katsnelson and A. I. Lichtenstein, Adsorbates on graphene: Impurity states and electron scattering, *Chem. Phys. Lett.* **476**, 125-134 (2009).

49. F. Schedin, A. K. Geim, S. V. Morozov, E. W. Hill, P. Blake, M. I. Katsnelson and K. S. Novoselov, Detection of individual gas molecules adsorbed on graphene, *Nat. Mater* **6**, 652-655 (2007).

50. S. Rumyantsev, G. Liu, M. S. Shur, R. A. Potyrailo and A. A. Balandin, Selective gas sensing with a single pristine graphene transistor, *Nano Lett.* **12**, 2294-2298 (2012).

51. R. Berdan, T. Prodromakis, I. Salaoru, A. Khiat and C. Toumazou, Memristive devices as parameter setting elements in programmable gain amplifiers, *Appl. Phys. Lett.* **101**, 243502 (2012).

52. Y. Shao, J. Wang, H. Wu, J. Liu, I. A. Aksay and Y. Lin, Graphene based electrochemical sensors and biosensors: a review, *Electroanalysis* **22**, 1027-1036 (2010).

53. Q. He, H. G. Sudibya, Z. Yin, S. Wu, H. Li, F. Boey, W. Huang, P. Chen and H. Zhang, Centimeter-long and large-scale micropatterns of reduced graphene oxide films: fabrication and sensing applications, *ACS Nano* **4**, 3201-3208 (2010).

54. T. V. Dubovik, A. I. Itsenko, L. P. Dolinskaya, N. S. Zyatkevich and L. Y. Levkov, Use of boron carbonitride in devices for rapid analysis of slag oxidation, *Powder Metall. Met. Ceram.* **33**, 100-102 (1995).

55. K. Pramoda, K. Moses, U. Maitra and C. N. R. Rao, Superior performance of a MoS_2-RGO composite and a borocarbonitride in the electrochemical detection of dopamine, uric acid and adenine, *Electroanalysis* **27**, 1892-1898 (2015).

56. H. Li, J. Wu, Z. Yin and H. Zhang, Preparation and applications of mechanically exfoliated single-layer and multilayer MoS_2 and WSe_2 nanosheets, *Acc. Chem. Res.* **47**, 1067-1075 (2014).

57. D. J. Late, Y.K Huang., B. Liu, J. Acharya, S. N. Shirodkar, J. Luo, A. Yan, D. Charles, U. V. Waghmare, V. P. Dravid and C. N. R. Rao, Sensing behavior of atomically thin-layered MoS_2 transistors, *ACS Nano* **7**, 4879-4891 (2013).

58. Y. Zhang, B. Zheng ,C. Zhu, X. Zhang, C. Tan, H. Li, B. Chen, J. Yang, J. Chen, Y. Huang, L. Wang and H. Zhang, single-layer transition metal dichalcogenide nanosheet-based nanosensors for rapid, sensitive, and multiplexed detection of DNA, *Adv. Mater* **27**, 935-939 (2015).

59. G. Chen, T. M. Paronyan and A. R. Harutyunyan, Sub-ppt gas detection with pristine graphene, *Appl. Phys. Lett.* **101**, 053119 (2012).

60. S. Lim, B. Kang, D. Kwak, W. H. Lee, J. A. Lim and K. Cho, Inkjet-printed reduced graphene oxide/poly(Vinyl Alcohol) composite electrodes for flexible transparent organic field-effect transistors, *J. Phys. Chem. C* **116**, 7520-7525 (2012).

61. C. N. R. Rao, K. Gopalakrishnan and A. Govindaraj, Synthesis, properties and applications of graphene doped with boron, nitrogen and other elements, *Nano Today* **9**, 324-343 (2014).

62. H. S. Lee, S.W. Min, Y.G. Chang, M. K. Park, T. Nam, H. Kim, J. H. Kim, S. Ryu and S. Im, MoS_2 nanosheet phototransistors with thickness-modulated optical energy gap, *Nano Letters* **12**, 3695-3700 (2012).

63. C. N. R. Rao, H. S. S. R. Matte and K. S. Subrahmanyam, Synthesis and selected properties of graphene and graphene mimics, *Acc. Chem. Res.* **46**, 149-159 (2013).

64. D. Jariwala, T. J. Marks and M. C. Hersam, Mixed-dimensional van der Waals heterostructures, *Nat. Mater.* **16**, 170-181 (2016).

65. L. L. Zhang and X. S. Zhao, Carbon-based materials as supercapacitor electrodes, *Chem. Soc. Rev.* **38**, 2520-2531 (2009).

66. Z. Wen, X. Wang, S. Mao, Z. Bo, H. Kim, S. Cui, G. Lu, X. Feng and J. Chen, Crumpled nitrogen-doped graphene nanosheets with ultrahigh pore volume for high-performance supercapacitor, *Adv. Mater.* **24**, 5610-5616 (2012).

67. K. Gopalakrishnan, A. Govindaraj and C. N. R. Rao, Extraordinary supercapacitor performance of heavily nitrogenated graphene oxide obtained by microwave synthesis, *J Mater. Chem. A* **1**, 7563-7565 (2013).

68. L. Sun, L. Wang, C. Tian, T. Tan, Y. Xie, K. Shi, M. Li and H. Fu, Nitrogen-doped graphene with high nitrogen level via a one-step hydrothermal reaction of graphene oxide with urea for superior capacitive energy storage, *RSC Adv.* **2**, 4498-4506 (2012).

69. L. Niu, Z. Li, W. Hong, J. Sun, Z. Wang, L. Ma, J. Wang and S. Yang, Pyrolytic synthesis of boron-doped graphene and its application as electrode material for supercapacitors, *Electrochim. Acta* **108**, 666-673 (2013).

70. Z. Zuo, Z. Jiang and A. Manthiram, Porous B-doped graphene inspired by Fried-Ice for supercapacitors and metal-free catalysts, *J. Mater. Chem. A* **1**, 13476-13483 (2013).

71. M. Acerce, D. Voiry and M. Chhowalla, Metallic 1T phase MoS_2 nanosheets as supercapacitor electrode materials, *Nat. Nanotechnol* **10**, 313-318 (2015).

72. M. Liu, Y. Song, S. He, W. W. Tjiu, J. Pan, Y.Y. Xia and T. Liu, Nitrogen-doped graphene nanoribbons as efficient metal-free electrocatalysts for oxygen reduction, *ACS Appl. Mater. Interfaces* **6**, 4214-4222 (2014).

73. L. Zhang and Z. Xia, Mechanisms of oxygen reduction reaction on nitrogen-doped graphene for fuel cells, *J. Phys. Chem. C* **115**, 11170-11176 (2011).

74. K. Moses, V. Kiran, S. Sampath and C. N. R. Rao, Few-layer borocarbonitride nanosheets: Platinum-free catalyst for the oxygen reduction reaction, *Chem. Asian J.* **9**, 838-843 (2014).

75. Y. Xue, D. Yu, L. Dai, R. Wang, D. Li, A. Roy, F. Lu, H. Chen, Y. Liu and J. Qu, Three-dimensional B,N-doped graphene foam as a metal-free catalyst for oxygen reduction reaction, *Phys. Chem. Chem. Phys.* **15**, 12220-12226 (2013).

76. K. Gopalakrishnan, K. Pramoda, U. Maitra, U. Mahima, M. A. Shah and C. N. R. Rao, Performance of MoS_2-reduced graphene oxide nanocomposites in supercapacitors and in oxygen reduction reaction, *Nanomater. and Energy* **4**, 9-17 (2015).

77. C. N. R. Rao, U. Maitra and U. V. Waghmare, Extraordinary attributes of 2-dimensional MoS_2 nanosheets, *Chem. Phys. Lett.* **609**, 172-183 (2014).

78. M. Pumera, Graphene-based nanomaterials for energy storage, *Energy Environ. Sci.* **4**, 668-674 (2011).

79. R. R. Haering, J. A. Stiles and K. Brandt, Lithium molybdenum disulphide battery cathode, U.S Patent 4224390 (1980).

80. C. Feng, J. Ma, H. Li, R. Zeng, Z. Guo and H. Liu, Synthesis of molybdenum disulfide (MoS_2) for lithium ion battery applications, *Mater. Res. Bull.* **44**, 1811-1815 (2009).

81. K. Chang and W. Chen, In situ synthesis of MoS_2/graphene nanosheet composites with extraordinarily high electrochemical performance for lithium ion batteries, *Chem. Comm.* **47**, 4252-4254 (2011).

82. D. Deng, K. S. Novoselov, Q. Fu, N. Zheng, Z. Tian and X. Bao, Catalysis with two-dimensional materials and their heterostructures, *Nat. Nanotechnol* **11**, 218-230 (2016).

83. M. Chhetri, S. Maitra, H. Chakraborty, U. V. Waghmare and C. N. R. Rao, Superior performance of borocarbonitrides, $B_xC_yN_z$, as stable, low-cost metal-free electrocatalysts for the hydrogen evolution reaction, *Energy Environ. Sci.* **9**, 95-101 (2016).

84. U. Gupta and C. N. R. Rao, Hydrogen generation by water splitting using MoS_2 and other transition metal dichalcogenides, *Nano Energy* **41**, 49-65 (2017).

85. M. Chhetri, U. Gupta, L. Yadgarov, R. Rosentsveig, R. Tenne and C. N. R. Rao, Effects of p- and n-type doping in inorganic fullerene MoS_2 on the hydrogen evolution reaction, *ChemElectroChem* **3**, 1937-1943 (2016).

86. D. Voiry, M. Salehi, R. Silva, T. Fujita, M. Chen, T. Asefa, V. B. Shenoy, G. Eda and M. Chhowalla, Conducting MoS_2 nanosheets as catalysts for hydrogen evolution reaction, *Nano Letters* **13**, 6222-6227 (2013).

87. N. K. Singh, A. Soni, R. Singh, U. Gupta, K. Pramoda and C. N. R. Rao, Remarkable photochemical HER activity of semiconducting $2H\text{-}MoSe_2$ and MoS_2 covalently linked to layers of 2D structures and of the stable metallic 1T phases prepared solvo- or hydro-thermally, *J. Chem. Sci.* **130(10)**, 131 (2018).

88. K. Pramoda, U. Gupta, M. Chhetri, A. Bandyopadhyay, S. K. Pati and C. N. R. Rao, "Nanocomposites of C_3N_4 with layers of MoS_2 and nitrogenated RGO, obtained by covalent cross-linking: synthesis, characterization, and HER activity, *ACS Appl. Mater. Interfaces* (2017).

Chapter 3

Opportunities and Challenges in Quantum Dots

Pradeep K. R. and Ranjani Viswanatha[*]

*New Chemistry Unit, International Centre for Materials Science and School of
Advanced Materials, Jawaharlal Nehru Centre for Advanced Scientific
Research, Jakkur, Bangalore 560064, Karnataka, India*
**rv@jncasr.ac.in*

Quantum dots (QDs) are typically semiconductor crystals in the size range of 2-20 nm. Due to the quantum confinement effects arising from their small size, they exhibit composition, shape and size-dependent electrical and optical properties. These highly tunable properties have driven the research for several decades into their understanding and subsequent commercial application in several fields especially for the optoelectronic devices and photovoltaics. II-VI QDs and more recently perovskite QDs have shown great promise as tunable light absorbing and/or emitting layers in LED devices and displays, in photovoltaics and imaging. Although research in this field has led to promising results, a better understanding of important factors like robustness, stability, processability, toxicity and cost effectiveness are necessary before implementing them for commercial purposes. In this chapter, the properties of QDs, studies towards commercialization for different applications and the future challenges are summarized.

1. Introduction

Quantum dots (QDs), with their unique properties, have given rise to a variety of new applications that have been very useful. These semiconductor nanoparticles have electronic properties that are significantly different from their bulk counterparts as a consequence of constrained electron and hole wavefunctions. This quantum size effect was first experimentally discovered in a study of CuCl nanocrystals

55

     K. R. Pradeep & R. Viswanatha

whose size varied from 2 to 31 nm[1] and has since demonstrated varied unexpected magnetic, optical and opto-electronic properties that can be traced to their sizes.[2,3] These properties are of interest both from a fundamental perspective as well as from the perspective of many practical applications.

While the early 1990s were marked by the preparation of quantum dots starting from complicated and expensive molecular beam epitaxy[4] to the development of colloidal synthesis,[5] the research in the past decade has been driven by practical applications. For example, most recently, the all inorganic cesium and organic hybrid lead halide perovskite QDs have dominated the field with their high quantum yields and high photo-voltaic efficiency.[6] In fact, the broad range of applications of QDs driven by their unique properties include functional nanomaterials for photovoltaics, light emitting diodes, backlit displays, lasers, photodetectors, sensing and bioimaging. Development in the synthesis of various type of heterostructures with an increasing complexity and integrating multiple functionality in QDs offers new applications in the field of magnetism and spintronics. However, challenges remain in reducing the toxicity of these QDs by bringing in more green chemistry in the synthesis techniques and the quest for a stable and improved performance of devices is keeping this area active. This review will focus on the properties driving the various applications of QDs of II-VI semiconductors and the newly emerging perovskite QDs in the area of optical and magnetic applications. The review concludes with a brief discussion on challenges and future perspectives of QD devices. Though a number of interesting but more specific and specialized topics have not been addressed here, we hope that this review provides an introduction to the field of exciting applications of QDs.

2. Synthesis of Colloidal QDs

The efficiency of applications based on colloidal QDs depends largely on the synthesis of QDs that are highly monodisperse with a narrow size distribution. Initial synthesis techniques of QDs ranged from complex,

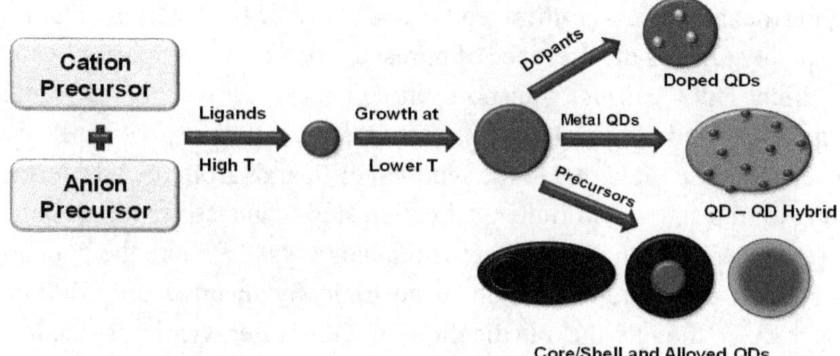

Fig. 1. A schematic showing the synthesis of nanocrystals starting from various precursors leading to complex structures of core/shell or hybrid or doped nanocrystals.

expensive methods like the molecular beam epitaxy to obtain ultrapure arrays of QDs to the solution processed micellar, hydrothermal or microwave assisted synthesis that gave rise to unstable QDs with a large size distribution. Eventually, the middle ground is stuck by the use of high temperature colloidal synthesis involving the rapid nucleation followed by the slow growth. In addition to the high quality of the sample with a good size distribution, this high temperature synthesis is also accommodative of the control over multiple reaction parameters like concentration of precursors and their reactivity, surfactants and temperature of the reaction that is useful in the synthesis of different heterostructures. A schematic showing a typical modern synthesis of colloidal QDs is shown in Figure 1. It is important to note here that this methodology of synthesis caters to various needs including core/shell QDs for surface passivation,[7,8] passivation tunability to alter the morphology and hence their optical properties as well as to dope transition metals into these QDs for magneto-optical and opto-electronic applications. The two most important parameters, namely the interface of the core/shell and the interface of the ligands with the QDs on the surface, play an important role in determining the properties of the QDs and can be accurately controlled within the high temperature synthesis route. While we discuss below the various properties driven by surface ligands, we discuss the engineering of the QD interface and their consequences on device efficiency in the later sections of this chapter.

Purification of as-synthesized QDs is also a critical step in obtaining high quality QDs as the presence of unreacted precursors or excess ligands can affect the carrier transport in devices. However, sequential precipitation and dispersion in suitable solvents can give rise to a reduced quantum yield due to the removal of ligands from the QD surface leading to trap states. Multiple purification steps can also lead to a change in the phase transformation of perovskite nanocrystals where the optically active cubic phase transforms in to an optically inactive orthorhombic phase[6]. Even though the purification of QDs after synthesis leads to complex degradation processes it is important in obtaining good performance in devices as the purification of QDs allows for obtaining good morphology by preparing pinhole free thin films.[9]

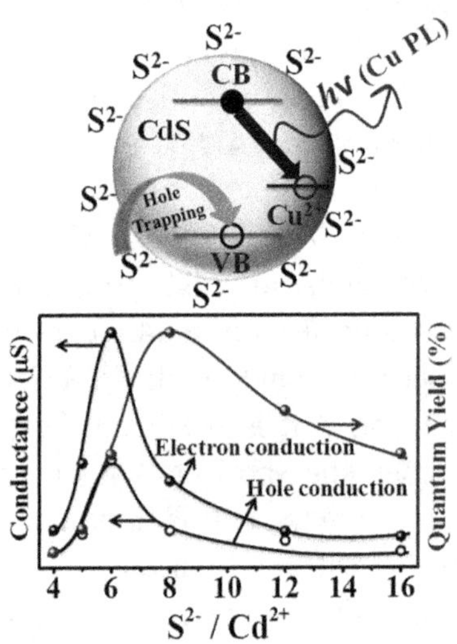

Fig. 2. Schematic showing ligand exchanged QDs of Cu doped CdS and variation of conductance and PL quantum yield for varying S^{2-} to Cd^{2+} ratio. Reproduced with permission from Ref. 10 . Copyright 2014 American Chemical Society.

As synthesized QDs are typically capped with long insulating ligands (Oleic acid or Oleylamine) as these ligands are important to obtain colloidally stable QDs. However, these long chain ligands introduce spacing between QDs which in turn lead to low density in QD films.[11] To obtain an improved carrier transport, ligands can be exchanged from long ligands to short as the interparticle spacing can be reduced by doing so. As shown in Figure 2 ligand free or sulfide capped Cu-doped CdS QDs have been synthesized to assist the transport in these QDs. These QDs are shown to be luminescent and exhibit significant conductivity.[10] To date, the most efficient reported QD solar cell has employed a combination of partial fusing with strong passivation of the remaining surface using MPA and chloride ligands.[12] Recently high performance $CsPbX_3$ perovskite QD light emitting devices have been prepared via solid state ligand exchange.[13] Similarly targeted ligand exchange chemistry on cesium lead halide perovskite QDs lead to high efficient photovoltaic devices.[14]

3. Optical Applications

Colloidal II-VI semiconductor QDs and inorganic perovskite QDs are emerging as promising materials for optoelectronic devices by virtue of their optical properties. The size and composition dependent bandgap tunability of QDs also make these materials practical for low cost, large scale devices. Here, we try to address the major advances in the area of optoelectronic applications of QDs.

3.1. *Photovoltaics*

The application of QDs in photovoltaic systems is one of the most important and promising avenues of research. The critical parameters which influence the performance of the QD photovoltaic devices are carrier mobility, doping density, trap density and diffusion length in films. The power conversion efficiency (PCE) of quantum dot photovoltaic (QD-PV) devices has increased rapidly up to 9.9 % in the past decade. The present QD-PV is currently dominated by devices based on Pb chalcogenides (PbS,[15] PbSe,[16] PbS_xSe_{1-x}) which exhibit a wide range of bandgap tunability in the range of NIR to visible region which is suitable for PV application. However there are reports on Cd chalcogenides QD-PV as well.[17]

Though QD-PV cells initially started as simple Schottky cells, heterojunction solar cells were developed to overcome the limitations of Schottky QD solar cells. A typical heterojunction solar cell consists of a highly doped n-type metal oxide in a p-n heterojunction with a p-type QD film. The typical substrate consists of a transparent base followed by a thin transparent conductive oxide layer, namely indium tin oxide (ITO). The most commonly used wide bandgap semiconductor as a n-doped junction layer is TiO_2. The p-type QD film is coated on top of a metal oxide layer and the device is coated with a back reflective contact with a deep work function metal such as gold and followed by silver or aluminum. Under this geometry, the external quantum efficiency (EQE) typically ranges between 15-25 % for a high performing device.[18]

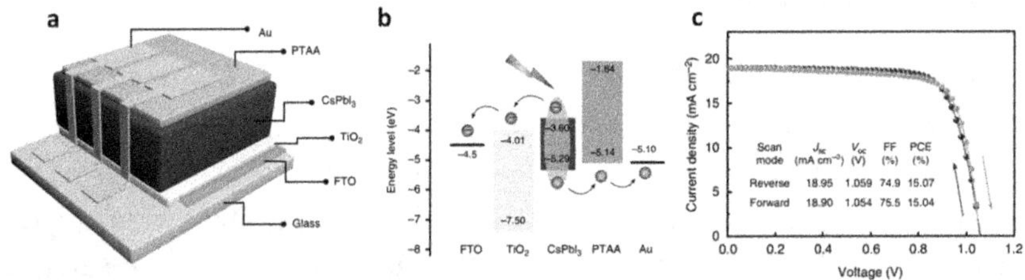

Fig. 3. (a) A schematic of a device structure and (b) Energy level diagram of the CsPbI₃ PV cell. (c) Current density–voltage curves of the device. Reproduced with permission from Ref. 19. Copyright 2018 The Nature Publishing Group.

However, these promising results are constrained by long term stability and spectral tunability with an ideal bandgap to absorb the solar energy. Recent studies have shown that these inorganic perovskite QDs synthesized by hot injection methods are capable of performing well in the area of QD-PV.[20] A schematic of a typical solar cell based on perovskite QDs are shown in Fig. 3(a) and (b) respectively. Fig. 3(c) shows the current density-voltage curve for the CsPbI₃ based device with PCE of 15.1%.[19] To avoid thermal/moisture induced QDs agglomeration which is well known in perovskite QDs Wang et al. used high mobility micrometre sized graphene sheets to crosslink CsPbI₃ QDs to achieve long term stability.[21] CsPbBrₓI₃₋ₓ based QDs are known to show improved stability under ambient conditions compared to CsPbI₃.[22]

However, the major drawback of these materials is the toxicity of Pb that hinders these devices from commercialization. Lead free $CsSnX_3$ QD based solar cells have been reported by Chen et al.[23] with a high PCE of 12.96 % to partially address this aspect. In summary, the performance of $CsPbX_3$ QD based solar cells typically have greater than 10% PCE with a large J_{sc} and V_{oc} of about 1 eV. As seen from the progress shown by all inorganic perovskite QDs based solar cells, $CsPbX_3$ based solar cells show promise as a strong contender for commercially available solar cells.

3.2. *Quantum Dot Light Emitting Diodes (QD-LEDs)*

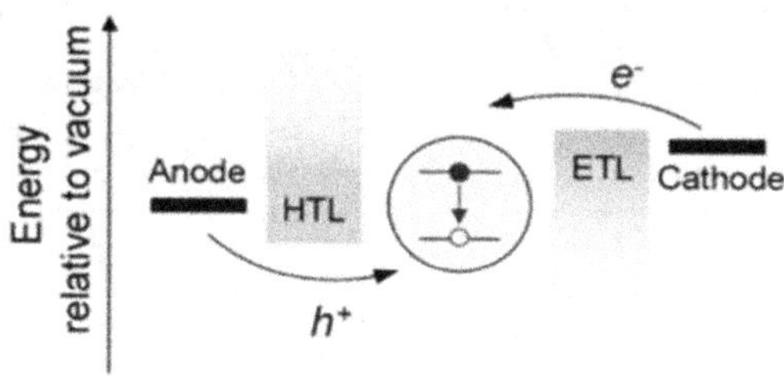

Fig. 4. Energy level diagram for a typical QD-LED.

Electroluminescence (EL) is a phenomenon of light emission from a material excited by the electric current. When charge carriers are injected into an emissive material, like QDs through contact electrodes they recombine radiatively giving rise to EL. Similar to light emitting diodes (LEDs) QD-LEDs typically have a p-i-n structure, which comprises of an anode, a hole transporting layer, a QD active layer and an electron transport layer. Figure 4 shows the energy level band diagram of a typical QD-LED. Electrons from a cathode and holes from an anode are transported by the carrier transport layers and are injected to the active QD layer where they radiatively recombine.

QD-LEDs were first prepared back in 1994[24] using cadmium selenide QDs. Since then the external quantum efficiency (EQE) of QD-LEDs, defined as the ratio of the number of photons extracted from the device to the number of injected e-h pairs, has substantially improved.

This is achieved by majorly focusing on high quantum yield active materials and various device architectures. The reason for low EQE devices reported in the literature is because of the low photoluminescence efficiencies (PL QY) of the QDs used as active materials.[25] Another step to improve the performance is to choose the suitable hole and electron transport materials with respect to the band alignment for smoother carrier transport.

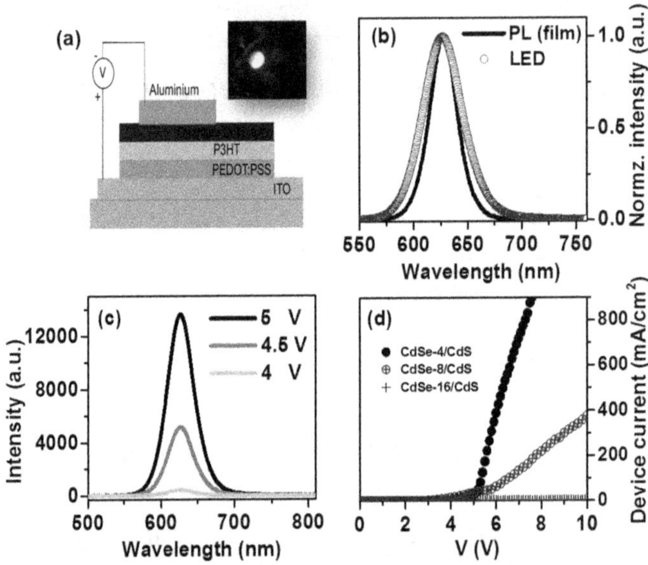

Fig. 5. (a) A schematic of the device architecture. Inset shows the photograph of QD-LED. (b) Spectra properties of the LED and comparison with PL. (c) LED emission spectrum with various driving voltages. (d) J-V characteristics of the device obtained from the QD-LEDs. Reproduced with permission from Ref. 26. Copyright 2013 American Chemical Society.

Vladimir Bulovic and co-workers have demonstrated QD-LEDs with an EL tunable over the entire visible spectrum by choosing different organic charge transport layers and various QDs with different bandgaps to tune the LEDs over a visible spectrum.[27] However, the device performance and EQE is limited by the PL QY of the active material. By carefully engineering the crystal defects, near unity QY microstructures of CdSe/CdS QDs can be synthesized demonstrating the role of interfaces in QD solids for application.[26] Simple unoptimized LEDs obtained from these QDs as their active layer demonstrate performances in excess of 7000 Cd/m^2 with a power conversion efficiency of 1.5 % lm/W. The device architecture used in the device is shown in Figure 5(a). This device is constructed without an electron transport layer and the hole transport layer used here is poly (3-hexyl thiophene) (P3HT). The equivalence of PL and EL indicates a common excitonic origin with the QD layer as the emission source as seen from Figure 5(b). As the bias voltage increases the emission intensity also increases and the emission wavelength remains unchanged as shown in Figure 5(c). The current voltage characteristics shown in Figure 5(d) exhibit typical diode like features. It is also important to note here that the efficiency of the device is non-linearly proportional to the absolute PL QY emphasizing the importance of high quality QDs.

Recently all inorganic perovskite CsPbX$_3$ QDs have emerged as a new class of optoelectronic materials with PL QY up to 90%.[6] Unlike hybrid organic inorganic halide perovskite materials inorganic perovskite QDs are highly stable at ambient conditions making them excellent candidates for high performance LEDs. In 2015 Song et al.[28] for the first time reported CsPbX$_3$ QDs based LEDs with a typical device structure composed of ITO/PEDOT:PSS/PVK/CsPbX$_3$ QDs/TPBi/LiF/Al. Here, the CsPbX$_3$ QDs were synthesized through the hot-injection method discussed earlier and the bandgap is tuned by the halide composition. The luminance and EQE values were reported to be 742,946, and 528 cd/m^2 and 0.07%, 0.12%, and 0.09% respectively for blue, green and orange CsPbX$_3$ QD-LEDs. However, besides the toxicity, one needs to consider crucial contribution from ligands to obtain high quality QD-LEDs with perovskite QDs as active layers. While a high concentration of ligands are needed for better passivation and hence high PL QY, environmental and phase stability, their presence hinders electric conductivity due to their long hydrophobic tail.[29] Hence it is essential to obtain optimal balance between surface passivation and charge carrier injection by the ligand density control as well as the ligand exchange by shorter ligands on CsPbX$_3$ QDs.[30] Yet another way of improving the electro/ photoluminescence efficiency of perovskite QDs is the doping of transition metals. Yao et al.[31] proposed a high-performance LED by Ce^{3+} doping in to the CsPbX$_3$ QDs. QD-LED based on these samples showed an improvement in EQE clocking up to 4.5% with almost a 100% increase compared to their undoped counterparts.

Along with the optimizations of the active layer, the HTL and ETL also play an important role in improving the performance of QD-LEDs. Subramanian et al.[32] reported a $CsPbBr_3$ QDs based LED with inverted LED architecture where ETL is Li doped TiO_2. Li doping facilitates the charge carrier balance between ETL and HTL leading to superior green emitting LED with enhanced EQE. Although $CsPbI_3$ based LED showed promising results, the stability of these QDs in an ambient environment still remains as a challenge. In fact, several challenges remain in the path for the LED application of perovskite QDs based LEDs, more efforts need to be put to tackle the challenges and exciting developments are expected to continue in the upcoming decade.

3.3. *Quantum dots for display applications*

One very important application of LEDs with excellent color purity is in display applications wherein mixed proportions of sources of primary colors (RGB) are required. The most economical and efficient structure for displays available in the market is constructed by the blue InGaN LED with a yellow phosphor (Ce-doped YAG). But the lack of color purity makes it inefficient to express the natural colours.[33] Hence currently, white LEDs are attracting a great deal of interest in the commercialization of displays, especially for backlights in liquid-crystal displays. There have been many efforts to develop QD color converters for LEDs for general purposes after Bawendi's report on QD-polymer composites.[34] Y. Kim and co-workers have synthesized highly luminescent multi-shell structured green CdSe//ZnS/CdZnS QDs and red CdSe/CdS/ZnS/CdZnS QDs which showed almost 100% quantum efficiency.[35] White LEDs for the display backlights were prepared by combining the above QDs with blue LEDs. The EQEs of green and red QD-LEDs reached up to 72% and 34% respectively and the QD-LED maintained a high efficiency and stability over a long period of time up to 2500 hrs.

Fig. 6. Schematic illustration of spin casting different QDs on to donor substrate and electroluminescence image of a 4-inch full colour QD-display using a TFT backplane with a 320*240-pixel array. Reproduced with permission from Ref. 36. Copyright 2011 Nature Publishing Group.

Figure 6 shows a schematic of the process of fabrication of spin cast QD LED display device and electroluminescence image of a prototype 4-inch full colour QD-display on a 320*240-pixel array of TFT backplane.[36] The development of these EL devices strongly relies on the precise band engineering of multi core/shell QDs. Alternative materials with a simple synthesis would be valuable for future development. Recent developments in the EL properties of perovskite QDs make them excellent contenders for II-VI semiconductors based EL devices for display applications. With all the advantages of wide colour tunability, inherent narrow band emission, high PLQY, and low-cost production, perovskite QDs are very promising candidates for display applications.[37] Recently H.C. Wang et al.[38] created mixed green and red perovskite QDs in silicon resin and subjected them to excitation using a blue InGaN chip. The green emitting $CsPbBr_3$ QDs were coated with mesoporous silica to avoid anion exchange with red emitting $CsPbI_3$ QDs. In the LED packaging, the red emitting $CsPb(Br/I)_3$ was mixed with this coated green $CsPbBr_3$ and then dropped in the blue LED chip. The obtained spectra were found to be pure with a NTSC value of 113 % which is more than 104 % NTSC value reported by II-VI semiconductor QDs earlier.[35] However the low stability and high toxicity are still concerns for perovskite QDs from commercialization and far away from practical applications for daily usage.

3.4. *Quantum dot lasing*

Lasing in QDs were observed in the early 90's where QDs were grown by epitaxial techniques using optical and electrical pumping.[39] The difficulty in achieving lasing on colloidal QDs is due to the high nonradiative carrier losses arising due to surface defects. An important milestone in the area of QD lasing was when the researchers realized that the optical gain in QDs mostly relies on emission not from single excitons but biexcitons and other multiexcitons of higher orders.[40] A simplified scheme of biexciton generation is depicted in Figure 7(a). In order to obtain optical gain, one must excite two elections to the conduction band and generate biexcitons as shown in the Figure. But due

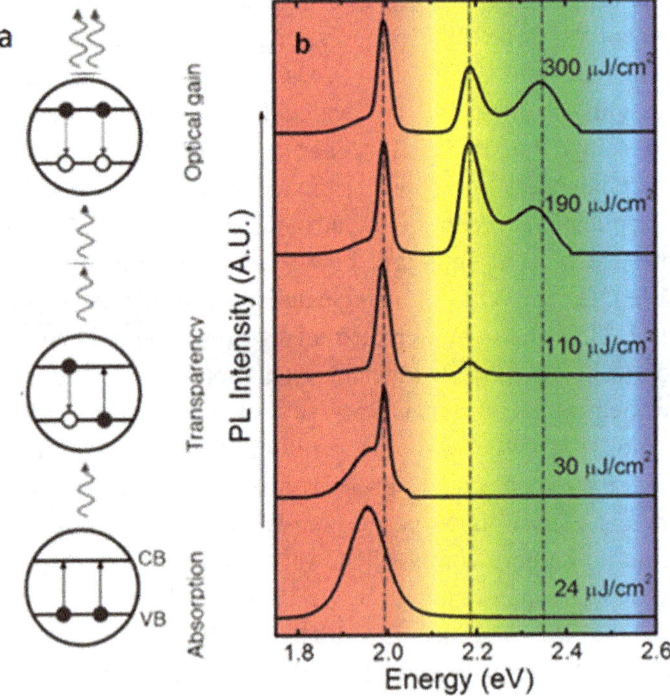

Fig. 7. (a) Simplified scheme illustrating biexciton generation leading to stimulated emission. (b) PL spectra showing first and second order laser emission in CdSe/CdS films recorded as a function of pump intensity. Reproduced with permission from Ref. 42. Copyright 2013 American Chemical Society.

to the highly efficient Auger recombination, the optical gain depletes which in turn leads to more complication especially when QDs are in a solution.[40] The experiment that lead to the first successful demonstration of the amplified stimulated emission (ASE) in QDs is in QD films and excited by femtosecond pulses.[40] Since then multiple efforts in improving the performance of QD lasers, mostly concentrating on improving the sample emission and excitonic generation by reducing the surface defects and improvisation by introducing thick shell and other architectures[41] are underway. A typical series of PL spectra showing the first and second order laser emission in CdSe/CdS films recorded as a function of pump intensity is shown in Figure 7(b).[42]

Recent advances and early research suggest that the perovskite QDs are emerging as a new research paradigm in photonics and will compete favorably with both conventional CdSe-based QDs and organic- inorganic hybrid halide perovskites as optical gain media.[43]

4. Magnetism in Quantum Dots

Magnetism at the nanoscale level has drawn tremendous attention for its significance in growing applications such as in magnetic storage devices

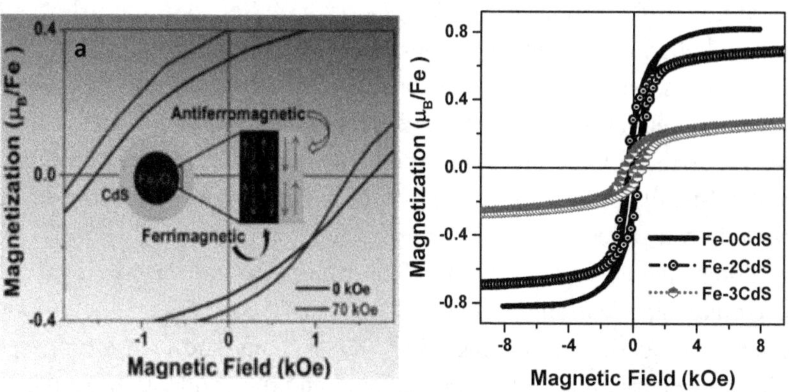

Fig. 8. (a) Formation of a Thin AFM Layer (FeS) at the Interface Giving Rise to FM/AFM Coupling; (b) M vs H hysteresis loop of different Fe_3O_4–nCdS QDs during initial shell growth, measured at 2 K. Reproduced with permission from Ref. 47. Copyright 2017 American Chemical Society.

and microelectronics.[44,45] Semiconductor QDs serving as a multi-component system doped with transition metals having unpaired d-electrons open on to the idea of dilute magnetic semiconducting quantum dots (DMSQD). DMSQDs with the three-dimensional quantum confinement provide high saturation magnetization with near room temperature Curie temperatures (T_c)[46] as well as an ability to manipulate opto-magnetic properties. This combination of storage as well as information processing in a single material provides a path for its potential application in spintronics and magneto-optics.

Owing to continuous efforts, surprising results of DMSQDs were accounted on exceptionally different nature of magnetism. Viswanatha et al., demonstrated co-doping of ZnO with Cu and Fe possessing high quality ferromagnetism with a high magnetic moment of 600 memu/g[47] in spite of the contrasting behavior of anti-ferromagnetism in individually doped Fe or Cu in ZnO. Recently, Saha et al., studied the interface of the magnetic/nonmagnetic Fe_3O_4/CdS QD heterostructures which showed the presence of a substantial exchange bias (as shown in Figure 8(a)) in a well-known optically active and magnetically inactive material CdS.[48] Furthermore, the effect of variation in size of the core on saturation magnetization was successfully demonstrated[49, 50] as shown in Figure 8(b).

Magneto-PL properties of Mn^{2+}-doped ZnSe/CdSe core-shell QDs were studied to understand spin- polarized Mn^{2+} emission from Mn doped QDs.[51] These studies reveal the process of excitation and emission of Mn^{2+} in colloidal QDs which are different compared to bulk materials. As the size of these QDs are very small, quantum confinement influences circular Mn PL polarization along with a large energy splitting in spin forbidden Mn states.

Photo-magnetization memory in timescales of hours is another spectacular application of DMSQDs. The study of magneto-optical property in Cu doped chalcogenide QD systems shows a strong spin-exchanged interaction in the valence band and the conduction band of the host and the Cu dopant. These photo-excitation phenomena were further studied using magnetic circular dichroism (MCD) which exhibited paramagnetism of up to 100% in Cu doped ZnSe/CdSe QDs under the UV excitation. Also, the optically controlled magnetism in Mn-doped

CdSe QDs has been reported, which can give rise to potential applications.[52] This showed a large Zeeman splitting in the absence of an external magnetic field due to enhanced dopant-carriers exchange field. Photo-excited carriers arising from surface defect states on a varying smaller size scale of QDs lead to carrier-mediated ferromagnetic interaction and is well accounted in Mn-doped CdSe QDs.[53] Recent developments in the understanding of the internal structure and magnetic property in nanocrystals will provide a stepping stone to understand a wide range of properties of magnetic nanoparticles.

5. Biomedical Imaging and Drug Delivery

QDs, with their dimensional similarities with biomolecules such as proteins and nucleic acids, size dependent optical properties and resistance against photo bleaching, make them potential candidates for imaging and theranostic applications.[43] Their improved signal brightness compared to the conventional organic dyes has been used for bio imaging.[54] However, their major drawback in biological applications is the presence of heavy elements as well as blinking. Extensive research has been carried out to prepare heavy-metal free biocompatible QDs possessing favorable fluorescent properties such as the long fluorescence lifetime, large stokes shift and the narrow emission band.[55] Recently, the synthesis and application of lead-free ternary I-III-VI chalcogenide QDs (Cu-In-S, Cu-In-Se) have taken centre stage owing to the multiple recombination channels present in their system.[56] Also, coating QDs with a wide band gap semiconductor shell like ZnS (~3.6 eV) forming a core-shell structure has been adapted efficiently not only to enhance the stability of the core but also to encapsulate the heavy-metal rich core like CdSe. In this aspect, Klimov et al. (2017), reported a $CuInS_2$/ZnS system achieving narrow emission line widths up to 60 meV in NIR I and NIR II regions extending its robustness, overcoming a serious limitation of a large photoluminescence (PL) line width (typically >300 meV).[57] Though $CuInS_2$ and $CuInSe_2$ systems have attracted notable attention, they suffer from fairly low PL QY compared to other QDs. However, a quaternary QD system $ZnAgInSe$[58] is known to be heavy metal free with high PL QY (~70%). In fact, in 2017, Deng et al., demonstrated ZnAgInSe/ZnS core-shell QD conjugated with sulfobetaine -poly(isobutylene-altmaleic anhydride)–histimine (SPH) polymer and a cyclic RGD peptide for its significant application in cancer imaging.[59]

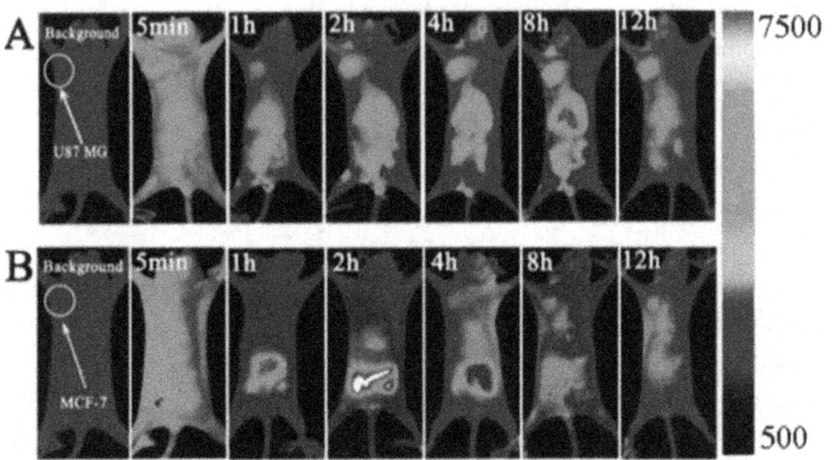

Fig. 9. Longitudinal distribution of RGD–SPH ZnAgInSe/ZnS QDs-clusters in nude mice bearing (A) glioblastoma (B) breast cancer, detected using NIR QD imaging (λ ex = 660 nm, λ em = 740 nm). Reproduced with permission from Ref. 59. Copyright 2017 American Chemical Society.

This was injected in glioblastoma or breast cancer tumor bearing mice; NIR images were taken up to 72h post-injection as shown in Figure 9. QD cores providing a structural scaffold for drug delivery is also well established in recent years. QDs possessing good biocompatibility, inertness towards drugs, loading capacity with high in-vivo residence time and a suitable particle size and shape with stability are excellent choices as successful drug carriers. Thus, this wide possibility of tuning material characteristics opens up an interesting platform to engineer the drug carrier materials with high selectivity and superior physical properties such as size, cell transfection, charge etc.[60]

Utilization of QDs as drug carriers for therapeutics range from cancer, neurological disorders, hypertension and also gene therapy.[61] QDs with Lipofectamine ™2000 and small interfering RNA (siRNA) as a gene therapeutic approach was discussed by Chen and his co-workers.[62] Another nano-vehicle for the anticancer drug was established in 2017, by L Zhang et al.[63] based on a self-assembled quantum dot DNA hydrogel system that exhibited both enzyme-responsive drug delivery of doxorubicin and cell-specific targeting.

6. Quantum Dots: Present Status and Outlook for the Future

Despite its history of more than 30-years, the science of colloidal QDs still represents an exciting area of research that appeals to scientists with a various range of backgrounds, including inorganic and colloidal chemistry, condensed matter physics, optoelectronics, biological and medical sciences. With the introduction of perovskite inorganic QDs the possibility is further increased by a recent demonstration of their applications in different technologies ranging from photovoltaics, QD-LEDS, lasers, bio-imaging etc. Colloidal QDs are rapidly generating attention not only in the scientific community but also engineering and entrepreneurial communities due to their potential in various applications.

QDs in real life applications are at developmental stages for some applications, some applications like QD based down-converting displays have already become a commercial reality. However, there are a lot of major issues and concerns that need to be addressed to make these applications to become part of our daily lives.

The stability of colloidal QDs and thin films are some of the major issues here. A brief exposure of QDs solid to oxygen or moisture may entirely modify the landscape of surface trapping state thus affecting its conductivity as well as the optical properties. In case of perovskite QDs a slight exposure to air, moisture, UV light can fasten the degradation of these cubic crystals to orthorhombic phase which is an optical inactive phase leading to inefficiency in these devices. Even though the cell encapsulation can reduce the ill effects of oxygen and water contamination, it is useless against thermodynamic instability.

The removal of long chain organic ligands accelerates degradation on densely packed QDs and eventually causes aggregation and transformation to optically inactive and inefficient stages. Stability issues are attended to an extent by the encapsulation of QDs in a polymer matrix like PMMA, or silica coating air stable QDs with a high performance needs to be engineered.

Additionally, the cost of manufacturing and the lack of scalability hinders some of these materials to come out of the laboratory scale to an industrial large-scale production. A majority of highly efficient colloidal QDs are synthesized either using heavy metals like Cd, Hg, Pb etc. The social and environmental concerns regarding toxicity of these heavy metals are still a huge concern in bringing these technologies to real life applications. Replacing heavy metals from these QDs using lesser toxic elements like Sn and Cu have been partially successful, however their optoelectronic performance compared to the native QDs have been really poor. For example, all inorganic perovskite Sn based $CsSnX_3$ are synthesized using high temperature colloidal synthesis but these QDs are highly unstable and exhibit a negligible PL QY. Hence for the commercialization of Cd or Pb based QD devices in the near future, efficient policies in recycling have to be first formulated and followed.

References

1. Ekimov, A. I. & Onushchenko, A. A. Quantum size effect in three-dimensional microscopic semiconductor crystals. *JETP Lett.* **34**, 345-349 (1981).

2. Norris, D. J. & Bawendi, M. Measurement and assignment of the size-dependent optical spectrum in CdSe quantum dots. *Phys. Rev. B* **53**, 16338 (1996).

3. Ekimov, A. I., Hache, F., Schanne-Klein, M., Ricard, D., Flytzanis, C., Kudryavtsev, I., Yazeva, T., Rodina, A. & Efros, A. L. Absorption and intensity-dependent photoluminescence measurements on CdSe quantum dots: assignment of the first electronic transitions. *J. Opt. Soc. Am.* **10**, 100-107 (1993).

4. Alferov, Z. I. The history and future of semiconductor heterostructures. *Semiconductors* **32**, 1-14 (1998).

5. Murray, C., Norris, D. J. & Bawendi, M. G. Synthesis and characterization of nearly monodisperse CdE (E= sulfur, selenium, tellurium) semiconductor nanocrystallites. *J. Am. Chem. Soc* **115**, 8706-8715 (1993).

6. Protesescu, L., Yakunin, S., Bodnarchuk, M. I., Krieg, F., Caputo, R., Hendon, C. H., Yang, R. X., Walsh, A. & Kovalenko, M. V. Nanocrystals of cesium lead halide perovskites ($CsPbX_3$, X= Cl, Br, and I): novel optoelectronic materials showing bright emission with wide color gamut. *Nano Lett.* **15**, 3692-3696 (2015).

7. Hines, M. A. & Guyot-Sionnest, P. Synthesis and characterization of strongly luminescing ZnS-capped CdSe nanocrystals. *J. Phys. Chem. A* **100**, 468-471 (1996).

8. Peng, X., Schlamp, M. C., Kadavanich, A. V. & Alivisatos, A. P. Epitaxial growth of highly luminescent CdSe/CdS core/shell nanocrystals with photostability and electronic accessibility. *J. Am. Chem. Soc.* **119**, 7019-7029 (1997).

9. Sanehira, E. M., Marshall, A. R., Christians, J. A., Harvey, S. P., Ciesielski, P. N., Wheeler, L. M., Schulz, P., Lin, L. Y., Beard, M. C. & Luther, J. M. Enhanced mobility $CsPbI_3$ quantum dot arrays for record-efficiency, high-voltage photovoltaic cells. *Sci. Adv.* **3** (2017).

10. Grandhi, G. K., Swathi, K., Narayan, K. & Viswanatha, R. Cu doping in ligand free CdS nanocrystals: conductivity and electronic structure study. *J. Phys. Chem. Lett.* **5**, 2382-2389 (2014).

11. Remacle, F. On electronic properties of assemblies of quantum nanodots. *J. Phys. Chem. A* **104**, 4739-4747 (2000).

12. Carey, G. H., Levina, L., Comin, R., Voznyy, O. & Sargent, E. H. Record charge carrier diffusion length in colloidal quantum dot solids via mutual dot-to-dot surface passivation. *Adv. Mater.* **27**, 3325-3330 (2015).

13. Suh, Y.-H., Kim, T., Choi, J. W., Lee, C.-L. & Park, J. High-performance $CsPbX_3$ perovskite quantum-dot light-emitting devices via solid-state ligand exchange. *ACS Appl. Nano Mater.* **1**, 488-496 (2018).

14. Wheeler, L. M., Sanehira, E. M., Marshall, A. R., Schulz, P., Suri, M., Anderson, N. C., Christians, J. A., Nordlund, D., Sokaras, D. & Kroll, T. Targeted ligand-exchange chemistry on cesium lead halide perovskite quantum dots for high-efficiency photovoltaics. *J. Am. Chem. Soc.* **140**, 10504-10513 (2018).

15. Speirs, M., Balazs, D., Fang, H.-H., Lai, L.-H., Protesescu, L., Kovalenko, M. V. & Loi, M. Origin of the increased open circuit voltage in PbS–CdS core–shell quantum dot solar cells. *J. Mater. Chem. A* **3**, 1450-1457 (2015).

16. Crisp, R. W., Kroupa, D. M., Marshall, A. R., Miller, E. M., Zhang, J., Beard, M. C. & Luther, J. M. Metal halide solid-state surface treatment for high efficiency PbS and PbSe QD solar cells. *Sci. Rep.* **5**, 9945 (2015).

17. Pan, Z., Zhao, K., Wang, J., Zhang, H., Feng, Y. & Zhong, X. Near infrared absorption of $CdSe_xTe_{1-x}$ alloyed quantum dot sensitized solar cells with more than 6% efficiency and high stability. *ACS Nano* **7**, 5215-5222 (2013).

18. Ip, A. H., Thon, S. M., Hoogland, S., Voznyy, O., Zhitomirsky, D., Debnath, R., Levina, L., Rollny, L. R., Carey, G. H. & Fischer, A. Hybrid passivated colloidal quantum dot solids. *Nat. Nanotechnol.* **7**, 577 (2012).

19. Wang, K., Jin, Z., Liang, L., Bian, H., Bai, D., Wang, H., Zhang, J., Wang, Q. & Shengzhong, L. All-inorganic cesium lead iodide perovskite solar cells with stabilized efficiency beyond 15%. *Nat. Commun.* **9**, 4544 (2018).

20. Park, B. W., Philippe, B., Zhang, X., Rensmo, H., Boschloo, G. & Johansson, E. M. Bismuth based hybrid perovskites $A_3Bi_2I_9$ (A: methylammonium or cesium) for solar cell application. *Adv. Mater.* **27**, 6806-6813 (2015).

21. Wang, C., Chesman, A. S. & Jasieniak, J. J. Stabilizing the cubic perovskite phase of $CsPbI_3$ nanocrystals by using an alkyl phosphinic acid. *ChemComm* **53**, 232-235 (2017).

22. Ghosh, D., Ali, M. Y., Chaudhary, D. K. & Bhattacharyya, S. Dependence of halide composition on the stability of highly efficient all-inorganic cesium lead halide perovskite quantum dot solar cells. *Sol. Energy Mater Sol. Cells* **185**, 28-35 (2018).

23. Chen, L.-J., Lee, C.-R., Chuang, Y.-J., Wu, Z.-H. & Chen, C. Synthesis and optical properties of lead-free cesium tin halide perovskite quantum rods with high-performance solar cell application. *J. Phys. Chem. Lett.* **7**, 5028-5035 (2016).

24. Colvin, V., Schlamp, M. & Alivisatos, A. P. Light-emitting diodes made from cadmium selenide nanocrystals and a semiconducting polymer. *Nature* **370**, 354 (1994).

25. Niu, Y.-H., Munro, A. M., Cheng, Y. J., Tian, Y., Liu, M. S., Zhao, J., Bardecker, J. A., Jen- La Plante, I., Ginger, D. S. & Jen, A. Y. Improved performance from multilayer quantum dot light- emitting diodes via thermal annealing of the quantum dot layer. *Adv. Mater.* **19**, 3371-3376 (2007).

26. Saha, A., Chellappan, K. V., Narayan, K., Ghatak, J., Datta, R. & Viswanatha, R. Near-unity quantum yield in semiconducting nanostructures: Structural understanding leading to energy efficient applications. *J. Phys. Chem. Lett.* **4**, 3544-3549 (2013).

27. Anikeeva, P. O., Halpert, J. E., Bawendi, M. G. & Bulovic, V. Quantum dot light-emitting devices with electroluminescence tunable over the entire visible spectrum. *Nano Lett.* **9**, 2532-2536 (2009).

28. Song, J., Li, J., Li, X., Xu, L., Dong, Y. & Zeng, H. Quantum dot light- emitting diodes based on inorganic perovskite cesium lead halides ($CsPbX_3$). *Adv. Mater.* **27**, 7162-7167 (2015).

29. Yang, D., Li, X. & Zeng, H. Surface chemistry of all inorganic halide perovskite nanocrystals: passivation mechanism and stability. *Adv. Mater. Interfaces* **5**, 1701662 (2018).

30. Pan, J., Sarmah, S. P., Murali, B., Dursun, I., Peng, W., Parida, M. R., Liu, J., Sinatra, L., Alyami, N. & Zhao, C. Air-stable surface-passivated perovskite quantum dots for ultra-robust, single-and two-photon-induced amplified spontaneous emission. *J. Phys. Chem. Lett.* **6**, 5027-5033 (2015).

31. Yao, J.-S., Ge, J., Han, B.-N., Wang, K.-H., Yao, H.-B., Yu, H.-L., Li, J.-H., Zhu, B.-S., Song, J.-Z. & Chen, C. Ce^{3+}-doping to modulate photoluminescence kinetics for efficient $CsPbBr_3$ nanocrystals based light-emitting diodes. *J. Am. Chem. Soc.* **140**, 3626-3634 (2018).

32. Subramanian, A., Pan, Z., Zhang, Z., Ahmad, I., Chen, J., Liu, M., Cheng, S., Xu, Y., Wu, J. & Lei, W. Interfacial energy-level alignment for high-performance all-inorganic perovskite $CsPbBr_3$ quantum dot-based inverted light-emitting diodes. *ACS Appl. Mater. Interfaces* **10**, 13236-13243 (2018).

33. Berns, R. S. *Billmeyer and Saltzman's Principles of Color Technology*. (Wiley New York, 2000).

34. Lee, J., Sundar, V. C., Heine, J. R., Bawendi, M. G. & Jensen, K. F. Full color emission from II–VI semiconductor quantum dot–polymer composites. *Adv. Mater.* **12**, 1102-1105 (2000).

35. Jang, E., Jun, S., Jang, H., Lim, J., Kim, B. & Kim, Y. White- light- emitting diodes with quantum dot color converters for display backlights. *Adv. Mater.* **22**, 3076-3080 (2010).

36. Kim, T.-H., Cho, K.-S., Lee, E. K., Lee, S. J., Chae, J., Kim, J. W., Kim, D. H., Kwon, J.-Y., Amaratunga, G. & Lee, S. Y. Full-colour quantum dot displays fabricated by transfer printing. *Nat. Photonics* **5**, 176 (2011).

37 Bai, Z. & Zhong, H. Halide perovskite quantum dots: potential candidates for display technology. *Sci. Bull.* **60**, 1622-1624 (2015).

38 Wang, H. C., Lin, S. Y., Tang, A. C., Singh, B. P., Tong, H. C., Chen, C. Y., Lee, Y. C., Tsai, T. L. & Liu, R. S. Mesoporous silica particles integrated with all-inorganic $CsPbBr_3$ perovskite quantum-dot nanocomposites (MP-PQDs) with high stability and wide color gamut used for backlight display. *Angew. Chem. Int. Ed* **55**, 7924-7929 (2016).

39. Kirstaedter, N., Ledentsov, N., Grundmann, M., Bimberg, D., Ustinov, V., Ruvimov, S., Maximov, M., Kop'ev, P. S., Alferov, Z. I. & Richter, U. Low threshold, large T/sub o/injection laser emission from (InGa) As quantum dots. *Electron. Lett.* **30**, 1416-1417 (1994).

40. Klimov, V. I., Mikhailovsky, A. A., McBranch, D., Leatherdale, C. A. & Bawendi, M. G. Quantization of multiparticle Auger rates in semiconductor quantum dots. *Science* **287**, 1011-1013 (2000).

41. Pietryga, J. M., Park, Y.-S., Lim, J., Fidler, A. F., Bae, W. K., Brovelli, S. & Klimov, V. I. Spectroscopic and device aspects of nanocrystal quantum dots. *Chem. Rev.* **116**, 10513-10622 (2016).

42. García-Santamaría, F., Chen, Y., Vela, J., Schaller, R. D., Hollingsworth, J. A. & Klimov, V. I. Suppressed auger recombination in "giant" nanocrystals boosts optical gain performance. *Nano Lett.* **9**, 3482-3488 (2009).

43. Wang, Y., Li, X., Song, J., Xiao, L., Zeng, H. & Sun, H. All-inorganic colloidal perovskite quantum dots: a new class of lasing materials with favorable characteristics. *Adv. Mater.* **27**, 7101-7108 (2015).

44. Mandal, S. K., Mandal, A. R. & Banerjee, S. High ferromagnetic transition temperature in PbS and PbS: Mn nanowires. *ACS Appl. Mater. Interfaces* **4**, 205-209 (2011).

     K. R. Pradeep & R. Viswanatha

45. Zhou, Y., Liu, K., Xiao, H., Xiang, X., Nie, J., Li, S., Huang, H. & Zu, X. Dehydrogenation: a simple route to modulate magnetism and spatial charge distribution of germanane. *J. Mater. Chem. C* **3**, 3128-3134 (2015).

46. Ottaviano, L., Continenza, A., Profeta, G., Impellizzeri, G., Irrera, A., Gunnella, R. & Kazakova, O. Room-temperature ferromagnetism in Mn-implanted amorphous Ge. *Phys. Rev. B* **83**, 134426 (2011).

47. Viswanatha, R., Naveh, D., Chelikowsky, J. R., Kronik, L. & Sarma, D. D. Magnetic properties of Fe/Cu codoped ZnO nanocrystals. *J. Phys. Chem. Lett.* **3**, 2009-2014 (2012).

48. Saha, A. & Viswanatha, R. Magnetism at the interface of magnetic oxide and nonmagnetic semiconductor quantum dots. *ACS Nano* **11**, 3347-3354 (2017).

49. Saha, A. & Viswanatha, R. Volume and concentration scaling of magnetism in dilute magnetic semiconductor quantum dots. *J. Phys. Chem. C* **121**, 21790-21796 (2017).

50. Saha, A., Shetty, A., Pavan, A., Chattopadhyay, S., Shibata, T. & Viswanatha, R. Uniform doping in quantum-dots-based dilute magnetic semiconductor. *J. Phys. Chem. Lett.* **7**, 2420-2428 (2016).

51. Viswanatha, R., Pietryga, J. M., Klimov, V. I. & Crooker, S. A. Spin-polarized Mn^{2+} emission from mn-doped colloidal nanocrystals. *Phys. Rev. Lett.* **107**, 067402 (2011).

52. Pandey, A., Brovelli, S., Viswanatha, R., Li, L., Pietryga, J., Klimov, V. & Crooker, S. Long-lived photoinduced magnetization in copper-doped ZnSe–CdSe core–shell nanocrystals. *Nat. Nanotechnol.* **7**, 792 (2012).

53. Zheng, W., Kumar, P., Washington, A., Wang, Z., Dalal, N. S., Strouse, G. F. & Singh, K. Quantum phase transition from superparamagnetic to quantum superparamagnetic state in ultrasmall $Cd_{1-x}Cr(II)_xSe$ quantum dots? *J. Am. Chem. Soc.* **134**, 2172-2179 (2011).

54. Zrazhevskiy, P., Sena, M. & Gao, X. Designing multifunctional quantum dots for bioimaging, detection, and drug delivery. *Chem. Soc. Rev.* **39**, 4326-4354 (2010).

55. Weissleder, R. & Pittet, M. J. Imaging in the era of molecular oncology. *Nature* **452**, 580 (2008).

56. Zhong, H., Bai, Z. & Zou, B. Tuning the luminescence properties of colloidal I–III–VI semiconductor nanocrystals for optoelectronics and biotechnology applications. *J. Phys. Chem. Lett.* **3**, 3167-3175 (2012).

57. Zang, H., Li, H., Makarov, N. S., Velizhanin, K. A., Wu, K., Park, Y.-S. & Klimov, V. I. Thick-shell $CuInS_2$/ZnS quantum dots with suppressed "Blinking" and narrow single-particle emission line widths. *Nano Lett.* **17**, 1787-1795 (2017).

58. Zhang, J., Xie, R. & Yang, W. A simple route for highly luminescent quaternary Cu-Zn-In-S nanocrystal emitters. *Chem. Mater.* **23**, 3357-3361 (2011).

59. Deng, T., Peng, Y., Zhang, R., Wang, J., Zhang, J., Gu, Y., Huang, D. & Deng, D. Water-solubilizing hydrophobic ZnAgInSe/ZnS QDs with tumor-targeted cRGD-

sulfobetaine-PIMA-histamine ligands via a self-assembly strategy for bioimaging. *ACS Appl. Mater. Interfaces* **9**, 11405-11414 (2017).

60. Xiao, S., Zhou, D., Luan, P., Gu, B., Feng, L., Fan, S., Liao, W., Fang, W., Yang, L. & Tao, E. Graphene quantum dots conjugated neuroprotective peptide improve learning and memory capability. *Biomaterials* **106**, 98-110 (2016).

61. Jia, N., Lian, Q., Shen, H., Wang, C., Li, X. & Yang, Z. Intracellular delivery of quantum dots tagged antisense oligodeoxynucleotides by functionalized multiwalled carbon nanotubes. *Nano Lett.* **7**, 2976-2980 (2007).

62. Chen, A. A., Derfus, A. M., Khetani, S. R. & Bhatia, S. N. Quantum dots to monitor RNAi delivery and improve gene silencing. *Nucleic Acids Res.* **33**, e190-e190 (2005).

63. Zhang, L., Jean, S. R., Ahmed, S., Aldridge, P. M., Li, X., Fan, F., Sargent, E. H. & Kelley, S. O. Multifunctional quantum dot DNA hydrogels. *Nat. Commun.* **8**, 381 (2017).

Chapter 4

Advances in Heterostructure Metamaterials for Solid-State Energy Conversion

Shashidhara Acharya, Dheemahi Rao and Bivas Saha[*]

International Centre for Materials Science and School of Advanced Materials,
Jawaharlal Nehru Centre for Advanced Scientific Research
Jakkur, Bangalore 560064, Karnataka, India
[*]bsaha@jncasr.ac.in

Artificially structured materials in the form of heterostructure and superlattices are one of the most celebrated classes of materials, having led to the discovery of many novel scientific concepts and device technologies. Semiconductor heterostructures are the most prominent among such "man-made" crystals and have been utilized for over six decades in the search for exotic new physics, such as quantum confinement of carriers, quantum Hall effect, mini-band formation in superlattices etc., and for the development of advanced electronic and optoelectronic device technologies, such as light-emitting diodes, quantum-well lasers, quantum-cascade lasers etc. However, with the changing time and to address the contemporary grand challenges of our society, recent research and development in heterostructure metamaterials have focused on topics such as solid-state energy conversion, energy efficient computing, secure information processing, imaging, sensing, etc. Similarly, complementing the semiconductor heterostructures including the layered 2D materials, the single crystalline metal/semiconductor superlattice heterostructure has emerged as a new addition to the epitaxial heterostructure. Starting with the basic concepts of the heterostructure development and a brief description of the semiconductor heterostructures, here, we will discuss the epitaxial metal/semiconductor heterostructure superlattices and their application in thermoelectricity and thermionic-emission devices, hot-electron plasmonic energy conversion, (solar) photovoltaics and neuromorphic computing devices.

78

1. Introduction

Technologically and commercially, solid-state energy conversion materials and devices are one of the most important areas of research and development in physics, chemistry, materials science and device engineering. Such materials and devices are necessary and important not only for the conversion of unconventional and wasted sources of energies into effective power, such as solar energy conversion, thermoelectric energy conversion, piezoelectric electronic energy harvest, etc. but also to develop modern energy efficient electronic devices, such as transistors, sensors, switches, etc. These materials have evolved over the last several generations based on the ever-changing industrial requirements and the underlying physical mechanism. At the dawn of the modern electronics era in the early 1950s, such materials and devices were developed to perform energy efficient computing and communications in transistors and integrated circuits, while the next three-to-four decades have seen their widespread application in almost all types of optoelectronic devices such as in solar cells, light emitting diodes, lasers, sensors etc. Most of these devices employ semiconductor heterostructures having homo and/or heterojunctions, along with the idea of quantum confinement of carriers in the active device regions. While such remarkable progress in the semiconductor heterostructure research is still continuing albeit with reduced rate, the last few years have also seen the emergence of several novel type of heterostructures such as layered 2D materials, metal/semiconductor superlattice heterostructures, neuromorphic computing heterostructures, and their applications in unexplored research fields such as in plasmonic and thermionic energy conversions, hot-electron based devices, neuromorphic computing, etc.

In this chapter, we will emphasize on the concepts behind the heterostructure and metamaterial development, and their evolution with industrial requirements and functionalities. After a brief description of the semiconductor heterostructure and 2D materials, we will discuss about the development of metal/semiconductor heterostructure and superlattices, which have emerged as new classes of "man-made" crystals. Physical properties of such heterostructure superlattices and

                    S. Acharya, D. Rao & B. Saha

their usefulness in several contemporary device technologies will also be addressed.

2. Idea Behind Superlattice and Heterostructure Metamaterials

The principle idea behind the use of heterostructure metamaterials is the utilization of interfaces that separate materials having different physical properties such as mass density, carrier concentrations, band structure, refractive index, etc. In addition, when the thickness of the individual layers in a heterostructure becomes less than or comparable to the mean free path of the energy carriers, the quantum confinement of such carrier particles can be achieved and utilized for device applications. Since the 1950s, a large number of electronic devices such as transistors, detectors, LEDs, lasers, etc. were developed based on *p-n* junctions, where two different parts of an interface being doped with hole (*p*-type) and electron (*n*-type) carriers. If the *p-n* junction is merely an interface

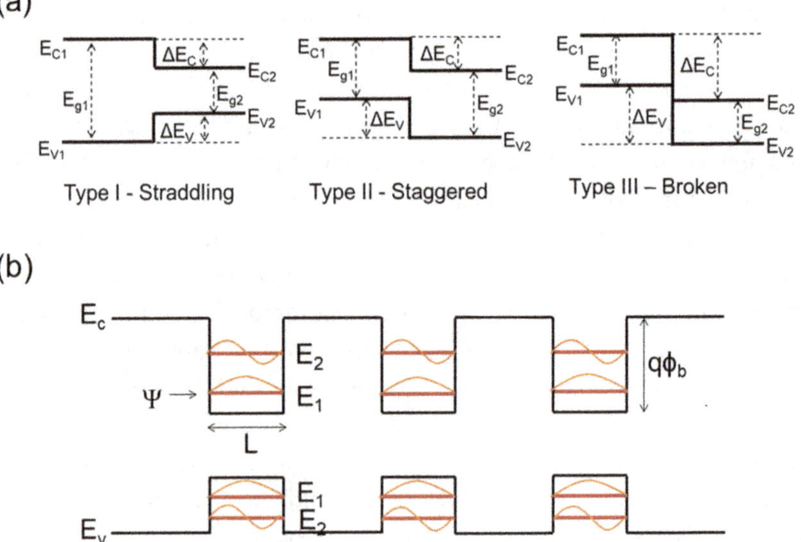

Fig. 1. (a) Three types of energy band alignments and (b) energy band diagram of the heterostructure with multiple quantum wells.

between the *n*-type and *p*-type doped regions of the same material, then such junctions are called homojunctions. Though such homojunctions are still an essential part of several modern devices, present-day advanced device technologies such as room temperature continuous-wave laser, high-electron-mobility transistors or high-efficiency solar energy converters would have been difficult to achieve with only homojunctions due to the same bandgap, refractive index, etc. throughout the device regions. Homojunctions also limit the ability to control unidirectional carrier injections, thus increasing the electrical and optical losses leading to less efficient devices. On the other hand, heterogeneous structures built from two or more materials can overcome such limitations and can lead to greater device efficiencies.

The most important feature of a heterojunction device is the discontinuity in energy levels at the hetero-interface creating conduction and valence band offsets. The band offset is a critical parameter, which dictates the performance and even the feasibility of the device operation for a given material system to a specific application. Band offsets allow us to optimize carrier injection and/or separation, carrier confinement, and barrier height adjustments. In terms of the electronic properties, heterostructures can be classified in three types based on the band alignment: (a) straddling gap or type I, (b) staggered gap or type II and (c) broken (misaligned) gap or type III (see Figure 1(a)). The band alignment of type I heterostructure leads to the accumulation of charge (both type) carriers in narrow bandgap semiconductors, whereas the type II heterostructure separates the charge carriers at the interface. Hence the type I heterostructure is used mostly in light emission applications (LED and laser), while the type II heterostructure is mostly used for solar energy conversion or detector applications.

The idea of the heterostructure has been further extended by introducing periodic layers (periodic potentials) of two different semiconductors stacked one over another, called as superlattices, that result in a series of quantum wells (see Figure 1(b)). The periodicity of alternating layers is typically maintained at a length scale less than the mean free path of electrons, which results in the quantum confinement of charge carriers and leads to novel quantum electronic devices. The period thickness is an important parameter to observe the quantum effects such as resonant tunneling, quantum cascade lasing, etc. and the crystal quality, sharpness of the interface, effective mass of the carriers, etc. plays a dominant role.

The types of heterostructures and metamaterials discussed here do not occur naturally and are traditionally prepared using various sophisticated thin film deposition techniques such as molecular beam epitaxy (MBE), metal-organic chemical vapor deposition (MOCVD), etc. Such growth techniques along with an advanced in-situ growth monitoring capability allow researchers to deposit ultra-thin layers (as small as a monolayer) with planarity, compositional and other controls. However, with the advances in growth technologies and materials science, several other methods such as ultra-high vacuum magnetron sputtering, hybrid vapor phase epitaxy, remote epitaxy (or layer transfer technique), etc. are currently explored for the materials and device fabrications.

3. Semiconductor Heterostructure: 50 years of Remarkable Progress

From a historical perspective, advances in heterostructure device technologies progressed on three main aspects: (a) device concepts (b) availability of suitable materials and (c) technologies to grow thin films of such materials, as well as device fabrication process. To realize the heterostructure's devices, a wide range of high-quality materials and expertise to engineer hetero-interfaces with minimized structural and electronic defects were required. The development of MBE and MOCVD have largely addressed the growth challenge and defect free heterostructures are regularly deposited with atomic scale precisions. Similarly, the development of the compound semiconductor (particularly III-V, II-VI classes of semiconductors) thin films and their alloys have immensely contributed to heterostructures with engineered properties.

The idea of heterojunctions to control electronic properties emerged along with the development of the transistor in 1951 when Shockley et al. proposed using a wide bandgap emitter to attain one-way injection.[1] However, the most significant contribution and the general design principle for heterojunction was first described by Kroemer et al. in 1957.[2] It was hypothesized that the heterojunction can give high unidirectional carrier injection efficiencies in comparison to homojunctions, which later became the precursor for the development of the heterostructure bipolar transistor (HBT) by Kroemer et al. and the double-heterostructure (DH) based laser invented independently by Alferov et al. and Kroemer et al.[3,4] In the case of DH, the feasibility of achieving a high-density carrier injection, accumulation and population inversion in the active region were predicted in order to achieve lasing action. Even though the DH was predicted in 1963, none of the immediate efforts could achieve continuous-wave (CW) room temperature laser action primarily due to the unavailability of a suitable lattice-matched heterostructure. However, with the development of a relatively defect-free lattice-matched GaAs/AlGaAs heterostructures, the room temperature lasing action was achieved. In 1970, Alferov et al. demonstrated the CW room temperature laser (Figure 2(a)), which later lead to the fabrication of several optoelectronic devices such as highly-efficient LEDs, heterostructure solar cells, thyristor, etc. Meanwhile, on the materials front, a wide variety of compound semiconductor solid-solutions were developed, which allowed the lattice constant and energy band gap to be varied independently. All such developments lead to the fabrication of the heterostructure with different types of band alignment, and hence, devices with new phenomena and light emission with extended spectral range became possible.

The development of the DH laser is one of the most remarkable technological achievements, which gave rise to several novel device concepts such as quantum wells and superlattice based electronic and optical devices. In 1970, Esaki and Tsu et al., while investigating the electron transport in ultra-thin quantum well structure, established the theory of resonant tunneling diodes and proposed the idea of superlattice heterostructures.[5] During the same time period, the growth of ultra-thin layers was made possible leading to the fabrication of such quantum structures. In such superlattices, with a varying composition in a period of a few 100 angstroms, the parabolic bands split into minibands that had a small bandgap, and the Brillouin zone is dictated by the period of the superlattice. In the early 1970s, Esaki et al. demonstrated the first experimental resonant tunneling in AlGaAs-GaAs based quantum well and superlattice structures,[5] which was an important milestone in achieving high-speed electronics applications.

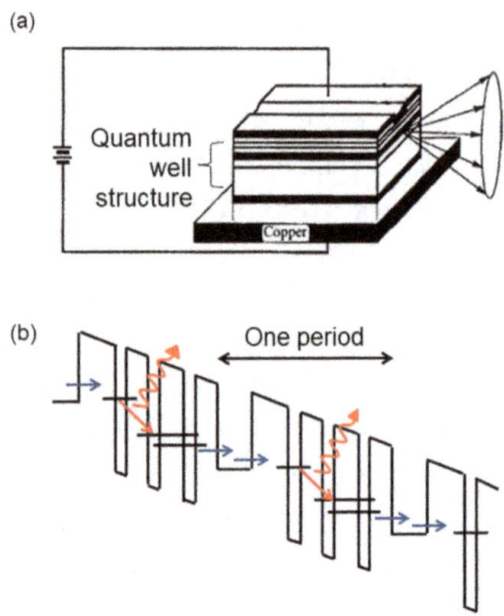

Fig. 2. Schematic representation of the (a) structure of the DH laser operating in the CW regime at room temperature and (b) the conduction band of a quantum cascade laser.

A study of the electron mobility in modulation-doped superlattice in 1978, showed for the first time that the mobility in such structures can be enhanced in comparison to the bulk semiconductors, which stimulated the research on 2D electron gas with a high mobility and gave rise to a new type of device called high-electron-mobility transistor (HEMT) based on AlGaAs/GaAs heterostructures for microwave applications. Following this, many other systems, for instance, AlGaN/GaN based HEMT have been studied for the 2D electron motion and demonstrated for high power and high-frequency applications.

Similar to electron transport, the superlattice heterostructure has a similar influence on the optical properties. Quantum well structures were used for light emission, in particular, for lasing applications. At the same time, Trang et al. from Bell communication laboratory demonstrated the advantage of the quantum well structure by fabricating an optimized structure with a separate confinement and smooth variation of the refractive index in the waveguiding region, which reduced the threshold current density considerably.[6] The threshold current density was further reduced from a few hundred A/cm^2 for DH laser to a few tens of mA/cm^2 with the development of the quantum well with a short-period superlattice structure. Similarly, the stimulated emission in the superlattice structure proposed by Kazarinov and Suris[7] shortly after the breakthrough work of Esaki and Tsu et al. on the superlattice, gave birth to the quantum cased laser (Figure 2(b))and was first demonstrated by F. Capasso et al. in 1994.[8]

4. 2D Semiconductor Heterostructure

Prior to 2004, based on few theoretical predictions, it was believed that it is not feasible to stabilize a material in an atom-thick layer at room temperature until Geim et al. mechanically exfoliated a monolayer of carbon, called graphene.[9] It paved the way for the exploration of various 2D materials and their heterostructures for several applications. Along with graphene, several other 2D materials such as boron nitride (BN), transition metal dichalcogenides (TMD) such as MoS_2, $MoSe_2$, WS_2, WSe_2, etc. have been studied extensively for the last 10-15 years. The 2D materials are in the limelight also due to their novel optical bandgap, strong interaction with electromagnetic radiations, electron-hole confinement, flexibility, transparency, and large specific surface area. Such materials generally possess strong in-plane covalent bonds and weaker van der Waals force between the layers and can be formed either laterally or vertically. Lateral heterostructures consist of different 2D crystals bonded in a single atomic layer, while vertical heterostructures consist of stacking of different 2D layers one over the other. Among the well-known 2D materials, graphene possesses semimetal, h-BN demonstrate insulator and the TMDs mostly exhibit semiconducting properties, though the TMD family also has metallic materials such as TiS_2, $TiSe_2$ VSe_2, $NbSe_2$, TaS_2, etc. Such diversity in electronic properties and nano-fabrication techniques enable the formation of heterostructures that can have applications in numerous fields, including photovoltaics and thermoelectrics.

                            S. Acharya, D. Rao & B. Saha

In 2010, Dean et al. introduced the first atomically thin 2D heterostructures, in which graphene was placed on top of thin h-BN layers.[10] Following this study, several semimetal/semiconductors and semiconductor heterostructures were developed (see Figure 3).

TMD based semiconductor/semiconductor heterojunctions fabricated on silicon were explored for solar cells application.[11] With plasma doping to create p-n junction, Wi et al. developed ITO/n-MoS_2/p-MoS_2/Au stacked heterostructure and reported the photoconversion-efficiency (PCE) of 2.8% and external-quantum-efficiency (EQE) of 37 to 78% for the wavelength range of 300nm-700nm.[12] Further research demonstrated a PCE of 5.23% by Tsai et al. in the CVD grown MoS_2/p-Si, where the built-in electric field near the interface facilitates photogenerated charge separation.[13]

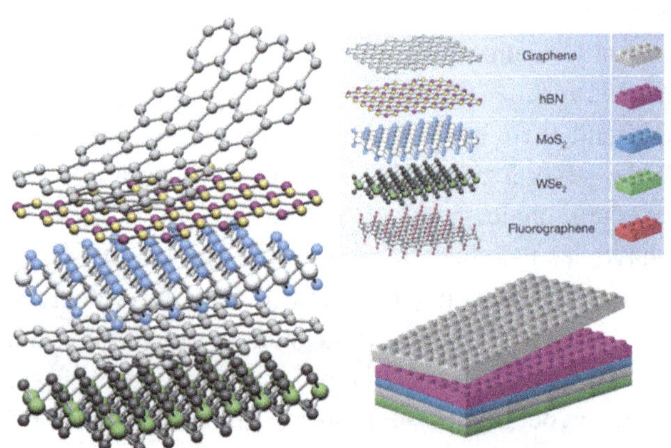

Fig. 3. Building 2D (van den Waals) heterostructures. Adapted by permission from AK Geim & IV Grigorieva Nature 499, 419-425.[14] Copyrights 2013 Springer Nature.

In one of the first reports of semimetal/semiconductor heterostructure, Shanmugam et al. developed a graphene/WS_2 heterostructure and photoelectric conversion efficiency of about 3.3% due to better absorptivity of WS_2 layer in the visible range.[15] Results also showed that stacking multiple layers of graphene improved the Schottky contact, while an effective absorption by WS_2 remained almost unaffected.

Due to the layered stacking of 2D heterostructures with van der Waals bonding along with the out of plane directions, thermal conductivity across the layers are usually very low, which allows one to maintain a temperature gradient across few layers of materials. Chen et al. reported a thermoelectric behavior of graphene/h-BN/graphene heterostructure. Though the conversion efficiency was rather small, the relatively high Seebeck coefficient –99.3 μV/K and temperature drop (a drop of 39 K for almost 4mV) suggested the possibility of using them as thermoelectric generators. Significant research in this direction, however, needs to be performed to evaluate the 2D materials' usefulness for such applications.

5. Metal/Semiconductor Heterostructure: A New Paradigm in Solid-State Devices

For over six-decades, semiconductor heterostructures have been the backbone of the modern technological revolution and have immensely had an impact on the discovery of novel and functional materials, as well as energy efficient electronic and optoelectronic devices. However, some of the current major societal requirements such as energy conversion, energy-efficient computing, secure information processing, imaging, sensing, etc. demand materials and devices that are in many instances beyond the reach of conventional semiconductor heterostructures, and in this regard, the recent development of metal/semiconductors heterostructures have the potential to overcome such challenges. In addition, metal/semiconductor heterostructures have also opened up new avenues, where one can manipulate not only the transport of current (electrons) but also the heat and light propagation in or through materials in devices. For example, plasmonic devices, hyperbolic metamaterials, thermionic energy conversion devices, etc. that were not possible with previously developed semiconductor heterostructures, are now a reality with metal-semiconductor metamaterials. In the following sections, we will discuss how the metal/semiconductors heterostructures and metamaterials are playing a crucial role in energy conversion and transport research fields, as well as energy efficient electronic device applications.

5.1. *Hot electrons – photochemical reactions*

Metal/semiconductor heterostructures are gaining increasing importance and attention from researchers working on photocatalysis, where optical (light) energy is used to carry out chemical reactions, such as water splitting to generate hydrogen fuel. Conventional photocatalytic reactions use semiconductors as photo-electrodes, which absorb light to generate electron-hole pairs and subsequently drive oxidation-reduction reaction on a liquid medium. However, the efficiency of such chemical reactions is smaller mainly due to the limitations of semiconductors, i.e. (a) the spectral range of absorption of a semiconductor is limited by its bandgap, leading to a small absorption range of the solar spectra to be useful for photocatalysis (in near UV regions for most cases) (b) thickness of the semiconductor though increases the absorbance, it also decreases the possibility of extracting electrons and holes for photocatalysis. However, in a metal-semiconductor based photo-electrode, light absorption and carrier generation or separation are enhanced by the surface plasmon resonance and the Schottky barriers at the metal-semiconductor interfaces. It leads to a heterostructure generating photocurrent in a wider band of the solar spectrum that comprises UV, visible and near IR regions, which in turn, increases the efficiency and lowers the cost of such devices with fewer materials required.

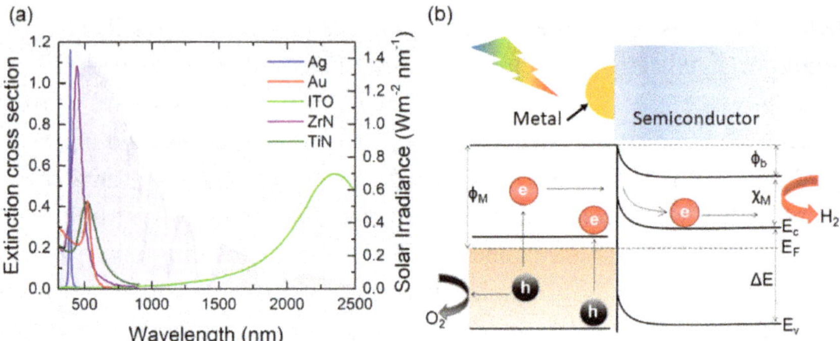

Fig. 4. (a) Extinction cross-section of metals (Ag, Au), transition metal nitrides (TiN, ZrN) and conducting oxides (ITO) plasmonic nanostructures compared with the solar irradiance spectrum is plotted in the background and (b) schematic representation of hot-electron mediated photochemical reaction where ϕ_M is the work function of the metal, ϕ_b is the barrier height and χ_M is the electron affinity of the semiconductor.

Plasmonic nanostructures are developed to trap, confine and manipulate light at the nanoscale. Metals having a large carrier concentration and plasmonic response are deposited or patterned on semiconductor surfaces, which generate energetic carriers (hot electrons and holes) due to the decay of surface plasmon (Landau Damping) upon excitation with light. Tian et al. showed that the hot carriers can be separated in an Au/TiO_2 photoelectrode due to the Schottky barrier at the metal/semiconductor interfaces (see Figure 4(b)).[16] The enhancement in the photocurrent due to the Au surface plasmon decay was observed in the visible spectral region around 520 nm (Au plasmonic resonance band) and at the UV region (~380 nm, TiO_2 bandegde). In addition, a 20× enhancement in the incident photon to current conversion efficiency (ICPE) was achieved in the presence of the suitable donor solution in contact with metal.[17] Results showed that for the continuous photogeneration and transfer process, a subsequent regeneration and injection of electrons in metal from the donor solution was required. Following this exciting demonstration, several researchers investigated Au or Ag nanostructures in contact with TiO_2 having various structural configurations that exhibited an extended lifetime of carriers generated in the visible region, and an enhancement in the photochemical conversion by several orders of magnitude.[18-23] Furthermore, various other semiconductors and metal combinations were also investigated; for example, Pt nanoparticle-TiO_2 film, Au nanoparticle-ZnO nanorod, Au nanoparticle-CeO_2, Au nanoprism-WO_3, AgBr/Ag nanoparticle on Al_2O_3, CdS/Au nanoparticle on TiO_2 nanorod, etc.[17,24]

Several studies were focused on understanding the mechanism of carrier generation and transfer pathways from the metal to the semiconductor layer. Theoretical investigations have shown that plasmons excited at metal surfaces can decay nonradiatively either through the inter-band (direct) or through intra-band (indirect) transitions, depending on, whether its energy is above or below the threshold energy, respectively. The intra-band (indirect) transitions are mediated mostly by phonons to conserve the momentum.[17] Further, density function theory based studies have shown that the size and shape of the plasmonic nanostructures also significantly control the decay pathway.[25] When the nanostructure size was decreased below 20 nm,

 S. Acharya, D. Rao & B. Saha

geometry (physical confinement) assisted transition dominated over the phonon-assisted transition. The time scale of such transitions spans from a few femtoseconds to several picoseconds depending on the plasmon energy and relaxation mechanism. Therefore, the hot carrier dynamics (generation, relaxation, and injection across the interface) are generally characterized using an ultrafast visible-pump/infrared-probe femtosecond transient absorption spectroscopy. Using such a pump-probe technique, Furube et al. found that the hot-electron generation and the injection process required less than 50 fs in Au/TiO$_2$ system,[26] and the injection efficiency was around 40% at 550 nm excitation. Results also showed that the hot carrier can inject back to a nanoparticle in the absence of a donor solution in contact with metal nanoparticles.

For efficient device applications, a metal-semiconductor photo-catalyst needs to generate a larger number of hot-electrons and should be able to inject them efficiently into semiconductors. As discussed before, reducing the particle size can increase the hot carrier generation, but will also reduce the absorption strength. Hence, a balanced approach needs to be adapted. For example (a) the surface plasmon polaritons (SPP) with adiabatic compression and hot spots for increased hot carrier generation rates and (b) carrier compensation by using a suitable donor or acceptor, etc. can be used. In addition, metal nanoparticles having narrow absorptions in the high-frequency regime of the visible spectrum due to their large carrier concentrations can be replaced with well-known transparent conducting oxides such as aluminum doped ZnO, ITO, and RuO$_2$ with a lower carrier concentration and higher optical losses. Therefore, the plasmonic structures made from such materials can have broader absorption spectra extending into the IR region and can increase the photon-to-electron conversion efficiency significantly. Similarly, the transition metal nitrides such as TiN, ZrN, etc., which exhibit interesting plasmonic behavior (close to that of noble materials, see Figure 4(a)) are also promising to replace noble metals.

5.2. *Thermionic energy conversion*

Thermionic energy conversion devices were conceptualized and developed in the 1950s to convert waste heat energy directly into electricity by means of thermionic emission. In its simplest form, the thermionic energy converter (TEC) has two parallel plates separated by a small gap (either vacuum or filled with some dielectric medium). One of the electrodes called *emitter* or *cathode* is heated to emit electron thermionically into the vacuum (or medium filling the gap), which will be collected at the cold electrode called *collector* or *anode*. The anode develops a negative potential due to the charge accumulation which eventually stops the electron flow. If two plates are connected through an electric load, the accumulated charge will flow through the load generating output voltage and current. In principle, the efficiency of such devices can approach up to 90% of the Carnot efficiency. A TEC composed of a vacuum diode having an inter-electrode gap of about 100 μm demonstrated a device efficiency of 10-15%. However, the use of TECs has been limited to high temperature domains such as space applications. The efficiency of such a device is limited by the work functions of the electrodes and the space charge effect at the interelectrode region. In order to achieve maximum efficiency, the collector work function should be minimum and that of the emitter should be at least 1eV more than the collector. In addition, the emitter electrode materials should have high temperature stability.

Several approaches have been adapted to mitigate such challenges, for example using an alkali metal intercalation to reduce the work function of the host materials, but with limited success. Recently, Lee et al. have developed a micro-fabricated suspended cathode based on SiC to reduce the inter-electrode distance, which would reduce the space charge effect.[27] But the conversation efficiency was still low due to the high work function of SiC.

In the late 1990s, Shakouri and Bowers developed the idea of a metal/semiconductor based solid-state TEC, where the vacuum gap (vacuum diodes in previously used TEC) was replaced with a solid, usually a semiconductor.[28] The motivation was to create a solid-state refrigeration or cooling device, which operated at a reverse bias in comparison to a TEC power generation. Such devices have several advantages over the conventional TEC such as (a) the device application can be extended to room temperature (b) thermionic emission in such metal/semiconductor heterostructures depends on the Schottky barrier height, which can be tailored with changes in material properties and (c) the space charge effect can be reduced by doping. However, one main disadvantage of such a solid-state TEC is the heat leakage by conduction, which is severe in comparison to the radiative heat transfer at a relatively lower temperature. To overcome this challenge, Mahan and Woods et al. proposed metal-semiconductor multilayer/superlattice structures (see Figure 5) and suggested that the increase in thermal resistance due to multiple interfaces can increase the energy conversion efficiency by

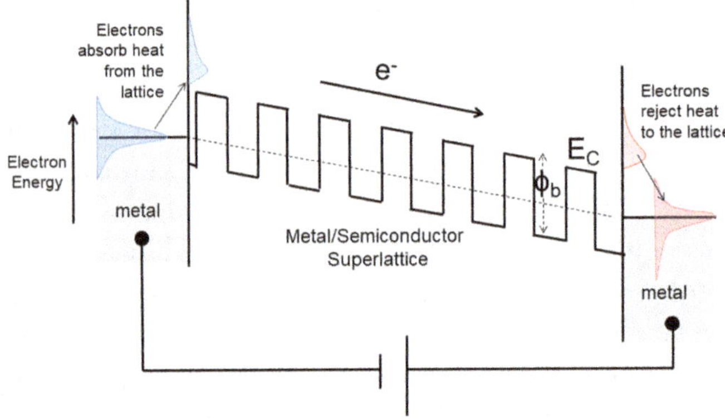

Fig. 5. Metal-semiconductor based multilayer TEC having multiple Schottky barriers to filter and ballistically transport hot electrons and efficiently block cold electrons.

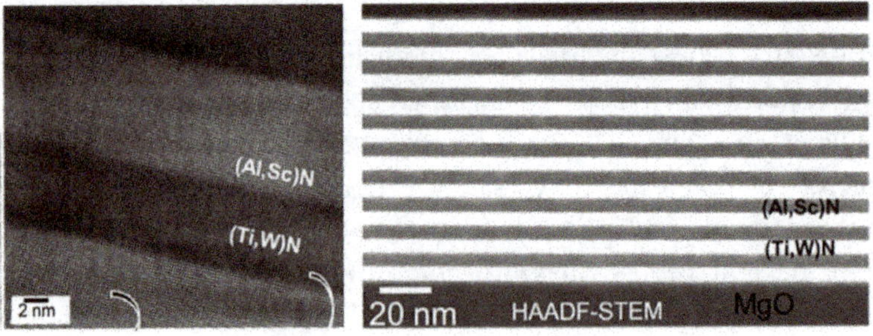

Fig. 6. HRTEM and HAADF-STEM image of the superlattices (Ti,W)N/(Al,Sc)N. Reproduced with permission from Saha et al., *Phys. Rev.* B 93, 045311 (2016).[31] Copyright 2016 American Physical Society.

more than a factor of two.[29] Results also showed that the material with a lower thermal conductivity would be better suited for such heterostructures and metal-semiconductor Schottky barriers can lead to desired low barrier heights for achieving high energy conversion efficiencies. *Vashaee et al.* showed that the effective thermoelectric *figure-of-merit* can be increased to more than 5 at room temperature with metal-semiconductor superlattices.[30]

Such theoretical predictions of thermionic refrigeration and power generation based on metal-semiconductors superlattices has inspired researchers to design epitaxial defect free heterostructures. In this regard, single crystalline and lattice-matched TiN/(Al,Sc)N metal/semiconductor superlattices have been developed recently that exhibit abrupt and sharp interfaces. Semiconducting rocksalt ScN was alloyed with AlN to stabilize the cubic-(Al,Sc)N and was lattice-matched with metallic TiN on MgO substrates for such a metal/semiconductor superlattice development. High-resolution electron microscopy images (see Figure 6) confirmed the epitaxial pseudomorphic cube-on-cube crystal growth and very little intermixing of atoms at the interfaces. Thermal transport measurements by the time domain thermoreflectance (TDTR) measurement exhibited a thermal conductivity of several orders of lower magnitude in superlattices with respect to the individual layers due to the increased interface scattering of phonons. A minimum room temperature of 1.7 W/m-K has been achieved with (Ti,W)N/(Al,Sc)N superlattices, that is suitable for thermionic energy conversion applications.[31] Similarly, the electrical transport properties across the cross-plane directions also have been investigated albeit with limited success due to the difficulty in defect and shunt free device fabrication. For ZrN/ScN metal/semiconductor multilayers, a Schottky barrier height of 0.28 eV has been achieved from the temperature dependent current-voltage (I-V) measurements. With future research and engineering of the work function and electron affinity of metal and semiconductor layers respectively, the Schottky barrier height can be tuned for specific applications, that will further the potential of such superlattices in diverse applications.

5.3. *Hyperbolic metamaterials*

Advances in nanofabrication techniques have given rise to several novel classes of artificial metamaterials that exhibit optical properties that are remarkably different from those that are observed in naturally occurring materials. In a pioneering work in the late 1960s, Veselago predicted that materials with a simultaneous negative permittivity and permeability should exhibit optical properties that are markedly different from normal dielectric or metallic materials, such as the negative refractive index.[32] However, the fabrication of such a material was made possible only in the 2000s, when Pendry theoretically predicted that the "negative refraction makes a perfect lens".[33] Such exotic optical properties of metamaterials arise primarily from the specific sub-wavelength structures and interactions of electromagnetic waves in such systems. Along with the demonstration of such negative refractive index materials at various spectral regions, the last 20 years have also seen research and development of several other types of metamaterials exhibiting interesting optical properties such as negative refraction, superlens, sub-wavelength imaging, etc.

Hyperbolic metamaterials are a new addition to the optical metamaterials class and exhibit several novel optical properties. As the name would suggest, the transverse magnetic (TM) wave in hyperbolic metamaterials exhibits hyperbolic dispersion of its iso-frequency surfaces. Such hyperbolic dispersion leads to a giant enhancement in the densities of photonic states, negative refraction, sub-wavelength imaging also known as hyperlens and the possibility of super-Planckian higher photonic thermal emission and thermal conductivity in materials.

The conventional demonstration of hyperbolic metamaterials employs polycrystalline elemental metal/dielectric multilayers with an alternative sign of the dielectric permittivity in metallic and semiconducting layers, for example, Au/TiO_2, Ag/Si, etc. Though such a structure exhibits hyperbolic metamaterial properties, they are not amenable to the atomic scale control of interfaces as well as doping, alloying and other quantum size effects. Given such limitations, epitaxial single crystalline TiN/(Al,Sc)N metal/dielectric superlattices have been developed recently that exhibit hyperbolic dispersion in a visible to near IR spectral range.[34] Moreover, such epitaxial superlattices have also exhibited an enhancement in the densities of photonic states. Due to the ability to control atomic scale parameters in such metamaterials, researchers were also enabled to modify and control the HMM properties of such structures.

5.4. *Solar-thermophotovoltaics*

Solar thermophotovoltaic (STPV) is an effective strategy for solar energy conversion. The STPV conversion devices mainly consist of the absorber, filter, emitter and a photovoltaic cell. The principle involved in such devices is similar to that of a typical photovoltaic (PV) cell except for the fact that the photons are not received directly from the sun but from the high temperature emitters that capture thermal energy from the sunlight (see Figure 7). The hot emitter emits photons in a narrow spectral range (in the IR region) and by choosing a PV cell with a band gap matching the spectral emission of the emitter, most of the solar energy can be converted into electricity. Hence STPV based devices are expected to have a superior efficiency in comparison to the conventional PV, where the efficiency is lowered due to the inability to absorb the full solar spectrum.

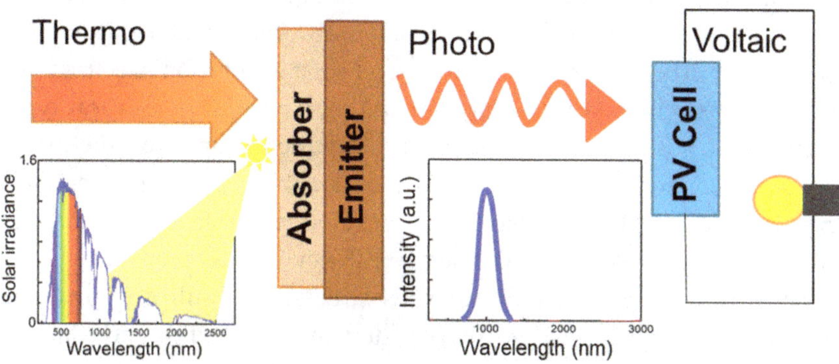

Fig. 7. Schematic representation of the working principle of a thermophotovoltaic cell.

The efficiency of a STPV device is defined by the ratio of the electric power generated by the cell over the net radiative power exchanged between the source and the cell. However, the overall efficiency of such devices depends on the efficiencies of the constituent components of the STPV system, such as the light concentration efficiency of the concentrating optics, the thermal transfer efficiency of the absorber/emitters and finally the heat-to-electricity conversion efficiency of the PV cell. Various strategies have been adapted over the course of several years of STPV development in order to maximize the net radiative flux with photon energy above the absorption edge of the PV cell.

Based on the theoretical predictions of Würfel et al., one of the initial strategies adopted to improve the STPV efficiency was to use lower bandgap semiconductors in PV cell, which could reduce the sub-bandgap losses.[35] They predicted that the up to 65% efficiency can be achieved with an absorber band gap of 0.8 eV. Several absorber materials have been investigated for PV cell application in STPV systems such as indium gallium arsenide (InGaAs), tellurium doped gallium antimonide (GaSb), silicon, germanium, etc.[36] Among such candidates, InGaAs is the most attractive material as it offers flexibility in the peak wavelength with a composition modulation. However, in order to reduce the cost, other alternative materials are being greatly investigated.

The next important approach that was adopted to reduce the sub-bandgap losses, was to develop selective absorbers and emitters. The selective absorbers must have a broadband solar absorption spectrum and a low thermal emittance at a large operating temperature. Several types of selective absorbers such as metal-dielectric composites (cermets), metal-semiconductor tandems, plasmonic absorbers and photonic crystals (1D/2D/3D PhCs) have been developed. Using these strategies up to 86%-92 % of thermal transfer efficiency has been achieved.[36] However, the thermal stability of such structures is still a challenging issue. On the other hand, selective emitters are required to emit thermal photons at or above the energy bandgap of the PV cell. In addition, they should have an optimum band width of photons above the bandgap and should be thermally stable for a long period of time. These requirements were well satisfied by rare-earth oxides such as erbium oxide, ytterbium oxide, etc. Hence, such materials have been the first choice for selective emission applications. Apart from that, photonic crystal and metamaterial based selective emitters were also developed, as such structures allow the tailoring of the emission band and can strongly suppress the undesired emission. In order to suppress the sub-bandgap emission further, the STPV systems include filters and photon recycling that selectively returns or reflects the sub-bandgap photons to the emitter.

In spite of the significant progress made over the last two decades to improve the design of the selective absorber, emitter, and PV cell, the efficiency of STPV devices remains significantly low. Also, the previous designs were based on far-field energy transfer, and hence, the efficiency has been dictated by black body limit. It led to the recent development of the near-field STPV, where the separation between the absorber/emitter assembly and the PV cell is reduced to a micro or nanoscale range. In such devices, the energy transfer takes place by coupling of evanescent surface modes and the radiative heat exchange can exceed several order magnitudes than in the large distance separation TPVs. The efficiency of such devices can reach the Carnot limit, which is well beyond the classical TPVs (limited by thermodynamics or Shockley–Queisser limit). However, near-field STPVs are still far from ideal conversion systems. They are affected by several losses such as radiative losses, electrical losses, and thermal losses. Several studies addressing these issues suggested using the thermal analog of quantum well or hyperbolic metamaterials, which may improve the efficiency further[36].

6. Neuromorphic Computing Heterostructure

Neuromorphic computing refers to systems containing electronic circuits that mimic neuro-biological architectures present in the human nervous system. In recent times, neuromorphic computing is actively researched for the artificial neural network design such as vision-system, hand-eye system, autonomous systems, etc. and utilize oxide-based memristors, spintronic memories, threshold switches, transistors, etc. Most of these hardware elements utilize some versions of heterostructure metamaterials, however, memristive based devices are the most prominent among them.

A memristor is a resistive device with an inherent memory and has a simple form of a two-terminal structure, where a total of only three layers- two electrodes, to send and receive electrical signals and a storage layer in between is needed (see Figure 8). Unlike static resistors, however, the storage layer can be dynamically configured by electrical inputs. The storage layer is typically a few nanometers thick, thus even a small voltage drop across the layer can generate enough electric field to drive the ionic process to alter the ionic configuration of the material. Oxidation, migration, reduction of the cation or anion species in the storage layer can lead to changes in the local conductivity. Such changes can be either abrupt (binary) or gradual (analog) with the different timescales of operation that lead to rich device physics in such heterostructure materials. A detailed discussion on the neuromorphic computing heterostructures are beyond the scope of this chapter and could be found at Ref. 37.

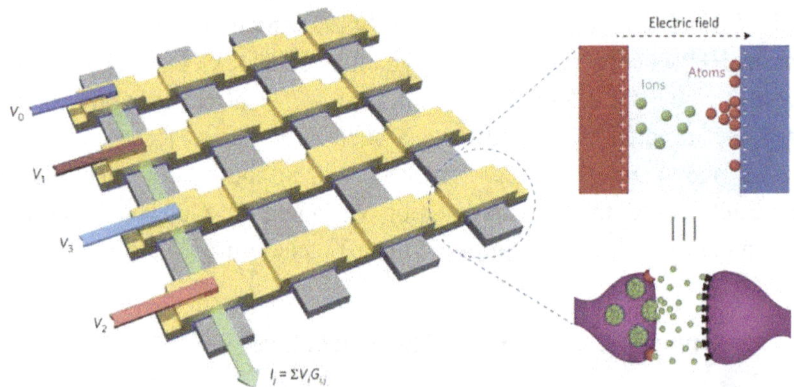

Fig. 8. Hardware implementation of artificial neural networks in a memristor crossbar. A memristor is formed at each crosspoint and can be used to simultaneously store data and process information. In this approach, vector-matrix multiplication can be obtained through Ohm's law and Kirchhoff's law through a simple read operation. Reprinted by permission from M. A. Zidan et al., Nature Electronics 1, 22 .[37] Copyrights 2018 Springer Nature.

7. Conclusion and Outlook

In conclusion, from the early years of the semiconductor heterostructure development to the modern era of the layered 2D semiconductor and epitaxial lattice-matched metal/semiconductor superlattices, heterostructure metamaterials have impacted many branches of basic science and device engineering properties, especially the energy transport and conversion research fields. Today the research and development in hot-electron based photochemical reactions to generate H_2 and other fuels, solar-photovoltaics and thermophotovoltaics, thermoelectrics and thermionic energy conversion devices, etc. extensively utilize various types of natural or "man-made" heterostructure metamaterials for improved efficiencies. Apart from the electronic and optoelectronic properties achieved with semiconductor heterostructures, the development of metal/semiconductor superlattices have extended their application in plasmonics and thermal energy transport and conversion research fields with unprecedented possibilities. There is no doubt that devices for the future such as low voltage electronic switches, neuromorphic computing hardware, etc. will use heterostructure materials in the coming days and further vitalize heterostructure materials and devices for the next few decades.

References

1.   W. Shockley, Circuit Element Utilizing Semiconductive Material, *US Patent* **2,569,347**, 1-13 (1951).
2.   H. Kroemer, Theory of a wide-gap emitter for transistors. *Proc. IRE* **45**(11), 1535–1537 (1957).
3.   H. Kroemer, Nobel lecture: Quasielectric fields and band offsets: Teaching electrons new tricks. *Rev. Mod. Phys.* **73**(3), 783–793 (2001).
4.   Z. I. Alferov, The history and future of semiconductor heterostructures. *Semiconductors* **32**(1), 1–14 (2002).
5.   L. Esaki, A bird's-eye view on the evolution of semiconductor superlattices and quantum wells. *IEEE J. Quant. Electron.* **22**(9), 1611–1624 (1986).
6.   W. T. Tsang, Extremely low threshold (AlGa)As graded-index waveguide separate-confinement heterostructure lasers grown by molecular beam epitaxy. *Appl. Phys. Lett.* **40**(3), 217–219 (1982).
7.   R. F. Kazarinov and R. A. Suris, Possibility of the amplification of electromagnetic waves. *Sov. Phys. Semicond.* **5**(4), 707–709 (1971).
8.   J. Faist, F. Capasso, D. L. Sivco, C. Sirtori, A. L. Hutchinson, and A. Y. Cho, Quantum cascade laser. *Science* **264**(5158), 553–556 (1994).
9.   K. S. Novoselov, A. K. Geim, S. V. Morozov, D. Jiang, Y. Zhang, S. V. Dubonos, I. V. Grigorieva, and A. A. Firsov, Electric field in atomically thin carbon films. *Science* **306**(5696), 666–669 (2004).

10. C. R. Dean, A. F. Young, I. Meric, C. Lee, L. Wang, S. Sorgenfrei, K. Watanabe, T. Taniguchi, P. Kim, K. L. Shepard, and J. Hone, Boron nitride substrates for high-quality graphene electronics. *Nat. Nanotechnol.* **5**(10), 722–726 (2010).

11. M. Y. Li, C. H. Chen, Y. Shi, and L. J. Li, Heterostructures based on two-dimensional layered materials and their potential applications. *Mater. Today* **19**(6), 322–335 (2016).

12. S. Wi, H. Kim, M. Chen, H. Nam, L. J. Guo, E. Meyhofer, and X. Liang, Enhancement of photovoltaic response in multilayer MoS2 induced by plasma doping. *ACS Nano* **8**(5), 5270–5281 (2014).

13. M. L. Tsai, S. H. Su, J. K. Chang, D. S. Tsai, C. H. Chen, C. I. Wu, L. J. Li, L. J. Chen, and J. H. He, Monolayer MoS2 heterojunction solar cells. *ACS Nano* **8**(8), 8317–8322 (2014).

14. A. K. Geim and I. V Grigorieva, Van der Waals heterostructures. *Nature* **499** (7459), 419–425 (2013).

15. M. Shanmugam, R. Jacobs-Gedrim, E. S. Song, and B. Yu, Two-dimensional layered semiconductor/graphene heterostructures for solar photovoltaic applications. *Nanoscale* **6**(21), 12682–12689 (2014).

16. Y. Tian and T. Tatsuma, Mechanisms and applications of plasmon-induced charge separation at TiO2 films loaded with gold nanoparticles. *J. Am. Chem. Soc.* **127**(20), 7632–7637 (2005).

17. M. Kim, M. Lin, J. Son, H. Xu, and J. M. Nam, Hot-electron-mediated photochemical reactions: principles, recent advances, and challenges. *Adv. Opt. Mater.* **5**(15), 1–21 (2017).

18. D. B. Ingram, P. Christopher, J. L. Bauer, and S. Linic, Predictive model for the design of plasmonic metal/semiconductor composite photocatalysts. *ACS Catal.* **1**(10), 1441–1447 (2011).

19. D. Gong, W. C. J. Ho, Y. Tang, Q. Tay, Y. Lai, J. G. Highfield, and Z. Chen, Silver decorated titanate/titania nanostructures for efficient solar driven photocatalysis. *J. Solid State Chem.* **189**, 117–122 (2012).

20. A. Sousa-Castillo, M. Comesaña-Hermo, B. Rodríguez-González, M. Pérez-Lorenzo, Z. Wang, X. T. Kong, A. O. Govorov, and M. A. Correa-Duarte, Boosting hot electron-driven photocatalysis through anisotropic plasmonic nanoparticles with hot spots in Au-TiO2 nanoarchitectures. *J. Phys. Chem. C* **120**(21), 11690–11699 (2016).

21. K. Qian, B. C. Sweeny, A. C. Johnston-Peck, W. Niu, J. O. Graham, J. S. Duchene, J. Qiu, Y. C. Wang, M. H. Engelhard, D. Su, E. A. Stach, and W. D. Wei, Surface plasmon-driven water reduction: Gold nanoparticle size matters. *J. Am. Chem. Soc.* **136**(28), 9842–9845 (2014).

22. Q. Zhang, D. T. Gangadharan, Y. Liu, Z. Xu, M. Chaker, and D. Ma, Recent advancements in plasmon-enhanced visible light-driven water splitting. *J. Mater.* **3**(1), 33–50 (2017).

23. J. Lee, S. Mubeen, X. Ji, G. D. Stucky, and M. Moskovits, Plasmonic photoanodes for solar water splitting with visible light. *Nano Lett.* **12**(9), 5014–5019 (2012).

24. C. Clavero, Plasmon-induced hot-electron generation at nanoparticle/metal-oxide interfaces for photovoltaic and photocatalytic devices. *Nat. Photonics* **8**(2), 95–103 (2014).

25. A. M. Brown, R. Sundararaman, P. Narang, W. A. Goddard, and H. A. Atwater, Nonradiative plasmon decay and hot carrier dynamics: Effects of phonons, surfaces, and geometry. *ACS Nano* **10**(1), 957–966 (2016).

26. L. Du, A. Furube, K. Hara, R. Katoh, and M. Tachiya, Ultrafast plasmon induced electron injection mechanism in gold-TiO2 nanoparticle system. *J. Photochem. Photobiol. C Photochem. Rev.* **15**(1), 21–30 (2013).

27. J. H. Lee, I. Bargatin, T. O. Gwinn, M. Vincent, K. A. Littau, R. Maboudian, Z. X. Shen, N. A. Melosh, and R. T. Howe, Microfabricated silicon carbide thermionic energy converter for solar electricity generation. *Proc. IEEE Int. Conf. Micro Electro Mech. Syst.* No. February, 1261–1264 (2012).

28. A. Shakouri and J. E. Bowers, Heterostructure integrated thermionic coolers. *Appl. Phys. Lett.* **71**(9), 1234–1236 (1997).

29. G. D. Mahan and L. M. Woods, Multilayer thermionic refrigeration. *Phys. Rev. Lett.* **80**(18), 4016–4019 (1998).

30. D. Vashaee and A. Shakouri, Improved thermoelectric power factor in metal-based superlattices. *Phys. Rev. Lett.* **92**(10), 10–13 (2004).

31. B. Saha, Y. R. Koh, J. Comparan, S. Sadasivam, J. L. Schroeder, M. Garbrecht, A. Mohammed, J. Birch, T. Fisher, A. Shakouri, and T. D. Sands, Cross-plane thermal conductivity of (Ti,W)N/(Al,Sc)N metal/semiconductor superlattices. *Phys. Rev. B* **93**(4), 1–11 (2016).

32. V. G. Veselago, The electrodynamics of substances with simultaneously negative values of ε and μ. *Sov. Phys. Uspekhi* **10**(4), 509–514 (1968).

33. J. B. Pendry, Negative refraction makes a perfect lens. *Phys. Rev. Lett.* **85**(18), 3966–3969 (2000).

34. B. Saha, G. V. Naik, S. Saber, C. Akatay, E. A. Stach, V. M. Shalaev, A. Boltasseva, and T. D. Sands, TiN/(Al,Sc)N metal/dielectric superlattices and multilayers as hyperbolic metamaterials in the visible spectral range. *Phys. Rev. B* **90**(12), 125420 (2014).

35. P. Würfel and W. Ruppel, Upper limit of thermophotovoltaic solar-energy conversion. *IEEE Trans. Electron Devices* **27**(4), 745–750 (1980).

36. Z. Zhou, E. Sakr, Y. Sun, and P. Bermel, Solar thermophotovoltaics: reshaping the solar spectrum. *Nanophotonics* **5**(1), 1–21 (2016).

37. M. A. Zidan, J. P. Strachan, and W. D. Lu, The future of electronics based on memristive systems. *Nat. Electron.* **1**(1), 22–29 (2018).

Chapter 5

Self-forming Templates and Nanofabrication

Indrajit Mondal,[a] Bharath B.[b] and Giridhar U. Kulkarni[a,b,*]

aCentre for Nano and Soft Matter Sciences, Jalahalli, Bangalore 560013, Karnataka, India
bChemistry & Physics of Materials Unit and School of Advanced Materials, Jawaharlal Nehru Centre for Advanced Scientific Research, Jakkur, Bangalore 560064, Karnataka, India
**kulkarni@jncasr.ac.in*

Material concepts pertaining to the nanoscale are realized into technology applications through nanofabrication. As regards to large area applications, self-forming templates have taken a central stage in the fabrication as the conventional lithography tools though capable, prove impractical and expensive in terms of the process cost and time. The chapter provides an introductory reading interlaced with a brief historical account, covering essential aspects of self-forming template-based methods focusing in particular, on the close packing of nanospheres, desiccated crack networks and phase segregation in polymer layers. Many diverse applications derived from these methods have been dealt with as case examples while citing the relevant literature. The merits and demerits of each technique have also been brought out.

1. Introduction

Fabrication at any length scale relates to assembling component materials following design and deriving a functional aspect from the assembled device. Fabrication involving small components has become attractive as it brings down the material cost, overall size and weight of the device, energy required to operate and derive the function, be it a mechanical, electronic, optical or a hybrid device. The last few decades have seen tremendous progress in micrometer level fabrication culminating into highly miniaturized devices which we enjoy.[1,2] Further, miniaturization would only mean fabrication at the nano level; the latter is of interest not only for miniaturization but more for exploiting the emergent material properties of the nanoscale. As Feynman remarked, there is indeed 'plenty of room at the bottom'.[3] And presently, the efforts are towards exploiting the 'room at the bottom'.

At nanoscale, matter behaves differently. Discretization of energy bands, quantum confinement of charge carriers, enhancement in surface bound effects, intensification of electric fields at sharp corners and so on- each sets in for a nanoscale object depending sensitively on its size and shape. In essence, the properties- electrical, optical, magnetic, thermal, mechanical, reactive etc. tend to be size-dependent. This explains the unprecedented interest and excitement in nanoscience and the associated technology. Over the last three decades, a wide variety of materials have been realised as nanomaterials, thanks to the novel synthetic techniques. Interestingly, this excitement owes much to nanomaterials of varied dimensionalities, hitherto unknown. Controlled manipulation of nano objects and their collective behavior have brought out new phenomena and added a great flavor to this branch of science. Nanofabrication is the key to translate this excitement to up-scaling, prototyping and finally, next generation devices and products.

Although nanofabrication concepts are around since the advent of advanced theoretical approaches half a century ago,[4] the developments in nanofabrication had to wait for the relevant technology tools to mature. Thus, several microscopy techniques such as scanning electron microscope (SEM), transmission electron microscope (TEM), ion beam microscope which enable visualization at the nano and sub-nano level have also become tools of nanofabrication. Since 1990s, atomic force microscope (AFM), scanning tunneling microscope (STM) and a variety of stamping and molding techniques have been developed offering such high resolutions that they are finding a wide usage in nanofabrication.[5,6] Today the world is poised with a plethora of lithography related tools to carry out material organization at various length scales be it in 1D, 2D or in 3D.[7,8]

2. Nanofabrication

Nanofabrication in contrast to microfabrication often involves high-end instrumentation mandated to serve at such a fine length scale. Most works related to producing materials in patterns or cutting and sizing of nano-objects are done using established methods of lithography. A large variety of methods are available starting from optical lithography to deal with micron-submicron features, to e-beam and other charged beam based lithography for finer features down to a few nanometers. When it comes to low cost, fast and large area micro or nanofabrication, the conventional top-down lithography techniques such as optical lithography, soft lithography, nanoimprint lithography, charged beam-based lithography are less desired. Although the methods are highly reliable, the working area is limited, a few mm^2 at best, and therefore, are ideally suited to produce high-density circuits and memory devices. However, devices such as energy devices or touch panels given their functionality, are meant to be of a large area and there exists a wide gap where the conventional lithography techniques are not applicable in practice. In this gap is the emergence of a new range of nanofabrication techniques that rely on self- assembling or self-forming templates, which is the subject matter of this chapter.

Self-assembly as the term indicates refers to the assembly of entities such as (i) molecules specially designed to interact with the surface of the chosen substrate (ii) nanoparticles and nano-spheres and (iii) copolymers formed by clustering two or more monomers. Molecules such as alkanethiols are known to form self-assembled monolayers (SAMs) on metal substrates via thiol chemisorption assisted by the steric interactions prevalent among the hydrocarbon chains.[9] The molecules remain well ordered over large areas to form a 2D-layer of molecules. SAMs are used as etch masks in combination with other lithography methods and are not dealt with here separately.[10,11] A system of nanospheres has the ability to form a closed pack or a superlattice structure on a liquid or a solid surface via entropy minimization, assisted by the capillary and hydrodynamic forces.[12] The area not shadowed by the packed particles is of interest for patterning and the method is termed as Nanosphere lithography (NSL). Block copolymers (BCPs) get into blocks of repeating units to spontaneously form well-defined structures or patterns due to phase segregation under the annealing conditions.[13] In both cases, NSL and BCP patterning, segregated regions or simply gaps formed at the nanometric level serve as templates. In contrast to the above self-assembling systems, the templet formation with self-forming is more physical; here, one would rely on desiccation of a colloidal thin film to release the stress to form cracks of controlled dimensions.[14]

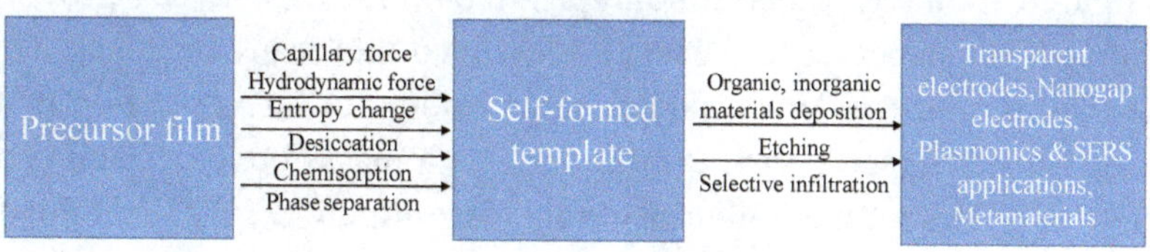

Fig. 1. Self-forming templates in nanofabrication.

The template process in any of the processes begins with a precursor film on a substrate which either undergoes chemically driven assembly, phase segregation or crack formation to form a template having random, interconnected grooves or even periodic gaps, used as a stencil for filling up a metal or some active material. On washing away the template, the material deposited inside of the grooves is left behind on the host substrate which depending on the connectivity, shape and the length scale can form a plasmonic array, a transparent conducting mesh or a set of nanogap electrodes. In this sense, the methods are subtractive in that, the material in predefined patterns is only retained while the rest gets removed. Importantly, these are easily realizable over large areas without serious implications on the cost if the material to be deposited is chosen carefully.

2.1. *Fabrication methods*

2.1.1. *Nanosphere lithography*

Nanosphere lithography (NSL) is being widely used in nanofabrication as it is a simple, low cost process yet reliable. It is useful in producing periodic nanopatterns over cm^2 areas in a short time. In NSL, nanospheres of polystyrene (PS), silica or such uniformly sized nanoparticles with smooth surfaces are spread over a substrate by dip coating, spin coating or by transferring in the form of a Langmuir film, to form close packed arrays leaving curved triangular regions in between. These open cavities are deposited with a desired material thus creating a regular nanostructured array of the material (see Figure 2(a)-(e)).[15] In spite of its simplicity, the method allows a great control on the periodicity, shape and size of the nanostructures and that explains its widespread use as compared to many other lithography techniques. NSL has grown to become sophisticated and indeed, has evolved into many branches. Thus, one may produce ordered nanospacings by in situ size reduction of the spread particles (see Figure 2(f)) for use as a mask to etch Si and create arrays of pillars with a high aspect ratio (see Figure 2(g)).[16,17] Following well defined steps- further reducing particle diameter, depositing Ni layer, removing the nanospheres and finally etching the exposed Si surface, one may even obtain an array of 3D hollow pillars or tubes as shown in the SEM image in Figure 2(h).[17] NSL has been used to create a Si membrane which upon decoration with Pd nanoparticles has worked as an efficient hydrogen gas sensor importantly at an ambient temperature without requiring heating.[12] Instead of the line of site deposition a tilted metal deposition can produce an interesting variety of nanostructures. In one such instance, oriented line pattern could be obtained in the place of disjoined features which served as a unidirectional transparent electrode.[18] Controlled texturing of surface using NSL has produced solid surfaces with varying degree of wettability.[19] The NSL technique is not only limited to fabricating simple periodic nanostructures, but also applicable in developing more complex structures. One such example is the Moire nanosphere lithography (M-NSL). It involves a sequential transfer and stacking of nanosphere monolayers at different angles resulting in the formation of moiré patterns (see Figure 2(i)). As shown in Figures 2(j) and (k), complementary patterns have been fabricated by depositing gold on a moiré-patterned template. This example clearly demonstrates the effectiveness of the M-NSL towards creating metasurfaces with a tailor-made optical response.[20] While NSL can produce such sophisticated patterns, the presence of defects such as holes restricts its routine usage, particularly for a large area patterning.

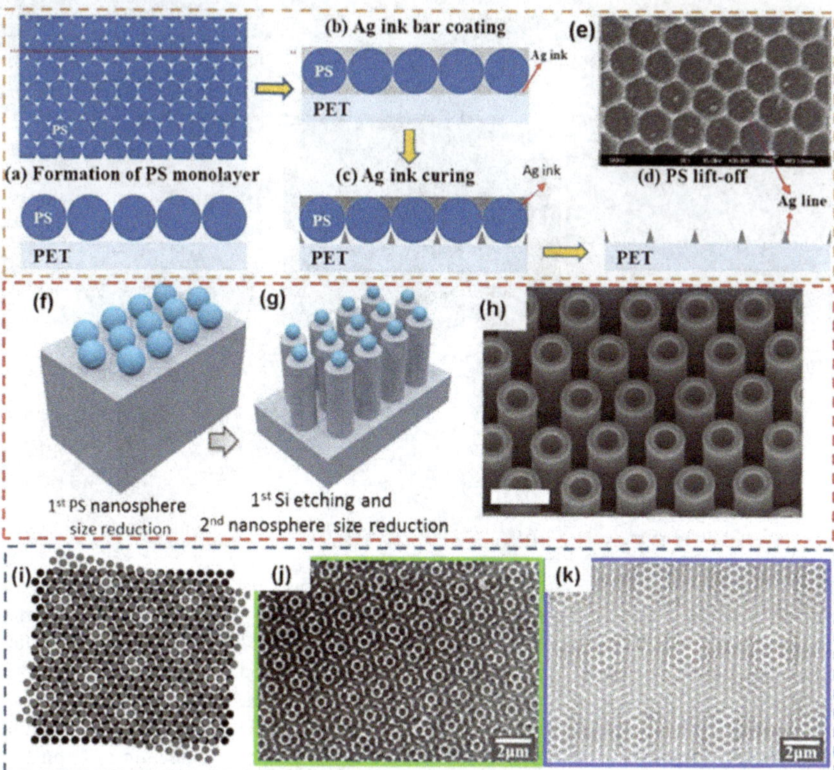

Fig. 2. Nanosphere lithography: Schematic diagram of (a) the top and side views of a closely packed polystyrene (PS) monolayer formed by the floating transfer method, (b) Ag ink bar coating and (c) curing followed by (d) PS layer removal. (e) SEM image of the as-fabricated Ag honeycomb mesh pattern.[15] (f) Size reduction of nanospheres to create a defined spacing between the particles and the use of the nanosphere array as a mask for Si etching, (g) After the first etching, further size reduction of nanospheres and the successive steps of Ni deposition, nanosphere removal and deep reactive ion etching to create a 3D tube pattern of Si as shown in (h) SEM image. Scale bar, 1μm.[17] (i) Schematic of a moire pattern generation by depositing two PS layers, the second layer rotated with respect to the bottom one. The spacing shown between the spheres is due to the size reduction induced by ion etching. (j) and (k) The experimental patterns fabricated by depositing Cr (5 nm) + Au (70 nm) metal on the nanoire-patterned template by e-beam deposition.[20] Adapted from Refs. [15, 17, 20].

2.1.2. *Self-forming*

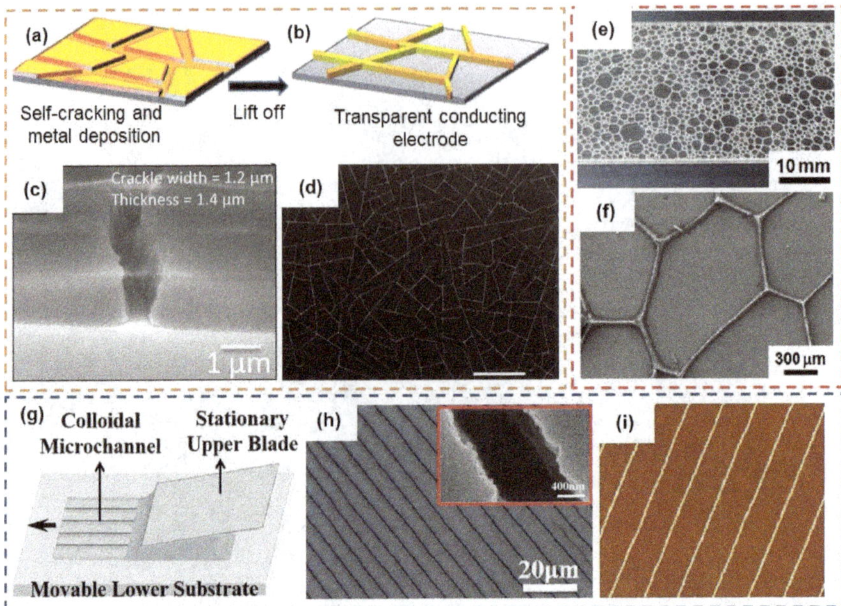

Fig. 3. Self-forming: Schematic of (a) a metal deposition on a random crack network formed after the drying of a crackle precursor thin film and (b) a metal network formed after the lifting off of the sacrificed crackle layer.[21] (c) Tilt SEM image of a desiccated crack down to the substrate.[14] (d) SEM image of a as-fabricated metal network (scale bar: 200µm). (e) Photograph of bubbles containing Ag nanowires (AgNWs) between two glass substrates to form continuous polygonal structure. (f) FESEM of AgNWs bundle network after drying the bubble template.[23] (g) schematic of the production of large area parallel microchannels by controlling the drying direction of a PS latex particle suspension on a computer-controlled movable substrate.[25] (h) Low and high (inset) magnified SEM images of the parallel cracks. (i) AFM image of Au stripes formed after the metal deposition on the cracked template (the stripe width is 600 nm).[24] Adapted from Refs. [14, 21, 22, 23, 24, 25].

The self-forming methods are applicable in producing large area template patterns with relatively simpler process conditions. An example of such a template is a desiccated crack layer formed as a way of releasing stress in the precursor colloidal layer film arising due to the solvent evaporation (Figures 3(a) and (c)).[14,21,22] By optimizing the associated physical parameters, highly interconnected micro and nano-cracks can be realised in the form of a single network, which by nature is random but do cover indefinitely large area substrates, uniformly. When used as a template for metal deposition, it can produce a large area connected metal network (see Figures 3(b) and (d)) and can be used in transparent conducting electrode applications.[14,21,22] In another example, a polygon network of bundled silver nanowires was formed along the ridges of a surfactant bubble template (see Figures 3(e) and (f)). In this case, the nanowires were first dispersed in bubbled water and the dispersion was then sandwiched between two glass substrates.[23]

Generally, the desiccated cracks are randomly oriented to form a mesh structure due to the randomness in drying; however a little control over the drying direction can produce aligned crack patterns. In one instance, a dispersion of nanoparticles was kept between two nearly parallel plates and by dragging one against another, manually or by a computer controlled instrument (as shown in Figure 3(g)), drying at the liquid-solid-air interface makes it directional to produce long parallel cracks (see Figure 3(h) and its inset). The cracks after metal deposition gave rise to parallel metal strips as shown in Figure 3(i).[24,25] There are other examples of such parallel and radial crack templates in colloidal films caused by enhanced drying from the directionally thinned front, the latter induced either by the gravity flow or by the dip coating methods.[26,27] The metal grating systems derived from such templates find applications as transparent conducting electrodes and photonic sensors.

2.1.3. *Self-assembly*

Self-assembly of block copolymers (BCPs) has attracted great attention in the context of sub 100 nm scalable patterning, complementary to expensive methods. BCPs, which comprise of two or more different polymers, are known to micro-phase separate to form periodic nano-patterns when they are thermal or solvent annealed, due to the decrement in the segregation strength resulting in an increase in the inter polymer chain diffusion. By proper surface modification of the substrate, choosing required volume fraction and molecular weight of the BCPs, the pattern morphology could be varied as described in the theoretical phase diagram in Figure 4(c). Typically, the patterns are obtained in the form of cylinders (Figure 4(a)), lamellae (Figure 4(b)) or other structures (gyroids or

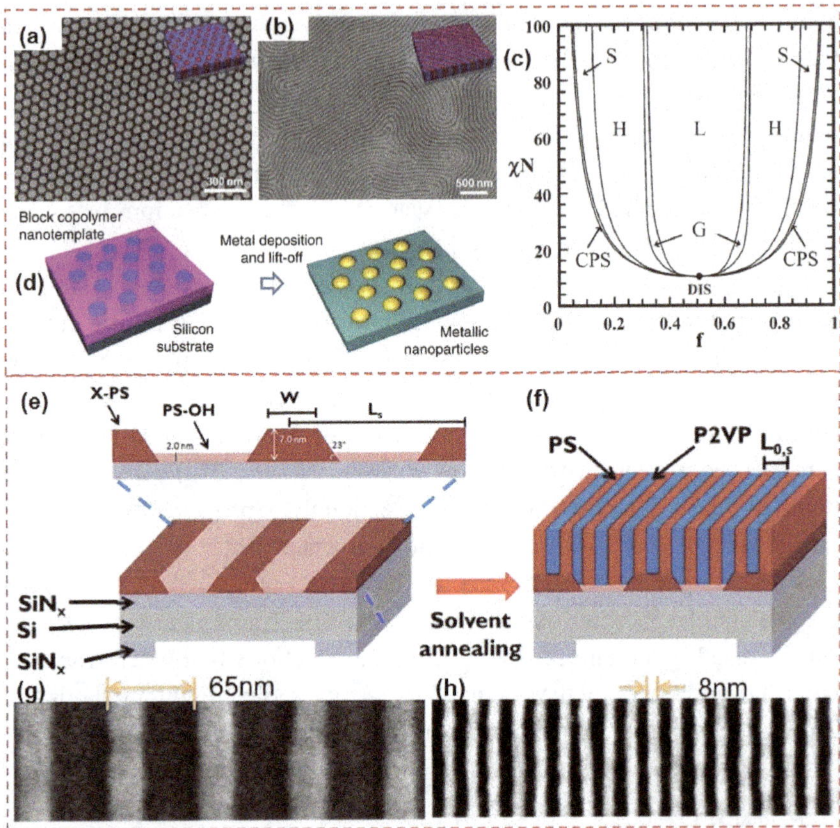

Fig. 4. **Self-assembly**: SEM images of self-assembled (a) cylindrical and (b) lamellae phases di-block copolymer (BCP) after a micro phase separation by solvent annealing (insets are the schematic of the same).[28] (c) Theoretical phase diagram of a di-BCP. χN is the segregation strength (χ: Flory–Huggins interaction parameter, N: degree of polymerization) and f is the fill factor of one of the two blocks. At various equilibriums, due to the minimization of the thermodynamic free energy, there forms different morphologies such as L: lamellae, G: the double gyroid, S: body-centred cubic packed spheres, H: hexagonally packed cylinders, CPS: close-packed spheres and DIS: disordered.[13] (d) Schematic of metallic nanoparticles array fabrication using a template of phase separated BCPs film after the selective removal of a single polymer.[34] **Directed self-assembly**: Schematic of (e) a pattern template of PS-OH brush and cross-linked polystyrene (X-PS) guiding stripes fabricated by e-beam lithography and plasma etching followed by the brush coating, (f) directed self-assembly of P2VP-*b*-PS-*b*-P2VP BCPs on the template after acetone vapor solvent annealing,[39] SEM images of (g) the photoresist pattern by e- beam lithography and (h) self-assembled BCPs pattern having four times higher line density in comparison to that obtainable with the e-beam pattern.[36] Adapted from Refs. [13, 28, 34, 36, 39].

spheres).[13] In recent times, an enormous amount of effort has been put to create nanopatterns by utilizing the controlled morphologies of BCPs. One general way to create patterns is to first form the desired nanopatterns followed by a selective removal of a particular block chemically or using plasma etching to use it as a template for material deposition (Figure 4(d)) or for masking for a future step.[13,28-34]

While BCP self-assembly produces periodic patterns locally in the nanometer range, the unfavorable domain orientations is the main concern when it comes to a large area nanofabrication. To address this, a guided technique called directed self-assembly is implemented to produce a long range ordering either by an external applied field or by pre-patterning the substrate.[35-42] In the example described in Figures 4(e) and (f), a polystyrene pre-patterned using e-beam lithography (Figure 4(g)) is selectively treated with oxygen plasma to serve as a guide for the BCPs (P2VP-b-PS-b-P2VP). The lamellae structure of BCPs hosts 4 times higher line density (Figure 4(h)) ideally suited for high density information storage.[36,38]

2.2. *Device Fabrication and Applications*

Self-forming templating methods can be process integrated relatively easily and that explains their extensive usage to produce device quality components. Below are some examples of materials in patterns produced using such methods often in combination. Applications in optoelectronics, molecular electronics, plasmonics, etc are provided.

2.2.1. *Transparent conducting electrodes*

In applications where light is required to pass through or interact with an active material beneath an electrode, the electrode is ought to be transparent to allow photons to reach the active material and such electrodes are called transparent conducting electrodes (TCEs). TCE is an essential component of all the opto-electronic devices (e.g. solar cells, LEDs, smart windows, displays, touch panels etc.) and also of non-optoelectronic applications such as defoggers, defrosters, transparent heaters and EMI shields.[21,43-47] Currently, tin-doped indium oxide (ITO) is the widely used TCE material at the industrial level due to its high transparency (90%) and low sheet resistance (10 Ω) at 300 nm thickness; but the cost, lower stability and brittleness combined with the scarcity of indium hinder the applicability of ITO in the next-generation optoelectronic and energy devices, whose sizes as well as usage are constantly on the increase. This being the scenario, there is an urge to reduce the overall cost of the devices, particularly that of the TCE itself. In this context, TCEs produced using self-formed templates have gained great attention worldwide. As shown in Figure 5(a), a large roll of PET could be converted to a flexible TCE by making it host a metal mesh, produced using a cracked template.[48] The optoelectronic properties of such electrodes are indeed comparable to a typical ITO film (see Figure 5(b)), and in many respects, the former, outperforms in terms of cost effectiveness, flexibility and stability.[21,44] TCEs obtained using self-formed templates are also superior to other network based TCEs formed by the brute-force deposition of nanowire dispersions, where the wire connectivity cannot be assured and the loosely bound junctions tend to act as hotspots.[49] There are applications uniquely realizable with metal mesh TCEs made from self-forming templates, which are otherwise not possible with ITO coatings. A Au mesh TCE can be employed as a transparent heater in air with temperature reaching beyond 600 °C at nominal power (see Figure 5(c)).[21] Such heaters can stand a large number of heating cycles without any performance degradation. In smart window applications, gentle heating of the active layer using a transparent heater can cause switching of its transparency as shown in Figure 5(d) and (e).[44] The self-formed templating method has been exploited to make a random network of interconnected nanowire bundles and used it as a flexible TCE in a touch panel (Figure 5(f)).[43] In another example, the random metal mesh was semi-embedded in an elastomeric substrate for use as a strain sensor with ultrahigh sensitivity.[45] The sensor action is derived from reproducible make-break nanogaps; higher the strain, larger will be the number of break junctions in wire segments and in turn, higher will be the overall resistance. It works for both tensile and compressive strains. Its application as a biomedical device is demonstrated as shown in Figure 5(g) and (h), where the movement of jaws while eating is monitored.

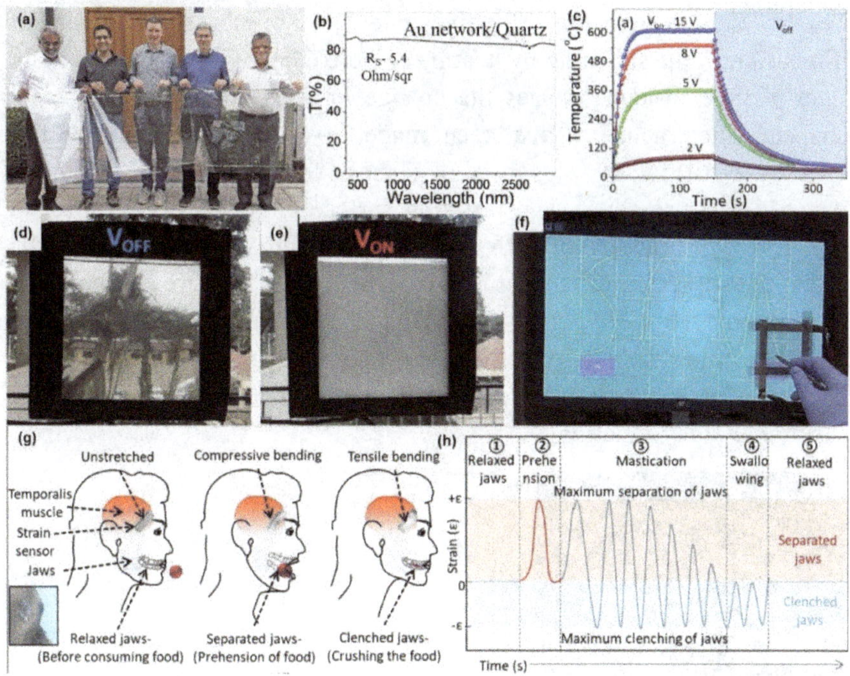

Fig. 5. (a) Photograph showing the team members holding a large area PET sheet deposited with a random Al mesh produced by a roll-to-roll process. The sheet exhibited visible transparency of 84% and sheet resistance of 40 Ω.[48] (b) Transmittance plot of a Au wire mesh based TCE on quartz substrate.[21] Applications of the TCE as (c) transparent heater,[21] (d) and (e) hydrogel based smart window where hydrogel switches its transparency after a certain temperature,[44] (f) touch sensor,[43] and (g) and (h) strain sensor where the crack formation in the random metal wire network due to an applied strain gives rise to a resistance change.[45] Adapted from Refs. [21, 43, 44, 45].

2.2.2. *Applications from BCP patterning*

In addition to crack templates, BCPs also have been employed in fabricating TCEs.[28] The method relies on simultaneous macro and micro phase separations in a BCP and then following the general method to create a template for metal deposition. BCP patterning is being used in a wide variety of contexts. As demonstrated in Figure 4(d), a metal nanodots array is fabricated on a Si substrate by BCP self-assembly and transferred on a stretchable PMMA substrate (Figure 6(a)).[34] The as-fabricated metasurfaces made of different metals or alloy nanoparticles can be used to tune the refractive index just by shrinking the lateral dimension of the substrate by heating (Figure 6(b)). The decrement in the inter-particle distance causes the change in the refractive index. Cone shaped nanostructures have been made in Si (Figure 6(c)) by first creating a Au dot array using a BCP mask followed by the etching of exposed Si vertically.[33] The resulting surface could easily enhance the antireflection property of a Si solar cell in a broadband (Figure 6(d)). Thus, the short circuit current density is enhanced (Figure 6(e)) and the power conversion efficiency is increased from 8.7% to 13.1%. Another example is the fabrication of a microporous honeycomb structure that consists of an array of spherical cavity which was used as a template for functional dots formation.[31]

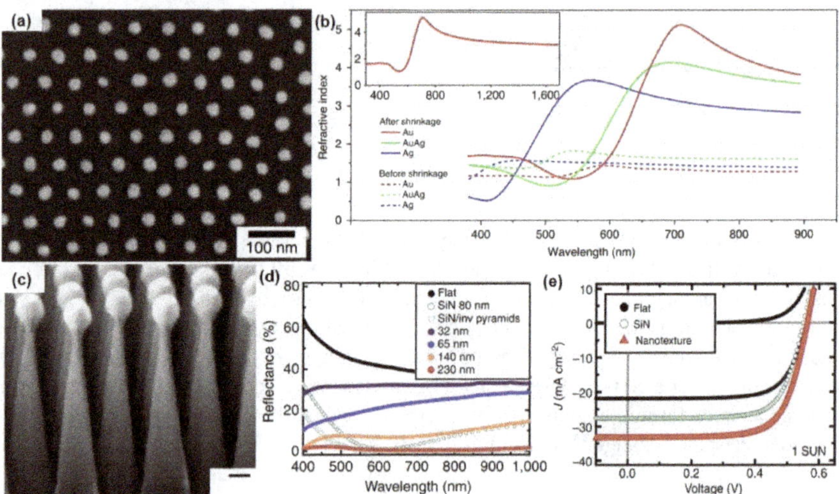

Fig. 6. (a) SEM image of a hexagonal Au nano-dots array fabricated on a stretchable PMMA substrate by BCP self-assembly with tunable refractive index. The process is extended to other metal and alloy nanoparticles such as Ag, Au, Au-Ag etc.[34] (b) Ellipsometry measurements of refractive indices of the as-fabricated metasurfaces made of different nanoparticles before (dashed lines) and after (solid lines) substrate shrinkage up to 40%.[34] (c) SEM image of Si nanocones fabricated by etching a Si wafer while masking with a alumina nanopattern formed by infiltrating a PS-b-PMMA BCP in its cylindrical phase, for enhanced antireflection in a Si solar cell. (Scale bar: 20 nm).[33] (d) Reflectance vs wavelength plot for flat Si film, a SiN film (80 nm thick) and nanotextured Si film with nanocone spacing 52 nm and height ranging from 32 to 230 nm.[33] (e) Current– voltage (J–V) characteristics of flat Si solar cell (black) and similar cells coated with 80 nm thick SiN antireflection coating (green open circles) and a surface nanotexture (red).[33] Adapted from Refs. [33, 34].

2.2.3. *Nanogap arrays*

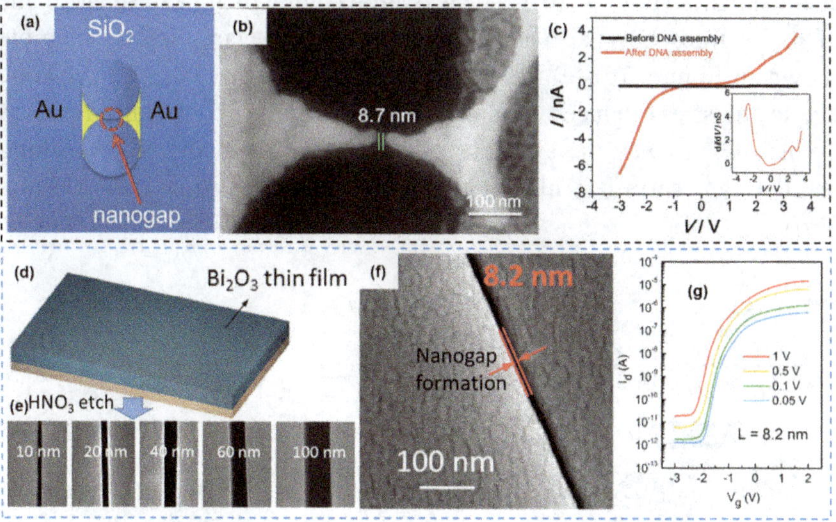

Fig. 7. **Nanogap arrays**: (a) Schematic and (b) SEM image of a nanogap fabricated by nanosphere lithography for template formation followed by metal deposition and nanospheres lift-off. (c) I-V characteristics of a DNA device fabricated by assembling DNA into the nanogap of the electrode.[50] (d) Schematic of bismuth trioxide film on a silicon substrate, (e) SEM images of nanocracks formed in the Bi_2O_3 film after the annealing and HNO_3 etching. Crack width could be increased with the increasing etching time. (f) SEM image of a monolayer MoS_2 FET on the nanocracked film (before the top gate fabrication) and (g) its transfer characteristics.[51] Adapted from Ref. [50, 51].

One of the main challenges in molecular electronics pertains to the fabrication of gap electrodes, gaps being the order of molecular length. Although there exists a range of sophisticated techniques for obtaining fine features, such as e-beam lithography, ion beam lithography and scanning probe lithography, these processes are tedious and expensive. Here, self-assembly or self-forming methods bring a ray of hope, particularly in the context of a large area, multigap electrode systems. As regards NSL, the process is similar to that described in the example in Figure 2, where by properly controlling the etching time of the nanospheres, one can achieve adesired nanogaps of the metal deposited (Figure 7(a), (b)).[50] In this example, the gap was as small as 10 nm enough to host individual DNA strands- 10.4 nm long poly (G)-poly(C)

DNA oligomers containing 30 base pairs, for obtaining electrical characteristics (Figure 7(c)).[50] Cracks propagating along a cleavage plane of oxide films have also been exploited as nanogaps. In the example shown in Figure 7(d)-(e), cracks get formed in β-Bi$_2$O$_3$ on Si substrate due to the stress induced by annealing which is synergistically released by dissolving away the affected region in a dilute acid.[51] The resulting parallel gaps served as ultrashort channels for MoS$_2$ transistors (Figure 7(g)).

2.2.4. *SERS*

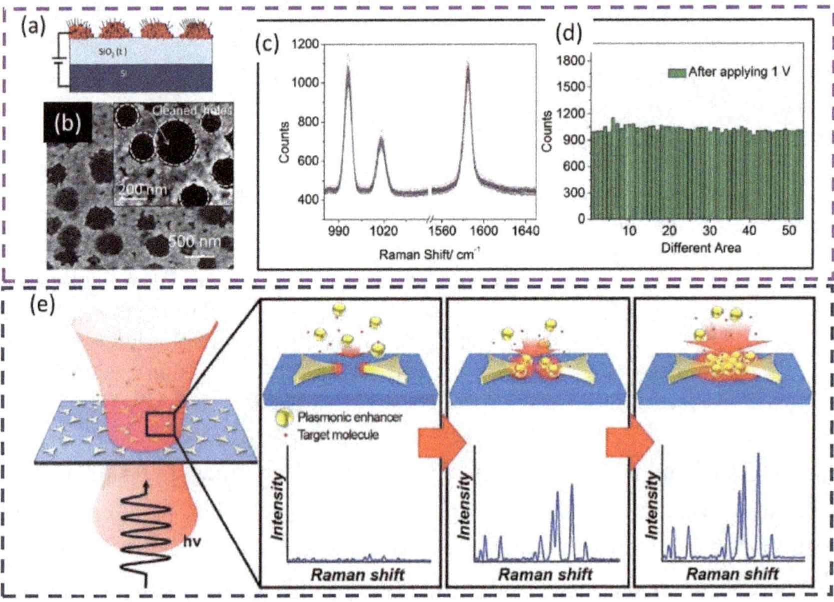

Fig. 8. Nanofabrication for SERS application. (a) Clean random nanohole array over the large area is formed by thermally annealing the Ag ink coated SiO$_2$/Si substrate followed by electrical annealing (1V) as shown in the cross-section schematic along with thiophenol molecules anchored to the Ag surface. (b) SEM image showing these well-defined random nanoholes. (c) Observed enhancement in the Raman signal collected from different regions of the large area (1 cm^2) substrate. (d) Intensity distribution of SERS signal at 996 cm-1 which suggests the high uniformity of the SERS substrate.[53] (e) Schematic illustration of autoenhanced Raman spectroscopy. 10 µM R6G solution mixed with gold nanoparticle is dispersed on a glass substrate with nanogap and irradiated with laser. With time (0, 120 and 240 min), the collected Raman spectra displayed enhancement in the signal due to trapped hotspots.[54] Adapted from Refs. [53, 54].

A major part of such an enhancement is often due to the sharp nanosized features on metal surfaces arising due to roughness, the latter sometimes by design. Although many complicated recipes have been proposed for preparing SERS active substrates, self-formed templating methods offer simple solutions.

Early studies on Ag nanohole arrays initiated by Van Duyne and coworkers using NSL are quite popular.[52] Even simpler methods have been developed wherein a Ag organic ink spin coated and annealed produces ~ 500 nm nanoholes in arrays.[53] By hosting the layer on a SiO_2/Si substrate followed by joule annealing, a cm^2 area platform with uniform SERS factors (10^7 to 10^8) has been realized (Figures 8(a)-(d)). In another study using NSL, R6G as analyte was loaded on 50 nm Au nanoparticles and the dispersion was dropped over the Au nanogap array. Due to the newly generated plasmonic effect hotspots, a significant increase in the SERS factor was observed (Figure 8(e)).[54] The recipe allowed the detection of R6G molecule in ultralow concentrations, ~100 pM.

3. Summary

The Chapter has dealt with various nanofabrication techniques that rely on self-forming templates. Depending on the process, the templates host a variety of nanogaps. Those formed by dissolving away a polymer component from a phase segregated mixture or by allowing a wet colloidal layer desiccate to dryness and crack, are typically random while the gaps that arise when the polymer nanospheres are organised into closed pack monolayers are periodic with a definite shape. These templates when used as a stencil produce materials in desired patterns - random networks metal wires, parallel lines as gratings, nanogap electrodes etc., are some examples mentioned above. Self-forming templates thus avoid complicated lithography approaches in producing patterns, a feature that is of great significance in the context of large area applications such as energy devices and smart windows. Indeed, these nanofabrication methods hold a great promise in futuristic technology development.

References

1. M. Campbell-Kelly, *Computer, Student Economy Edition: A History of the Information Machine*. Routledge (2018).

2. C. Mead and L. Conway, *Introduction to VLSI systems* (Vol. 1080). Reading, MA: Addison-Wesley (1980).

3. S. Abbott, J. Gribbin and M. Gribbin, *Richard Feynman: A Life in Science*. Math. Gaz., 1998.

4. K. E. Drexler, *Engines Of Creation: The Coming Era of Technology*. Anchor Books, Doubleday (1986).

5. A. A. Tseng, A. Notargiacomo and T. P. Chen, Nanofabrication by scanning probe microscope lithography: A review, *J. Vac. Sci. Technol. B: Microelectron. Nanometer Structur. Process. Measur. Phenomena* **23**(3), 877-894 (2005).

6. H. Schift, Nanoimprint lithography: An old story in modern times? A review, *J. Vac. Sci. Technol. B: Microelectro. Nanometer Structur. Process. Measur. Phenomena.* **26**(2), 458-480 (2008).

7. M. Stepanova and S. Dew, *Nanofabrication: Techniques and Principles*. Springer Science & Business Media (2011).

8. T. Pradeep, *A Textbook of Nanoscience and Nanotechnology*. Tata McGraw-Hill Education (2003).

9. D. K. Schwartz, Mechanisms and kinetics of self-assembled monolayer formation, *Ann. Rev. Phys. Chem.* **52**(1), 107-137 (2001).

10. J. P. Leeand and M. M. Sung, A new patterning method using photocatalytic lithography and selective atomic layer deposition, *J. Am. Chem. Soc.* **126**(1), 28-29 (2004).

11. I. Pang, J.H. Boo, H. Sohn, S.S. Kim and J.G. Lee, Amino-functionalized alkylsilane SAM-assisted patterning of poly (3-hexylthiophene) nanofilm robustly adhered to SiO_2 substrate, *Bull. Korean Chem. Soc.* **29**(7), 1349-1352 (2008).

12. M. Gao, M. Cho, H.J. Han, Y.S. Jung and I. Park, Palladium-decorated silicon nanomesh fabricated by nanosphere lithography for high performance, room temperature hydrogen sensing, *Small.* **14**(10), 1703691 (2018).

13. J.A. Dolan, B.D. Wilts, S. Vignolini, J.J. Baumberg, U. Steiner and T.D. Wilkinson, Optical properties of gyroid structured materials: from photonic crystals to metamaterials, *Adv. Opt. Mater.* **3**(1), 12-32 (2015).

14. K.D.M. Rao, R. Gupta and G.U. Kulkarni, Fabrication of Large Area, High-performance, transparent conducting electrodes using a spontaneously formed crackle network as template, *Adv. Mater. Interf.* **1**(6), 1400090 (2014).

15. N. Kwon, K. Kim, S. Sung, I. Yi and I. Chung, Highly conductive and transparent Ag honeycomb mesh fabricated using a monolayer of polystyrene spheres, *Nanotechnology.* **24**(23), 235205 (2013).

16. L. Li, Y. Fang, C. Xu, Y. Zhao, K. Wu, C. Limburg, P. Jiang and K.J. Ziegler, Controlling the geometries of Si nanowires through tunable nanosphere lithography, *ACS Appl. Mater. Interf.* **9**(8), 7368-7375 (2017).

17. X. Xu, Q. Yang, N. Wattanatorn, C. Zhao, N. Chiang, S.J. Jonas and P.S. Weiss, Multiple-patterning nanosphere lithography for fabricating periodic three-dimensional hierarchical nanostructures, *ACS Nano.* **11**(10), 10384-10391 (2017).

18. V. Oddone and M. Giersig, Nanosphere lithography with variable deposition angle for the production of one-directional transparent conductors, *Phys. Status Solidi Rapid Res. Lett.* **11**(3), 1700005 (2017).

19. P.C. Kuo, N.W. Chang, Y. Suen and S.Y. Yang, Fabrication of biomimetic dry-adhesion structures through nanosphere lithography, *Appl. Phys. A.* **124**(3), 236 (2018).

20. K. Chen, B.B. Rajeeva, Z. Wu, M. Rukavina, T.D. Dao, S. Ishii, M. Aono, T. Nagao and Y. Zheng, Moiré nanosphere lithography, *ACS nano.* **9**(6), 6031-6040 (2015).

21. K.D.M. Rao and G.U. Kulkarni, A highly crystalline single Au wire network as a high temperature transparent heater, *Nanoscale.* **6**(11), 5645-5651 (2014).

22. B. Han, Q. Peng, R. Li, Q. Rong, Y. Ding, E.M. Akinoglu, X. Wu, X. Wang, X. Lu, Q. Wang and G. Zhou, Optimization of hierarchical structure and nanoscale-enabled plasmonic refraction for window electrodes in photovoltaics, *Nat. Commun.* **7**, 12825 (2016).

23. T. Tokuno, M. Nogi, J. Jiu, T. Sugaharaand and K. Suganuma, Transparent electrodes fabricated via the self-assembly of silver nanowires using a bubble template, *Langmuir.* **28**(25), 9298-9302 (2012).

24. W. Han, B. Li and Z. Lin, Drying-mediated assembly of colloidal nanoparticles into large-scale microchannels, *ACS nano.* **7**(7), 6079-6085 (2013).

25. B. Li, B. Jiang, W. Han, M. He, X. Li, W. Wang, S.W. Hong, M. Byun, S. Lin and Z. Lin, Harnessing colloidal crack formation by flow-enabled self-assembly, *Angew. Chem. Int. Ed.* **56**(16), 4554-4559 (2017).

26. I. Mondal, A. Kumar, K.D.M. Rao and G.U. Kulkarni, Parallel cracks from a desiccating colloidal layer under gravity flow and their use in fabricating metal micro-patterns, *J. Phys. Chem. Solids.* **118**, 232-237 (2018).

27. O. Dalstein, E. Gkaniatsou, C. Sicard, O. Sel, H. Perrot, C. Serre, C. Boissière and M. Faustini, Evaporation-directed crack-patterning of metal–organic framework colloidal films and their application as photonic sensors, *Angew. Chem. Int. Ed.* **56**(45), 14011-14015 (2017).

28. J.Y. Kim, H.M. Jin, S.J. Jeong, T. Chang, B.H. Kim, S.K. Cha, J.S. Kim, D.O. Shin, J.Y. Choi, J.H. Kim and G.G. Yang, Bimodal phase separated block copolymer/homopolymer blends self-assembly for hierarchical porous metal nanomesh electrodes, *Nanoscale.* **10**(1), 100-108 (2018).

29. R. Shenhar, E. Jeoung, S. Srivastava, T.B. Norsten and V.M. Rotello, Crosslinked nanoparticle stripes and hexagonal networks obtained via selective patterning of block copolymer thin films, *Adv. Mater.* **17**(18), 2206-2210 (2005).

30. S. Park, J.Y. Wang, B. Kim and T.P. Russell, From nanorings to nanodots by patterning with block copolymers, *Nano Lett.* **8**(6), 1667-1672 (2008).

31. B. de Boer, U. Stalmach, H. Nijland and G. Hadziioannou, Microporous honeycomb-structured films of semiconducting block copolymers and their use as patterned templates, *Adv. Mater.* **12**(21), 1581-1583 (2000).

32. Q. Peng, Y.C. Tseng, S.B. Darling and J.W. Elam, Nanoscopic patterned materials with tunable dimensions via atomic layer deposition on block copolymers, *Adv. Mater.* **22**(45), 5129-5133 (2010).

33. A. Rahman, A. Ashraf, H. Xin, X. Tong, P. Sutter, M.D. Eisaman and C.T. Black, Sub-50-nm self-assembled nanotextures for enhanced broadband antireflection in silicon solar cells, *Nat. Commun.* **6**, 5963 (2015).

34. J.Y. Kim, H. Kim, B.H. Kim, T. Chang, J. Lim, H.M. Jin, J.H. Mun, Y.J. Choi, K. Chung, J. Shin and S. Fan, Highly tunable refractive index visible-light metasurface from block copolymer self-assembly, *Nat. Commun.* **7**, 12911 (2016).

35. H.S. Suh, P. Moni, S. Xiong, L.E. Ocola, N.J. Zaluzec, K.K. Gleason and P.F. Nealey, Sub-10-nm patterning via directed self-assembly of block copolymer films with a vapour-phase deposited topcoat, *Nat. Nanotechnol.* **12**(6), 575 (2017).

36. S. Xiong, L. Wan, Y. Ishida, Y.A. Chapuis, G.S. Craig, R. Ruiz and P.F. Nealey, Directed self-assembly of triblock copolymer on chemical patterns for sub-10-nm nanofabrication via solvent annealing, *ACS Nano.* **10**(8), 7855-7865 (2016).

37. A.P. Lane, X. Yang, M.J. Maher, G. Blachut, Y. Asano, Y. Someya, A. Mallavarapu, S.M. Sirard, C.J. Ellison and C.G. Willson, Directed self-assembly and pattern transfer of five nanometer block copolymer lamellae, *ACS Nano.* **11**(8), pp.7656-7665 (2017).

38. C.C. Liu, E. Franke, Y. Mignot, R. Xie, C.W. Yeung, J. Zhang, C. Chi, C. Zhang, R. Farrell, K. Lai and H. Tsai, Directed self-assembly of block copolymers for 7 nanometre FinFET technology and beyond, *Nat. Electron.* **1**(10), 562 (2018).

39. T. Segal-Peretz, J. Ren, S. Xiong, G. Khaira, A. Bowen, L.E. Ocola, R. Divan, M. Doxastakis, N.J. Ferrier, J. de Pablo and P.F. Nealey, Quantitative three-dimensional characterization of block copolymer directed self-assembly on combined chemical and topographical prepatterned templates, *ACS Nano.* **11**(2), pp.1307-1319 (2017).

40. H. Hu, M. Gopinadhan and C.O. Osuji, Directed self-assembly of block copolymers: a tutorial review of strategies for enabling nanotechnology with soft matter, *Soft Matter.* **10**(22), 3867-3889 (2014).

41. H.M. Jin, D.Y. Park, S.J. Jeong, G.Y. Lee, J.Y. Kim, J.H. Mun, S.K. Cha, J. Lim, J.S. Kim, K.H. Kim and K.J. Lee, Flash light millisecond self-assembly of high χ block copolymers for wafer-scale sub-10 nm nanopatterning, *Adv. Mater.* **29**(32), 1700595 (2017).

42. X. Zhang, Q. He, Q. Chen, P.F. Nealey, and S. Ji Directed self-assembly of high χ poly (styrene-b-(lactic acid-alt-glycolic acid)) block copolymers on chemical patterns via thermal annealing, *ACS Macro Lett.* **7**(6), 751-756 (2018).

43. Y.D. Suh, S. Hong, J. Lee, H. Lee, S. Jung, J. Kwon, H. Moon, P. Won, J. Shin, J. Yeo and S.H. Ko, Random nanocrack, assisted metal nanowire-bundled network fabrication for a highly flexible and transparent conductor, *RSC Adv.* **6**(62), 57434-57440 (2016).

44. A.K. Singh, S. Kiruthika, I. Mondal and G.U. Kulkarni, Fabrication of solar and electrically adjustable large area smart windows for indoor light and heat modulation, *J. Mater. Chem. C.* **5**(24), 5917-5922 (2017).

45. N. Gupta, K.D.M. Rao, K. Srivastava, R. Gupta, A. Kumar, A. Marconnet, T.S. Fisher and G.U. Kulkarni, Cosmetically adaptable transparent strain sensor for sensitively delineating patterns in small movements of vital human organs, *ACS Appl. Mater. Interf.* **10**(50), 44126-44133 (2018).

46. J.Y. Lee, S.T. Connor, Y. Cui and P. Peumans, Solution-processed metal nanowire mesh transparent electrodes, *Nano Lett.* **8**(2), 689-692 (2008).

47. H. Liu, V. Avrutin, N. Izyumskaya, Ü. Özgür and H. Morkoç, Transparent conducting oxides for electrode applications in light emitting and absorbing devices, *Superlattice microst.* **48**(5), 458-484 (2010).

48. http://igstc.org/images/newsletters/152653527320180517.pdf

49. G.U. Kulkarni, S. Kiruthika, R. Gupta and K.D.M. Rao, Towards low cost materials and methods for transparent electrodes, "*Curr. Opin. Chem. Eng.* **8**, 60-68 (2015).

50. D. Ji, T. Li and H. Fuchs, Nanosphere lithography for sub-10-nm nanogap electrodes, *Adv. Electron. Mater.* **3**(1), 1600348 (2017).

51. K. Xu, D. Chen, F. Yang, Z. Wang, L. Yin, F. Wang, R. Cheng, K. Liu, J. Xiong, Q. Liu and J. He, Sub-10 nm nanopattern architecture for 2D material field-effect transistors, *Nano Lett.* **17**(2), 1065-1070 (2017).

52. C. Fenzl, T. Hirsch and O.S. Wolfbeis, Photonic crystals for chemical sensing and biosensing, *Angew. Chem. Int. Ed.* **53**(13), 3318-3335 (2014).

53. R. Gupta, S. Siddhanta, G. Mettela, S. Chakraborty, C. Narayana and G.U. Kulkarni, Solution processed nanomanufacturing of SERS substrates with random Ag nanoholes exhibiting uniformly high enhancement factors, *RSC Adv.* **5**, 85019–85027 (2015).

54. S. Hong, O. Shim, H. Kwon and Y. Choi, Autoenhanced Raman spectroscopy via plasmonic trapping for molecular sensing, Anal. Chem. **88**(15), 7633-7638 (2016).

Chapter 6

Towards Precision and Adaptive Supramolecular Materials

Shikha Dhiman and Subi J. George[*]

New Chemistry Unit and School of Advanced Materials,
Jawaharlal Nehru Centre for Advanced Scientific Research, Jakkur,
Bangalore 560064, Karnataka, India
[*]*george@jncasr.ac.in*

The advances in supramolecular chemistry, which deals with non-covalent interactions between the molecules, have ignited the field of supramolecular materials exhibiting a highly dynamic, adaptive and reversible stimuli responsive behavior. Due to these structural properties, supramolecular materials have realized their applications in a wide and varied area of research such as in mechanical, biological and opto-electronic materials challenging the conventional covalent polymers and inorganic materials. Recent interest in smart programmable supramolecular materials, inspired by natural systems, has directed the field towards temporal programming over the structure and functions of supramolecular materials. Thus, this Chapter covers the evolution of supramolecular chemistry to supramolecular materials and a further advancement from passive and simpler equilibrium materials to programmable, active and smart non-equilibrium supramolecular materials.

1. Introduction

In 1978, Nobel laureate Prof. Lehn introduced "supramolecular chemistry" as the "chemistry beyond the molecule" that deals with two or more molecules interacting via intermolecular non-covalent interactions organizing them into complex species.[1] By virtue of the dynamic and reversible nature of the weak or moderate non-covalent interactions, typically of the order of 1-5 kcal/mol, such as hydrogen

123

bonding, electrostatic, π-π stacking, metal ion coordination, van der Waals forces etc., the supramolecular species exhibit thermodynamic stability, are kinetically more labile with the ability to adapt to changes in the environment and subsequently undergo reorganization as compared to the covalent polymers. Based on their distances and angles, these interactions have a different magnitude of strength and directionality. However, the directional and collective interactions lead to highly stable extended organizations of the molecules also called as supramolecular polymers which are analogous to covalent polymers. These supramolecular polymers, generally one-dimensional, are referred as supramolecular materials when used for a functional outcome (Figure 1). If the lifetime of the intermolecular bonds is too short in the range lower than a few μs, a weak and fragile supramolecular polymer is obtained, in contrast to the lifetime in the minute regime which produces supramolecular materials with less dynamicity.[2] The intermediate spectrum of bond lifetimes offer unique properties such as adaptability, stimuli-responsiveness, self-heal, self-repair, etc. Thus, careful designing of the chemical structure of molecules can provide tunable binding affinities, strength and a balanced dynamicity.

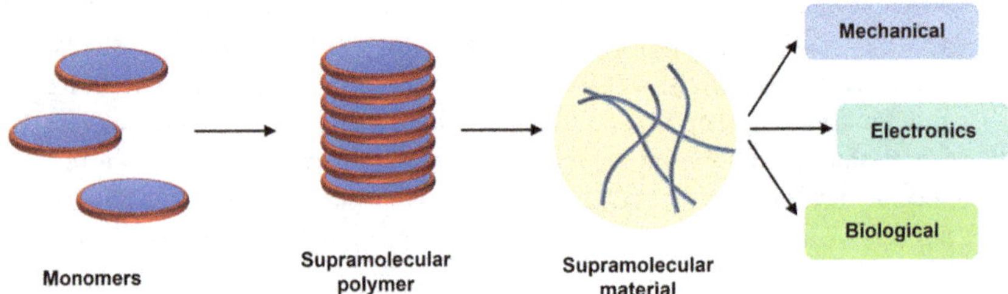

Fig. 1. Supramolecular polymerization of monomers to form supramolecular materials that exhibit mechanical, electronic and biological applications.

The intermolecular interactions due to their strength and directionality lead to a selective and specific binding of two different molecules termed as substrate and receptor, yielding a supramolecular species involving a molecular recognition process. The key examples include the lock and key concept in proteins, host-guest, metal-ligand and enzyme-substrate complexes. Hence, the chemical structure of the molecules carry a molecular information that is a crucial principle that determines its supramolecular organization and resultant properties.[3] These interactions can lead to small discrete supramolecular complexes composed of two or three molecules or extended supramolecular assembly/polymer composed of a large number of molecules.

The initial challenge in this field to design materials, was to achieve self-or hetero-complimentary groups or motifs with a high association constant ($>10^6$) to achieve polymeric like behavior with a large number of monomers or building blocks that are arranged in a well-ordered manner. In a break through, Meijer's groups have introduced the quadruple H-bonding and self-complimentary 2-ureido-4-pyrimidone (UPy) motif, after extensive research in this area to balance the secondary electrostatic interactions between the H-bonding donor and acceptor atoms.[4] The donor-donor-acceptor-acceptor type of hydrogen bonding interactions yield attractive secondary electrostatic interactions giving rise to an association constant for dimerization of more than 10^6 M^{-1} in $CHCl_3$. Subsequently, many such motifs and host-guests interactions with very high association constants have been developed to yield a variety of supramolecular polymers with a significantly high degree of polymerization.

The field also administers the organization of different structural and functional components into single complex multicomponent species for a modular functional outcome. As a consequence, supramolecular chemistry has expanded as an interdisciplinary field of research with implications in chemical, physical and biological sciences.

The next challenge the field addressed was the creation of supramolecular materials. Supramolecular materials composed of these reversible and dynamic non-covalent interactions have a unique reversibility, adaptive behavior and sensitivity to the environment granting these materials intriguing properties. The key features exhibited by supramolecular materials are (1) facile to fabricate multiple components into a single material, (2) reversibility due to the dynamic destruction and reconstruction at equilibrium with an ability to self-repair, (3) adaptive to external stimuli and environment resulting in the reconfiguration of structural and functional states and (4) ordering at the nanometer level for macroscopic materials.[5] These exceptional characteristics of supramolecular materials have seen applications in mechanical materials, soft optical and electronic nanomaterials and biomaterials.[2]

Most of the supramolecular architecture and materials are formed as a result of thermodynamically driven self-assembly, they reside in equilibrium and hence are temporally passive.[6] This is unlike the natural systems that work out-of-equilibrium and are the key inspiration for the growth and development of supramolecular chemistry. The out-of-equilibrium systems are fueled by an energy source that determine their temporal characteristics.[7] Also, a higher complexity with a simplistic strategy to form multicomponent supramolecular block copolymers is needed. Hence, the grand challenge for the supramolecular chemists is to design supramolecular precision materials which are more complex, are composed of multi-components and function out-of-equilibrium.

In this chapter, we shall first discuss the design and applications of supramolecular materials in the field of mechanical, optical, electronic and biology (Figure 1). Our main focus is on supramolecular materials composed of 1-dimensional supramolecular polymers. We shall then discuss the current trends in the field of supramolecular chemistry to build temporally programmed supramolecular materials with a well-defined structure, and temporally regulated functions.[8]

2. Passive Supramolecular Materials

Conventional supramolecular materials are composed of molecules that undergo an energetically downhill self-assembly process in appropriate environmental conditions such as solvent composition, temperature, pH etc. to form the most thermodynamically stable structure. Once these materials attain thermodynamic equilibrium, they are kinetically inert to undergo any alteration in spatial molecular organization with time and hence are termed as passive materials. Most of these materials are dynamic due to the interplay of weak intermolecular interactions and can undergo stimuli-responsive changes in structure and functions. Supramolecular materials, thus are good candidates for various functional materials and following are the important examples signifying the same. In this chapter we limit our discussion only on supramolecular materials constructed from 1-D supramolecular polymers.

2.1. *Mechanical*

Supramolecular materials, unlike conventional covalent polymers can provide excellent mechanical properties along with a dynamic self-healing ability that elongates the lifespan of materials and increases its reusability. Although the strength of supramolecular interactions is far low due to the non-covalent bonding, but a clever designing of a molecular motif can provide highly robust materials with exceptional mechanical properties.

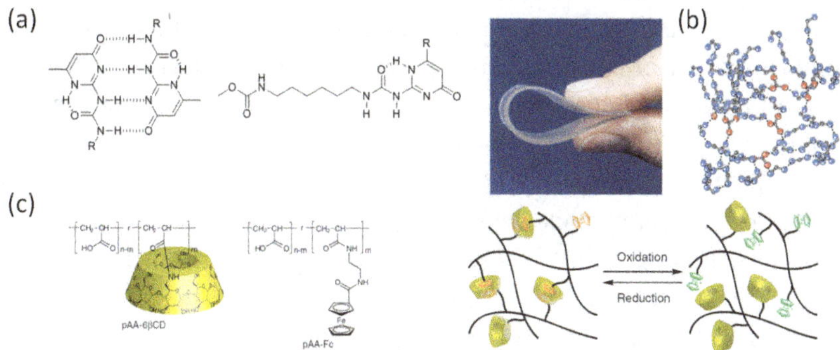

Fig. 2. (a) Quadruple H-bonding in self-complimentary 2-ureido-4-pyrimidone (UPy) forming supramolecular material with a high mechanical strength. (b) Schematic representation of cross-linking of directional interactions between ditopic (blue) and tritopic (red) molecules. (c) Chemical structures of poly(acrylic acid) polymer derivatives of β-Cyclodextrin (β-CD) host and Ferrocene guest. The host-guest polymer undergoes a redox responsive assembly-disassembly, yielding a self-healing property to the polymer. Adapted with permission from Refs. [9, 10, 11].

Meijer and coworkers introduced a supramolecular polymer composed of a self-complementary quadruple hydrogen bonding motif UPy, as discussed above, connected via an alkyl spacer (Figure 2(a)).[4] The lifetimes of these hydrogen bonds in solution at room temperature are in the order of 0.1 to 1 s. The polymer obtained has a molecular weight of more than 500 kDa at room temperature with unprecedented mechanical properties for supramolecular polymers in the solid state (bulk). Even at a high temperature where the lifetime of hydrogen bonds decrease significantly, these exhibit viscous nature and do not disassemble completely. Further hydroxy-telechelic poly(ethylene butylene)molecule (PEB) end functionalized with these UPy groups undergo crosslinking to form a rubbery matrix that exhibits self-healing characteristics.[9]

Leibler and coworkers synthesized a supramolecular reversible network comprising of ditopic and tritopic molecules that can interact via directional hydrogen bonds to form crosslinking networks (Figure 2(b)).[10] On fracturing, the two surfaces when brought into contact for some time allows a self-healing of material to its original strength as well as elasticity due to the dynamic nature of these hydrogen bonding interactions.

A supramolecular material composed of host-guest interactions form a class of materials that can be reversibly bound and unbound using stimuli that can affect the chemical structure or electronic property of guest molecules reducing their binding affinity towards the guest entity. Harada and coworkers synthesized a supramolecular hydrogel consisting of two poly(acrylic acid) polymers, one functionalized with the host functionality beta-cyclodextrin (CD) and other with guest motif ferrocene (Figure 2(c)).[11] The host and guest polymers undergo crosslinking via the host-guest complexation with an association constant of 1.1×10^3 M^{-1} forming a viscous hydrogel. Ferrocene converts from hydrophobic neutral guest Fe(II) to the hydrophilic charged guest Fe(III) on exposure to an oxidizing agent. This expels the guest from the hydrophobic pocket of the host and therefore destroys the crosslinking and the hydrogel. On reduction, the changes are reverted and a new crosslinking occurs retrieving the hydrogel and its mechanical properties. Since these host-guest interactions are reversible and stimuli-responsive, the hydrogel when cut into two halves can undergo self-healing when the cut surface is first subjected to an oxidizing agent sodium hypochlorate to destroy host-guest complexes present at the surface and subsequently adheres the surfaces via the formation of new complexes when reduced with glutathione.

Thus, clever designing of supramolecular materials have resulted in excellent mechanical properties and dynamic self-healing ability.

2.2. *Electronics*

Supramolecular electronics is the research field that bridges the gap between molecular electronics and plastic/polymer electronics.[12] Molecular electronics deals with the Ångstrom length range and polymer electronics higher than a 100 nm regime. Supramolecular electronics contain self-assembled nanowires of 5-100 nm length scale formed via supramolecular polymerization of π-conjugated systems.[13,14] The stacking of semiconducting chromophores allow the transport of charge carriers unidirectionally through the stacking direction over long distances. Supramolecular electronics addressed two key aspects; first is the synthesis of conducting supramolecular organic fibers and the second is to improve its order and alignment on the device.[15]

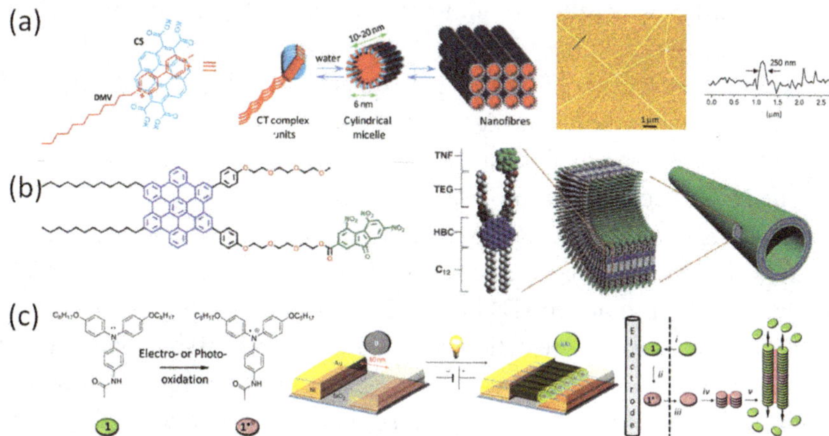

Fig. 3. (a) Supramolecular polymerization of semiconducting charge-transfer complex of coronene tetracarboxylate (CS) and dodecyl methyl viologen (DMV) into nanofibers utilized for Field-effect transistor (FET) devices. (b) Supramolecular assembly of amphiphilic hexabenzocoronene (HBC) derivatives to form radial p-n junction. (c) Supramolecular electro- or photo-polymerization of triarylamine (TAA) for aligned nanofibers bridging the electrodes (photo) and on the electrode surface (electro). Adapted with permission from Refs. [21, 23, 24, 25].

During the last two decades various p- or n-type chromophores were organized into a one-dimensional fiber using supramolecular interactions. Feringa and coworkers symmetrically functionalized p-type mono and bithiophene with urea derivatives for their supramolecular polymerization into nanofibers.[16] The hydrogen bonding assisted close packing of thiophenes facilitated an efficient pathway for charge transport. Oligo(p-phenylenevinylene)[17] derivatives are well-studied as p-type supramolecular nanostructures, via π-stacking interactions supported by functionalization with self-complimentary quadruple hydrogen bonding UPy groups, or with dendritic structures.[18]

To achieve a high charge carrier mobility, extended π-conjugated chromophores are functionalized with supramolecular motifs for their ordered one-dimensional arrangement for field effect transistors (FET). Amphiphilic design with asymmetric functionalization of HBC with dodecyl and tetraethylene glycol facilitated the formation of a graphite-like hollow nanotubular supramolecular polymer with a 14 nm diameter, 3 nm wall thickness and high aspect ratio was shown by Aida's group.[19]

In order to ensure a uniform and an inherent doping of the supramolecular fibers, our group has introduced charge-transfer (CT) nanofibers, with an alternating sequence of donor and acceptor chromophores. Supramolecular copolymers comprising of donor and acceptor molecules exhibit an inherent directional dipole moment that facilitates the conduction process. In order to construct these CT fibers, we have used a non-covalent amphiphilic CT monomer, such as the co-facial CT complex of coronene tetracarboxylate (CS) donor and acceptor dodecyl methyl viologen (DMV).[20] These non-covalent amphiphiles are stabilized by both electrostatic and CT interactions (association constant of 10^6 M^{-1}) and hence self-assemble into micrometer long 1-D structures with a cylindrical micellar organization (6 nm diameter) of the monomers (Figure 3(a)). The nanowires showed a high field effect mobility of 4.4 cm^2/V.s.[21] Using an amphiphilic approach, we self-assembled Oligo(p-phenylenevinylenes) (OPV) known for their performance in optoeletronic devices. An asymmetric amphiphilic derivatization of OPV by a long alkyl chain and tetraethylene glycol resulted in a highly ordered bilayer 2D sheets. A high degree of order in the charge transport direction due to the amphiphilic design resulted in a better electronic performance.[22] A similar amphiphilic approach has been used by Aida and co-workers to form co-axial, orthogonal radial p-n junctions by the functionalization of the terminal tetraethylene glycol of the amphiphilic HBC with an electron acceptor trinitrofluorenone. This has resulted in coaxial nanotubular structures with an electron-donating graphitic layer of π-stacked HBCs covered by an electron-accepting trinitrofluorenone layer (Figure 3(b)).[23] This arrangement facilitates the photochemical production of spatially separated charge carriers yielding a fast photoconductive response on irradiation with white light.

The next obvious challenge that the field needed to address was the device fabrication and soft-lithography for realistic applications of these semi-conducting supramolecular polymers. Triarylamine (TAA) is an important class of molecules exhibiting a high hole-transport mobility and it administers extensive applications in optoelectronic devices. Giuseppone's lab demonstrated the formation of supramolecular nanowires of a TAA derivative on visible light irradiation of the solution (Figure 3(c)).[24] On irradiation of TAA in an electron acceptor chlorinated solvent, the catalytic amount of TAA radical cations (TAA$^{•+}$) are generated. These TAA$^{•+}$ form charge-transfer interaction with neutral TAA. As a consequence, the supramolecular polymerization facilitated by π-π stacking, the hydrogen bonding and charge-transfer (CT) interaction occurs to form supramolecular nanofibers. This strategy was used for the in situ growth and alignment of supramolecular nanofibers between electrodes in the presence of an electric field. Recently, the study was extended for supramolecular electropolymerization of TAA between the electrodes.[25] This is an interesting step towards the fabrication of soft-supramolecular fibers between electrodes, where conventional lithographic techniques fail.

Thus, semiconducting supramolecular polymers have proven to be good candidates for supramolecular electronics. However, current challenges are the in situ synthesis of supramolecular polymers and the alignment of these fibers for efficient conduction.

2.3. *Biological*

Supramolecular materials form an important class of biomaterials with an ability to sense, adapt and respond to biological cues in order to mimic or affect biological structures and their functions.[26] The possibility of a rational yet a modular design of molecules using a tunable and a reversible non-covalent interaction with a feasibility for infusion of multiple functional motifs into a single entity allows the synthesis of unprecedented biomaterials with an enhanced activity and response. Owing to this, supramolecular biomaterials grant a multifarious toolbox that can address applications from signaling cells, drug delivery, tissue and immune-engineering to regenerative medicine. Supramolecular biomaterials composed of water soluble small molecules, mostly with a peptide residue can undergo assembly-disassembly-reassembly to fit the structural geometry required in the cell and degrade into non-toxic waste products and is minimal invasive. The monomers are coded with chemical structures (functional groups) that can bind cell receptors with a high affinity. A self-assembling amphiphilic peptide molecule encoded with laminin epitope IKVAV, which is a glycoprotein constituent of the connective tissue basement membrane responsible to promote cell adhesion is shown to aid in the spinal cord injury (SCI).[27] The formation of nanofibers in the extracellular spaces of the spinal cord triggered by the ionic strength of the environment, promotes the neurite outgrowth and axon elongation and regeneration.

To promote the growth of new blood vessels, a biopolymer that interacts with angiogenic growth factors, was synthesized. This was a cationic peptide amphiphile with the LRKKLGKA sequence having a high binding affinity ($\sim 10^7$ M^{-1}) for an anionic heparin.[28] By virtue of charge-neutralization, heparin stimulates the nucleation of supramolecular polymerization of the peptide amphiphile into nanofibers that display heparin chains to direct cell signaling proteins for the fast growth of new blood vessels as compared to the absence of the peptide amphiphile.[29]

Bone, being a complex composite of organic and inorganic nanophases makes the synthesis of materials that resemble it, uneasy. A clever chemical structure of the peptide amphiphile containing regions of (1) cysteine residues, that can be polymerized on oxidation via disulfide linkage, (2) flexible glycine residues, (3) phosphorylated serine that interact strongly with the calcium ion to promote the hydroxyapatite mineralization (an integral part of bone) and (4) Arg-Gly-Asp (RGS) sequence to direct the integrin-mediated cell adhesion, was synthesised.[30] These PA nanofibers via reversible crosslinking form a scaffold similar to the extracellular matrix with a tunable structural integrity.[31] Similar to the alignment between collagen fibrils and hydroxyapatite crystals found in bone, this scaffold allows a direct mineralization of hydroxyapatite to form a composite material.

The articular cartilage regeneration is shown by nanofibers composed of a co-assembly of peptide amphiphiles one acting as a nonbioactive filler and the other comprising of the transforming growth factor β-1 (TGF) binding epitope.[31] The supramolecular biomaterials have shown potential in various biological conditions.[26,32,33]

Thus, peptide amphiphiles have shown ability to sense, adapt and respond to biological cues and promise high potential in biomedical applications.

3. Active Supramolecular Materials

Biological systems exhibit a spatiotemporal control over their organization directed by a complex network of chemical cues and enzymatic transformations.[34] Owing to this, natural systems behave as active, adaptive and autonomous systems. Natural supramolecular polymers such as cytoskeleton proteins: actin and microtubules undergo fuel-driven (adenosine and guanosine triphosphate, ATP and GTP, respectively) kinetically controlled as well as out-of-equilibrium supramolecular polymerization. The transient change in the length/degree of polymerization modulates their functional attributes i.e. the cell motility. In contrast, synthetic materials, as discussed in the previous section, are simpler systems with a thermodynamically driven self-assembly. Although the system might exhibit a dynamic and stimuli responsive behavior, it doesn't demonstrate any temporal control.[35] The urge towards a bioinspired system and a complex control over the organization, has initiated the production of supramolecular polymerization with temporal control for narrow dispersity, controlled length and transient functions.[36] Active supramolecular materials are defined as the materials in which the complexity of self-organization changes with time.

3.1. *Living supramolecular polymerization*

Extensive studies on mechanistic features by Meijer's group on the supramolecular polymerization process of have defined two major types of mechanisms.[37] The isodesmic mechanism is defined by a continuous energy downhill process of the addition of monomers on the growing supramolecular polymer. The Gibbs free energy of addition of the monomer doesn't depend on the length of the supramolecular polymer. The cooperative mechanism is a two-stage mechanism where the initial step is an energy uphill nucleation (K_n) phase that comprises of the formation of a critical size oligomer referred as the nucleus and the subsequent energy downhill elongation step (K_e) where the monomers tend to associate with this nucleus to form a growing supramolecular polymer. These mechanisms are analogous to the step growth and chain-growth polymerization in covalent polymers, respectively.[37,38]

For the advancement in the performance of supramolecular materials their structural control i.e. the dispersity and degree of polymerization are very essential. The development of the living polymerization in covalent polymers for the synthesis of polymers with a narrow dispersity, controlled molecular weight and multi-component block copolymers have initiated the parallel field in supramolecular chemistry to achieve the same. In covalent polymers, living chain polymerization reactions required a kinetic control over the propagation step to minimize the bimolecular termination processes and has been successfully addressed by generating the dormant monomer states for propagating chains. On the other hand, mechanistic understandings in the analogous nucleation-growth polymerization revealed that an uncontrolled nucleation due to the low activation energy is the challenge to be addressed to achieve living supramolecular polymerization. A nucleation-elongation growth with a slow nucleation step and faster elongation rate can result in a controlled dispersity and degree of polymerization. The key to address this is, to logically design monomers that undergo a nucleation-elongation growth and create their dormant state that can retard/delay the nucleation phase.[39]

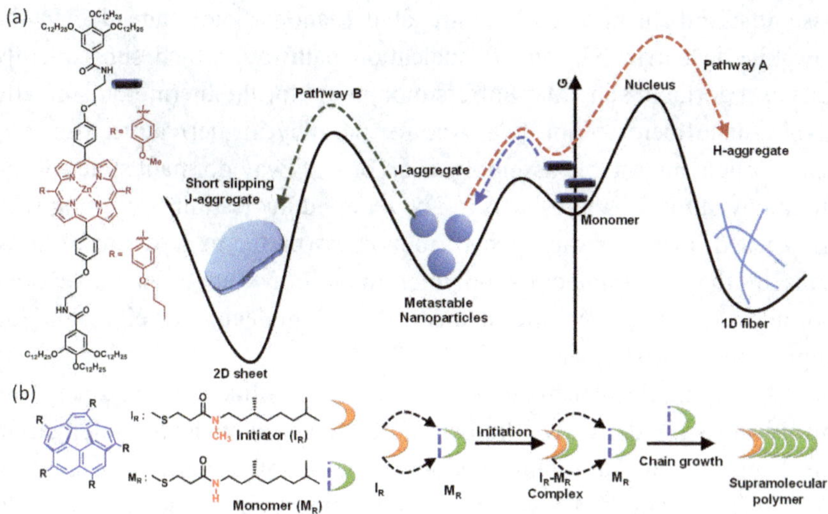

Fig. 4. (a) Molecular structures of porphyrin molecules undergoing self-assembly into one-dimensional fiber via pathway A and into two-dimensional sheets via pathway B. Monomers undergo formation of a metastable nanoparticle assembly prior to self-assembly. G is Gibbs Free Energy. (b) Molecular structures of non-hydrogen bonded initiator and hydrogen-bonded inactive monomers undergoing in chain-growth supramolecular polymerization. Adapted with permission from Ref. [8].

Monomers with low dynamicity can get trapped in a metastable assembled state. This state has a low energy barrier at experimental conditions and the system tends to escape the trap to form a thermodynamic assembled state. Since, the relaxation of the metastable to a thermodynamic state is non-instantaneous due to the energy barrier, a delayed nucleation process for the thermodynamic self-assembled state occurs giving a controlled dispersity and size. The metastable state here acts as a reservoir for a slow and a continual buffering of monomers. This pathway complexity of the process of supramolecular polymerization can be manipulated by system conditions such as the solvent and temperature as suggested by comprehensive investigations performed by Meijer's group.[40]

Sugiyasu and Takeuchi's group explored the metastable states as dormant reservoirs for monomers to kinetically retard the nucleation process (Figure 4(a)). They synthesized porphyrin molecule that assembles into a metastable aggregated (nanoparticle) state due to the low kinetic barrier via an off-nucleation pathway, which subsequently acts as a dormant state that buffers monomers for the thermodynamically stable nanofiber assembly.[41] Another porphyrin derivative has this nanoparticle metastable assembly as an on-pathway dormant state for the thermodynamic 2D nanosheets.[42] The retarded nucleation step along with the seeded characteristics (where monomers prefer to grow on existing ends of the supramolecular polymer than form their own nucleation) provides control over the dispersity and molecular weight of the supramolecular polymer.

An intramolecular hydrogen bonding was utilized to trap a perylene bisimide (PBI) derivative known to undergo a nucleation-elongation mechanism into a conformational metastable dormant state by Würthner's group.[43] To undergo 1D organization, monomers need to undergo a transition from an intramolecular hydrogen bonding closed state to intermolecular hydrogen bonding open state which controls the nucleation phase to yield a living supramolecular polymer.

Miyajima and Aida's group demonstrated the first example of a chain growth living supramolecular polymerization employing a rationally designed intramolecular hydrogen bonding mediated conformationally dormant corannulene monomer (Figure 4(b)).[44] They utilized the concept of the conformational dormant state of monomer which is unable to polymerize on its own. Another molecule synthesized by methylating the amide groups in order to destroy its ability to form any intramolecular hydrogen bonding acts as an initiator that forms an intermolecular hydrogen bonding with the dormant monomers and activate them to undergo supramolecular polymerization. The variation of the initiator to the monomer ratio resulted in a precise control over dispersity and molecular weight.

The above stated examples are a breakthrough in the field of the living supramolecular polymerization but the main disadvantage is that they are highly dependent on the molecular structure, lack in a generic strategy and have no external control over kinetics. On the contrary, natural systems, being an inspiration to supramolecular systems utilize a thermodynamic dormant state for monomers with a negligible potency to assemble and trigger its supramolecular polymerization by a specific chemical fuel. For instance, the G-actin, a cytoskeleton protein, is unable to polymerize on its own and stays in a dormant state. On the addition of a biofuel ATP, G-actin undergoes a conformational change to F-actin that activates it for its fuel-driven kinetically controlled polymerization into actin filaments. This polymerization occurs via a cooperative mechanism that facilitates a seeded characteristic to attain control over their structural and functional attributes.[45]

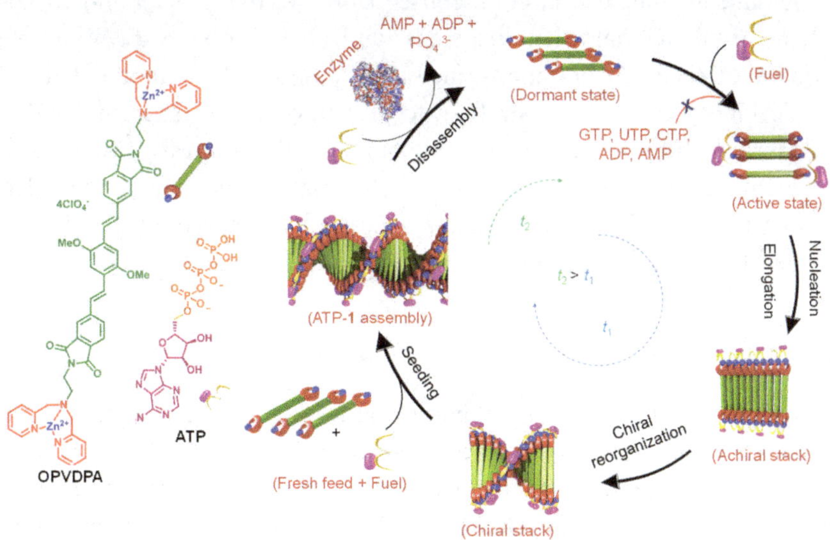

Fig. 5. Molecular structures of monomer OPVDPA and fuel ATP. Schematic representation of ATP-driven living and transient supramolecular polymerization. Adapted with permission from Ref. [46].

In this context, our group has introduced the concept of a bio-inspired fuel-driven strategy as a generic approach to create living supramolecular polymers. The essential criterion is to build thermodynamic dormant states for the monomers, which can be triggered via a fuel to undergo a kinetically controlled growth rate which will be determined by the rate of the fuel binding or its generation.

In an attempt to realize a fuel-driven kinetically governed supramolecular polymerization, we rationally designed an extended π-conjugated monomer, oligo (p-phenylenevinylene) derivative (OPVDPA), appended with a cationic phosphate receptor group: zinc dipicolylamine (ZnDPA).[46] The molecule has an acceptor–donor–acceptor (A–D–A) electronic structure that traps the monomers in an intermolecular charge-transfer (CT) slipped state facilitated by the electrostatic repulsion between the cationic terminal ZnDPA group (Figure 5). On electrostatic interaction with ATP, a charge neutralization occurs that drives a conformational change from a slipped CT state to π-π stacked extended supramolecular polymer. This kinetic barrier, retards a spontaneous nucleation and consequently a living supramolecular polymer with a controlled dispersity and size is obtained. With the variation of ATP concentrations, the kinetics of growth are modulated.

We have extended the strategy to a charge-transfer supramolecular polymerization by a fuel-driven in situ amphiphile synthesis.[47] Sugiyasu, Takeuchi's group also presented a photo-regulated living supramolecular polymerization.[48] Thus, a bioinspired strategy of a fuel-driven living supramolecular polymerization proves its generic nature.

3.2. *Transient supramolecular polymerization*

Biological systems have a temporal control over their structural and functional characteristics. Coupled chemical reactions signal the proteins to undergo structural variations and activate their functional state for a programmed time and subsequently switch back to the initial state. For instance G-actin on the addition of ATP undergoes a conformation change into the F-actin form that undergoes a supramolecular polymerization.[45] On subsequent hydrolysis/consumption of ATP, the polymer undergoes disassembly. These changes due to their temporary nature are referred as transient changes.[49] The concept requires a thermodynamic equilibrium state that is pushed to a higher energy non-equilibrium state on the addition of fuel which due to the energy consumption/dissipation can relax to the same equilibrium state or a state that is different from the initial state (Figure 6(a)). In order to construct synthetic transient materials, the rate of formation should be significantly higher than the rate of decay where the lifetime of the transient state is determined by the amount of fuel. The strategy employs a monomer lacking in its ability to assemble which is activated by fuel, a (chemical) agent that modifies its chemical structure, as a consequence of it, the altered monomer undergoes a spontaneous supramolecular polymerization. Concurrently, an opposite reaction reverts these structural changes thus reducing their capability to assemble and destroy the supramolecular polymer and resets back to equilibrium. An important requisite is the system should have high dynamicity and adaptiveness so that chemical transformation can determine the state of system.

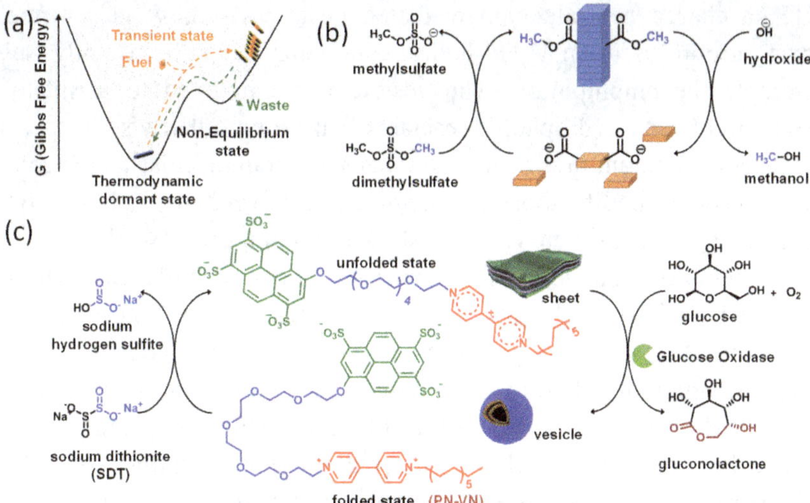

Fig. 6. Fuel-driven transient supramolecular polymerization. (a) Energy landscape depicting the pathway of transient assembly. (b) Dimethyl sulfate fueled transient self-assembly of dicarboxylate molecules (orange). (c) ATP fueled transient vesicle (nanoreactor) formation of an amphiphile in the presence of ATP-hydrolyzing enzyme apyrase. (d) Redox mediated transient morphological switching of a charge-transfer amphiphilic (PN-VN) molecule. Adapted with permission from Ref. [8].

van Esch's group coupled opposite yet orthogonal methylation–demethylation reactions to transiently self-assemble carboxylate appended monomers in an alkaline medium (Figure 6(b)).[50] Methylation converts this anionic carboxylate state to a neutral hydrophobic ester state in alkaline media which is subsequently hydrolyzed via hydroxyl ions to the initial carboxylate water soluble derivative.[51] At a higher concentration these transient supramolecular polymers form self-abolishing hydrogels with tunable lifetimes determined by the fuel concentration. When this transient supramolecular polymerization was observed under confocal microscopy, a tread-milling assembly and disassembly was seen similar to actin filaments. The strategy was extended to the polymeric system for transient clustering of colloids.[52]

An amphiphilic molecule functionalized with a cationic zinc(II) based ATP receptor motif by virtue of the electrostatic repulsion is monomeric in the aqueous solution but on feeding with an anionic fuel ATP, a charge neutralization mediated vesicle assembly gets formed. Prins' group used an ATP hydrolyzing enzyme to transiently self-assemble the amphiphile.[53] The vesicle was employed to facilitate a reaction between hydrophobic reactants that do not otherwise happen in an aqueous medium thus acting as transient supramolecular nanoreactors The transient modification of monomers catalyzed by a biocatalytic agent α-chymotrypsin enzyme was shown by Ulijn and coworkers.[54]

A transient change in the properties of the supramolecular material can also be influenced by altering their environmental factors such as the pH temporally. Walther's group employed urease catalyzed hydrolysis of urea into ammonium hydroxide or base aided lactone/ester hydrolysis for a temporal change in the pH of the solution.[55] This was applied for the transient gelation, temporal control for the photonics application.[56] Our group also utilized this strategy for transient regulation of ion transport.[57]

Our lab utilized viologen redox chemistry to modulate a charge-transfer interaction between the donor-acceptor system in a pyranine-viologen amphiphilic foldamer molecule to attain a transient conformational response i.e. unfolding of foldamer for affecting a supramolecular morphological transition from the vesicle to the sheet (Figure 6(d)).[58] The foldamer was unfolded by a reducing agent, sodium dithionite capable of reducing viologen to viologen radical cation in order to destroy its CT interaction with the pyranine, unfolding the foldamer. Viologen on oxidation by dissolved oxygen or enzymatically by glucose oxidase, refolds the foldamer with the retrieval of the CT interaction temporally. Thus transient conformational change affecting a transient morphological switching was obtained.

As a step towards the biomimetic system, our group carefully chose zinc(II)-diethylenedipicolylamine (ZnDPA) phosphate receptor, appended to naphthalene diimide chromophore (NDPA) (Figure 7).[59,60] The molecule is highly dynamic and adaptive and exhibits high selectivity and differentiability to adenosine phosphates. It undergoes a supramolecular polymerization into the right- and left-handed helical assembly on interacting with the ATP and the ADP/AMP respectively, however, doesn't assemble with inorganic phosphate.[61]

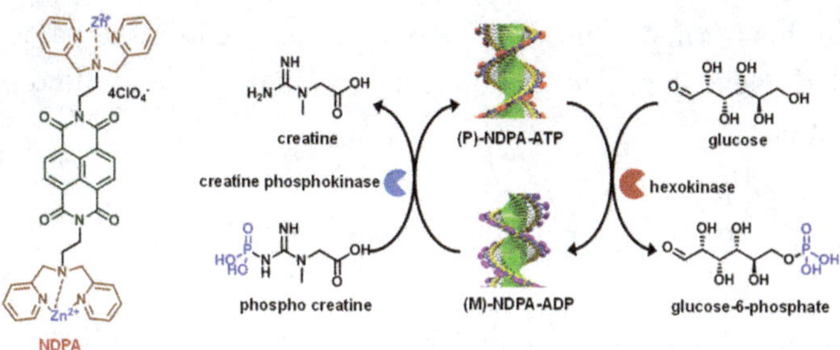

Fig. 7. The chemical structure of NDPA and ATP-fueled transient helical conformational switching by enzymes in tandem (creatine phosphokinase and hexokinase). Adapted with permission from Ref. [8].

A bioinspired "enzymes in tandem" strategy, with two orthogonal and opposite transphoryl enzymes, creatine phosphokinase and hexokinase were utilized to transiently convert ATP to ADP in the presence of their substrates, phosphocreatine and glucose respectively. This transformation in the presence of the NDPA results in a transient helical inversion from the left to the right to the left helicity (Figure 7).[62] The rates of the helical transformation and the lifetime of the transient right handed helical state was modulated with a variation of an enzyme and their substrate concentrations. The complexity of the system was increased to attain an unprecedented multi non-equilibrium state transient supramolecular polymerization.[63] The enzymes, creatine phosphokinase and phosphatase (alkaline phosphatase and apyrase) were coupled in appropriate concentrations to convert ADP→ATP→ADP/AMP→Pi temporally. This resulted in transient self-assembly that underwent three helical transient state in its polymerized form prior to complete disassembly. The various chemical cues showed an effect on the rate and extent of these transient states to present a modular multi non-equilibrium transient supramolecular polymerization.

Hermans' group coupled protein kinase A (PKA) and λ-protein phosphatase to transiently phosphorylate serine residue of a peptide derivative of perylenediimide (PDI) via consumption of ATP.[64] Using a continuous flow device allowing a continuous injection of fuel and removal of waste resulted in a non-equilibrium steady left handed helical state of the PDI.

So far fundamental aspects of transient supramolecular polymerizations and proof-of-concept applications have been investigated. Next step is to increase the complexity via multicomponent self-assembly.

4. Outlook

Supramolecular materials by virtue of their dynamic and adaptive nature have shown immense potential in mechanical, biological and electronic applications. The recent shift of interest from supramolecular materials functioning at a thermodynamic equilibrium to out-of-equilibrium materials have introduced time as a new attribute. This temporal dimension has paved the way to precisely control the supramolecular structural property such as the length and dispersity and to synthesize programmable functional materials. The field is still under infancy and promises high potential to unveil unprecedented properties.

The next challenge is to move towards a higher complexity. Advancing from naive single component systems to complex multicomponent systems, that either communicate with each other or are orthogonal, can result in extraordinary properties. The coordination of multiple components in the temporal regime can result in the synthesis of supramolecular block co-polymers for p-n junctions, catalysis in tandem and microscopic motion.

Multicomponent out-of-equilibrium systems shall result in autonomous and a cascade of transformations. With the incorporation of feedback loops and the network of chemical conversions, oscillating, self-replicating and autonomous supramolecular materials can be prepared similar to the living organisms.

In conclusion, supramolecular interactions are highly dynamic and adaptive which on careful designing can result in materials with varied properties. Thus, supramolecular chemists should work towards the simultaneous orchestration of the fuel-driven temporal control to multicomponent self-assembling materials for unprecedented properties.

References

1. J. M. Lehn, Supramolecular chemistry, *Science* **260**, 1762-1763 (1993).
2. T. Aida, E. W. Meijer and S. I. Stupp, Functional supramolecular polymers. *Science* **335**, 813-817 (2012).
3. J. M. Lehn, Perspectives in supramolecular chemistry—from molecular recognition towards molecular information processing and self-organization, *Angew. Chem. Int. Ed.* **29**, 1304-1319 (1990).
4. R. P. Sijbesma, H. Felix, L. B. Beijer, J. B. Brigitte, J. H. K. Folmer, H. Ky, R. F. M. Lange, J. K. L. Lowe and E. W. Meijer, Reversible polymers formed from self-complementary monomers using quadruple hydrogen bonding, *Science* **278**, 1601-1604 (1997).
5. K. Liu, Y. Kang, Z. Wang and X. Zhang, 25th anniversary article: Reversible and adaptive functional supramolecular materials: "Noncovalent interaction" matters, *Adv. Mater.*, **25**, 5530-5548 (2013).
6. D. B. Amabilino, D. K. Smith and J. W. Steed, Supramolecular materials, *Chem. Soc. Rev.* **46**, 2404-2420 (2017)
7. J. F. Lutz, J. M. Lehn, E. W. Meijer and K. Matyjaszewski, From precision polymers to complex materials and systems, *Nat. Rev. Mater.*, **1**, 16024 (2016).
8. S. Dhiman and S. J. George, Temporally controlled supramolecular polymerization. *Bull. Chem. Soc. Jpn.* **91**, 687-699 (2018).
9. B. J. Folmer, R. P. Sijbesma, R. M. Versteegen, J. A. J. van der Rijt and E. W. Meijer, Supramolecular polymer materials: Chain extension of telechelic polymers using a reactive hydrogen- bonding synthon, *Adv. Mater.*, **12**, 874-878 (2000).
10. P. Cordier, T. François, S. -Z. Corinne and L. Leibler, Self-healing and thermoreversible rubber from supramolecular assembly, *Nature* **451**, 977 (2008).
11. M. Nakahata, Y. Takashima, H. Yamaguchi and A. Harada, Redox-responsive self-healing materials formed from host–guest polymers, *Nat. Commun.* **2**, 511 (2011).
12. E. W. Meijer, and A. P. H. J. Schenning, Chemistry: Material marriage in electronics, *Nature* **419**, 353 (2002).
13. A. P. H. J. Schenning and E. W. Meijer, Supramolecular electronics, nanowires from self-assembled π-conjugated systems, *Chem. Commun.* **26**, 3245-3258 (2005).

14. A. P. H. J. Schenning, P. Jonkheijm, F. J. M. Hoeben, J. van Herrikhuyzen, S. C. J. Meskers, E. W. Meijer, L. M. Herz, C. Daniel, C. Silva, R. T. Phillips and R. H. Friend, Towards supramolecular electronics, *Synthetic Metals*, **147**, 43-48 (2004).

15. A. Jain and S. J. George, New directions in supramolecular electronics, *Materials Today*, **18**, 206-214 (2015).

16. F. S. Schoonbeek, J. H. van Esch, B. Wegewijs, D. B. Rep, M. P. De Haas, T. M. Klapwijk, R. M. Kellogg and B. L. Feringa, Efficient intermolecular charge transport in self- assembled fibers of mono- and bithiophene bisurea compounds, *Angew. Chem. Int. Ed.*, **38**, 1393-1397 (1999).

17. P. Jonkheijm, F. J. Hoeben, R. Kleppinger, J. van Herrikhuyzen, A. P. H. J. Schenning and E. W. Meijer, Transfer of π-conjugated columnar stacks from solution to surfaces, *J. Am. Chem. Soc.*, **125**, 15941-15949 (2003).

18. A. P. H. J. Schenning, J. van Herrikhuyzen, P. Jonkheijm, Z. Chen, F. Würthner and E. W. Meijer, Photoinduced electron transfer in hydrogen-bonded oligo (p-phenylene vinylene)− perylene bisimide chiral assemblies, *J. Am. Chem. Soc.*, **124**, 10252-10253 (2002).

19. J. P. Hill, W. Jin, A. Kosaka, T. Fukushima, H. Ichihara, T. Shimomura, K. Ito, T. Hashizume, N. Ishii and T. Aida, Self-assembled hexa-peri-hexabenzocoronene graphitic nanotube, *Science*, **304**, 1481-1483 (2004).

20. K. V. Rao, K. Jayaramulu, T. K. Maji and S. J. George, Supramolecular hydrogels and high-aspect-ratio nanofibers through charge- transfer- induced alternate coassembly, *Angew. Chem. Int. Ed.*, **49**, 4218-4222 (2010).

21. A. A. Sagade, K. V. Rao, U. Mogera, S. J. George, A. Datta and G. U. Kulkarni, High- mobility field effect transistors based on supramolecular charge transfer nanofibers, *Adv. Mater.*, **25**, 559-564 (2013).

22. B. Narayan, S. P. Senanayak, A. Jain, K. S. Narayan and S. J. George, Self-assembly of π- conjugated amphiphiles: free standing, ordered sheets with enhanced mobility, *Adv. Funct. Mater.* **23**, 3053-3060 (2013).

23. Y. Yamamoto, T. Fukushima, Y. Suna, N. Ishii, A. Saeki, S. Seki, S. Tagawa, M. Taniguchi, T. Kawai and T. Aida, Photoconductive coaxial nanotubes of molecularly connected electron donor and acceptor layers, *Science*, **314**, 1761-1764 (2006).

24. V. Faramarzi, F. Niess, E. Moulin, M. Maaloum, J. F. Dayen, J. B. Beaufrand, S. Zanettini, B. Doudin and N. Giuseppone, Light-triggered self-construction of supramolecular organic nanowires as metallic interconnects, *Nat. Chem.*, **4**, 485 (2012).

25. T. K. Ellis, M. Galerne, J. J. Armao IV, A. Osypenko, D. Martel, M. Maaloum, G. Fuks, O. Gavat, E. Moulin and N. Giuseppone, Supramolecular electropolymerization, *Angew. Chem. Int. Ed.*, **57**, 15749-15753 (2018).

26. M. J. Webber, E. A. Appel, E. W. Meijer and R. Langer, Supramolecular biomaterials, *Nat. Mater.* **15**, 13. (2016)

27. V. M. Tysseling-Mattiace, V. Sahni, K. L. Niece, D. Birch, C. Czeisler, M. G. Fehlings, S. I. Stupp and J. A. Kessler, Self-assembling nanofibers inhibit glial scar formation and promote axon elongation after spinal cord injury, *J. Neurosci.*, **28**, 3814-3823 (2008).

28. K. Rajangam, H. A. Behanna, M. J. Hui, X. Han, J. F. Hulvat, J. W. Lomasney and S. I. Stupp, Heparin binding nanostructures to promote growth of blood vessels, *Nano Lett.*, **6**, 2086-2090 (2006).

29. E. D. Spoerke, S. G. Anthony and S. I. Stupp, Enzyme directed templating of artificial bone mineral, *Adv. Mater.*, **21**, 425-430 (2009).

30. J. D. Hartgerink, E. Beniash and S. I. Stupp, Self-assembly and mineralization of peptide-amphiphile nanofibers, *Science*, **294**, 1684-1688 (2001).

31. R. N. Shah, N. A. Shah, M. M. D. R. Lim, C. Hsieh, G. Nuber and S. I. Stupp, Supramolecular design of self-assembling nanofibers for cartilage regeneration, *Proc. Natl. Acad. Sci.*, **107**, 3293-3298 (2010).

32. O. J. Goor, S. I. Hendrikse, P. Y. Dankers and E. W. Meijer, From supramolecular polymers to multi-component biomaterials, *Chem. Soc. Rev.*, **46**, 6621-6637 (2017).

33. J. Boekhoven and S. I. Stupp, 25th anniversary article: supramolecular materials for regenerative medicine, *Adv. Mater.*, **26**, 1642-1659 (2014).

34. R. Merindol and A. Walther, Materials learning from life: concepts for active, adaptive and autonomous molecular systems, *Chem. Soc. Rev.*, **46**, 5588-5619 (2017).

35. S. Dhiman, A. Sarkar and S. J. George, Bioinspired temporal supramolecular polymerization, *RSC Adv.*, **8**, 18913-18925 (2018).

36. A. Sorrenti, J. Leira-Iglesias, A. J. Markvoort, T. F. de Greef and T. M. Hermans, Non-equilibrium supramolecular polymerization, *Chem. Soc. Rev.*, **46**, 5476-5490 (2017).

37. T. F. A. de Greef, M. M. J. Smulders, M. Wolffs, A. P. H. J. Schenning, R. P. Sijbesma and E. W. Meijer, Supramolecular polymerization, *Chem. Rev.* **109**, 5687–5754 (2009).

38. C. Kulkarni, S. Balasubramanian and S. J. George, What molecular features govern the mechanism of supramolecular polymerization? *ChemPhysChem*, **14**, 661 − 673 (2013).

39. P. A. Korevaar, S. J. George, A. J. Markvoort, M. M. Smulders, P. A. Hilbers, A. P. H. J. Schenning, T. F. de Greef, E. W. Meijer, Pathway complexity in supramolecular polymerization, *Nature*, **481**, 492-496 (2012).

40. P. Jonkheijm, P. van der Schoot, A. P. H. J. Schenning and E. W. Meijer, Probing the solvent-assisted nucleation pathway in chemical self-assembly, *Science* **313**, 80-83 (2006).

41. S. Ogi, K. Sugiyasu, S. Manna, S. Samitsu and M. Takeuchi, Living supramolecular polymerization realized through a biomimetic approach, *Nat. Chem.* **6**, 188-195 (2014).

42. T. Fukui, S. Kawai, S. Fujinuma, Y. Matsushita, T. Yasuda, T. Sakurai, S. Seki, M. Takeuchi and K. Sugiyasu, Control over differentiation of a metastable supramolecular assembly in one and two dimensions, *Nat. Chem.*, **9**, 493 (2017).

43. S. Ogi, V. Stepanenko, K. Sugiyasu, M. Takeuchi and F. Würthner, Mechanism of self-assembly process and seeded supramolecular polymerization of perylene bisimide organogelator, *J. Am. Chem. Soc.*, **137**, 3300-3307 (2015).

44. J. Kang, D. Miyajima, T. Mori, Y. Inoue, Y. Itoh and T. Aida, A rational strategy for the realization of chain-growth supramolecular polymerization, *Science* **347**, 646-651 (2015).

45. M. Kasai, S. Asakura and F. Oosawa, The cooperative nature of GF transformation of actin, *Biochim. Biophys. Acta*, **57**, 22-31 (1962).

46. A. Mishra, D. B. Korlepara, M. Kumar, A. Jain, N. Jonnalagadda, K. K. Bejagam, S. Balasubramanian and S. J. George, Biomimetic temporal self-assembly via fuel-driven controlled supramolecular polymerization, *Nat. Commun.*, **9**, 1295 (2018).

47. A. Jain, S. Dhiman, A. Dhayani, P. K. Vemula and S. J. George, Chemical fuel-driven living and transient supramolecular polymerization, *Nat. Commun.* **10**, 450 (2019).

48. M. Endo, T. Fukui, S. H. Jung, S. Yagai, Takeuchi, M. and Sugiyasu, K., Photo-regulated living supramolecular polymerization established by combining energy landscapes of photoisomerization and nucleation–elongation processes, *J. Am. Chem. Soc.* **138**, 14347–14353 (2016).

49. F. della Sala, S. Neri, S. Maiti, J. L. Y. Chen, L. J. Prins, Transient self-assembly of molecular nanostructures driven by chemical fuels, *Curr. Opin. Biotechnol.* **46**, 27-33 (2017).

50. J. Boekhoven, A. M. Brizard, K. N. K. Kowlgi, G. J. M. Koper, R. Eelkema, J. H. van Esch, Dissipative self-assembly of a molecular gelator by using a chemical fuel, *Angew. Chem. Int. Ed.*, **49**, 4825-4828 (2010).

51. J. Boekhoven, W. E. Hendriksen, G. J. Koper, R. Eelkema and J. H. van Esch, Transient assembly of active materials fueled by a chemical reaction, *Science*, **349**, 1075-1079 (2015).

52. B. G. van Ravensteijn, W. E. Hendriksen, R. Eelkema, J. H. van Esch and W. K. Kegel, Fuel-mediated transient clustering of colloidal building blocks, *J. Am. Chem. Soc.*, **139**, 9763-9766 (2017).

53. S. Maiti, I. Fortunati, C. Ferrante, P. Scrimin and L. J. Prins, Dissipative self-assembly of vesicular nanoreactors, *Nat. Chem.* **8**, 725–731 (2016).

54. S. Debnath, S. Roy and R. V. Ulijn, Peptide nanofibers with dynamic instability through nonequilibrium biocatalytic assembly, *J. Am. Chem. Soc.* **135**, 16789–16792 (2013).

55. L. Heinen, T. Heuser, A. Steinschulte and A. Walther, Antagonistic enzymes in a biocatalytic pH feedback system program autonomous DNA hydrogel life cycles, *Nano Lett.*, **17**, 4989-4995 (2017).

56. T. Heuser, R. Merindol, S. Loescher, A. Klaus and A. Walther, Photonic devices out of equilibrium: transient memory, signal propagation, and sensing, *Adv. Mater.*, **29**, 1606842 (2017).

57. K. P. Sonu, S. Vinikumar, S. Dhiman, S. J. George, M. Eswaramoorthy, Bio-inspired temporal regulation of ion-transport in nanochannels, *Nanoscale Adv.*, **1**, 1847-1852 (2019).

58. K. Jalani, S. Dhiman, A. Jain and S. J. George, Temporal switching of an amphiphilic self-assembly by a chemical fuel-driven conformational response, *Chem. Sci.*, **8**, 6030-6036 (2017).

59. M. Kumar, M. D. Reddy, A. Mishra and S. J. George, The molecular recognition controlled stereomutation cycle in a dynamic helical assembly, *Org. Biomol. Chem.*, **13**, 9938-9942 (2015).

60. M. Kumar and S. J. George, Homotropic and heterotropic allosteric regulation of supramolecular chirality, *Chem. Sci.*, **5**, 3025-3030 (2014).

61. M. Kumar, P. Brocorens, C. Tonnelé, D. Beljonne, M. Surin and S. J. George, A dynamic supramolecular polymer with stimuli-responsive handedness for in situ probing of enzymatic ATP hydrolysis, *Nat. Commun.*, **5**, 5793 (2014).

62. S. Dhiman, A. Jain and S. J. George, Transient helicity: fuel- driven temporal control over conformational switching in a supramolecular polymer, *Angew. Chem.*, **129**, 1349-1353 (2017).

63. S. Dhiman, A. Jain, M. Kumar and S. J. George, Adenosine-phosphate-fueled, temporally programmed supramolecular polymers with multiple transient states, *J. Am. Chem. Soc.*, **139**, 16568-16575 (2017).

64. A. Sorrenti, J. Leira-Iglesias, A. Sato and T. M. Hermans, Non-equilibrium steady states in supramolecular polymerization, *Nat. Commun.*, **8**, 15899 (2017).

Chapter 7

Porous Materials: Recent Developments

K. P. Sonu, Subhajit Laha, Muthusamy Eswaramoorthy[*]
and Tapas Kumar Maji[†]

Chemistry and Physics of Materials Unit, New Chemistry Unit and School of Advanced Materials, Jawaharlal Nehru Centre for Advanced Scientific Research, Jakkur, Bangalore 560064, Karnataka, India
[*]*eswar@jncasr.ac.in;* [†]*tmaji@jncasr.ac.in*

Porous materials have influenced humanity ever since their initial applications on catalysis and adsorption. In the last few decades, porous materials have been subjected to extensive research that have widened their application arena to energy storage, sorption and separation, catalysis, light emission and drug delivery. In this chapter, we highlight some of the recent developments of porous materials related to the aforementioned applications. We emphasize how various strategies of pore size engineering and attachment of desired surface functionalities resulted in rationally designed porous materials with specific applications. Further, the current research directions and challenges on the further developments on various porous materials are summarized.

1. Introduction to Porous Materials

"Empty vessels make the most noise"— the proverb subtly intends that one who is 'hollow' makes more noise and is therefore unimportant. In contrast, 'hollow' (meaning porosity) in a nanostructured system is so important that materials with empty spaces are indispensable in our daily lives. The empty spaces generated at the nanoscale by the arrangement of matter in space do make "beneficial noise" in the field of catalysis, energy storage, separation and drug delivery.

Materials containing pores in the form of cavities, channels and interstices are defined as porous materials. According to IUPAC norms, they are classified (based on the pore size) into micro (< 2 nm), meso (between 2 -50 nm) and macroporous (> 50 nm) materials.[1] The pores are ordered or disordered in nature and the skeletons are made up of organic[2,3], inorganic[4-6] or hybrid building blocks.[7-10] Some of the important classes of porous materials known today are zeolites[11-13], mesoporous oxides[14,15], porous carbons[16-18], metal-organic frameworks (MOFs)[19-22], covalent-organic frameworks (COFs)[3,23], conjugated microporous polymers (CMP)[24], pillared clays[25,26] etc. A scheme for classification of different porous materials is given in Figure 1.

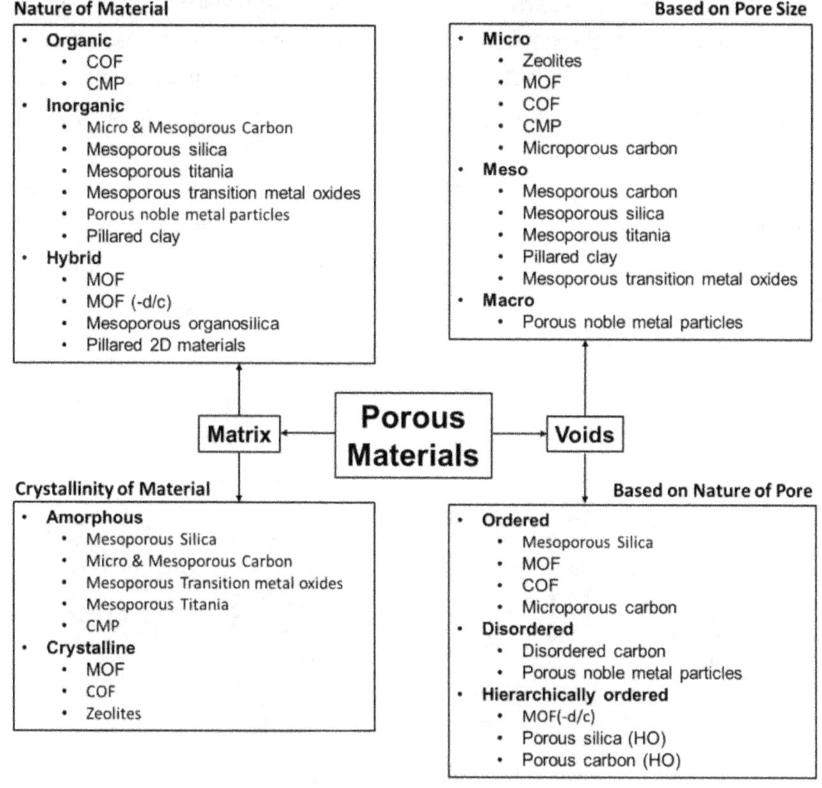

Fig. 1. Schematic representation showing different classifications of porous materials. Representative examples are included for each category.

As the size, shape and the nature of pores have a direct bearing on the performance of a porous material,[27-29] tailoring these parameters is of paramount importance for many specific applications. Significant advances have been made in the last few decades to obtain materials with a tunable pore size, geometries and functionalities.[2,3,13,14,17,21,23,30-32] In this chapter, we highlight the recent developments of porous materials related to energy storage, sorption and separation, catalysis, light emission and drug delivery.

2. Energy Storage (Supercapacitor Applications)

The emergence of low-carbon economy as an alternative to fossil fuel not only depends on the generation of energy from renewable sources but also its storage. Supercapacitors represent an important class of energy storage devices having a high specific power density (more than 10 kW/kg) and a long durability (over 10^6 cycles). There are two types of supercapacitors: (i) the electrical double layer capacitor (EDLC) where only electrostatic interaction takes place between the charged (polarized) surface and ions (Fig. 2(a)) and (ii) pseudo-capacitors where chemical reactions in addition to electrostatic interactions contribute to the capacitance (Fig. 2(b)).

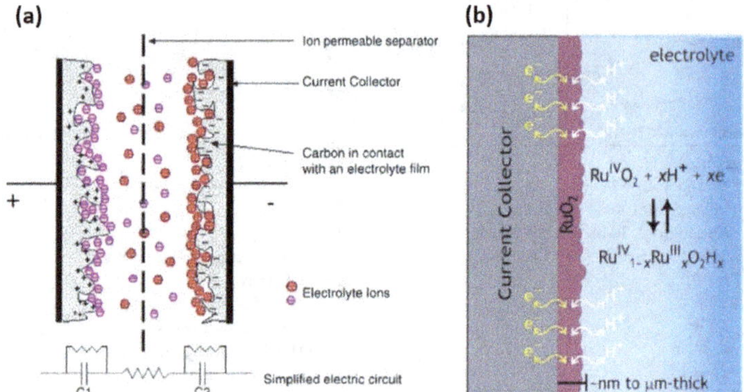

Fig. 2. (a) Principle of a single cell double layer capacitor. The capacitance is originating from the charge developed at the interface of active material (carbon). Higher the surface area carbons have for the charge accumulation, higher the resultant capacitance. (b) Schematic representation of redox pseudocapacitance taking RuO_2 as an example. Reproduced with permission from Ref. 40 and Ref. 38.

Porous materials with high surface areas, huge pore volumes and tunable pore sizes are much sought after materials for supercapacitor applications. Many porous solids such as metals[33-35], metal oxides[36-38], carbons[39,40] and conducting polymers[41-43] are being explored in this direction. Among them, carbon based materials have gained significant attention owing to their high electrical polarizability, tunable surface area, easy processability and low cost economy. They have a wide range of operation temperatures and are stable in extreme pH (from strongly basic to strongly acidic).[39]

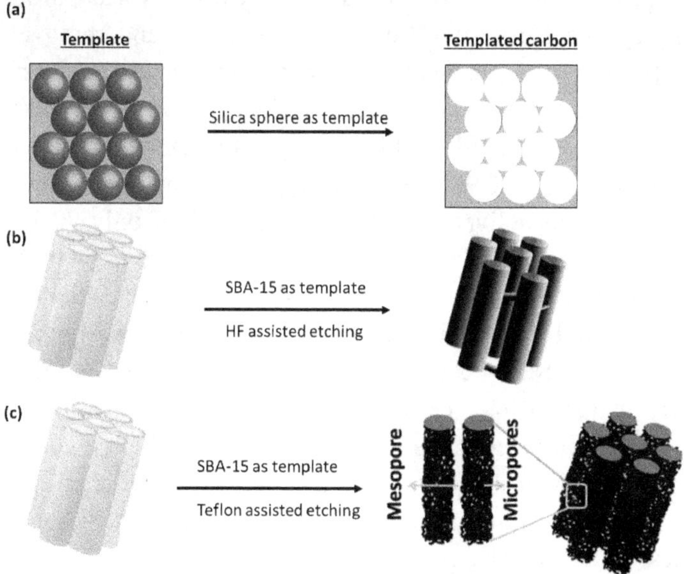

Fig. 3. Porous carbons using silica spheres as template. (a) Porous carbons obtained using colloidal crystal template- stober spheres. (b) Ordered mesoporous carbon obtained using SBA-15 as template and (c) Micro/mesoporous (hierarchical) carbons using SBA-15 as template.

Activated carbons (AC) having a high surface area in the range of 1000 to 3500 cm^2/g are extensively employed in EDL capacitors. In general, depending upon the surface nature of carbons the double layer capacitance varies between 15 and 50 $\mu F/cm^2$ (which could be translated to 150 - 500 F/g if the surface area of carbon is counted as 1000 m^2/g). However, the capacitance values of high surface area microporous carbons are often limited to a few tens of F/g due to the restricted access of its surface by the large electrolyte ions.[44] Therefore, it is necessary to design porous carbons with easily accessible pores (without compromising the surface area) to further improve the capacitor performance.

In this direction, templated porous carbons with a high specific surface area and hierarchal pore structures have generated a lot of interest. The general strategy to synthesis porous carbons through templating techniques involves the infiltration of carbon precursor (sucrose, glucose etc.) into the pores of the template (mesoporous silica, colloidal crystals etc.) followed by carbonization and removal of the template (Figure 3). For example, ordered nanostructured carbon, CMK-3,is prepared using SBA-15(mesoporous silica) as a hard template.[18] The resultant carbon shows fiber –like hexagonal stacking structures with a specific surface area of 900 m^2/g and a pore diameter of 3.9 nm. This material showed to deliver a specific capacitance of only 90 F/g.[45,46] A high surface area (2545 m^2/g), mesostructured carbon with bi-modal (micro-meso) pore size was reported to show a high capacitance of 292 F/g which is well above the CMK-3 having only mesopores.[48] The carbon prepared (SBA-15 as the template and sucrose, as the carbon precursor) by a single step carbonization and silica removal using Teflon as the fluorine source contains a lot of micropores which could be easily accessed by the electrolyte through the mesopores. The larger mesopores facilitates the mass transport and the micropores provide a huge surface area necessary for the charge accumulation.

In addition to EDLC, the facile faradic processes in porous metal oxides such as RuO_2, Fe_3O_4, Co_3O_4, NiO, MnO_2, porous polymers and surface modified carbons find application in *pseudo*-capacitors owing to their remarkable energy storage capabilities.[47] The term *pseudo*-capacitance originated because it involved faradaic charge transfer reactions.[48] RuO_2 containing porous architecture showed a pseudo capacitance value of 700 F g^{-1} [49] which was far higher than the film form. The charge storage mechanism involves the "double insertion" of electrons and protons to the structure as expressed below.

$$RuO_x(OH)_y + \delta H^+ + \delta e \rightarrow RuO_{x-\delta}(OH)_{y+\delta}$$

During this process, the oxidation state of ruthenium would change between Ru^{4+}, Ru^{3+} and Ru^{2+}. Later a number of transition metal oxides (MnO_2, TiO_2, Nb_2O_5, NiO, Co_3O_4, SnO_2, and Fe_2O_3) were also reported to possess a *pseudo*-capacitive behavior.[50] In addition to redox responsive transition metal oxides, surface engineered porous carbons have also emerged as an important class of *pseudo*-capacitor materials.[51,52] In particular, the carbon enriched with oxygen and nitrogen groups[52] and functionalized with facile redox motifs such as quinones[53] have gained interest owing to their lower cost and easy fabrication compared to transition metal oxide counterparts.

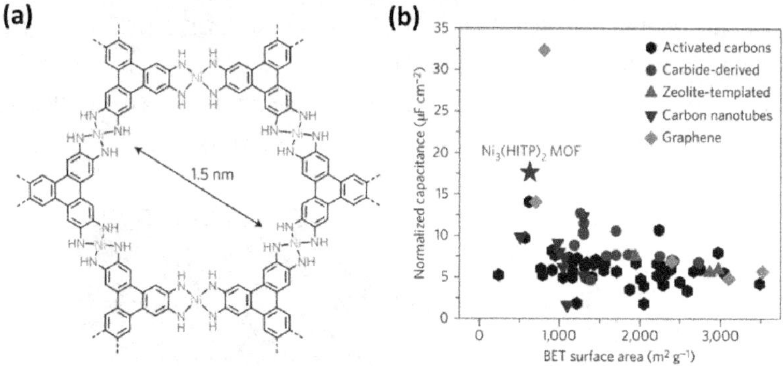

Fig. 4. (a) Molecular structure of Ni$_3$(HITP)$_2$. (b) Comparison of Ni$_3$(HITP)$_2$ areal capacitance to that of various materials normalized relative to their BET surface areas. Reproduced with permission from Ref. 54.

Metal organic frameworks (MOFs) materials represent an important class of high surface area materials. However, despite their high surface area, their applications as electrode materials for supercapacitors are limited due to the insulating nature. Recently, Dincă and co-workers reported a Ni$_3$(hexaiminotriphenylene)$_2$ (Ni$_3$(HITP)$_2$) MOF electrode with high surface area and intrinsic electrical conductivity (Figure 4).[54] The double layer capacitance was tested in a two electrode setup and the capacitance was as high as 110 F/g comparable to that of AC. It is worth noting that the electrodes were made from Ni$_3$(HITP)$_2$ alone and no binders or conductive carbon additives were added. Molecular porous systems with energy storage properties are rare and developments in these directions are expected to offer many design and implementation opportunities.

3. Adsorption and Separation

Porous materials have been extensively studied in the last few decades as active materials for sorption and separation owing to their high surface area and tunable pore size and versatile surface chemistry.

Adsorption and separation of molecules based on their size/shape have been achieved using materials containing pores in the range of molecular size.[55] For example, the separation of para-xylene from the ortho- and meta- isomers, one of the long standing problems in the chemical industry was possible by size selective separation using zeolites. For instance, Ba^{2+} and K^+ ion-exchanged X zeolite showed a higher adsorption capacity and selectivity for para-xylene.[56]

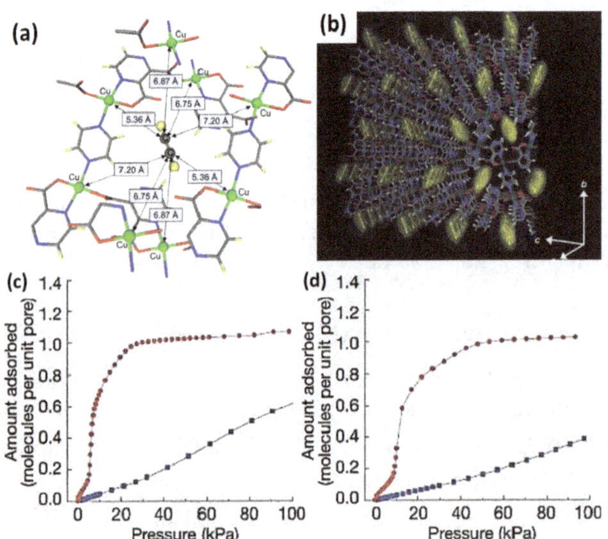

Fig. 5. (a) A representation of the pore structure and the distances between acetylinc carbon atoms and copper atoms (Cu–C distances). Crystal structure of complex 1 with C_2H_2 at 170 K from MEM/Rietveld analysis. Cu, green; O, red; C, grey; N, blue; H, white. (b) Orthographic views down the c axis. Adsorption isotherms for C_2H_2 and CO_2 temperatures are at (c) 300 K and (d) 310 K. Reproduced with permission from Ref. 58.

The hallmark characteristics of metal-organic framework (MOF) such as the ultrahigh porosity (highest reported Langmuir surface area 10,000 m^2g^{-1}), controllable pore size (a few Angstroms to ~10 nm) make MOFs of interest for potential application in high-capacity adsorbents, storage and separation. MOFs have shown promising applications in adsorption based separation of different gases (CO_2/H_2, CO_2/N_2, CO_2/CH_4, C_2H_2/C_2H_4, CH_4/C_2H_6 etc.), geometrical isomers (xylene isomers) and in the purification of several industrially important raw materials.[57] For instance, the removal of CO_2 is an essential step to obtain ultrapure acetylene (C_2H_2); which is hardly possible by zeolites and the activated carbon due to their similar physical and structural properties. Mita and co-workers developed a porous structure ($[Cu_2(pzdc)_2(pyz)]_n$; pzdc is pyrazine-2,3-dicarboxylate and pyz is pyrazine) which selectively adsorbs C_2H_2 over CO_2 at a density that is 200 times of a safe compression limit upto 1 atm (Figure 5).[58]

In recent days, downsizing the MOF to nanoscale has attracted significant attention due to its enhanced textural porosity and active external surface compare to the bulk analogue. For example, Kitagawa *et al.* have demonstrated the crystal downsizing of a two-fold interpenetrated framework which regulates the structural flexibility and induces a shape-memory effect (shows flexible sorption phenomena by a nonporous closed phase and a guest-included open phase) in the coordination frameworks.[59] Similarly, Maji et al. reported a way of downsizing a mixed-linker based 3D MOF to the nano/mesoscale by a coordination modulation method using the n-dodecanoic acid as a modulator.[60] The smaller hexagonal MOF nanoparticles showed a high BET surface area with higher mass transfer kinetics which was reflected in a better CO_2/N_2 separation efficiency in a breakthrough experiment under an ambient condition.

Whilst microporous materials were largely successful for small molecules, the separation and sequestration of larger motifs such as proteins and biomolecules requires materials with a larger pore size. Besides, microporous materials are not effective in cases where the sequestration of small molecules and ions necessitate the chemical modification of pores. Mesoporous materials with a pore size in the range of 2-50 nm are often functionally modified and are being explored for the capture of larger molecules and ions through electrostatic, hydrophobic or covalent interactions.[61,62]

Recently, the charge transfer interactions (C-T) were also being explored to separate the molecules based on the binding abilities. For instance, Pawan et al. demonstrated a selective adsorption of chromophores in an ordered mesoporous silica though charge transfer interactions.[63] The viologen (acceptor) modified porous silica showed an enhanced uptake for donor molecules such as perylene tetracarboxylate (PS) and coronene tetracarboxylate owing to the formation of the C-T complex. The difference in the binding constants of these donors with the viologen accepter was utilized for the breakthrough column separation of donors (Figure 6).

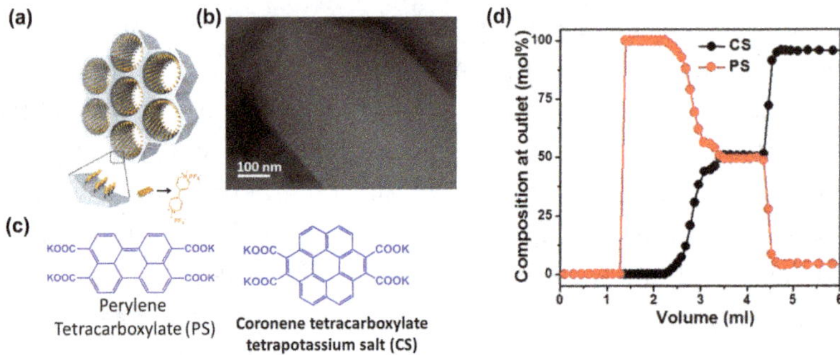

Fig. 6. (a) Schematic representation of viologen functionalized mesoporous silica. (b) TEM image of mesoporous silica. (c) Chemical structure of PS and CS. (d) Experimental breakthrough curves for an equimolar mixture of CS and PS flowing through a bed of SBA-V, exhibiting clear separation and high selectivity. Reproduced with permission from Ref. 63.

In order to adsorb and separate macromolecules such as proteins and biopolymers, the porous materials should possess sufficiently larger pores and the porous surface should be engineered to have specific interactions with macromolecules. For example, the larger pore size mesoporous silica was studied for the adsorption of larger biomolecules. Adsorption of larger molecules like proteins on as-synthesized mesoporous silica usually occurs through the electrostatic interactions between the charged silica surface and proteins which could be optimized by varying the pH of the solution. For example, as-synthesised mesoporous silica shows a negatively charged surface above pH 3.0. Hence, the maximum protein adsorption (via electrostatic attraction) occurs between pH 3.0 and the pI (isoelectric point) of the protein. It is also possible to separate the biomolecules based on the hydrophobicity of the molecules. For example, hydrophobic proteins such as myoglobin have greater interactions with mesoporous silica with hydrophobic interiors, whilst hydrophilic biopolymer shows the least interactions with the hydrophobic mesoporous silica. The difference in interactions of protein with the silica surface have been successfully employed to separate these macromolecules from a mixture.[63]

4. Catalysis

Porous materials have been traditionally used in catalysis ever since the advent of synthetic zeolites in the late 1940s (owing to the pioneering work of Barrer and Miton).[64,65] In 1962, synthetic faujasites containing crystalline aluminosilicate framework were used on an industrial scale in the fluid catalytic cracking (FCC) of heavy petroleum distillates (an important chemical process worldwide).[66] Currently, zeolites are very extensively used in a variety of chemical processes such as hydrocracking of heavy petroleum distillates, improvement of octane number of light gasoline and many more. It is interesting to note that catalysis is the single most important application of zeolite in terms of the financial market size.[67] The basic principles of zeolite chemistry (their surface acidity and shape selectivity) and catalysis is well-documented elsewhere.[68,69]

On the other hand, an ordered mesoporous silica generally contains an amorphous silica wall and has been used as a catalytic support to immobilize organometallic complexes, noble metal nanoparticles, transition metal oxide particles, nano-alloys etc.[70,71] Owing to the high thermal and chemical stability of mesoporous silica, it acted as as good catalytic support. The immobilization of metal nanoparticles on the mesoporous silica improved the stability of these particles by preventing them from leaching which lead to an enhanced performance. Functionally modified mesoporous silica with catalysts were employed for several chemical reactions like acid catalysed reactions (alkylation reactions, fluid catalytic cracking, oligomerization processes, Friedel-Crafts acylation, cycloaddition of CO_2 into epoxides, ring opening of epoxides etc.), base catalysed reactions, oxidation reactions (by transition metal doped mesoporous silica), hydrogenation etc.[72,73]

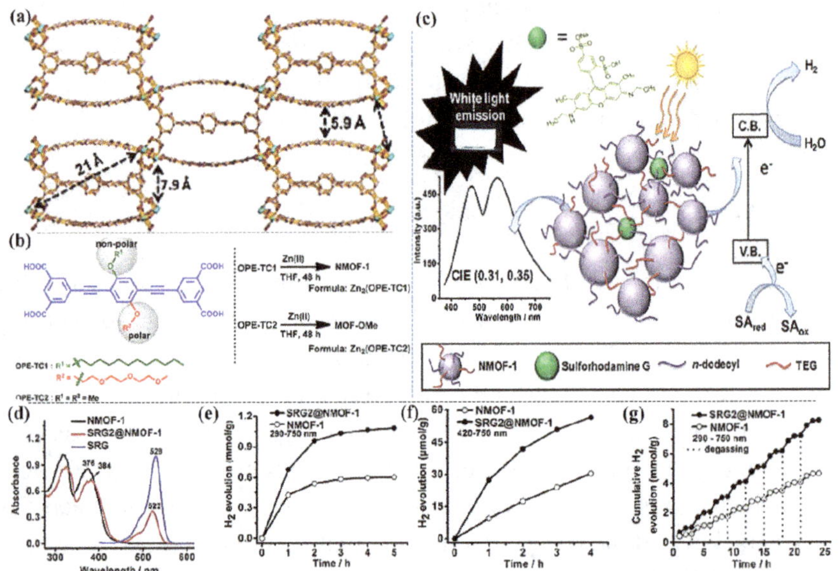

Fig. 7. (a) Nanovesicular MOF with Omniphilic Porosity: Bimodal Functionality for White-Light Emission and Photocatalysis by Dye Encapsulation. (b) Ligands **OPE-TC1** and **OPE-TC2** and Their Self-Assembly with Zn^{2+}, (c) Graphical Representation of the Bimodal Functionality; White-light emission and photocatalysis in the present work, (d) UV−vis spectra of SRG, NMOF-1, and **SRG2@NMOF-1** as a water dispersion. Photocatalytic activity of NMOF-1 and **SRG2@ NMOF-1** using (e) full range light (290−750 nm) and (f) visible light (420−750 nm). (g) Cumulative H_2 production upon intermittent degassing by NMOF-1 and **SRG2@NMOF-1** (290−750 nm). Reproduced with permission from Ref. 77.

In recent years, MOFs have been gaining significant attention in the field of catalysis owing to the presence of coordinatively unsaturated sites (CUS), structural defects along with high surface area and porosity. Besides these tunable properties, the nature of ligands also affects the catalytic activities by acting as Bronsted acid sites (sulfonic acid group) or basic sites (amine groups). The effective interaction between reactants and the unsaturated metal centre of the MOF which acts as a Lewis acid site have shown an increase in the electrophilicity of the reactant and the corresponding catalytic activity in different organic reactions. Further, the pore size and shape also play an important role in introducing the selectivity in the catalytic reactions. Lin *et al.*, have shown an enantioselective catalytic activity of a hydrogenation reaction based on a chiral MOF.[74] In the year 1994, Fujita and co-workers first performed the cyanosilylation of aldehydes by using a 2D MOF ([Cd(4,4'-bpy)$_2$(NO$_3$)$_2$], (bpy = bipyridine)).[75]

MOFs have also begun to be recognized as suitable candidates in the field of photocatalysis and electrocatalysis. The performance efficiency of the electrochemical reaction (such as oxygen evolution reaction (OER), hydrogen evolution reaction (HER), oxygen reduction reaction (ORR), CO$_2$ reduction reaction (CO$_2$RR) etc.) strongly depends upon the structure, morphology, chemical stability, electro and redox activity of the catalytic electrodes. The MOFs cannot be used directly in electrocatalysis due to its poor electrical conductive nature. To overcome this most of the MOFs are converted to highly conductive carbons by a controlled pyrolysis in a high temperature under an inert atmosphere. These MOF derived active materials show tremendous potential in electrocatalysis. For example, Maji *et al.* recently reported an efficient and a stable bi-functional electrocatalyst derived from a 3D metal-organic framework (([Co(bpe)$_2$(N(CN)$_2$)]N(CN)$_2$.5H$_2$O)) (Co-MOF), (bpe = 1,2-bis(4-pyridyl) ethane and N(CN)$_2^-$ = dicyanamide).[76] Two different catalysts were prepared by pyrolyzing at 800 °C under Ar and H$_2$/Ar atmosphere, and resulted in Co-NCNT-Ar and Co-NCNT-H$_2$ (NCNT= nitrogen doped carbon nanotube), respectively. Co$_3$O$_4$@Co/NCNT composite material has been synthesized by further calcination of the sample Co-NCNT-H$_2$ in air at 250 °C. Both catalysts showed an outstanding bi-functional activity for the ORR and OER. However, the Co$_3$O$_4$@Co/NCNT nanostructure exhibited a superior electro catalytic activity for ORR with a potential of 0.88V at a current density of 1 mA cm^{-2} and the OER with a potential of 1.61 V at 10 mA cm^{-2}.

By appropriately choosing the organic linker of MOF systems, optical properties can be fine-tuned. Therefore, MOFs based photocatalysts can be rationally designed. For example, a novel asymmetric bola-amphiphilic linker, oligo-(p-phenyleneethynylene) tetracarboxylate (OPE-TC), containing dodecyl (non-polar) and triethyleneglycol-monomethylether (TEG, polar) side chains in the backbone, combined with Zn^{2+} resulted in a nanoscale MOF $(Zn_2(OPE-TC))_n$ (NMOF-1) (Figure 7(a)-(d).[77] The N_2 isotherms reveal an additional porosity in the mesoscale which is further supported by the morphology characterization and the dye encapsulation (sulforhodamine G, SRG) studies. Both NMOF-1 and SRG encapsulated NMOF-1 (SRG@NMOF-1) showed a significant photocatalytic activity for H_2 evolution from water and opened up the door for OPE-based materials as water-splitting photocatalysts without using any metal co-catalyst (Figure 7(e)-(g).

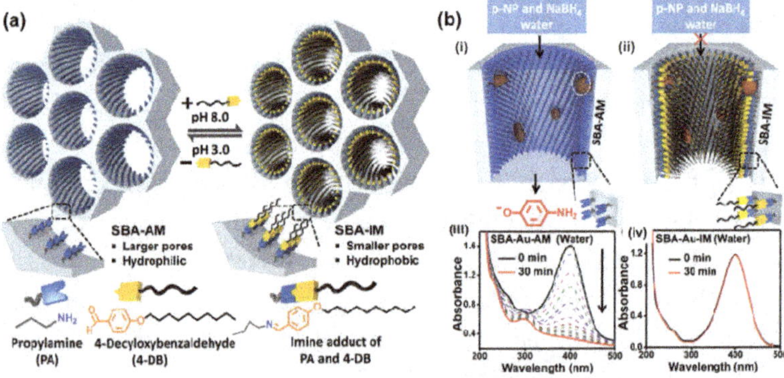

Fig. 8. (a) Illustration depicting reversible engineering of the pore size and philicity of mesoporous SBA via dynamic covalent chemistry triggered by changes in pH. (b) Catalysis in water medium. (i) Schematic showing gold nanoparticles (encircled in white) immobilized in the pores of SBA-Au-AM, carrying out catalytic reduction of p-nitrophenol. (ii) UV-Vis spectra indicating complete reduction of p-nitrophenol to p-aminophenol within 30 min by SBA-Au-AM (arrow indicating the decrease in intensity with time). (iii) Illustration showing the absence of catalytic activity due to de-wetting of the pores caused by hydrophobicity of SBA-Au-IM. (iv) Corresponding UV-Vis spectra indicating no catalytic reduction. Reproduced with permission from Ref. 82.

High surface area, porous metal nanostructures have been used as efficient catalysts for the direct synthesis H_2O_2, methanol/ethanol electrooxidation, oxygen reduction reaction, formic acid oxidation etc.[78-

[81] Their high surface area originating from the interconnected particles lend them superior catalytic activity compared to their bulk counterparts.

Besides, conventional catalytic applications, porous materials (with rationally designed properties) are now emerging as smart catalysts that are capable of performing a switchable catalytic activity in response to stimuli such as pH, light, temperature, magnetic field etc. For example, D. K Singh *et al.* reported a pH switchable catalysis based on the functionalized mesoporous silica (Figure 8).[82] The smart catalyst consisted of mesoporous silica with gold nanoparticles inside the pores and the pores were functionalised with propylamine groups (inside the pores) (Au-SBA-NH$_2$). They have carried out the reduction of p-nitrophenol in an aqueous phase as a model reaction. The protonated amine makes the pores hydrophilic in the aqueous medium which in turn allows the reactant p-nitrophenol (and the reducing agent borohydride) to get into the pores and reduced to aminophenol on the gold surface. However, when the pores were reversibly modified by alkyl chains through the dynamic covalent bond between the amine and aldehyde (imine chemistry), the pores became hydrophobic. The reactant dissolved in water (and sodium borohydride) now could not reach the gold nanoparticles through the hydrophobic channels and hence no reduction. The pore philicity can be switched between hydrophilic and hydrophobic in a pH responsive manner as the imine bonds are labile to an acidic pH demonstrating the pH responsive catalysis within the porous silica.

5. Drug Delivery

Porous materials have gained a lot of importance in drug delivery owing to their high pore volume, their ease of surface modification and biocompatibility. In the last few decades, significant developments have been made in the drug delivery platform towards cancer therapy due to the severe side effects of anti-cancer drugs. For instance, Feng and co-workers reported mesoporous silica having an acidic pH responsive release.[83]

In the early twenties, Férey and co-workers first proposed that MOFs can be used as drug carriers, owing to their considerable loading capacities and their controlled release behavior. They utilized two well-known MOFs (MIL-100(Cr) and MIL-101(Cr)), which exhibited a remarkable IBU adsorption with 0.35 g g^{-1} of IBU/dehydrated MIL-100 and 1.4 gg^{-1} of IBU/dehydrated MIL-101.[84] Wang *et al.* have reported a chiral Zn-based MOF assembled from 5,5′5″-(1,3,5-triazine-2,4,6-triyl) tris (azanediyl) triisophthalate (TATAT) and Zn^{2+} for the delivery of an anticancer drug 5-fluorouracil (5-Fu) with a loading efficiency of 0.5 gg^{-1}.[85] Recently, Maji and his coworkers have designed a unique luminescent organic molecule that contains polar chemical groups on one side of the backbone and non-polar groups on the other side.[86] This asymmetric organic linker self-assembled with Zn^{2+} to form three reversibly shape shifting and non-toxic NMOFs with nanovesicle (**NMOF-1**), inverse nanovesicle (**NMOF-2**) and nanoscroll (**NMOF-3**)

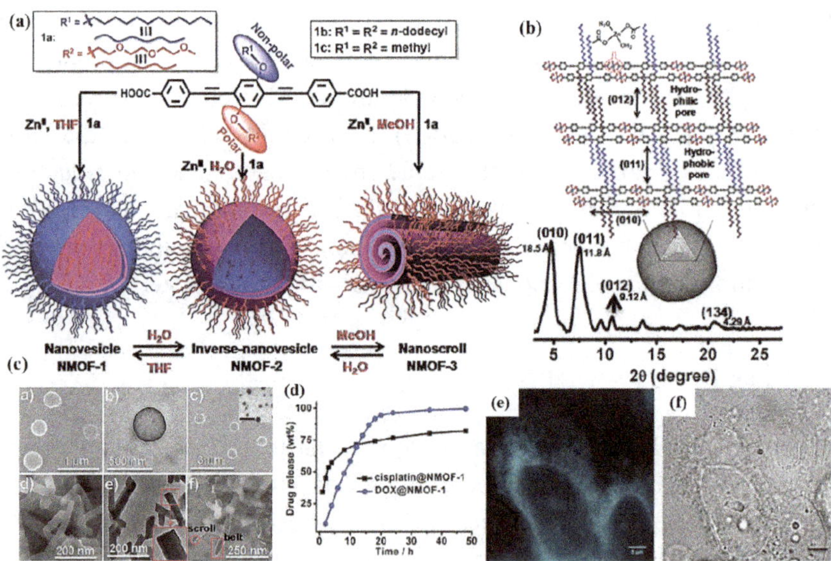

Fig. 9. (a) Schematic representation of synthesis and morphology transformation of NMOFs. (b) Possible packing and PXRD pattern of the NMOF-1. (c) FESEM images and TEM images of NMOF-1 (a and b), NMOF-2 (c and d), and NMOF-3. (e and f). (d) Drug release profile of cisplatin@NMOF-1 and DOX@N-MOF-1in 1×PBS buffer. (e-f) SR-SIM image of HeLa cells after treatment with cisplatin@NMOF-1 (scale bar: 5 μm). Reproduced with permission from Ref. 86.

morphologies by a varying solvent polarity (Figure 9(a)-(d)). The emissive property of NMOFs encourages its application in bio-imaging. The interior part of nanovesicles has been decorated with polar groups which facilitate the successful accommodation of the anti-cancer drugs Cisplatin and Doxorubicin. The process of cell permeation and drug release is quite facile for this MOF due to its nanoscale architecture and non-toxicity. Cisplatin (14.4 wt%) and doxorubicin (4 wt%) were encapsulated in flexible NMOF by a non-covalent interaction and, in vitro and in vivo drug releases were studied. The 67 % drug was released within 8 h at pH 7.4 ($1 \times$PBS buffer) and the release kinetics became slow providing 82% of the drug release in 48 h (Figure 9(e)). Furthermore, the delivery potential of cisplatin@NMOF-1has been studied in HeLa cell with a micromolar cytotoxicity.

6. Light Emission and Sensing Applications

The past two decades have witnessed an enormous growth of design and synthesis of inorganic-organic hybrid materials, like the coordination polymer or MOFs and the organic porous materials and realized for their broad range of potential application in light emitting, harvesting, display, sensing, chiral recognition and optical devices.[87] However, the interest in organic luminescent materials has been primarily inspired by their usages in organic light emitting diodes (OLEDs).[88]

MOFs, being a hybrid system, luminescence can either arise from direct π-chromophoric organic linkers (highly conjugated ligands) or from metal-centered emission (widely observed in lanthanide MOFs through the so-called antenna effect), and also from the charge-transfer such as ligand-to-metal charge transfer (LMCT) and metal-to-ligand charge transfer (MLCT) or as a result of the guest's luminescence in the MOF. Till date, hundreds of luminescent MOFs have been reported.[87] Liu *et al.* first reported a red emissive lanthanide MOF (Na[EuL $(H_2O)_4]_3.2H_2O$) (L = 1,4,8,11-tetraazacyclo-tetradecane-1,4,8,11-tetrapropionic acid) and have exploited for sensing different cations such as Cu^{2+}, Ag^+, Zn^{2+}, Cd^{2+}, and Hg^{2+} by encapsulating inside the MOF.[89] Maji *et al.* for the first time reported a tunable emission in MOFs based

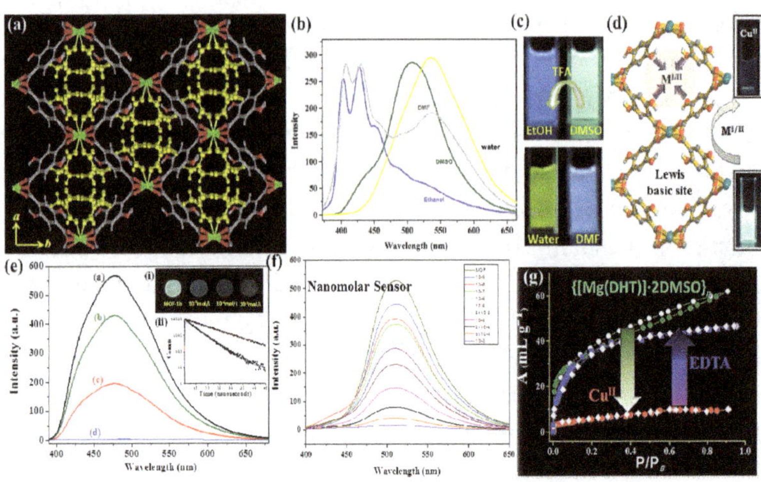

Fig 10. (a) View of 3D framework of 1 bridged by DHT along the c-axis (coordinated DMF molecules are shown in yellow color; a = 2 x, y, 3/2 z; b = 2 x, y, 1 z; c = x, y, 1/2 + z); (b) Emission spectra of 1 in different solvents ethanol, DMSO, water and DMF (c) with corresponding photograph (d) View of the 3D framework of 1a along the crystallographic c axis showing Lewis basic pendant −OH groups and photographs under UV light corresponding to free 1a and after its Cu^{2+} immobilization in DMSO. (e) Luminescence spectra of solid (Cu@Mg-MOF) obtained after the incorporation of different concentrations of Cu^{2+} in 1a in DMSO: (a) 1a; (b) 10^{-5}M Cu^{2+}; (c) 10^{-4} M Cu^{2+}; (d) 10^{-3} M Cu^{2+}. Insets: (i) Photographs of the corresponding samples under UV light; (ii) lifetime data for 1a and (Cu@Mg-MOF) (10^{-3} M). (f) Comparison of luminescence intensities of $(Mg(DHT)_n$ 1a in DMSO after incorporation of different concentrations of copper ion (10^{-3} -10^{-9} M). (g) CO_2 adsorption (closed) and desorption (open symbol) isotherms at 195 K for desolvated 1b, (Cu@Mg-MOF) and the compound obtained from (Cu@Mg-MOF) after treatment with EDTA. Reproduced with permission from Ref. 90 and Ref. 91.

on an excited state proton transfer (ESIPT) responsive linker 2,5-dihydroxy terephthalate, which is a bifunctional salicylic acid derivative, well studied for ESIPT properties. The MOF $(Mg(DHT)(DMF)_2)_n$ is synthesized by the solvo-thermal condition and the 3D structure is composed of Mg^{2+} and DHT linker.[90,91] Each carboxylate group of DHT linker connects two Mg^{2+} centres through the *syn-anti* bridging mode and each DHT acts as a tetradentate ligand. DHT linker binds 1D chains of $Mg(CO_2)_2$ along the three directions resulting in a 3D framework with 1D channels (5.3 x 5.3 Å^2) occupied by the two coordinated DMF molecules which fulfill the octahedral geometry of the Mg^{2+} center. The uncoordinated OH groups of the DHT linkers remain free in the channels and are H-bonded with the coordinated carboxylate oxygen atoms. The permanent porosity of the desolvated framework is established by the type I CO_2 uptake profile at 195 K. The detailed fluorescence spectroscopic studies of $(Mg(DHT)(DMF)_2)_n$ and its comparison with the linker have been reported. The solid MOF showed the Stokes shifted bright green emission which is attributed to the ESIPT process of the linker. Then the emission of the MOF is reported in different solvents of varied polarity. The high ionic character of Mg^{2+} facilitates the partial solubility of the MOF in polar solvents. The MOF showed a multi-colour emission ranging from blue and yellow depending on the solvent. In DMSO a red shifted green emission at 508 nm was observed whereas in the ethanol solution the MOF exhibited a blue emission with a maxima at 429 nm. This high energy blue and red shifted green emission can be attributed to the enol and keto isomers of the DHT linker, respectively. As mentioned previously the framework $(Mg(DHT)(DMF)_2)_n$ **(1)** contains free pendent –OH groups on the pore surface (Fig. 10(a)). The desolvated framework contains four –OH group. Each channels with a distance of 5.44, 7.11 and 7.17 Å. This framework further exploited for capture and sensing application for specific metal ions. The desolvated compound **1a** immersed in a 10^{-3} M solution containing different metal ions such as Li^+, Na^+, Ca^{2+}, Mn^{2+}, Co^{2+}, Ni^{2+}, Cu^{2+}, Zn^{2+} and Cd^{2+} in DMSO. The fluorescence spectroscopic study of the sample of (M@Mg-MOF) exhibited the quenching of green emission at 508 nm due to the presence of transition metal ions Co^{2+}, Ni^{2+}, Cu^{2+} in particular the quenching effect of Cu^{2+} is most significant (Fig. 10(b)-(f)). Interestingly, alkali, alkali-earth and transition metal ion with filled d-shells (Zn^{2+}, Cd^{2+}) shows no effect on luminescent after encapsulation into the pores. The quenching of the emission with transition metal ion was attainted to the energy or charge transfer process through the artially filled d-orbitals based on ligand field transitions (d-d) or due to the re-adsorption of emission energy for the d-d transition.

7. Conclusion and Future Prospects

In this chapter, we have featured some of the recent and exciting research directions of porous materials. Traditionally porous materials have been utilized in catalysis and adsorption. The initial focus was in fine tuning the pore size and distribution to suit the demands of the respective catalytic processes. However, the improvements of chemical tools and better molecular level understanding led to the exponential growth of various porous materials. This resulted in employing porous materials in various applications other than catalysis and adsorption. Apart from conventional materials like zeolites, mesoporous silica and porous carbons, new types of porous materials such as MOF, COF, CMP etc. have emerged. We have discussed some of the recent developments in the field of porous materials in terms of their applications such as adsorption and separation, photoluminescence and sensing, drug delivery, catalysis and electro-/photo-catalysis. The different strategies of pore engineering (in terms of both pore size tuning and surface functionalization) that have made them suitable for respective applications have been summarized. The challenges in the field of porous materials are specific to the type of materials. For example, one of the major concerns about of the MOFs is their instability in aqueous environments and hence their limited applicability. Of late, this problem is being addressed. Another concern with respect to MOFs is their scalability. The synthesis of many MOFs on a large scale is very limited. Therefore, the commercialization is hindered. An important application of porous materials where intense research is going on (to tackle certain major challenges) is drug delivery. One particular challenge (irrespective of the materials used) is the protein corona formation *in vivo*. Although, the designed delivery system behaves in the expected manner (in terms of release) in simulated fluids, the release would be severely different in biological systems (*in vivo*) due to the non-specific adsorption of proteins around the delivery particles. However, solutions to these challenges are not out of reach with the existing sophisticated ability to tailor the materials at a molecular level and the current research interest in porous materials.

References

1. J. Rouquerol, D. Avinir, C. W. Fairbridge, D. H. Everett, J. H. Haynes, N. Pernicone, J. D. F. Ramsay, K. S. W. Sing, K. K. Unger, Recommendations for the characterization of porous solids (Technical Report). *Pure Appl. Chem.* **66** (8), 1739 (1994).

2. S. Das, P. Heasman, T. Benand S. Qiu. Porous organic materials: Strategic design and structure–function correlation. *Chem. Rev* **117** (3), 1515 (2017).

3. X. Feng, X. Ding and D. Jiang. Covalent organic frameworks. *Chem. Soc. Rev.* **41** (18), 6010 (2012).

4. C. J. Brinker. Porous inorganic materials. *Curr. Opin. Solid State Mater. Sci.* **1** (6), 798 (1996).

5. W. P. Ruren Xu, Jihong Yu, Qisheng Huo, Jiesheng Chen *Chemistry of Zeolites and Related Porous Materials: Synthesis and Structure*; 1 ed.; John Wiley & Sons, Inc, (2007).

6. M. E. Medina, A. E. Platero-Prats, N. Snejko, A. Rojas, A. Monge, F. Gándara, E. Gutiérrez-Puebla and M. A. Camblor. Towards inorganic porous materials by design: looking for new architectures. *Adv. Mater.* **23** (44), 5283 (2011).

7. T. Asefa, M. J. MacLachlan, N. Coombs and G. A. Ozin. Periodic mesoporous organosilicas with organic groups inside the channel walls. *Nature* **402**, 867 (1999).

8. T. Asefa, C. Yoshina-Ishii, M. J. MacLachlan and G. A. Ozin. New nanocomposites: putting organic function "inside" the channel walls of periodic mesoporous silica. *J. Mater. Chem.* **10** (8), 1751 (2000).

9. A. Sayari and S. Hamoudi. Periodic mesoporous silica-based organic–inorganic nanocomposite materials. *Chem. Mater.* **13** (10), 3151 (2001).

10. F. Xie, Z. Xie, P. Liu, Y. Yuan, J. Li, S. Chen, F. Guo, J. Chen and Z. Gu. A highly organic functionalized three-connected periodic mesoporous silica by Co-condensation with hydridosilica. *Microporous Mesoporous Mater.* **266**, 177 (2018).

11. M. E. Davisand R. F. Lobo. Zeolite and molecular sieve synthesis. *Chem. Mater.* **4** (4), 756 (1992).

12. B. M. Weckhuysen and J. Yu. Recent advances in zeolite chemistry and catalysis. *Chem. Soc. Rev.* **44** (20), 7022 (2015).

13. S. I. Zones and M. E. Davis. Zeolite materials: recent discoveries and future prospects. *Curr. Opin. Solid State Mater. Sci.* **1** (1), 107 (1996).

14. F. Tang, L. Li and D. Chen. Mesoporous silica nanoparticles: synthesis, biocompatibility and drug delivery. *Adv. Mater.* **24** (12), 1504 (2012).

15. Y. Wang, Q. Zhao, N. Han, L. Bai, J. Li, J. Liu, E. Che, L. Hu, Q. Zhang, T. Jianget al. Mesoporous silica nanoparticles in drug delivery and biomedical applications. *Nanomedi.: Nanotechnol., Biol. Med.* **11** (2), 313 (2015).

16. M. R. Benzigar, S. N. Talapaneni, S. Joseph, K. Ramadass, G. Singh, J. Scaranto, U. Ravon, K. Al-Bahily and A. Vinu. Recent advances in functionalized micro and

mesoporous carbon materials: synthesis and applications. *Chem. Soc. Rev.* **47** (8), 2680 (2018).

17. J. Lee, J. Kim and T. Hyeon. Recent progress in the synthesis of porous carbon materials. *Adv. Mater.* **18** (16), 2073 (2006).

18. R. Ryoo, S. H. Joo, M. Kruk and M. Jaroniec. Ordered mesoporous carbons. *Adv. Mater.* **13** (9), 677 (2001).

19. L. Zhang, Z. Kang, X. Xin and D. Sun. Metal–organic frameworks based luminescent materials for nitroaromatics sensing. *CrystEngComm* **18** (2), 193 (2016).

20. J.-R. Li, J. Sculley and H.-C. Zhou. Metal–organic frameworks for separations. *Chem. Rev* **112** (2), 869 (2012).

21. L. E. Kreno, K. Leong, O. K. Farha, M. Allendorf, R. P. Van Duyne and J. T. Hupp. Metal–organic framework materials as chemical sensors. *Chem. Rev* **112** (2), 1105 (2012).

22. J. Lee, O. K. Farha, J. Roberts, K. A. Scheidt, S. T. Nguyen and J. T. Hupp. Metal–organic framework materials as catalysts. *Chem. Soc. Rev.* **38** (5), 1450 (2009).

23. S.-Y. Ding and W. Wang. Covalent organic frameworks (COFs): from design to applications. *Chem. Soc. Rev.* **42** (2), 548 (2013).

24. Y. Xu, S. Jin, H. Xu, A. Nagai and D. Jiang. Conjugated microporous polymers: design, synthesis and application. *Chem. Soc. Rev.* **42** (20), 8012 (2013).

25. A. Gil, S. A. Korili and M. A. Vicente. Recent advances in the control and characterization of the porous structure of pillared clay catalysts. *Catal Rev.* **50** (2), 153 (2008).

26. K. K. R. Datta, D. Jagadeesan, C. Kulkarni, A. Kamath, R. Datta and M. Eswaramoorthy. Observation of pore-switching behavior in porous layered carbon through a mesoscale order–disorder transformation. *Angew. Chem.* **123** (17), 4015 (2011).

27. T. Gebäck, M. Marucci, C. Boissier, J. Arnehed and A. Heintz. Investigation of the effect of the tortuous pore structure on water diffusion through a polymer film using lattice boltzmann simulations. *J. Phys. Chem. B* **119** (16), 5220 (2015).

28. C. S. Kong, D.-Y. Kim, H.-K. Lee, Y.-G. Shul and T.-H. Lee. Influence of pore-size distribution of diffusion layer on mass-transport problems of proton exchange membrane fuel cells. *J. Power Sources* **108** (1), 185 (2002).

29. A. Walcarius. Mesoporous materials and electrochemistry. *Chem. Soc. Rev.* **42** (9), 4098 (2013).

30. Q.-L. Zhu and Q. Xu. Metal–organic framework composites. *Chem. Soc. Rev.* **43** (16), 5468 (2014).

31. B. G. Trewyn, I. I. Slowing, S. Giri, H.-T. Chen and V. S. Y. Lin. Synthesis and functionalization of a mesoporous silica nanoparticle based on the sol–gel process and applications in controlled release. *Acc. Chem. Res.* **40** (9), 846 (2007).

32. S.-H. Wu, C.-Y. Mou and H.-P. Lin. Synthesis of mesoporous silica nanoparticles. *Chem. Soc. Rev.* **42** (9), 3862 (2013).

33. N. Kobayashi, T. Sakumoto, S. Mori, H. Ogata, K. C. Park, K. Takeuchi and M. Endo. Investigation on capacitive behaviors of porous Ni electrodes in ionic liquids. *Electrochim. Acta* **105**, 455 (2013).

34. X. Y. Lang, H. T. Yuan, Y. Iwasa and M. W. Chen. Three-dimensional nanoporous gold for electrochemical supercapacitors. *Scripta Mater.* **64** (9), 923 (2011).

35. Y. Ding and Z. Zhang *Nanoporous Metals for Supercapacitor Applications*, Springer International Publishing: Cham, (2016).

36. L. Wang, H. Ji, S. Wang, L. Kong, X. Jiang and G. Yang. Preparation of Fe3O4 with high specific surface area and improved capacitance as a supercapacitor. *Nanoscale* **5** (9), 3793 (2013).

37. X. Lang, A. Hirata, T. Fujita and M. Chen. Nanoporous metal/oxide hybrid electrodes for electrochemical supercapacitors. *Nat. Nanotechnol* **6**, 232 (2011).

38. J. W. Long, D. Bélanger, T. Brousse, W. Sugimoto, M. B. Sassin and O. Crosnier. Asymmetric electrochemical capacitors—Stretching the limits of aqueous electrolytes. *MRS Bull.* **36** (7), 513 (2011).

39. L. L. Zhang and X. S. Zhao. Carbon-based materials as supercapacitor electrodes. *Chem. Soc. Rev.* **38** (9), 2520 (2009).

40. A. G. Pandolfo and A. F. Hollenkamp. Carbon properties and their role in supercapacitors. *J. Power Sources* **157** (1), 11 (2006).

41. X. Lang, L. Zhang, T. Fujita, Y. Ding and M. Chen. Three-dimensional bicontinuous nanoporous Au/polyaniline hybrid films for high-performance electrochemical supercapacitors. *J. Power Sources* **197**, 325 (2012).

42. Y. Hou, L. Chen, L. Zhang, J. K ang, T. Fujita, J. Jiangand M. Chen. Ultrahigh capacitance of nanoporous metal enhanced conductive polymer pseudocapacitors. *J. Power Sources* **225**, 304 (2013).

43. K. Wang, H. Wu, Y. Meng and Z. Wei. Conducting polymer nanowire arrays for high performance supercapacitors. *Small* **10** (1), 14 (2014).

44. K. Kierzek, E. Frackowiak, G. Lota, G. Gryglewicz and J. Machnikowski. Electrochemical capacitors based on highly porous carbons prepared by KOH activation. *Electrochim. Acta* **49** (4), 515 (2004).

45. H.-Q. Li, J.-Y. Luo, X.-F. Zhou, C.-Z. Yu and Y.-Y. Xia. An ordered mesoporous carbon with short pore length and its electrochemical performances in supercapacitor applications. *J. Electrochem. Soc.* **154** (8), A731 (2007).

46. H. Zhou, S. Zhu, M. Hibino and I. Honma. Electrochemical capacitance of self-ordered mesoporous carbon. *J. Power Sources* **122** (2), 219 (2003).

47. V. Augustyn, P. Simon and B. Dunn. Pseudocapacitive oxide materials for high-rate electrochemical energy storage. *Energy Environ. Sci* **7** (5), 1597 (2014).

48. S. Trasatti and G. Buzzanca. Ruthenium dioxide: A new interesting electrode material. Solid state structure and electrochemical behaviour. *J. Electroanal. Chem.* **29** (2), A1 (1971).

49. J. W. Long, K. E. Swider, C. I. Merzbacher and D. R. Rolison. Voltammetric characterization of ruthenium oxide-based aerogels and other RuO2 solids: the nature of capacitance in nanostructured materials. *Langmuir* **15** (3), 780 (1999).

50. G. Wang, L. Zhang and J. Zhang. A review of electrode materials for electrochemical supercapacitors. *Chem. Soc. Rev.* **41** (2), 797 (2012).

51. Y. Zhao, W. Ran, J. He, Y. Song, C. Zhang, D.-B. Xiong, F. Gao, J. Wu and Y. Xia. Oxygen-rich hierarchical porous carbon derived from artemia cyst shells with superior electrochemical performance. *ACS Appl. Mater. & Interf.* **7** (2), 1132 (2015).

52. Y. Song, L. Li, Y. Wang, C. Wang, Z. Guo and Y. Xia. Nitrogen-doped ordered mesoporous carbon with a high surface area, synthesized through organic–inorganic coassembly, and its application in supercapacitors. *ChemPhysChem* **15** (10), 2084 (2014).

53. A. Le Comte, T. Brousse and D. Bélanger. Simpler and greener grafting method for improving the stability of anthraquinone-modified carbon electrode in alkaline media. *Electrochim. Acta* **137**, 447 (2014).

54. D. Sheberla, J. C. Bachman, J. S. Elias, C.-J. Sun, Y. Shao-Horn and M. Dincă. Conductive MOF electrodes for stable supercapacitors with high areal capacitance. *Nat. Mater.* **16**, 220 (2016).

55. Z. Wu and D. Zhao. Ordered mesoporous materials as adsorbents. *Chem. Commun.* **47** (12), 3332 (2011).

56. L. T. Furlan, B. C. Chaves and C. C. Santana. Separation of liquid mixtures of p-xylene and o-xylene in X-zeolites: the role of water content on the adsorbent selectivity. *Ind. Eng. Chem. Res* **31** (7), 1780 (1992).

57. J.-R. Li, R. J. Kuppler and H.-C. Zhou. Selective gas adsorption and separation in metal–organic frameworks. *Chem. Soc. Rev.* **38** (5), 1477 (2009).

58. R. Matsuda, R. Kitaura, S. Kitagawa, Y. Kubota, R. V. Belosludov, T. C. Kobayashi, H. Sakamoto, T. Chiba, M. Takata, Y. Kawazoeet et. al. Highly controlled acetylene accommodation in a metal–organic microporous material. *Nature* **436** (7048), 238 (2005).

59. Y. Sakata, S. Furukawa, M. Kondo, K. Hirai, N. Horike, Y. Takashima, H. Uehara, N. Louvain, M. Meilikhov, T. Tsuruokaet et. al. Shape-memory nanopores induced in coordination frameworks by crystal downsizing. *Science* **339** (6116), 193 (2013).

60. N. Sikdar, M. Bhogra, Umesh V. Waghmare and T. K. Maji. Oriented attachment growth of anisotropic meso/nanoscale MOFs: tunable surface area and CO2 separation. *J. Mater. Chem. A* **5** (39), 20959 (2017).

61. K. Y. Ho, G. McKay and K. L. Yeung. Selective adsorbents from ordered mesoporous silica. *Langmuir* **19** (7), 3019 (2003).

62. Z. Yan, S. Tao, J. Yin and G. Li. Mesoporous silicas functionalized with a high density of carboxylate groups as efficient absorbents for the removal of basic dyestuffs. *J. Mater. Chem.* **16** (24), 2347 (2006).

63. B. V. V. S. P. Kumar, K. V. Rao, T. Soumya, S. J. George and M. Eswaramoorthy. Adaptive pores: charge transfer modules as supramolecular handles for reversible pore engineering of mesoporous silica. *J. Am. Chem. Soc.* **135** (30), 10902 (2013).

64. R. M. Barrer. Zeolites and their synthesis. *Zeolites* **1** (3), 130 (1981).

65. R. M. Barrer. 435. Syntheses and reactions of mordenite. *J. Chem. Soc.*, DOI:10.1039/JR9480002158 10.1039/JR9480002158(0), 2158 (1948).

66. E. T. C. Vogt and B. M. Weckhuysen. Fluid catalytic cracking: recent developments on the grand old lady of zeolite catalysis. *Chem. Soc. Rev.* **44** (20), 7342 (2015).

67. J. Čejka, R. E. Morris. *Zeolites in Catalysis: Properties and Applications*, Royal Society of Chemistry, (2017).

68. L. P. Jens Weitkamp *Catalysis and Zeolites: Fundamentals and Applications*, Springer Science & Business Media, (17-Apr-2013).

69. J. Weitkamp. Zeolites and catalysis. *Solid State Ion.* **131** (1), 175 (2000).

70. G. Zhan and H. C. Zeng. Integrated nanocatalysts with mesoporous silica/silicate and microporous MOF materials. *Coord. Chem. Rev.* **320-321**, 181 (2016).

71. R. Jin, D. Zheng, R. Liu and G. Liu. Silica-supported molecular catalysts for tandem reactions. *ChemCatChem* **10** (8), 1739 (2018).

72. G. Martínez-Edo, A. Balmori, I. Pontón, A. Martí del Rio and D. Sánchez-García. Functionalized ordered mesoporous silicas (MCM-41): Synthesis and applications in catalysis. *Catalysts* **8** (12), 617 (2018).

73. E. Doustkhah, J. Lin, S. Rostamnia, C. Len, R. Luque, X. Luo, Y. Bando, K. C.-W. Wu, J. Kim, Y. Yamauchiet et. al. Development of sulfonic-acid-functionalized mesoporous materials: synthesis and catalytic applications. *Chem.: Eur. J.* **25** (7), 1614 (2019).

74. L. Ma, C. Abney and W. Lin. Enantioselective catalysis with homochiral metal–organic frameworks. *Chem. Soc. Rev.* **38** (5), 1248 (2009).

75. M. Fujita, Y. J. Kwon, S. Washizu and K. Ogura. Preparation, clathration ability, and catalysis of a two-dimensional square network material composed of Cadmium(II) and 4,4'-Bipyridine. *J. Am. Chem. Soc.* **116** (3), 1151 (1994).

76. N. Sikdar, B. Konkena, J. Masa, W. Schuhmann and T. K. Maji. Co3O4@Co/NCNT Nanostructure derived from a dicyanamide-based metal-organic framework as an efficient bi-functional electrocatalyst for oxygen reduction and evolution reactions. *Chem.: Eur. J.* **23** (71), 18049 (2017).

77. D. Samanta, P. Verma, S. Roy and T. K. Maji. Nanovesicular MOF with omniphilic porosity: bimodal functionality for white-light emission and photocatalysis by dye encapsulation. *ACS Appl. Mater. & Interf.* **10** (27), 23140 (2018).

78. S. Maity and M. Eswaramoorthy. Ni–Pd bimetallic catalysts for the direct synthesis of H2O2 – unusual enhancement of Pd activity in the presence of Ni. *J. Mater. Chem. A* **4** (9), 3233 (2016).

79. H.-C. Shin and M. Liu. Copper foam structures with highly porous nanostructured walls. *Chem. Mater.* **16** (25), 5460 (2004).

80. L. Wei, K. Goh, Ö. Birer, H. E. Karahan, J. Chang, S. Zhai, X. Chen and Y. Chen. A hierarchically porous nickel–copper phosphide nano-foam for efficient electrochemical splitting of water. *Nanoscale* **9** (13), 4401 (2017).

81. S. Fu, J. Song, C. Zhu, G.-L. Xu, K. Amine, C. Sun, X. Li, M. H. Engelhard, D. Duand Y. Lin. Ultrafine and highly disordered Ni2Fe1 nanofoams enabled highly efficient oxygen evolution reaction in alkaline electrolyte. *Nano Energy* **44**, 319 (2018).

82. D. K. Singh, B. V. V. S. Pavan Kumar and M. Eswaramoorthy. Reversible control of pore size and surface chemistry of mesoporous silica through dynamic covalent chemistry: philicity mediated catalysis. *Nanoscale* **7** (32), 13358 (2015).

83. R. Liu, Y. Zhang, X. Zhao, A. Agarwal, L. J. Mueller and P. Feng. pH-responsive nanogated ensemble based on gold-capped mesoporous silica through an acid-labile acetal linker. *J. Am. Chem. Soc.* **132** (5), 1500 (2010).

84. P. Horcajada, C. Serre, M. Vallet-Regí, M. Sebban, F. Taulelle and G. Férey. Metal–organic frameworks as efficient materials for drug delivery. *Angew. Chem. Int. Ed.* **45** (36), 5974 (2006).

85. C.-Y. Sun, C. Qin, C.-G. Wang, Z.-M. Su, S. Wang, X.-L. Wang, G.-S. Yang, K.-Z. Shao, Y.-Q. Lan and E.-B. Wang. Chiral nanoporous metal-organic frameworks with high porosity as materials for drug delivery. *Adv. Mater.* **23** (47), 5629 (2011).

86. D. Samanta, S. Roy, R. Sasmal, N. D. Saha, P. K R, R. Viswanatha, S. S. Agasti and T. K. Maji. Solvent adaptive dynamic metal-organic soft hybrid for imaging and biological delivery. *Angew. Chem. Int. Ed.* **58**, 5008 (2019),

87. Y. Cui, Y. Yue, G. Qian and B. Chen. Luminescent functional metal–organic frameworks. *Chem. Rev.* **112** (2), 1126 (2012).

88. S.-C. Lo and P. L. Burn. Development of dendrimers: macromolecules for use in organic light-emitting diodes and solar cells. *Chem. Rev.* **107** (4), 1097 (2007).

89. W. Liu, T. Jiao, Y. Li, Q. Liu, M. Tan, H. Wang and L. Wang. Lanthanide coordination polymers and their Ag+-modulated fluorescence. *J. Am. Chem. Soc.* **126** (8), 2280 (2004).

90. K. Jayaramulu, P. Kanoo, S. J. George and T. K. Maji. Tunable emission from a porous metal–organic framework by employing an excited-state intramolecular proton transfer responsive ligand. *Chem. Commun.* **46** (42), 7906 (2010).

91. K. Jayaramulu, R. P. Narayanan, S. J. George and T. K. Maji. Luminescent microporous metal–organic framework with functional lewis basic sites on the pore surface: specific sensing and removal of metal ions. *Inorg. Chem.* **51** (19), 10089 (2012).

Chapter 8

Development of Biomolecule Integrated Materials and their Biological Applications

Lakshmi P. Datta, Shivaprasad Manchineella and Thimmaiah Govindaraju[*]

New Chemistry Unit and School of Advanced Materials (SAMat), Jawaharlal Nehru Centre for Advanced Scientific Research, Jakkur, Bangalore 560064, Karnataka, India
[*]tgraju@jncasr.ac.in

The utilization of biological molecular systems for the development of bioactive materials has great potential in the field of biomaterials science. Advancements in the synthetic approaches to appropriately integrate the biological components have led to the generation of biohybrid artificial systems to precisely mimic natural systems, tissues and organs of the human body. In this chapter, we provide an overview of biomolecule integrated materials systems and their assimilation within the synthetic systems to realize numerous biomedical applications. Various design approaches with the focus on diverse biomaterial systems generated by integrating proteins, peptides, nucleic acids, carbohydrates and lipids have been discussed.

1. Introduction

Innovations and assimilation of knowledge on the diverse class of materials systems have been the perennial source of advancements in the exciting area of biomaterials science (Figure 1(a)).[1,2] Since ancient times, biomolecules are ubiquitous in clinical practice whether by employing corals and wood as dental implants or silk fabrics as sutures.[1,3] The utilization of biomolecules in the past for medicinal applications has inspired material chemists to design synthetic biomaterial systems with diverse functional features (Figure 1(b)).[4] A myriad of synthetic

     L. P. Datta, S. Manchineella & T. Govindaraju

materials, for example, polymers, ceramics, metals, and composites are successfully employed as scaffolds for the incorporation of biomolecules.[5,6] The conjugation of distinct biomolecules within synthetic systems are particularly challenging, since, the heterogeneous structural and chemical functionalities of biomolecules invoke a unique design approach for individual biomolecules. A plethora of techniques have been explored by researchers to integrate synthetic biomimetic materials and biomolecules that exhibit a remarkable resemblance to the complexities of nature.[7,8]

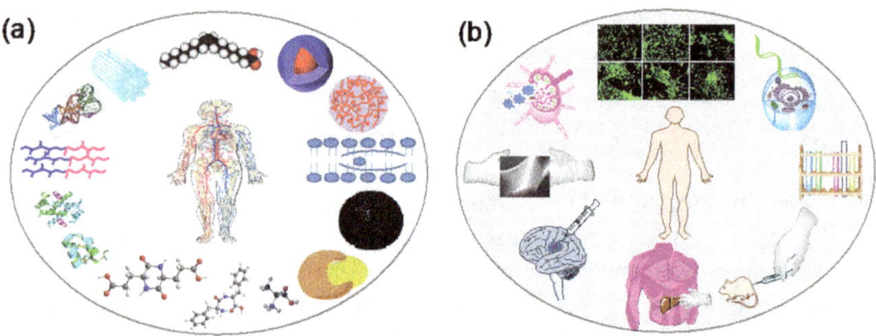

Fig. 1. (a) Numerous biomolecules and their derived structures used for biomaterials applications. (b) Diverse biomedical applications of biological molecules integrated biomaterials. The inset image was taken from Ref. [12].

Ontogenesis of next-generation biomaterials must maintain the architectural and functional synergism with biological systems.[8,9] Apart from the molecular-level interactions, the successful design and production of biomimetic materials demand the contribution of chemical and mechanical engineering arsenal. In this context, the utilization of amino acids, peptides, nucleobases and oligonucleotides as sophisticated building blocks and auxiliaries is essential to impart greater structural and functional integrity to resultant hybrid biomaterials.[10] In this chapter, we present an overview of recent advancements in the field of biomolecules integrated materials systems with potential value in the biomedical engineering and therapeutics. The perspective of this emerging field is assembled to highlight the major advances in the design and development of biomolecules integrated biomaterials for several biomedical applications.

2. Protein Based Materials

Proteins are the most abundant biomolecules in the human body with a diverse molecular architecture and characteristic properties responsible for the structural integrity and biological functions. Structural and fibrous proteins such as silk, collagen, elastin, keratin and gelatin are regarded as natural building blocks and accordingly greater attention has been devoted at using these proteins for the construction of biomaterials.[11] We have reported the successful fabrication of silk fibroin biomaterials with promising tissue engineering and regenerative medicine applications. In one of the studies, silk fibroin (SF) scaffolds were fabricated and surface-functionalized with short peptide epitopes YIGSR (L1) and GYIGSR (L2) derived from integrin binding domain of laminin, in a reductionist approach.[12] Human mesenchymal stem cell (hMSC) maintenance, proliferation and on-demand differentiation were demonstrated using these surface functionalized SF scaffolds. This study was aimed at evaluating the potential of surface modified SF films for stem cell-based *in vitro* and *in vivo* tissue engineering and regenerative applications. The cytocompatibility and suitability of the SF films was evaluated using cell proliferation and live/dead assays. The hMSC cells were cultured on the silk films and stained with fluorescein diacetate (FDA) and propidium iodide (PI) to ascertain the cytocompatibility of the silk scaffolds. The data shown in Figure 2(a) suggests a high cytocompatible nature of the laminin epitope modified surface functionalized silk films. Remarkably, hMSCs on the L2 modified SF scaffolds showed a transdifferentiation into the neuronal lineage. Figure 2(b) depicts the presence of a laterally elongated homogeneous cell population with an interconnected network. The geometric transformation of hMSC cells from spindle like morphology to the interconnected cellular network is a clear indication of the differentiation of stem cells into neuronal-like cells on the peptide (L2) modified SF films. The observed qualitative phenotypic change in the cell morphology has been further validated by investigating the quantitative neuronal-specific gene expression analysis. The neuronal-specific gene (NES, TUBB3, NEFL, MAP2) expression of hMSCs cultured on SF scaffolds were evaluated using RT-PCR followed by quantification. Upregulation of mature neuronal marker genes such as MAP2 for the hMSCs cultured on functionalized SF scaffolds (L2) further validated the suitability of rationally designed scaffolds for stem cell therapy and regenerative medicine applications.

Skeletal muscle tissue engineering (SMTE) is another important field in biomaterial science which requires biomaterial scaffolds with inherent electroactive properties and extracellular matrix (ECM) mimicking topography to induce myogenic differentiation of myoblasts to myotubes.[13] In a contemporary approach, we developed ECM mimicking fibrous electrospun scaffolds of melanin incorporated silk (SM), also termed as pigmented silk, with inherent electroactive and antioxidant properties to stimulate the myogenesis.[14,15] An in-house designed robust electrospinning setup was employed to prepare the aligned fiber mats of SM under ambient conditions. The incorporation of natural pigment melanin into silk bestowed the fabricated composite scaffold with inherent conducting and antioxidant properties. FESEM micrograph of electrospun SM composite mat indicates the presence of bead-free, aligned fiber networks as shown in Figure 2(c). The observations regarding adhesion, proliferation and differentiation of murine skeletal myoblast C2C12 cells into myotubes on the SM composite scaffolds affirmed the crucial role of the surface topography and conductivity in modulating the formation of myotubes (Figure 2(d)). The fluorescence staining of cultures on SM scaffolds further confirmed the *in vitro* transformation of myoblast cells to myotubes through myogenesis (Figure 2(e)). The introduction of electrical cues to impart conductivity property within the bioactive scaffold is a fascinating and a multifaceted neuroregenerative approach that has been successfully implemented in our laboratory.[15,16] SM aligned nanofibrous composite was exploited to assess the neurogenic potential using human neuroblastoma cell line (SH-SY5Y). The *in vitro* cellular studies ascertained the potential of antioxidant electrospun SM fiber mats for the nerve tissue engineering applications (Figure 2(f)).

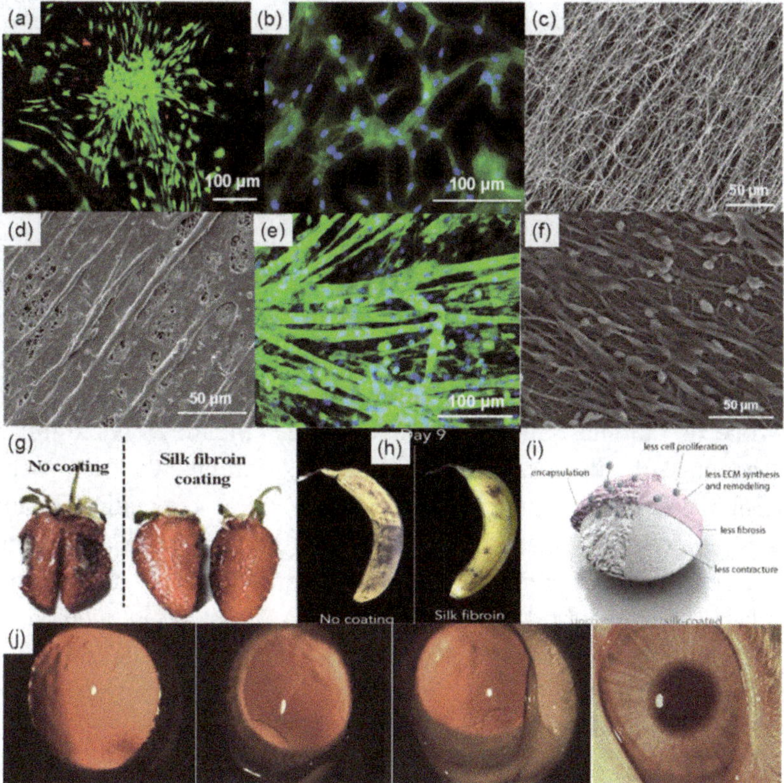

Fig. 2. Protein-based biomaterials and applications. (a) Fluorescence live/dead cell assay image showcasing the cytocompatibility of surface functionalized silk films for hMSC. (b) Fluorescence image shows the on-demand neuron-like trans-differentiation of hMSCs cultured on surface functionalized silk films, (c) FESEM image of SF mat displaying the bead free, ECM-mimicking aligned fibers. (d) FESEM micrograph showcasing the aligned myotube formation on SM fiber mat. (e) Fluorescence image showing the myogenesis of C2C12 myoblasts cultured on SM fiber mat into multinucleated myotubes. (f) FESEM image shows the adhesion and orientation of Sh-SY5Y along the direction of SM aligned fibers. Edible silk fibroin coating protecting (g) strawberries and (h) banana stored under ambient conditions. (i) Schematic representation of silk coating over silicone implant, and (j) slit lamp images of the rabbit corneas at different time points after implantation of gelatin-based hydrogels as ophthalmologic implantations to recover the vision. Adapted from Refs. [12, 14, 16 18, 19, 20].

Generally, the impermeability and fragile nature of the films of several proteins, carbohydrates and lipids impede their use as edible coating.[17] The quintessential properties of silk i.e. polymorphism, conformability and hydrophobicity have been exploited by Kaplan and coworkers for developing edible silk coatings to extend the shelf-life of fruits by preventing the rancidity.[18] For instance, coating freshly picked strawberries and bananas with edible silk suspension results in a significant protection over a week's time of storage under ambient conditions (Figure 2(g) and 2(h)). The introduction of implants as foreign objects within the human body leads to several consequences such as the generation of immunogenicity and nonspecific protein adsorption with a high risk of infection. Coating of silicone implants with recombinant silk protein by deep coating or oxygen plasma technique leads to reduced non-specific protein adsorption and immunogenic response along with a low rate of fibroblast and histiocyte coverage (Figure 2(i)).[19] Similarly, the coating of silicone catheters with silk inhibits the non-specific adhesion and proliferation of cells followed by preventing biomass related infections. Gelatin, another robust protein material has been engineered by an appropriate material modification to treat corneal blindness.[20] The hybrid (physical and UV) crosslinking method was shown to endow a several fold higher strength to the transparent hydrogel system. The human corneal endothelial cells that were grown over the hydrogel surface showed an improved expression of zona-occludin cells (Figure 2(j)).

3. Peptide Based Materials

One of the simple and robust approaches to design functionally complex biomaterials with wide structural diversities is to create peptide-based biomaterials. The minimalistic approach also makes them attractive from a design perspective which helps in avoiding the formation of artifacts in macrostructures. Peptide amphiphiles (PAs) based nano- and micro-structures are the most fascinating category of peptide-based materials with the ability to form tunable higher ordered architectures.[21] As a result, several research groups across the globe have designed and developed various PAs with versatile properties and functions. Stupp and coworkers have developed a supramolecular liquid crystal delivery vehicle synthesized using PAs, which facilitates the encapsulation of cells along with the growth factors within the unidirectional nanofiber environment.[22] The sequence-dependent stiffness of the peptide nanofibers modulates the degree of cell alignment. The injectable scaffolds also assists the adhesion and proliferation of myogenic stem- and progenitor cells followed by their differentiation and maturation. In a separate work, they have reported *in vivo* use of self-assembling PAs carrying Tenascin-C signal (E2Ten-C PA) for the redirection of endogenous neuroblasts in the rodent brain.[23] The study of directed cell alignment has potential application in tissue engineering and regenerative medicine which demands directed cell growth and formation of the cell wire. In this context, a thermoresponsive mechanism was devised to transform the isotropic solutions of peptides to liquid crystals which resulted in the formation of aggregated bundled nanofiber filaments. The monodomain fibrous gels allow the formation of macroscopically aligned cells. Amphiphilic peptides containing the amino acids $V_3A_3E_3(COOH)$ capped with a C_{16} alkyl chain at the N-terminus forms monodomain noodle like hydrogels under the influence of ion and temperature stimuli (Figure 3(a)). The uniform birefringence along the length of the noodle like string structure was visualized by dragging the noodle like string over the salty medium (Figure 3(b)). The viability of hMSCs during the whole process of the string formation indicates excellent biocompatibility of the PA hydrogels (Figure 3(c)). An interesting enzyme responsive targeted drug delivery approach which takes advantage of the overexpressed active enzymes in the disease locality was reported by Ulijn and coworkers.[24] The incorporation of short β-sheet peptide xFFyG, where x is either glycine or phenylacetyl group and y is leucine or alanine, within the MMP substrate significantly enhances the MMP activity towards the scaffolds. The entrapment of the anticancer drug doxorubicin within the hydrophobic pocket of the MMP substrate conjugated peptides results in the formation of drug loaded nanofibers as a metastasis marker (due to MMP overexpression). The confocal microscopy image showed the proximity region of MDA-MB-231-luc-D3H2LN breast cancer cells which reveals the presence of peptide delivery vectors in the cytoplasm and nucleus of the malignant cell along with the presence of the aggregation of the nanofibers outside the cell (Figure 3(d)). The MMP responsive peptide-based cargo vehicle was found to effectively deliver the payload in a sustained manner.

Hartgerink and coworkers have developed the VEGF mimic biodegradable peptide-based hydrogel network as a potential therapeutic agent for hind limb ischemia treatment.[25] The *in vitro* experiments were performed with human umbilical vein endothelial cells (HUVECs) which confirmed that peptides can activate the cellular signaling pathways. The observations indicate that the angiogenic peptide is able to activate VEGFR1, VEGFR2, and NP-1 receptors followed by the activation of the vasculogenic receptor. The *in vivo* rodent experiments further corroborated the *in vitro* observed results. The study also showed that the peptide hydrogel matrix exhibited good cytocompatibility with the cells and the formation of micro-vessels were observed over a period of 7 days. The therapeutic investigation on a peripheral artery disease mouse model showed a tissue regeneration and retention of functions after 28 days of treatment (Figure 3(e) and 3(f)). Tirrell and coworkers have demonstrated the design of protein mimic PAs through controlled self-assembly pathways.[26] They have decorated a protein-resembling templated nanosphere (PRTN) from dendrimer templates and PA building blocks (Figure 3(g)). The nanofibers resembling the protein structure exhibit the multivalent nature and multifunctional features and interact with biomolecules like DNA. The cellular internalization study and the biocompatibility feature of this supramolecular protein biomaterial system has shown a high stability and potentiality as a biomimetic system.

The photothermolysis efficiency of gold nanostructures was stimulated by decorating the gold nanostars with TAT peptide (Figure 3(h)).[27] The cellular uptake efficiency of TAT capped gold nanostars were significantly higher than the corresponding PEGylated nanostructures. The mechanism of uptake was reported to be based on the actin-driven lipid raft-mediated micropinocytosis, wherein the cargo bodies are deposited in the macropinosomes and then leached into the cytoplasm. In a majority of the cases laser irradiation used for the photothermolysis treatment is higher than the maximal permissible exposure (MPE) of skin per ANSI regulation. In this case, photothermolysis was evaluated using 850 nm pulsed laser under 0.2

W/cm^2 irradiation which is the minimal value utilized for the photothermolysis treatment, specifically lower than the MPE of skin. The multi-photon microscopy image of photothermolysis depicts a significant damage of cells at an irradiation dose lower than the MPE and has confirmed that the system is an efficient photothermolysis system (Figure 3(i)).

Dipeptides are minimalistic models of peptide systems with attractive structural and functional features. Gazit group has reported a rationally designed diphenylalanine based self-assembled nanostructural formulation with a potential biocidal activity.[28] The mechanism of the antibacterial action was found to be initiated by the up-regulation of stress response genes that causes membrane depolarization followed by a significant corrugation to the bacterial cell morphology and finally leading to bacterial cell death (Figure (3j)). The design has a potential in treating bacterial infections owing to its simplistic dipeptide preparation cost-effectively on an industry scale. The retroverted correlation between Alzheimer's disease and cancer has encouraged the use of amyloids to impede cancer. This conceptual phenomenon was investigated by the same group wherein naphthalene-conjugated diphenylalanine nanofibrils were found to inhibit the progression of glioblastoma cells over neuronal cells.[29] Aromatic interactions emanated from the assimilation of naphthalene and phenyl groups in the dipeptide derivative induced molecular self-assembly in water followed by the formation of fibrils (Figure 3(k)). The *in vivo* studies in rodent models showed a significant decrease in the tumor volume after 19th day of treatment (Figure 3(l)).

Cyclic dipeptides (CDPs) are the simplest cyclic peptides renowned for their fascinating structural and functional features.[30-33] CDPs are well-known for their characteristic hydrogen bonding-driven assembly into either molecular chains or layers leading to the formation of discrete functional materials.[32] Our group has explored the efficient molecular self-assembly of CDPs and the derived molecular systems and materials for a variety of applications. For example, we have reported the synthesis and self-assembly of asymmetric *cyclo*(Gly-*L*-Lys) ε-amino derivatives and their self-assembly into molecular organo- and hydrogels.[32] The organo- and hydrogels of CDPs were found to exhibit a higher order morphology with fibrillar network and exemplary mechanical properties

(Figure 3(m)). The encapsulation efficiency and potential application of the CDP molecular gels for drug delivery applications was demonstrated using curcumin as a model drug and the hydrophobic dye rhodamine B, a dye extensively used in molecular biology (Figure 3(n)). In a recent work, we developed CDP-based ambidextrous supergelators (*cyclo*(*L*-Tyr-*L*-Glu(OtBu), *cyclo*(*L*-Phe-*L*-Glu(OtBu)) and demonstrated their use as injectable gels to create a drug depot for the controlled drug release applications.[34] The *in situ* hydrogelation of curcumin loaded CDP solution under simulated physiological conditions has been successfully achieved, and a representative image is shown in Figure 3(o). The *in vitro* cytotoxicity studies of the CDP based supergelators showed good cytocompatibility with excellent viability of the cells in the presence of CDP gelators. Furthering the application of rationally designed CDP molecules, a set of CDP incorporated SF-based electrospun biomaterial scaffolds were fabricated with excellent antioxidant properties.[35] The rationally designed CDPs containing unnatural amino acid dopamine imparted an outstanding radical scavenging property to the electrospun SF-scaffolds. The CDP-incorporated bead-free ECM mimicking SF-fiber mats fabricated is shown in Figure 3(p).

The radical scavenging activity of the CDP-incorporated electrospun scaffold was evaluated using 2,2′-diphenyl-1-picrylhydrazyl (DPPH) radicals under an ambient condition (Figure 3(p) and inset). In particular, the antioxidant property of CDPs (*cyclo*(*L*-Tyr-*L*-DOPA) and *cyclo*(*L*-Phe-*L*-DOPA)) makes SF-mats as excellent radical scavenging biomaterial scaffolds with potential applications in chronic wound healing and tissue engineering.

In a unique design strategy, we developed CDP-based cell penetrating peptidomimetics, wherein the incorporation of CDP units in peptide backbone imparted high serum stability and cellular uptake property (Figure 3(q)).[36] The lysine-aspartic acid based CDP (kd) was incorporated into the lysine containing peptide at alternate positions. This design takes the advantage of lysine (as in cationic peptides), rigidity of CDP (similar to the rod-like structure of polyproline) and the possible engagement of hydrogen bonding interactions of CDPs with cell membrane. In fact, our study revealed that not only the electrostatic interaction (of lysine units) rather the hydrogen bonding interaction

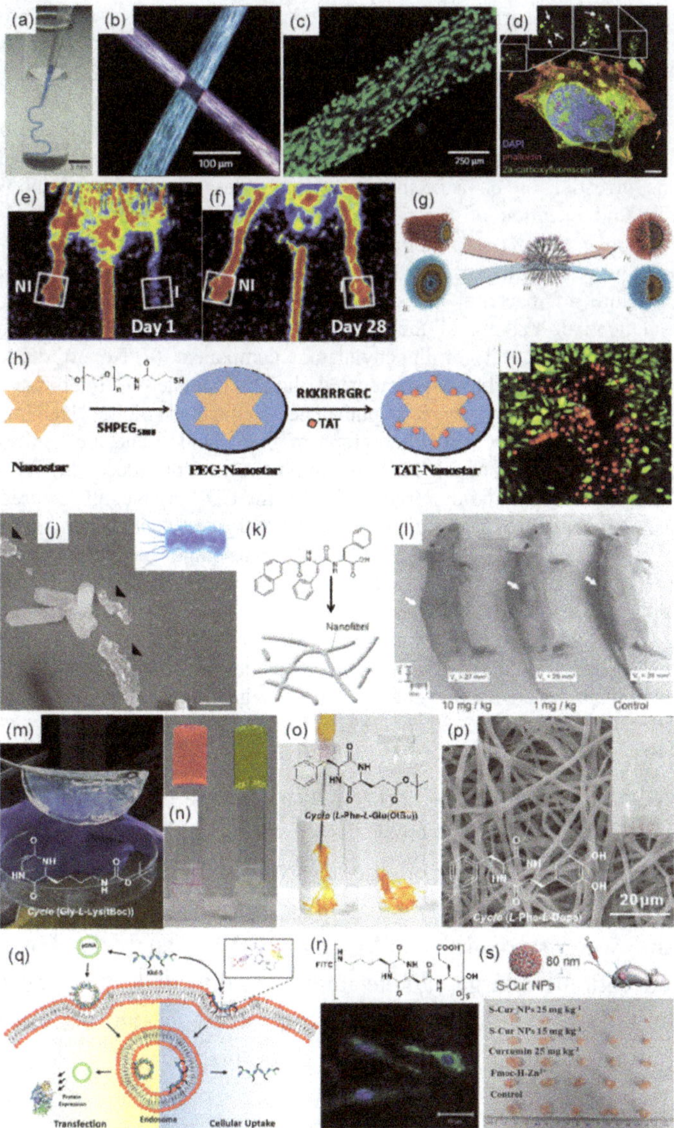

Fig. 3. Peptide-based biomaterials and their applications. (a) Trypan blue loaded peptide amphiphile (PA) solution in PBS forming peptide noodle structures. (b) The uniform alignment of the noodle-like peptide strings were observed by exciting light at the cross polars of the crosspoint of two noodles. (c) Visualization of calcein labeled cultured hMSC cells over the string. (e), (f) Hind limb ischemia treatment with VEGF-mimetic multi-domain peptide showcasing the recovery over a period of 28 days. (g) Schematic

representation of the dendrimers template micelle-forming and vesicle-forming PAs into spherical PRTNs. (h) A schematic of TAT coating over DNA to activate VEGFR1, VEGFR2, and NP-1 receptors followed by the activation of the vasculogenic receptor. The *in vivo* rodent experiments further corroborated the *in vitro* observed results. The study also showed that the peptide hydrogel matrix exhibited good cytocompatibility with the cells and the formation of micro-vessels were observed over a period of 7 days. Therapeutic investigation on a peripheral artery disease mouse model showed tissue regeneration and retention of functions after 28 coated gold nanoparticles and their biomedical applications. (i) Photothermolysis upon treatment of TAT coated gold nanoparticles which results in cell death, for therapeutic applications. (j) Corrugated surface structure of bacterial cells upon treatment with diphenylalanine derived antibacterial peptide, stalking of the bacterial cells with disrupted and compromised membrane denotes bioactivity of diphenylalanine component. (k) Nanofibrils formed by the self-assembly of naphthalene conjugated diphenylalanine. (l) Inhibition of tumor progression upon treatment with naphthalene conjugated diphenylalanine derived nanofibrils in a mice model. (m) Photograph of free-standing and mechanically stable cyclic dipeptide (CDP) organogel. (n) Photograph showing the successful encapsulation of model drug (curcumin) and dye (rhodamine) in the CDP organogels for drug delivery applications. (o) *In situ* hydrogelation of curcumin loaded CDPs upon injecting, into simulated biological fluid, PBS. (p) FESEM micrograph of antioxidant molecule incorporated silk/melanin (SM) electrospun mats. Inset shows the successful sequestration of DPPH radical by antioxidant SM mats. (q) Schematic representation of cellular uptake, DNA delivery and transfection study of CDP-based peptidomimetics. (r) Molecular structure and the corresponding cellular uptake of CDP peptidomimetic, (s) Histidine-curcumin based nano-agent for effective inhibition of glioblastoma growth. Adapted from Refs. [23, 24, 25, 26, 27, 28, 29, 32, 34, 35, 36, 40].

between the CDP unit and the polar head groups of the lipid membrane significantly contributes towards the observed cellular uptake of CDP-peptidomimetics (Figure 3(r)). The selective molecular level interactions between the CDP-core and polar head groups of lipids in the cell membrane were studied by NMR and SAXS analysis. Overall, the optimized cationic charge (lysine units), rigidity and hydrogen bonding (CDP units) of CDP-peptidomimetic resulted in an effective DNA complexation, cellular uptake and a delivery with an appreciable transfection efficiency.

Ghadiri et al reported the synthesis of a set of cyclic *D*, *L*-α-peptides and their structural and functional properties.[37] The abiotic structure of cyclic *D*, *L*-α-peptide makes them resistant to proteases and has been found to exhibit antiviral activity against the HCV virus. Similarly, Gazit and coworkers have shown that at a millimolar concentration, phenylalanine is capable of forming higher ordered fibrillar structures.[39]

The interaction of cystine with TPPS driven by Zn^{2+} mediated self-assembly led to the formation of three-dimensional nanorods, mimicking the grana organelle present in chloroplasts. In another study, they have used histidine, to improve the biological stability of the anticancer drug curcumin.[40] The metal binding activity of histidine has been exploited to construct effective curcumin-based nano-agents. The curcumin nano-agents were prepared in the presence of Fmoc-*L*-histidine in hexafluoroisopropanol (HFIP) and zinc chloride. The *in vivo* measurements showed a progressive decrease in the tumor volume which implies an increase in the bioactivity of curcumin by overcoming degradability and poor tissue penetration (Figure 3(s)).

4.   Nucleic Acid Based Materials

Nucleic acids are versatile biomolecules that serve as building blocks to construct a range of molecular systems and materials with a wide range of applications.[41,42] Nucleic acid-based architectonics is an emerging area in nanotechnology to mimic the natural systems.[42,43] The molecular architectonics of short nucleic acid structures essentially promise to drive the advancements in the area of molecular self-assembly exploiting the structural conformation of the DNA and RNA.[42-44] Nature's conception of designing functional bio-devices has been adapted by synthetic chemists to generate designer molecules with meticulously ordered structures mutually templated with nucleic acid chemistry.[42,45] In this context, we reported the design of small molecule (adenine conjugated naphthalenedimide: BNA) and oligothymidine (dTn) based hybrid DNA ensemble to achieve ultrasensitive detection of mercury in water.[46,47] The specific hydrogen bonding interaction between the BNA and oligothymidine led to the formation of functional 2D nanomaterials which were subsequently used to device an ultrasensitive sensor platform. The competitive binding interaction of mercury led to the strand displacement followed by the formation of metallo-DNA duplex through mercury-thymidine complexation which results in the displacement of BNA. The displacement of BNA led to significant changes in electrical conductivity and chiral properties thus aiding in the

  L. P. Datta, S. Manchineella & T. Govindaraju

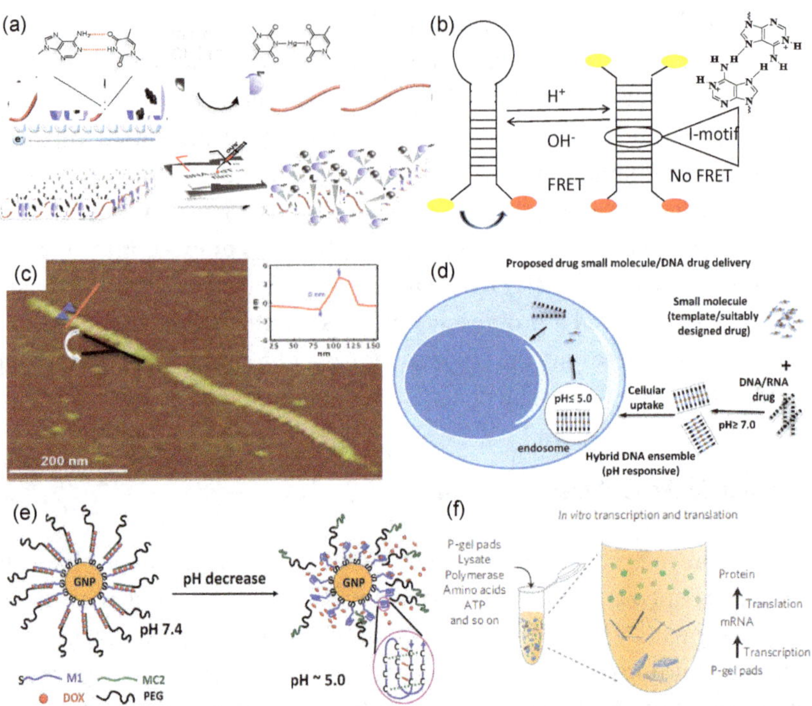

Fig. 4. Nucleic-acid based biomaterials and their applications. (a) Small molecule (BNA)-oligothymine (dTn) based hybrid DNA nano-architecture. Mercury (Hg^{2+}) mediated displacement of BNA from the DNA-BNA ensemble followed by the formation of a metallo-DNA duplex dT-Hg-dT$_n$, displacement of BNA and FET-based device to detect Hg. Adapted from Refs. [46, 49, 5, 52]. (b) FRET based DNA nanodevice (DNA switch) to acidic pH. (c) AFM micrograph of left handed DNA ensemble of dT20-APA-dT20. (d) A schematic of proposed DNA ensemble mediated drug delivery. (e) Schematic representation of pH responsive delivery of doxorubicin from a PEGylated gold nanoparticle system, under the acidic condition i-motif formation triggers the breakage of interaction between M1-MC2 DNA sequences followed by the release of cargo moieties. (f) A schematic of *in vitro* transcription and translation in the protein producing gel synthesized from an X shaped DNA. BNA: adenine conjugated naphthalenedimide, APA: adenine conjugated perylenediimide.

rapid and sub-nanomolar (0.1 nM, 0.02 ppb) detection of organic- and inorganic mercury (Figure 4(a)). The nanometer-sized architectures can be precisely tuned for site-selective functionalization and device fabrication.

In living organisms, several physiological functions have been regulated by an intracellular pH. Within a cell, there are regions and organelles with a variable pH controlled by both endocytic and exocytosis processes. The lysosomes (pH 4 to 5) and endosomes (pH 5 to 6) are acidic in nature which is required for the action of many enzymes. Similarly, cells with anomalous activities or under disease conditions like cancer are found to exhibit an acidic pH. Therefore, devising adequate, precise and sensitive approaches and means to monitor the change of pH in intracellular fluids and organelles is necessary. There are certain limitations with the conventional methods to employ them as monitoring pH sensitive electrodes which includes poor sensitivity to measure small changes in the acidic pH range. In this regard, designer DNA structures can be exploited as an alternative to build nanodevices and sensors for pH sensing.[45] The structural sensitivity of DNA to pH especially the protonation-deprotonation driven conformational changes is used to design molecular devices for pH sensing applications. We have fabricated a pH-responsive molecular beacon (MB)-based DNA nanoswitch as shown in Figure 4(b). [45, 48] For the construction of the DNA nanoswitch, MB was designed which contains an A-rich loop and a mixed sequence stem. Under variable acidic conditions (pH 5 to 3), the nanoswitch can transform between MB (closed state) and A-motif (open state) and a right-handed parallel duplex conformation. The fluorescence dye combination of Cy3 and Cy5 are positioned appropriately to facilitate FRET between them in a closed state which is disturbed in the open state. Upon lowering the pH, the protonation of adenine bases change the conformation from a close to an open state owing to the formation of A-motif supported by the reverse Hoogsteen $AH^+–H^+A$ hydrogen bonding interactions. The application of reversible protonation-deprotonation driven DNA nanoswitch for the sensing of acidic pH with small step-size (0.2-0.3) was demonstrated in an artificial vesicle as well as in live HeLa cells. In another distinct design, we reported the construction of a hybrid DNA ensemble of adenine conjugated perylenediimide (APA) and oligonucleotide (dBn) through double zipper helical assembly supported by unconventional hydrogen bonding between adenine of APA (small molecule template) and nucleobases of oligonucleotide sequence.[49] The AFM micrograph of $dT_{20}-(APA)_{20}-dT_{20}$ showed the formation of a DNA hybrid ensemble (Figure 4(c)). It is proposed that these hybrid DNA ensembles can be exploited for pH-dependent release of the gene and small molecules inside the cells under an acidic pH condition (Figure 4(d)). This minimalistic approach to construct a hybrid DNA ensemble of small molecule template (APA) and oligonucleotide may open up new possibilities in the area of DNA molecular architectonics and nanoarchitectonics and in the development of molecular devices.

Krishnan group has reported DNA based nanomachines as a pH sensor inside the endosomes.[50] Zhou group has reported a DNA based gold nanoparticle system for the pH-responsive release of drug moieties within cancerous cells.[51] The PEGylation of the DNA based gold nanoparticle system results in an increased serum stability, effectivity in delivering the entrapped anticancer drug doxorubicin and suppressed nonspecific protein adsorption (Figure 4(e)). The translation of RNA to proteins is one of the unique capabilities of the cellular machinery. In another report, a unique hydrogel system was constructed utilizing X shaped DNA molecules as crosslinkers.[52] The system is capable of transcribing the genes into RNA followed by translation into proteins. The transcription and translation process monitored via protein expression studies revealed that the surface area of cell-free gel system is an important parameter to modulate reaction conditions (Figure 4(f)). The present DNA based hydrogel biomaterial system for the acellular protein synthesis exemplifies the possibilities of designing novel biomaterials for biomedical applications.

5. Carbohydrate-Based Biomaterials

The strategic presence of carbohydrates in the structural organizations of the cell makes them crucial biomolecules with specific functions. Carbohydrates are also utilized as potential bio-cues in stem cell therapy and in the design of vaccine formulations.[53] Yang and Li coworkers have documented the conjugation of folic acid with dextran to generate a model antitumor delivery agent (Figure 5(a)).[54] Folic acid is over expressed in the cancer cells and the integration of dextran with folic acid results in the formation of stimuli-responsive and targeted self-assembled nanostructures. The anticancer drug doxorubicin has been encapsulated within the nano-assembly by exploiting the weak interactions between folic acid and doxorubicin. Along with a tailored anticancer activity, the system can alleviate the adverse side effects of doxorubicin and provides prolonged median survival rates as evident from the pharmacokinetic study. Glycans are an important class of carbohydrate-based biomolecule that play an important role in potentiating the signaling cascades and protect proteins from protease susceptibility. Stupp and coworkers reported decoration of supramolecular sulfated glycopeptide nanostructures, exhibiting a trisulfated monosaccharide on their surfaces that bind five critical proteins with different polysaccharide-binding domains.[55] The filamentous glycopeptide nanostructures exhibit high affinity to bind several heparin-binding proteins that act as important growth factors in the cellular signaling cascades. In comparison with heparin sulfate, the supramolecular interaction of the glycopeptide nanostructures have amplified the signaling of bone morphogenetic protein (BMP) and promoted the regeneration of the bone at a significantly lower dose of the protein. The immunosuppressive activity of mesenchymal stem cells (MSCs) makes them highly active therapeutic agents in regenerative medicine. For the efficient migration of MSCs at the inflammation site, the hyaluronic acid coated activation system has been generated to increase the expression of CD44 (hyaluronic acid receptor) on culture substrates.[56] Receptor activation leads to a significant increase in the homing potential of MSCs at the inflammation site. To design such an activation system, tissue culture plates were coated with hyaluronic acid (HA) and the CD44 expression was evaluated (Figure 5(b)). The immunohistochemistry analysis suggested that MSCs on HA-treated plates generated a strong anti-inflammatory response. For *in vivo* measurements, the lipopolysaccharide (LPS)-induced inflamed ear murine model was used by making inflammation in the ear of a mouse. The systematic administration of MSCs harvested from the HA coated tissue culture substrates at the inflammation site significantly increased the anti-inflammatory response validating the design efficacy. The mechanical and optoelectronic properties of single-walled carbon nanotubes (SWNTs) make them highly attractive materials for applications such as bioimaging, biosensing and cancer therapeutics.

However, the utility of SWNTs as a biomaterial is limited due to poor biocompatibility. In this context, the surface modification of SWNTs with a suitable biomolecule may impart the essential biocompatibility feature.

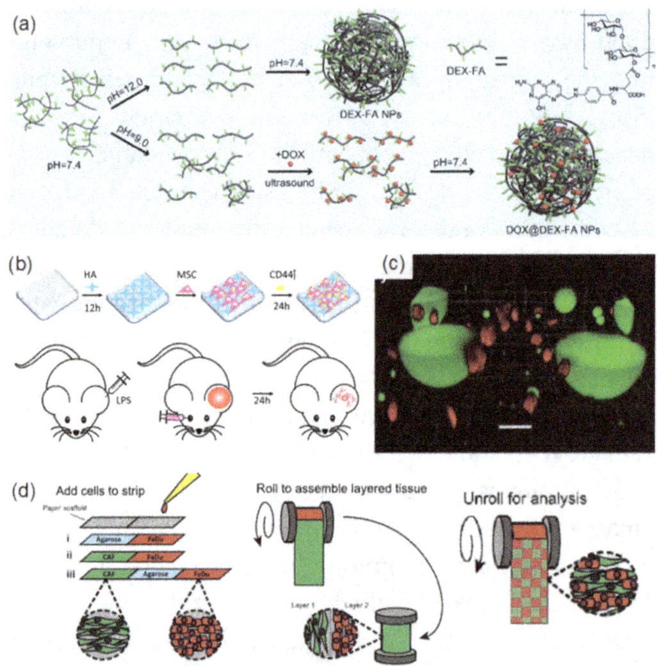

Fig. 5. Carbohydrate-based biomaterials and their applications. (a) Schematic illustration for the development of Dextran-folic acid NPs for the encapsulation and delivery of doxorubicin. Adapted from Refs. [54, 56, 59, 60]. (b) Schematic representation of hyaluronic acid coating, hMSC seeding, proliferation and CD44 expression study. (c) Confocal fluorescence microscopy images of FITC loaded PDDA/CMDX coacervate microdroplets after mixing with a dispersion of intact chloroplasts (red fluorescence). (d) A schematic of tissue roll for the analysis of cellular environment and response (TRACER) system for the *in vitro* 3D culture.

Bio-functionalization with glycopolymers is one of the highly promising approaches to improve the biocompatibility of SWNTs.[57] Functionalized SWNTs exhibit improved biocompatibility and cellular attachment behavior in comparison to pristine unmodified SWNTs. In recent years, decoration of carbohydrate-based compartmentalized proto-cellular structures using the microfluidic strategies has been documented by several research groups. Huck and Deng coworkers have fabricated PEG-dextran based coacervate microdroplets in monodisperse lysosomes utilizing the microfluidic strategy.[58] The potential of the coacervate microdroplets as artificial organelles have been further evaluated. In this context, sequestration of cells, biomolecules, organelles and maintenance of their activities within the artificial compartment is an important challenge to develop new technologies applicable in the area of bio-catalysis and regenerative medicine developments. Mann et al. reported a novel approach to sequester biological organelles within carbohydrate based coacervate microdroplets.[59]

The coacervate microdroplets were prepared by mixing both the synthetic polymer and biopolymer poly (diallyldimethylammonium chloride) (PDDA) and carboxymethyl-dextran (CMDX). The extracted chloroplast was integrated within the interior of the coacervate microdroplets employing weak electrostatic interactions. The surface wettability represents the physiological environment for the uptake and retention of intact chloroplast moieties within the carbohydrate-based coacervates. After sequestration, partitioning of the Hill reagent, 2,6-dichlorophenolindophenol (DPIP) within the chloroplast conjugated microdroplets was visualized (Figure 5(c)). The integrity and mechanism of the chloroplast mimetics were confirmed after exposure to light which resulted in the reduction of DPIP. The reduction of DPIP indicates that the electron transport chain is operational within the accumulated chloroplasts that facilitate the flow of electron from photosystem II to plastocyanin (photosystem I). McGuigan et al. introduced a cellulose hydrogel based novel tumor model platform termed as tissue roll for the analysis of cellular environment and response (TRACER) system to probe the tumor-stroma interactions.[60] In this particular system, the cells were seeded within cellulose scaffold hydrogel and the resulting hydrogel strip was rolled onto a cylindrical core which mimics ECM tissue environments in a 3D fashion (Figure 5(d)). The designed 3D culture system helps to quantify phenotypic changes in tumor cells upon cancer cell and cancer-associated fibroblast (CAF) interactions. In brief, the TRACER system helps to determine the effects of CAFs over the cancer cell progression by the cell mapping methodology.

6. Fatty Acid and Lipid Derived Biomaterials

Fatty acids and lipids are the class of amphiphilic biomolecules utilized in the development of functional biomaterials. Mann et al. contributed significantly to the development of fatty acid-based cell-like assemblies. In one of the designs, multi-compartmentalized structure was constructed by the encapsulation of glucose oxidase-containing proteinosomes in pH-responsive fatty acid based coacervate microdroplets.[61] The structural features of the nested proto-cell models were modulated by the internally generated enzymatic cascade. Cationic liposomes mainly synthesized from DC-Chol (3β-N-(N',N'-dimethyl amino ethane) carbamoylcholesterol), DOTAP (1,2-dioleoyl-3-trimethyl ammonium propane) and DOTMA (N-(2,3-dioleoyloxy)propyl-N, N, N-trimethylammonium chloride) are approved fatty acid-based gene delivery systems.[62] DOPE (3β-(N-(N',N'-dimethylaminoethane) carbarmoyl)-cholesterol)/dioleoylphosphatidyl ethanolamine is an important cationic liposome vector used for the sub-cutaneous injection to deliver cargo moieties. The complexation of allovectin-7 that encodes the gene of HLA-B7 and β2-microglobulin with a lipid mixture DMRIE (1,2-dimyristyloxypropyl-3-dimethyl hydroxyethyl ammonium bromide)/ DOPE generate an effective delivery combination which is under phase II clinical trial.[63] The hydroxyl groups in DMRIE facilitate the interactions with the DNA which results in the stabilization of DNA-liposome complexes and an improved uptake into the cell. For efficient delivery of drugs, the lipid formulation classification system (LFC) was introduced in the year of 2000 and further modified in 2006.[64, 65] Apart from liposomes and LFCs, solid lipid nanoparticles (SLNs) have been recently explored as lipid-based nanovectors, which consist of a solid lipid core stabilized by surfactants.[66] DOTAP based SLNs have been demonstrated as efficient vectors for the complexation of mRNA as well as to protect it from enzymatic degradation followed by targeted delivery. The lipid-based SLN complex was found to be aggregated in the presence of serum albumin and their efficient cellular uptake was monitored in Kupffer cells. Solid lipid nanocapsules (SLCs) are the second generation lipid-based vectors which consist of a liquid lipid core phase surrounded by a solid lipid shell at a physiological temperature.[67] SLCs are preferred delivery vehicles as they exhibit high encapsulation efficiency and controlled drug release characteristics.

7. Conclusion

The development of various biomolecule integrated materials systems and their biomedical applications have already been discussed in this chapter. A number of research groups around the globe are working towards such biomaterial development with a multitude of applications. Naturally occurring biomolecules are integrated within diverse synthetic materials like polymers, nanomaterials, ceramics or gel like macrostructures to develop smart materials with responsive features alongside the requisite compatibility and controlled degradation properties. In particular, the hurdles associated with implantation, controllable delivery and regenerative medicines demands the need for functional materials exhibiting good compatibility with the living system. Moreover, the field of biomaterials has evolved from being a mere inert material support to the current state of multifunctional smart materials towards the repair or replacement of malfunctioning tissues and organs. In this context, umpteen numbers of strategies have been developed in the last two decades for the development of functional biomaterials with the aid of cutting-edge technologies. Inspired by nature's elegant architectural design of biological systems employing discrete and a combination of biomolecules, synthetic biohybrid systems have been designed, synthesized and employed to fabricate smart biomaterials. As a result, a growing number of medical transplants, sensors, devices, delivery vehicles and biochips (lab-on-a-chip) have been developed based on "bioinert structures", which can dynamically interact with cells and tissues both *in vitro* and *in vivo*. In a definitive way, rational designs of biomolecule integrated materials systems are important in the ever-growing field to mimic natural systems with high throughput biomedical technologies.

Acknowledgments

We thank Prof. C. N. R. Rao FRS for his constant support and encouragement, JNCASR, the DST-Nanomission (grant: DST/SMS/4428 or SR/NM/TP-25/2016), the Government of India, and Sheikh Saqr Laboratory (SSL), ICMS-JNCASR for financial support.

References

1. M. W. Tibbitt, C. B. Rodell, J. A. Burdick, and K. S. Anseth. Progress in material design for biomedical applications, *Proc. Natl. Acad. Sci. U. S. A.* **112**, 14444-14451 (2015).
2. S.-B. Park, E. Lih, K.-S. Park, Y. K. Joung, and D. K. Han. Biopolymer-based functional composites for medical applications, *Prog. Polym. Sci.* **68**, 77-105 (2017).
3. F. G. Omenetto, and D. L. Kaplan. New opportunities for an ancient material, *Science* **329**, 528-531 (2010).
4. D. G. Anderson, J. A. Burdick, and R. Langer. Smart biomaterials, *Science* **305**, 1923-1924 (2004).
5. D. Hofmann, M. Entrialgo-Castaño, K. Kratz, and A. Lendlein. Knowledge-based approach towards hydrolytic degradation of polymer-based biomaterials, *Adv. Mater. (Weinheim, Ger.)* **21**, 3237-3245 (2009).
6. M. Geetha, A. K. Singh, R. Asokamani, and A. K. Gogia. Ti based biomaterials, the ultimate choice for orthopaedic implants–a review, *Prog. Mater. Sci.* **54**, 397-425 (2009).
7. A. V. Bryksin, A. C. Brown, M. M. Baksh, M. Finn, and T. H. Barker. Learning from nature–novel synthetic biology approaches for biomaterial design, *Acta Biomater.* **10**, 1761-1769 (2014).
8. T. Sun, G. Qing, B. Su, and L. Jiang. Functional biointerface materials inspired from nature, *Chem. Soc. Rev.* **40**, 2909-2921 (2011).
9. S. Cavalli, F. Albericio, and A. Kros. Amphiphilic peptides and their cross-disciplinary role as building blocks for nanoscience, *Chem. Soc. Rev.* **39**, 241-263 (2010).
10. M. B. Avinash, and T. Govindaraju. Nanoarchitectonics of biomolecular assemblies for functional applications, *Nanoscale* **6**, 13348-13369 (2014).
11. X. Wang, H. J. Kim, C. Wong, C. Vepari, A. Matsumoto, and D. L. Kaplan. Fibrous proteins and tissue engineering, *Mater. Today* **9**, 44-53 (2006).
12. S. Manchineella, G. Thrivikraman, B. Basu, and T. Govindaraju. Surface-functionalized silk fibroin films as a platform to guide neuron-like differentiation of human mesenchymal stem cells, *ACS Appl. Mater. Interf.* **8**, 22849-22859 (2016).

13. S. Levenberg, J. Rouwkema, M. Macdonald, E. S. Garfein, D. S. Kohane, D. C. Darland, R. Marini, C. A. Van Blitterswijk, R. C. Mulligan, and P. A. D'Amore. Engineering vascularized skeletal muscle tissue, *Nat. Biotechnol.* **23**, 879 (2005).

14. S. Manchineella, G. Thrivikraman, K. K. Khanum, P. C. Ramamurthy, B. Basu, and T. Govindaraju. Pigmented silk nanofibrous composite for skeletal muscle tissue engineering, *Adv. Healthcare Mater.* **5**, 1222-1232 (2016).

15. L.-M. Rauschendorfer. Pigments in tissue engineering, *ChemistryViews.* DOI: 10.1002/adhm.201501066.

16. M. Nune, S. Manchineella, T. Govindaraju, and K. Narayan. Melanin incorporated electroactive and antioxidant silk fibroin nanofibrous scaffolds for nerve tissue engineering, *Mater. Sci. Eng., C* **94**, 17-25 (2019).

17. M. Vargas, C. Pastor, A. Chiralt, D. J. McClements, and C. González-Martínez. Recent advances in edible coatings for fresh and minimally processed fruits, *Crit. Rev. Food Sci. Nutr.* **48**, 496-511 (2008).

18. B. Marelli, M. Brenckle, D. L. Kaplan, and F. G. Omenetto. Silk fibroin as edible coating for perishable food preservation, *Sci. Rep.* **6**, 25263 (2016).

19. P. H. Zeplin, N. C. Maksimovikj, M. C. Jordan, J. Nickel, G. Lang, A. H. Leimer, L. Römer, and T. Scheibel. Spider silk coatings as a bioshield to reduce periprosthetic fibrous capsule formation, *Adv. Funct. Mater.* **24**, 2658-2666 (2014).

20. M. Rizwan, G. S. Peh, H.-P. Ang, N. C. Lwin, K. Adnan, J. S. Mehta, W. S. Tan, and E. K. Yim. Sequentially-crosslinked bioactive hydrogels as nano-patterned substrates with customizable stiffness and degradation for corneal tissue engineering applications, *Biomaterials* **120**, 139-154 (2017).

21. X. Zhao, F. Pan, H. Xu, M. Yaseen, H. Shan, C. A. Hauser, S. Zhang, and J. R. Lu. Molecular self-assembly and applications of designer peptide amphiphiles, *Chem. Soc. Rev.* **39**, 3480-3498 (2010).

22. E. Sleep, B. D. Cosgrove, M. T. McClendon, A. T. Preslar, C. H. Chen, M. H. Sangji, C. M. R. Pérez, R. D. Haynes, T. J. Meade, H. M. Blau, and S. I. Stupp. Injectable biomimetic liquid crystalline scaffolds enhance muscle stem cell transplantation, *Proc. Natl. Acad. Sci.* **114**, E7919-E7928 (2017).

23. E. J. Berns, Z. Álvarez, J. E. Goldberger, J. Boekhoven, J. A. Kessler, H. G. Kuhn, and S. I. Stupp. A tenascin-C mimetic peptide amphiphile nanofiber gel promotes neurite outgrowth and cell migration of neurosphere-derived cells, *Acta Biomater.* **37**, 50-58 (2016).

24. D. Kalafatovic, M. Nobis, J. Son, K. I. Anderson, and R. V. Ulijn. MMP-9 triggered self-assembly of doxorubicin nanofiber depots halts tumor growth, *Biomaterials* **98**, 192-202 (2016).

25. V. A. Kumar, Q. Liu, N. C. Wickremasinghe, S. Shi, T. T. Cornwright, Y. Deng, A. Azares, A. N. Moore, A. M. Acevedo-Jake, N. R. Agudo, S. Pan, D. G. Woodside, P. Vanderslice, J. T. Willerson, R. A. Dixon, and J. D. Hartgerink. Treatment of hind limb ischemia using angiogenic peptide nanofibers, *Biomaterials* **98**, 113-119 (2016).

26. B. F. Lin, R. S. Marullo, M. J. Robb, D. V. Krogstad, P. Antoni, C. J. Hawker, L. M. Campos, and M. V. Tirrell. De Novo Design of bioactive protein-resembling nanospheres via dendrimer-templated peptide amphiphile assembly, *Nano Lett.* **11**, 3946-3950 (2011).

27. H. Yuan, A. M. Fales, and T. Vo-Dinh. TAT peptide-functionalized gold nanostars: enhanced intracellular delivery and efficient NIR photothermal therapy using ultralow irradiance, *J. Am. Chem. Soc.* **134**, 11358-11361 (2012).

28. L. Schnaider, S. Brahmachari, N. W. Schmidt, B. Mensa, S. Shaham-Niv, D. Bychenko, L. Adler-Abramovich, L. J. Shimon, S. Kolusheva, and W. F. DeGrado. Self-assembling dipeptide antibacterial nanostructures with membrane disrupting activity, *Nat. Commun.* **8**, 1365 (2017).

29. Y. Kuang, X. Du, J. Zhou, and B. Xu. Supramolecular nanofibrils inhibit cancer progression in vitro and in vivo, *Adv. Healthcare Mater.* **3**, 1217-1221 (2014).

30. S. Manchineella, and T. Govindaraju. Molecular self-assembly of cyclic dipeptide derivatives and their applications, *ChemPlusChem* **82**, 88-106 (2017).

31. T. Govindaraju. Spontaneous self-assembly of aromatic cyclic dipeptide into fibre bundles with high thermal stability and propensity for gelation, *Supramol. Chem.* **23**, 759-767 (2011).

32. S. Manchineella, and T. Govindaraju. Hydrogen bond directed self-assembly of cyclic dipeptide derivatives: gelation and ordered hierarchical architectures, *RSC Adv.* **2**, 5539-5542 (2012).

33. T. Govindaraju, M. Pandeeswar, K. Jayaramulu, G. Jaipuria, and H. S. Atreya. Spontaneous self-assembly of designed cyclic dipeptide (Phg-Phg) into two-dimensional nano- and mesosheets, *Supramol. Chem.* **23**, 487-492 (2011).

34. S. Manchineella, N. A. Murugan, and T. Govindaraju. Cyclic dipeptide-based ambidextrous supergelators: minimalistic rational design, structure-gelation studies, and in situ hydrogelation, *Biomacromolecules* **18**, 3581-3590 (2017).

35. S. Manchineella, C. Voshavar, and T. Govindaraju. Radical-scavenging antioxidant cyclic dipeptides and silk fibroin biomaterials, *Eur. J. Org. Chem.* **2017**, 4363-4369 (2017).

36. C. Madhu, C. Voshavar, K. Rajasekhar, and T. Govindaraju. Cyclic dipeptide based cell-penetrating peptidomimetics for effective DNA delivery, *Org. Biomol. Chem.* **15**, 3170-3174 (2017).

37. A. Montero, P. Gastaminza, M. Law, G. Cheng, F. V. Chisari, and M. R. Ghadiri. Self-assembling peptide nanotubes with antiviral activity against hepatitis C virus, *Chem. Biol.* **18**, 1453-1462 (2011).

38. D. M. Ryan, S. B. Anderson, F. T. Senguen, R. E. Youngman, and B. L. Nilsson. Self-assembly and hydrogelation promoted by F 5-phenylalanine, *Soft Matter* **6**, 475-479 (2010).

39. L. Adler-Abramovich, L. Vaks, O. Carny, D. Trudler, A. Magno, A. Caflisch, D. Frenkel, and E. Gazit. Phenylalanine assembly into toxic fibrils suggests amyloid etiology in phenylketonuria, *Nat. Chem. Biol.* **8**, 701 (2012).

40. X. Yan, Y. Li, Q. Zou, C. Yuan, S. Li, and R. Xing. Amino acid coordination-driven self-assembly for enhancing both biological stability and tumor accumulation of curcumin, *Angew. Chem.*, **130**(52), 17330-17334 (2018).

41. N. C. Seeman. Nucleic acid junctions and lattices, *J. Theor. Biol.* **99**, 237-247 (1982).

42. T. Govindaraju. (2019) Templated DNA nanotechnology — functional DNA nanoarchitectonics. Govindaraju, T. ed. New York: Pan Stanford.

43. B. Roy, D. Ghosh, and T. Govindaraju. (2019) Functional molecule templated DNA nanoarchitectures. In: *Templated DNA Nanotechnology: Functional DNA Nanoarchitectonics*. Govindaraju, T. ed. New York: Pan Stanford.

44. M. B. Avinash, and T. Govindaraju. Architectonics: design of molecular architecture for functional applications, *Acc. Chem. Res.* **51**, 414-426 (2018).

45. R. Madhu, B. Roy, and T. Govindaraju. (2019) DNA-based nanoswitches and devices. In: *Templated DNA Nanotechnology: Functional DNA Nanoarchitectonics*. Govindaraju, T. ed. New York: Pan Stanford.

46. A. K. Dwivedi, M. Pandeeswar, and T. Govindaraju. Assembly modulation of PDI derivative as a supramolecular fluorescence switching probe for detection of cationic surfactant and metal ions in aqueous media, *ACS Appl. Mater. Interfaces* **6**, 21369-21379 (2014).

47. P. Makam, R. Shilpa, A. E. Kandjani, S. R. Periasamy, Y. M. Sabri, C. Madhu, S. K. Bhargava, and T. Govindaraju. SERS and fluorescence-based ultrasensitive detection of mercury in water, *Biosens. Bioelectron.* **100**, 556-564 (2018).

48. N. Narayanaswamy, R. R. Nair, Y. V. Suseela, D. K. Saini, and T. Govindaraju. A molecular beacon-based DNA switch for reversible pH sensing in vesicles and live cells, *Chem. Commun.* **52**, 8741-8744 (2016).

49. N. Narayanaswamy, G. Suresh, U. D. Priyakumar, and T. Govindaraju. Double zipper helical assembly of deoxyoligonucleotides: mutual templating and chiral imprinting to form hybrid DNA ensembles, *Chem. Commun.* **51**, 5493-5496 (2015).

50. S. Modi, C. Nizak, S. Surana, S. Halder, and Y. Krishnan. Two DNA nanomachines map pH changes along intersecting endocytic pathways inside the same cell, *Nat. Nanotechnol.* **8**, 459 (2013).

51. L. Song, V. H. B. Ho, C. Chen, Z. Yang, D. Liu, R. Chen, and D. Zhou. Efficient, pH-triggered drug delivery using a pH-responsive DNA-conjugated gold nanoparticle, *Adv. Healthcare Mater.* **2**, 275-280 (2013).

52. N. Park, S. H. Um, H. Funabashi, J. Xu, and D. Luo. A cell-free protein-producing gel, *Nat. Mater.* **8**, 432 (2009).

53. J. Hu, P. H. Seeberger, and J. Yin. Using carbohydrate-based biomaterials as scaffolds to control human stem cell fate, *Org. Biomol. Chem.* **14**, 8648-8658 (2016).

54. Y. Tang, Y. Li, R. Xu, S. Li, H. Hu, C. Xiao, H. Wu, L. Zhu, J. Ming, and Z. Chu. Self-assembly of folic acid dextran conjugates for cancer chemotherapy, *Nanoscale* **10**, 17265-17274 (2018).

55. S. S. Lee, T. Fyrner, F. Chen, Z. Álvarez, E. Sleep, D. S. Chun, J. A. Weiner, R. W. Cook, R. D. Freshman, and M. S. Schallmo. Sulfated glycopeptide nanostructures for multipotent protein activation, *Nat. Nanotechnol.* **12**, 821 (2017).

56. B. Corradetti, F. Taraballi, J. O. Martinez, S. Minardi, N. Basu, G. Bauza, M. Evangelopoulos, S. Powell, C. Corbo, and E. Tasciotti. Hyaluronic acid coatings as a simple and efficient approach to improve MSC homing toward the site of inflammation, *Sci. Rep.* **7**, 7991 (2017).

57. P. Wu, X. Chen, N. Hu, U. C. Tam, O. Blixt, A. Zettl, and C. R. Bertozzi. Biocompatible carbon nanotubes generated by functionalization with glycodendrimers, *Angew. Chem. Int. Ed. Engl.* **47**, 5022-5025 (2008).

58. N. N. Deng, and W. T. Huck. Microfluidic formation of monodisperse coacervate organelles in liposomes, *Angew. Chem.* **129**, 9868-9872 (2017).

59. B. P. Kumar, J. Fothergill, J. Bretherton, L. Tian, A. J. Patil, S. A. Davis, and S. Mann. Chloroplast-containing coacervate micro-droplets as a step towards photosynthetically active membrane-free protocells, *Chem. Commun.* **54**, 3594-3597 (2018).

60. M. Young, D. Rodenhizer, T. Dean, E. D'Arcangelo, B. Xu, L. Ailles, and A. P. McGuigan. A TRACER 3D co-culture tumour model for head and neck cancer, *Biomaterials* **164**, 54-69 (2018).

61. N. Martin, J.-P. Douliez, Y. Qiao, R. Booth, M. Li, and S. Mann. Antagonistic chemical coupling in self-reconfigurable host–guest protocells, *Nat. Commun.* **9**, 3652 (2018).

62. I. Brigger, C. Dubernet, and P. Couvreur. Nanoparticles in cancer therapy and diagnosis, *Adv. Drug Delivery Rev.* **54**, 631-651 (2002).

63. A. T. Stopeck, A. Jones, E. M. Hersh, J. A. Thompson, D. M. Finucane, J. C. Gutheil, and R. Gonzalez. Phase II study of direct intralesional gene transfer of Allovectin-7, an HLA-B7/β2-microglobulin DNA-Liposome complex, in patients with metastatic melanoma, *Clin. Cancer Res.* **7**, 2285-2291 (2001).

64. H. Shrestha, R. Bala, and S. Arora. Lipid-based drug delivery systems, *J. Pharm.* **2014**, 801820-801820 (2014).

65. C. W. Pouton, and C. J. H. Porter. Formulation of lipid-based delivery systems for oral administration: Materials, methods and strategies, *Adv. Drug Delivery Rev.* **60**, 625-637 (2008).

66. R. H. Müller, M. Radtke, and S. A. Wissing. Solid lipid nanoparticles (SLN) and nanostructured lipid carriers (NLC) in cosmetic and dermatological preparations, *Adv. Drug Delivery Rev.* **54**, S131-S155 (2002).

67. W. Mehnert, and K. Mäder. Solid lipid nanoparticles: production, characterization and applications, *Adv. Drug Delivery Rev.* **64**, 83-101 (2012).

Chapter 9

The Colloidal Glass Transition

Manodeep Mondal[*] and Rajesh Ganapathy[†]

*Chemistry and Physics of Materials Unit, International Centre for
Materials Science and School of Advanced Materials
Jawaharlal Nehru Centre for Advanced Scientific Research
Jakkur, Bangalore 560064, Karnataka, India*
[*]manodeep@jncasr.ac.in [†]rajeshg@jncasr.ac.in

The microscopic underpinnings of how flowing liquids transform into
rigid glasses upon supercooling continues to elude our grasp. In this
chapter, we highlight contributions from experiments on colloidal suspensions that have helped advance our understanding of this problem.
This advance is in a large part due to our ability to probe and manipulate colloidal particles at the single-particle level. We begin with a
brief introduction on how the "Colloids as Model Atoms" paradigm has
been vital in literally bringing into focus an array of condensed matter phenomena. We then elaborate on the utility of these experiments
in probing glass transition phenomenology. We trace how these experiments, which started out more as an exploratory approach have now
reached a stage that allows us to discern between competing theories
of the glass transition. We conclude by emphasizing how dynamical
crossovers which correspond to quantitative changes in relaxation have
aided in this quest.

1. Introduction

Archaeological evidence suggests that the earliest known man-made glass
dates back to circa 3500 BC. Although substantial breakthroughs have been
made in manipulating the glassy state to suit our needs since then, the microscopic physics behind the liquid-glass transition is anything but clear.[1-4]
Like many other great unsolved problems in the natural sciences, the glass
transition problem is deceptively easy to state: why do glasses refuse to flow
like liquids despite being structurally similar to one? Unlike the other fa-

199

miliar phase transitions where a well-defined order parameter exists to help distinguish between the various phases, the liquid-glass transition lacks one or so it seems. A standard line of attack to tackle this problem has been to approach the glass form the liquid side.[2,4–9] When a liquid is cooled rapidly enough to bypass crystallization, it enters the supercooled regime where the liquid is metastable with respect to the crystal. Upon further cooling molecular relaxations become increasingly slow and is accompanied by a concomitant increase in viscosity. The relaxation time eventually exceeds the experimental duration at the glass transition temperature T_g and the liquid falls out-of-equilibrium. The viscosity of the liquid is very large for $T < T_g$ that for all practical purposes the material behaves essentially like a solid. This non-ergodic state is the glass.

Understanding glass formation is a formidable physics challenge since the rapid growth in relaxation times makes it impossible to equilibrate the supercooled liquid near the glass transition.[5] Measurements of the changes in various physical quantities like the specific-heat capacity at a constant pressure, c_P,[10] viscosity, η, and the relaxation time, τ,[7,11] on approaching T_g in atomic and molecular liquids have accumulated over the decades. In some cases the relaxation time data is available over nearly fourteen orders-of-magnitude.[12] An illuminating way to capture the how different liquids approach their glass transitions is through the Angell plot - a plot of $\log(\eta)$ versus T_g/T[8] first introduced by Austen Angell.

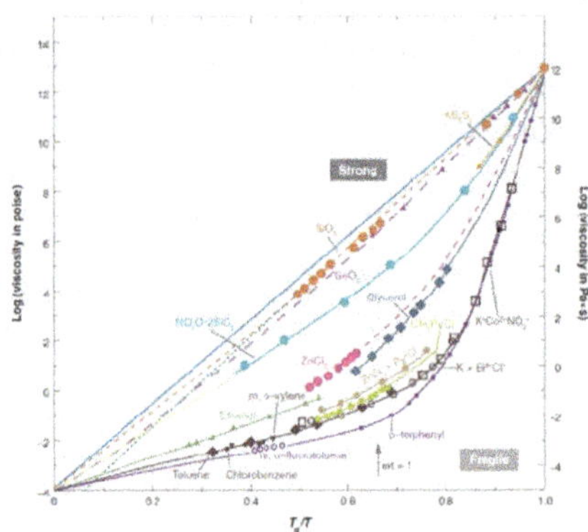

Fig. 1.   The Angell plot for various glass-forming liquids. Logarithm of the viscosity η plotted as a function of T_g/T, where T_g is the glass transition temperature. Adapted from.[13]

Figure 1 clearly shows that the rapidity with which the viscosity increases on approaching T_g for different liquids is remarkably different. In liquids like germanium oxide (GeO_2), the viscosity increases with supercooling in an almost Arrhenius fashion and these are called 'strong' glass formers. While in liquids like benzene or o-terphenyl the viscosity increases in a super-Arrhenius manner and these are called as 'fragile' glass formers. On a Angell plot, the slope of the viscosity curve at $T = T_g$ - termed the kinetic fragility m - quantifies the departure from Arrhenius behavior,[14]

$$m = \left(\frac{\partial \log_{10} \eta}{\partial (T_g/T)} \right)_{T=T_g}.$$ (1)

The dependence of viscosity on temperature can be described by an empirical expression of the form $\eta = \eta_0 \exp(E(T)/T)$, where $E(T)$ is a temperature-dependent activation barrier. In strong glass formers, $E(T)$ is almost temperature independent and molecular relaxations involve the crossing of a single activation barrier. Indeed, the activation barriers estimated from the Angell plot for prototypical strong glass formers is consistent with the covalent nature of bonds in these substances. For fragile glass formers on the other hand, $E(T)$ is strongly temperature-dependent and increases with supercooling. This suggests that the activation barrier in these materials involves the cooperative rearrangement of many molecules with the size of these cooperatively rearranging regions (CRRs) increasing with supercooling. In fact, the activation barrier at $T = T_g$ for fragile glass formers is substantially larger than typically measured bond energies in organic liquids. Although it remains highly debated whether the rapid growth of relaxation times is a direct consequence of the growing size of CRRs, it is indeed quite alluring to draw parallels between glass transition phenomenology and equilibrium critical phenomena. Thus, it comes as no surprise that the bulk of the research on the glass transition has been on fragile glass formers.

Given the wealth of available relaxation time data in atomistic and molecular glass formers, it is only natural to assume that this should suffice to distinguish between competing theories of the glass transition. This turns out to hardly be the case. In fact thermodynamic theories like the random first order transition theory (RFOT) which anticipates a finite temperature divergence of η and purely kinetic theories like dynamical facilitation (DF) that expect a divergence in η only at $T = 0$ fit the relaxation time data equally well (see Figure 2)! This clearly indicates that the growth in relaxation times is a manifestation of subtle changes in the local structure and/or dynamics that cannot be quantified directly in atomic and molecular experiments. Access to particle-resolved structure and dynamics is in fact the strength of computer simulations and colloid experiments.[15–17] Unfortunately, both simulations and colloid experiments have a limited dynamic range and can access only the first 5–6 decades of the growth in relaxation times unlike the fourteen orders of magnitude available in atomic experiments.

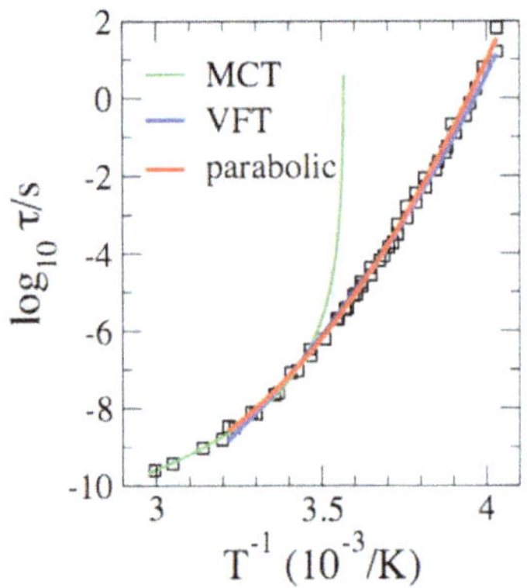

Fig. 2. Comparison of parabolic form predicted by DF: $\tau = \tau_0 \exp\left(J^2\left(\frac{1}{T} - \frac{1}{T*}\right)^2\right]$, Vogel-Fulcher-Tamman (VFT) form predicted by RFOT: $\eta = \eta_0 \exp\left(\frac{DT_0}{T-T_0}\right)$ and the mode-coupling theory (MCT) form: $\tau = \tau_0(T - T_c)^{-\gamma}$ fits to relaxation time data for ortho-terphenyl. Adapted from.[18]

We appear to have reached a deadlock. On the one hand relaxation time data alone is insufficient to discern between competing theoretical scenarios and on the other hand particle resolved studies do not have access to structure/dynamics over almost eight orders of magnitude en route to the glassy state. In this chapter we show there is still room for optimism. We will discuss how the access to dynamical crossovers - associated with quantitative changes in the nature of structural relaxation - that just about fall within the reach of colloid experiments and also simulations (which we will not focus on), can help prune theories of the glass transition. We begin with a brief introduction to colloids and the utility of colloid experiments. We then discuss the contribution from colloid studies to glass transition phenomenology with an emphasis on recent experiments that have helped critically assess various theories. We conclude by pointing out what we feel are promising future avenues.

2. Colloids as Model Atoms

2.1. *Basics*

A colloidal suspension comprises of particles typically in the nanometer to micrometer size range that remains suspended in a fluid due to the Brownian motion.[19,20] Colloidal suspensions are of immense technological relevance[21] and find applications in liquid body armor, protective coatings and paints and photonic bandgap materials to name a few.[22,23] From a fundamental physics viewpoint, due to thermalization by Brownian motion the statistical mechanics and phase behavior of colloidal suspensions share striking similarities with their atomic counterparts. This combined with their large size makes possible the probing of particle dynamics at the single-particle level and thus makes them excellent model systems to probe atomic scale phenomena.[24–26] Further, recent advances in the synthesis of colloidal particles with complex shapes and tailor-made interactions has not only allowed realizing novel self-assembled structures but also fine tuning the phase behaviour of colloidal suspensions to an extent that is impossible to mimic in atomic and molecular systems.[27–36]

The simplest model colloidal system where particles interact purely through excluded volume effects is a suspension of hard spherical particles. The interaction potential for hard spheres (HS) is given by

$$U_{HS} = \infty \quad \text{if} \quad 0 < r < \sigma \tag{2}$$

$$U_{HS} = 0 \quad \text{if} \quad r > \sigma. \tag{3}$$

Here, r is the inter-particle separation and σ is the particle diameter. Despite completely lacking an attractive component like in the Lennard-Jones potential between atoms, hard sphere colloids have proven invaluable in gaining insights into the physics of liquids, crystals and glasses. Since the repulsive part of the pair potential in conventional fluids varies far more rapidly with r than the attractive part and consequently determines its local structure, the HS model is a good approximation for real systems. The absence of attractions means that the phase diagram of HS is determined by entropy alone which solely depends on the colloid volume fraction

$$\phi = \frac{\pi \sigma^3 N}{6V} \tag{4}$$

where σ is the particle diameter, N is the number of particles and V is the volume.

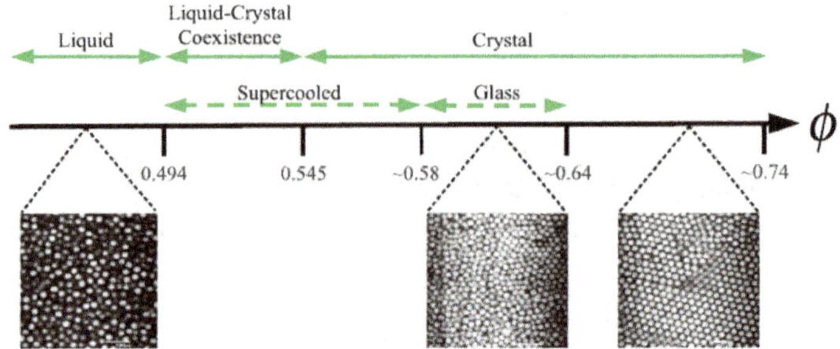

Fig. 3.   Phase diagram of a system interacting via hard sphere repulsions. The snapshots are confocal images corresponding to liquid, glass and crystal phases. Adapted from Ref. 16. For images of Bragg diffraction from 3D colloidal crystals, see Ref. 37 and 38.

Figure 3 shows the phase diagram of hard spheres. At $\phi < 0.495$, the system is a fluid. For $0.494 < \phi < 0.545$, there is a fluid-crystal coexistence and for $\phi > 0.545$, the minimum energy state is a crystal.[39,40] The first experimental realization of the HS phase diagram was by Pusey and van Megen[37] on a suspension of PMMA colloids that behave like near perfect hard spheres. Remarkably, the study revealed the presence of a disordered phase for at $0.58 <\sim \phi <\sim 0.64$ and subsequently set the stage for real-space experimental investigations of particle dynamics in dense disordered systems. The lack of an attractive component in the pair potential implies the system lacks a gas-liquid transition. Attractive interactions between colloids can be achieved through the well-known depletion attraction which involves the addition of non-adsorbing polymers, which are typically a fraction of the particle size, to the colloidal suspension. The form of the depletion attraction was calculated by Asakura and Oosawa[41] and is given by

$$U_{AO}(r) = -\frac{\pi k_B T N_d}{12V}[2(\sigma + \sigma_d)^3 - 3(\sigma + \sigma_d)^2 r + r^3], \quad \sigma \leq r \leq \sigma + \sigma_d$$
$$= 0, \quad \sigma + \sigma_d < r. \tag{5}$$

Here, σ_d and N_d are the diameter and number density of the depletant polymer, respectively, and the other symbols have their usual meaning. The ability to tune both σ_d and N_d independently has helped realize liquid-gas and the crystal-gas phase transitions, three phase coexistence, reentrant phase transitions and also glass-glass phase transitions in colloidal suspensions.[42–49]

2.2. *Brief summary of condensed matter phenomena probed using colloids*

The array of physical phenomena investigated using colloidal suspensions is too numerous to elaborate here and we highlight only a few key results. The success of these studies have for the large part also relied on simultaneous developments in confocal microscopic imaging and particle tracking algorithms.[50–53] In the context of crystallization, Weitz and coworkers have used colloidal suspensions to gain insights into crystal nucleation and growth.[54] Binary mixtures of positively and negatively charged particles have been used to realize NaCl and CsCl structures.[55] Non-close packed colloidal architectures such as the Kagome lattice has been achieved using colloids with patchy interactions.[56] Numerous strides have been made in the use of templated surfaces to steer crystal nucleation and growth.[57–59] These studies have not only helped shed light on the mechanisms that are at play during thin film growth[60,61] but have also aided in the fabrication of laterally ordered colloidal crystalline arrays that rival in complexity structures grown through atomic and molecular epitaxy.[62] In addition, colloidal particles whose size can be tuned through temperature have been instrumental in gaining insights into the vibrational properties of colloidal crystals with stiffness disorder,[63] premelting of colloidal crystals at defects and grain boundaries,[64] grain boundary dynamics[65] and solid-solid phase transitions[66] to name a few. Apart from these studies, the response of both single and polycrystalline colloidal materials to external fields such as shear and electric fields have also been investigated.[67,68] Real space visualization of dislocation nucleation and their subsequent dynamics have also been achieved[69–71] using a combination of laser diffraction microscopy[72] and confocal microscopy. In one of these ingenious experiments,[70] the authors indented a colloidal crystal with a sewing needle, essentially performing the colloid analogue of nano-indentation experiments on atomic crystals, to study the formation of dislocations, quantify their nucleation rate and map the associated strain field. Remarkably, these works have provided evidence that the defect dynamics in these systems can be well described by the continuum approach which is used to describe defect dynamics in atomic crystals.[73,74]

Colloid experiments have also made invaluable contributions to the physics of amorphous systems. Experiments on sheared colloidal glasses provided the first direct evidence for shear transformation zones.[75] Various aspects of amorphous solids such as the density of states,[76,77] aging,[78] yielding[79] and shear-induced melting[80] have also been investigated using colloids. In addition to all these studies, a large body of research on the glass transition using light scattering as well as microscopy techniques has accumulated over the years. These studies have contributed significantly to our understanding of important aspects of glass transitions phenomenology such as heterogeneous dynamics. The contributions from these studies is the subject for the remainder of this chapter. This chapter is an abridged version of the review article[9] and we encourage the interested reader to read this article to obtain a more detailed picture of the glass transition problem.

3. Supercooled Colloidal Liquids as a Testbed for Theories of Glass Transition

Since the seminal observation of a colloidal glass transition in PMMA hard-spheres by Pusey and van Megen[37,81] almost three decades ago, enormous strides have been made in understanding the glassy state and the approach to it. The literature is replete with numerous excellent reviews on the subject many of which are included in this chapter. Any attempt to include all the most spectacular results from colloid experiments is bound to fail given how numerous they are. We restrict ourselves to only those theories of the glass transition which have been put to test in colloid experiments by one of us (RG).

3.1. *Mode-coupling theory*

In the mid 1980's the mode coupling theory (MCT) of the glass transition was gaining impetus.[82–86] MCT, a purely dynamical theory, still remains the only first-principles theory of supercooled liquids and glasses. Using only the structure factor $S(q)$ of the liquid as an input, MCT anticipates that the non-linear feedback of density fluctuations on microscopic dynamics results in a complete structural arrest below a temperature T_c or equivalently above a volume fraction ϕ_c. The most appealing aspect of MCT from an experimentalist's viewpoint is that it predicts the precise form of the intermediate scattering function $F_s(q,t)$, which can be easily obtained from light-scattering experiments. Here q is the wave vector and t is time. The theory predicts three distinct relaxation regimes. At short times relaxation depends on the specific nature of microscopic dynamics. At intermediate times, also called the β-relaxation regime, particles are caged by their neighbors and this is manifest as a plateau in the $F_s(q,t)$. At longer times, particles break out of their cages and this corresponds to the α-relaxation regime. In this regime MCT predicts a stretched-exponential form for $F_s(q,t)$

$$F_s(q,t) \propto \exp\left(-(t/\tau_\alpha)^\beta\right) \tag{6}$$

with $\beta < 1$ and τ_α being the structural relaxation time. Further on approaching T_c or ϕ_c from the liquid side, $\tau_\alpha \sim (T-T_c)^{-\gamma}$ or $\tau_\alpha \sim (\phi-\phi_c)^{-\gamma}$ with the MCT prediction of the exponent $\gamma \sim 2.57$ for the case hard spheres.[87] The predictions of MCT were tested through static and dynamic light scattering measurements on colloidal liquids comprising of PMMA hard-spheres.[88] These measurements were made at the wave vector corresponding to the first peak of $S(q)$ and an excellent agreement with theoretical predictions was observed with a MCT singularity at $\phi_c = 0.557$ (Figure 4).

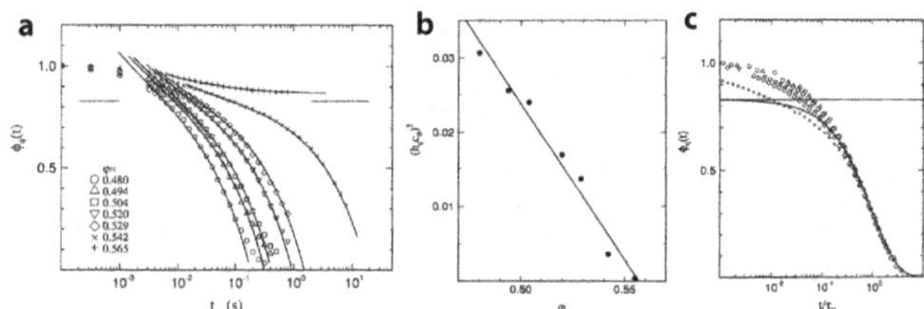

Fig. 4.   Verification of MCT predictions for hard sphere-like colloidal glass-formers (a) Normalized intermediate function $\phi_q(t) = F(q,t)/S(q)$ for different volume fractions φ. The solid curves are MCT fits. (b) Verification of the square root singularity predicted by MCT. The singularity occurs at $\varphi_c = 0.557$. (c) Scaling in the α relaxation regime. The solid curve is a Kohlrausch-Williams-Watts fit of the form $f_q\exp(-(t/\tau_\alpha)^\beta)$, with $\beta = 0.88$. Adapted from Ref. 89.

MCT has been extended to particles with attractive interactions as well. Depending on the strength of these interactions MCT predicts two distinct types of glasses - a repulsive glass at low attraction strengths and an attractive glass at higher strengths. At intermediate attraction strengths a ergodic liquid phase separates these two glasses and this phenomenon is called a re-entrant glass transition. The first direct evidence for this behavior came from DLS experiments by Poon and coworkers on collidal hard spheres in the presence of short-range depletion interactions.[48]

3.1.1. *Dynamical heterogeneity*

Perhaps the most important outcome of the above experiments is the observation of a stretched exponential relaxation at high degrees of supercooling. This is also consistent with the super-Arrhenius temperature dependence of η (Figure 1) in fragile glass formers which suggests the presence of a spectrum of relaxation times. While such a broad distribution of relaxation time could simply be due to an intrinsically non-exponential relaxation processes at the molecular level, there is indisputable evidence that relaxation in deeply supercooled liquids is both spatially and temporally heterogeneous. The spatial averaging of $F_s(q,t)$ over these different regions would then manifest itself as a stretched exponential relaxation.[90–94]

Around the same time, real-space imaging of colloidal particles using confocal imaging was making inroads. Kegel and van Blaaderen and Weitz and co-workers, independently provided the first direct experimental evidence of dynamical heterogeneities in colloidal glass-formers.[95,96] They first quantified the kurtosis of the particle displacement distribution through the non-Gaussian parameter

$$\alpha_2(t) = \frac{\langle \Delta x^4 \rangle}{3\langle \Delta x^2 \rangle^2} - 1 \tag{7}$$

where Δx is the particle displacement over t. In simple liquids, particle dynamics is Gaussian and $\alpha_2(t)$ is small. In supercooled liquids however $\alpha_2(t)$ is a maximum over the cage-breaking time t^* where the particles have to make anomalously large displacements. Further, the peak height grows of $\alpha_2(t)$ on approaching the glass transition. At longer times, owing to many cage-breaking events, the dynamics once again becomes Gaussian. In a dense liquid, breaking cages has to involve the correlated displacement of neighboring particles as well. The authors quantified these by determining the top 5% of the most-mobile particles over t^*. Interestingly these particles are spatially clustered and provided the first evidence of dynamical heterogenities. More importantly, as ϕ_c is approached from below, the size of these clusters grow. Beyond ϕ_c, however because the dynamics is essentially frozen and particles can only rattle in their cages, the cluser size drops (Figure 5). We point the interested reader to an excellent review on this topic by Hunter and Weeks.[16]

Subsequent studies also showed that just as the mobile particles are maximally clustered over t^*, the immobile ones are clustered over τ_α. Thus, the overall particle diffusivity D is dominated by the most-mobile particle

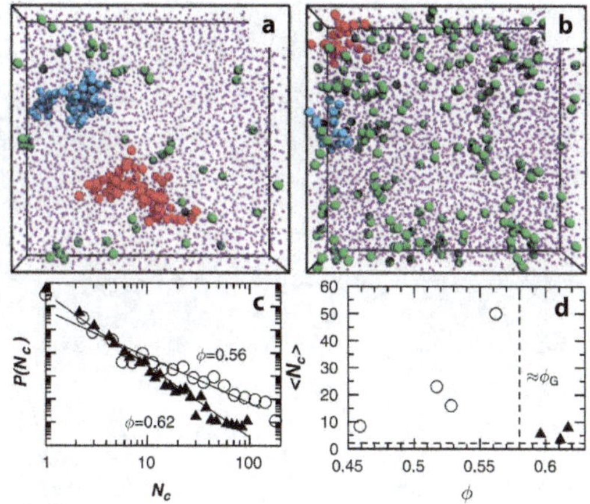

Fig. 5. Clusters of mobile particles. (a-b) Representative clusters formed by top 5% most mobile particles (green, red and blue) defined over t^* in (a) the supercooled liquid regime ($\phi = 0.56$) and (b) the amorphous regime ($\phi = 0.61$). In (a), the red cluster contains 69 particles and the blue cluster contains 50 particles. In (b), the largest (red) cluster contains 21 particles. (c) Distribution of cluster sizes $P(N_c)$ for a supercooled liquid ($\phi = 0.56$, open circles) and a glass ($\phi = 0.62$, filled triangles). The solid lines are power law fits to the data. (d) Average cluster size $\langle N_c \rangle$ as a function of ϕ. Data points in the supercooled liquid regime are shown as open circles and those in the glassy regime are shown as filled triangles. Adapted from Ref. 96.

and η is dominated by the most-immobile ones. An immediate consequence of these spatially distinct regions that govern D and η is that the Stokes-Einstein relationship $D = \frac{k_B T}{6\pi\eta a}$, a hallmark of simple liquids, breaks down in the supercooled regime. Here a is the particle radius and k_B is the Boltzmann constant.[97]

Due to advances in particle synthesis, MCT predictions could also be directly verified through real-space experiments on more complex particle shapes. For the case of ellipsoids where rotational as well as translational motion must be considered, MCT predicts that for aspect ratios $\alpha > 2.5$ a orientational glass transition precedes the translational one.[98–100] For $\alpha \leq 2.5$, there is a single ϕ_c at which both these degrees of freedom freeze. These predictions have been verified in 2D suspensions of colloidal ellipsoids by Han and Coworkers and Ganapathy and coworkers.[49,101,102] In fact in the latter study,[49] the authors also introduced short-range depletion interactions between colloidal ellipsoids of $\alpha = 2.1$. Although here MCT anticipates only a single glass transition, the presence of depletion attractions which depends on the local particle curvature substantially modifies it. Depletion interactions promote the lateral alignment of ellipsoids leading to the formation of local nematic domians. Such nematic ordering is a feature of high α ellipsoids where two glass transitions are observed. In fact, for intermediate attraction strengths, the authors observed a two-step glass transition with the orientational glass preceding the translational one (Figure 6).

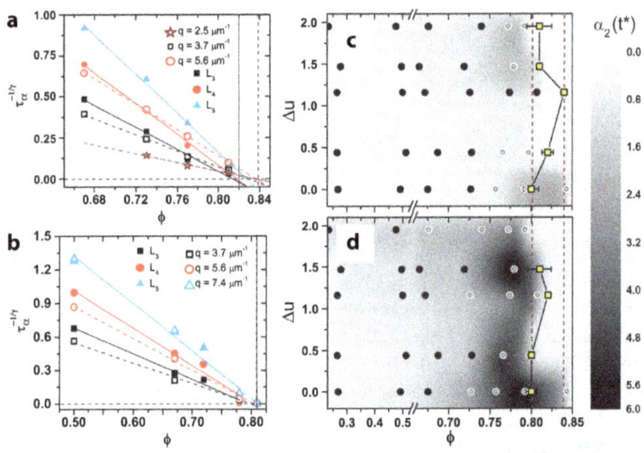

Fig. 6. Orientational and translational glass transitions in suspensions of colloidal ellipsoids of aspect ratio $\alpha = 2.1$ for various strengths Δu of attractive interactions. (a–b) $\tau^{-1/\gamma}$ vs ϕ for (a) $\Delta u = 1.16$ and (b) $\Delta u = 1.47$. The translational and orientational relaxation times have been extracted from $F_s(q,t)$ and orientational correlation function $L_n(t)$ for various values of q and n respectively. (c–d) Glass transition phase diagram in the Δu-ϕ plane for translational (c) and rotational (d) degrees of freedom. In (c) and (d), Black circles represent points at which relaxation functions decayed fully and white circles represent points where relaxation functions decayed partially. The yellow symbols denote glass transition points extracted from MCT scaling. The underlying colormap corresponds to the maximum value of the non-Gaussian parameter, $\alpha_2(t^*)$. Adapted from Ref. 49.

3.2. *Dynamical facilitation theory*

An alternate approach to MCT that is also purely kinetic is the dynamical facilitation (DF) theory of glasses.[103,104] The real-space DF theory traces its origins to a class of kinetically constrained models (KCMs) for spin systems developed in the 1980's (see review[105]) Despite their apparent simplicity KCMs captures the essential features of glassy slowing down. The basic idea of the DF approach is that a deeply supercooled liquid comprises of immobile regions which coexist with a few mobile ones. With time, mobility is transferred from the mobile to the immobile regions which subsequently themselves become immobile. Thus relaxation is 'facilitated' in the vicinity of a mobile region and system has a whole relaxes when a substantial portion of liquid has been visited by the mobility field. Just as in KCMs, DF further posits that the harbinger of mobilities are spatially and temporally localized objects called excitations. Upon supercooling, owing to a decrease in the concentration of these excitations it takes longer to facilitate relaxation in larger volumes, which naturally explains the observed growth in τ_α.

In a landmark numerical paper on atomistic glass-formers,[106] Chandler and coworkers laid out the recipe for identifying excitations and established strong parallels between atomistic glass formers and KCMs, in this case the East Model.[107] Subsequently, Gokhale et al., successfully applied this approach to colloid experiments and not only identified excitations but also found them to be both spatially and temporally localized objects as predicted by DF.[108] Although no explicit prediction for the excitation concentration, c_a, exists when ϕ, instead of T, is the control parameter, as expected within DF, c_a was indeed found to decrease on supercooling (Figure 7(a)).

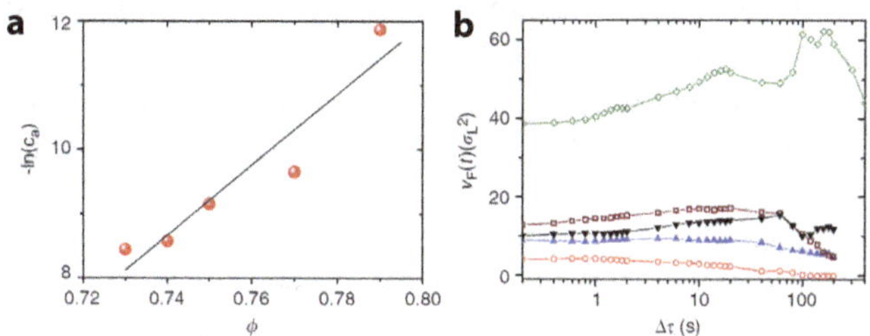

Fig. 7. Concentration of excitations and facilitation volumes. (a) Concentration of excitations c_a and (b) facilitation volume $v_F(t)$ for $a = 0.5\sigma_S$ for various area fractions ϕ. Adapted from Ref. 108.

Also, as predicted by DF, colloid experiments found the facilitation volume, $v_F(t)$, first increases with time, reaches a maximum value v_F^{max} at time t_{max} and then decreases at longer times (Figure 7(b)). Further, t_{max} as well as v_F^{max} increase with ϕ (Figure 7(b)). More recently, Mishra et al., have also extended these concepts to both repulsive and attractive liquids of colloidal ellipsoids.[109]

3.3. *Random first-order transition theory*

MCT and DF being purely dynamical theories, their very existence hinges on the experimental observation that the $S(q)$ of a supercooled liquid shows little or no change as it approaches the glass transition. It however is open to debate whether there is a subtle form of static order that is not immediately apparent in two-point density correlators. The random first-order transition theory (RFOT) a purely thermodynamic approach and one of the most prominent theories of the glass transition posits there is indeed such a form static order.[110] RFOT a mean field theory has its origins in a certain class of spin-glass models known as p-spin models.[111] In the mean field these models show two phase transitions: the first a purely dynamical transition that is associated with ergodicity breaking and is now believed to be the MCT singularity at T_c, and a second thermodynamic transition associated with the vanishing of the configurational entropy s_c at the Kauzmann temperature T_K.[112] In the $T_c > T > T_K$ regime, where the free energy landscape possesses a large number of metastable minima, the liquid is thought to exist as a patchwork of amorphous mosaics. Structural relaxation corresponds to hops between minima. While the gain in configurational entropy $Ts_c(T)\xi^d$ favors mosaic rearrangements, the surface free energy cost $\Gamma\xi^\theta$ due to a finite surface tension Γ between the nucleating phase and the frozen exterior abhors it. Here ξ is the mosaic size, d is the dimensionality and $\theta \leq d - 1$. The competition between these two terms yield the typical mosaic size $\xi^{*(d-\theta)} = \Gamma/Ts_c(T)$. On approaching T_K, $s_c(T)$ decreases and vanishes at T_K. The mosaic size thus grows and diverges at T_K resulting in a concomitant divergence in τ_α.

The presence of a finite Γ below T_c should favor a compact shape for cooperatively rearranging regions (CRRs). However numerical and experimental studies find the shapes of CRRs to be stringy, or fractal-like.[96,113] Wolynes and co-workers subsequently developed the 'Fuzzy Sphere Model' to account for this discrepancy.[114] Within this model, a CRR comprises of a compact core surrounded by a ramified shell. At high temperatures, the entropic term associated with the multiplicity of configurations for the ramified shell dominates and the CRR shape is predominantly stringy. At low temperatures, the energetic term assocaited with the breaking of favorable surface bonds of the CRR dominates and the shape is more compact. This change in shape is expected to coincide with the crossover from collisional to activated dynamics across T_c and is unique to the RFOT picture.

The precise physical meaning of the mosaic and a means of calculating its size was later provided by Bouchaud and Biroli.[115] Their method involved freezing of all particles outside of a circular cavity of radius R and investigating the evolution of particles within it. For $R < \xi^*$, the particles within the cavity remain frozen while for $R \geq \xi^*$, particles can reorganize. ξ^* is also called the point-to-set length ξ_{PTS}. Carrying out this procedure for many Rs and at many Ts, subsequently revealed a growing static length scale with supercooling. Apart from this pinning geometry, other pinning configurations have also been employed to probe static and dynamic correlations.[116] One of these is the amorphous wall geometry since it can not only probe the size of CRRs but also a change in their shape. The change is shape of CRRs is manifest as a non-monotonic evolution in the dynamic length scale ξ_{dyn}.[117] Recently, these predictions of RFOT have been verified by Manasa et al.,[118] who have use holographic optical tweezers to experimentally realize the amorphous wall geometry. By following the procedure described by Kob and coworkers,[117] the authors directly measured both a growing static and a non-monotonic dynamic length scale with supercooling. More importantly by the authors also provided direct evidence for a change in the shape of CRRs across T_c (Figure 8(c)). This study has been followed by a recent work by Ganapathy and coworkers who have devised a method to uncover both static and non-monotonic dynamic correlations even in the absence of external pinning. Most remarkably, the authors directly quantified Γ between amorphous-amorphous interfaces and observed a discontinuous jump across ϕ_c.[119]

More recently, an experimental study by Zhang and Cheng have also provided direct experimental evidence for a growing point-to-set length scale even in three dimensions using the spherical cavity pinning geometry.[120]

4. Using Dynamical Crossovers to Distinguish Between Competing Theories of Glass Formation

It should now become apparent to the reader why the glass transition problem continues to remain unsolved. Even within the three theoretical scenarios (MCT, DF and RFOT) considered here, colloid experiments clearly appear to support all three. The situation is not very different in numerical simulations. It is unclear if the theory of the glass transition exists at all. At first light such an endeavor appears hopeless since both colloid experiments and simulations have a very limited dynamical range corresponding

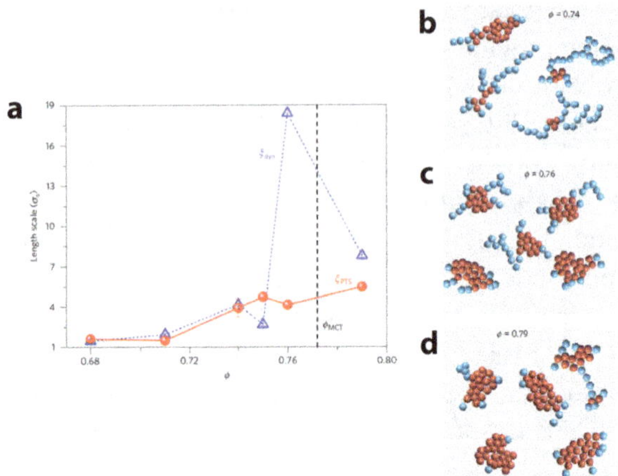

Fig. 8. (a) Point-to-set length scale, ξ_{PTS}, ($\bullet$) and dynamic length scale, ξ_{dyn}, ($\triangle$). The error bars have been obtained from the exponential fits. The dotted black line indicates the mode coupling crossover ϕ_{MCT}. (b–d) Representative 25-particle clusters of most mobile particles for $\phi = 0.74$, $\phi = 0.76$ and $\phi = 0.79$ respectively. Core-like particles are shown in red and string-like particles are shown in light blue. Adapted from Ref. 118.

to $T \geq T_c$ or equivalently $\phi \leq \phi_c$. In the recent past however, there has been growing evidence for dynamical crossovers in the vicinity of T_c (or ϕ_c). These dynamical crossovers correspond to quantitative changes in the nature of relaxation and can be used as a possible means to prune theories.

Perhaps the most well-known dynamical crossover en route to glass formation is the MCT transition. Colloid experiments[118,121,122] clearly have shown that the MCT singularity is avoided and the system continues to remain ergodic due to activated relaxation. This clearly rules out at least the idealized version of MCT as a plausible theory of the glass transition. While DF and RFOT could not be further apart in their formalisms and propose very different relaxation mechanisms, structural relaxation in DF is facilitated while in RFOT, where the landscape dominates, relaxation is activated. It is however not possible to tell if cooperative rearrangement of domains is activated based on dynamics alone. On the contrary, facilitated dynamics has a vectorial character (the transfer of mobility between mobile and immobile regions is directional) and is quite different from collective hopping. A way forward to distinguish between DF and RFOT is to evaluate the relative importance of facilitated dynamics and approaching the glass transition. Below we describe recent experiments that have exploited this strategy.

4.1. *The mobility transfer function*

One way to quantify the importance of facilitated dynamics in structural relaxation is through the mobility transfer function, $M(\Delta t)$, first introduced by Glotzer and coworkers.[123] Consider two successive time intervals of duration Δt, then $M(\Delta t)$ measures the excess probability that a mobile particle in the second interval is spatially close to a mobile particle in the first one.

$$M(\Delta t) = \frac{\int_0^{r_{min}} P_M(r, \Delta t)dr}{\int_0^{r_{min}} P_M^*(r, \Delta t)dr}. \tag{8}$$

Here, r_{min} is the first minimum of the radial pair distribution function $g(r)$. $P_M(r, \Delta t)$ is the probability that r is the minimum distance between a mobile particle in the second interval and the set of mobile particles in the first interval. Similarly, $P_M^*(r, \Delta t)$ is the probability that r is the minimum distance between a mobile particle in the second interval and a set of randomly chosen immobile particles in the first interval. Since mobile particles are highly correlated over the cage-breaking time t^*, $M(\Delta t)$ should be a maximum M_{max} for $\Delta t = t^*$. Since in DF mobile regions induce mobility in neighboring regions $M(\Delta t)$ should be large if facilitation is the dominant mechanism. In fact, Elmatad and Keys have shown that in the kinetically controlled East model, if one artificially introduces activated dynamics M_{max} first increases at high Ts where facilitation dominates and then decreases at low temperatures where activated processes kick in. In their colloid experiments, Nagamanasa et al.[118] computed M_{max} (Figure 9) and compared its evolution with ϕ with that of ξ_{dyn} (Figure 8(a)).

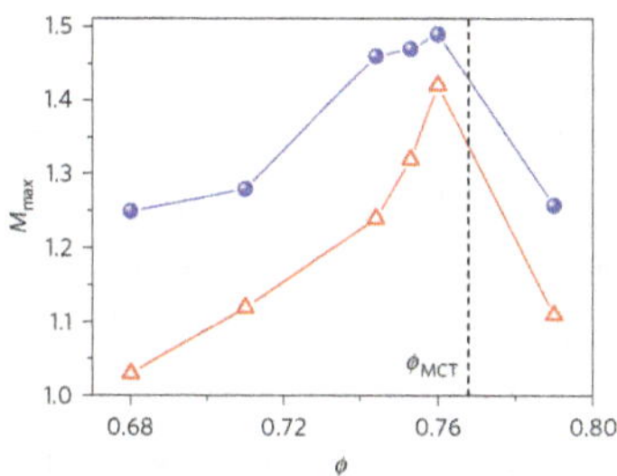

Fig. 9. The maximum value of the mobility transfer function, M_{max} as a function of area fraction ϕ for large (red triangles) and small (blue circles) particles in a binary colloidal glass-former composed of polystyrene particles. Adapted from Ref. 118.

Most remarkably, M_{max} exhibits a maximum at precisely the same $\phi \sim \phi_c$ where ξ_{dyn} is maximal. More importantly, this ϕ coincides with the change in shape of CRRs from string-like to compact which according to RFOT also occurs in the vicinity of ϕ_c. This clearly shows that while facilitation is the dominant mode of relaxation at low ϕs, at high ϕs relaxation is dominated by activated processes.

Following this study, Gokhale and coworkers[124] analyzed the spatial distribution of excitations within CRRs. Within RFOT, the compact CRR core is expected to undergo activated rearrangements while in the string-like shell relaxation is expected to be facilitated. If this picture were correct, excitations should be preferentially found in the shell as the CRR shape changes from string-like to compact with supercooling. This was indeed found to be the case in the colloid experiments.

5. Conclusions

The picture that emerges from the colloid experiments suggests that MCT and DF are valid over a rather limited dynamical range that extends from the onset of glassy dynamics (appearance of a plateau in $F_s(q,t)$) to the mode-coupling crossover. At higher degrees of supercooling where relaxation is dominated by hops between the minima in the free energy landscape and the RFOT theory appears to better capture experimental observations. More importantly, the recent attempts to directly quantify the surface tension of amorphous-amorphous interfaces and its observed evolution with ϕ further swing in favor of RFOT.[119] Clearly these experiments show that access to real-space dynamics in the $\phi > \phi_c$ regime is the way forward. Almost all real-space studies thus far have used μm-sized colloids where accessing dynamics beyond ϕ_c is a challenge owing to their large size and slow dynamics. The first steps towards circumventing this major limitation have already been taken by Royall and coworkers[122] who have used stimulated emission-depletion microscopy (STED) to image the real-space dynamics of sub-micrometer sized colloids. While this chapter is rather restricted in what it covers, there are other competing thermodynamic approaches such as those based on the concept of geometric frustration.[125,126] Designing experiments that utilize colloidal suspensions to critically evaluate these approaches promises to be an exciting future avenue.

References

1. C. Angell, The old problems of glass and the glass transition, and the many new twists, *Proc. Nat. Acad. Sci.* **92**(15), 6675–6682 (1995).
2. P. G. Debenedetti and F. H. Stillinger, Supercooled liquids and the glass transition, *Nature.* **410**(6825), 259–267 (2001).
3. K. Ngai, Why the glass transition problem remains unsolved?, *J. Non-Crystalline Solids.* **353**(8), 709–718 (2007).
4. L. Berthier and G. Biroli, Theoretical perspective on the glass transition and amorphous materials, *Rev. Mod. Phys.* **83**(2), 587 (2011).
5. A. Cavagna, Supercooled liquids for pedestrians, *Phys. Rep.* **476**(4), 51–124 (2009).
6. J. S. Langer, Theories of glass formation and the glass transition, *Rep. on Prog. in Phys.* **77**(4), 042501 (2014).
7. C. Angell, Perspective on the glass transition, *J. Phys. Chem. Solids.* **49**(8), 863–871 (1988).
8. C. A. Angell, Formation of glasses from liquids and biopolymers, *Science.* **267**(5206), 1924–1935 (1995).
9. S. Gokhale, A. Sood, and R. Ganapathy, Deconstructing the glass transition through critical experiments on colloids, *Adv. Phys.* **65**(4), 363–452 (2016).
10. N. O. Birge and S. R. Nagel, Specific-heat spectroscopy of the glass transition, *Phys. Rev. Lett.* **54**(25), 2674 (1985).
11. C. Angell, K. Ngai, and G. Wright, Relaxation in complex systems, *Naval Research Laboratory, Washington DC.* p. 3 (1984).
12. I. Chang, F. Fujara, B. Geil, G. Heuberger, T. Mangel, and H. Sillescu, Translational and rotational molecular motion in supercooled liquids studied by nmr and forced rayleigh scattering, *J. Non-crystalline Solids.* **172**, 248–255 (1994).
13. V. Lubchenko, Theory of the structural glass transition: a pedagogical review, *Adv. Phys.* **64**(3), 283–443 (2015).
14. V. Novikov, Y. Ding, and A. Sokolov, Correlation of fragility of supercooled liquids with elastic properties of glasses, *Phys. Rev. E.* **71**(6), 061501 (2005).
15. M. P. Allen and D. J. Tildesley, *Computer Simulation of Liquids.* Oxford University Press (1989).
16. G. L. Hunter and E. R. Weeks, The physics of the colloidal glass transition, *Rep. Prog. Phys.* **75**(6), 066501 (2012).
17. P. J. Lu and D. A. Weitz, Colloidal particles: crystals, glasses, and gels, *Annu. Rev. Condens. Matter Phys.* **4**(1), 217–233 (2013).
18. G. Biroli and J. P. Garrahan, Perspective: The glass transition, *J Chem Phys.* **138**, 12A301 (2013).
19. R. Jones, *Soft Condensed Matter.* Oxford Master Series in Physics, OUP Oxford (2002). ISBN 9780198505891. URL `https://books.google.co.in/books?id=V5nunQEACAAJ`.
20. J. Israelachvili, *Intermolecular and Surface Forces.* Intermolecular and Surface Forces, Elsevier Science (2010). ISBN 9780080923635.
21. T. Cosgrove, *Colloid Science: Principles, Methods and Applications.* John

Wiley & Sons (2010).

22. S.-H. Kim, S. Y. Lee, S.-M. Yang, and G.-R. Yi, Self-assembled colloidal structures for photonics, *NPG Asia Mater.* **3**(1), 25–33 (2011).

23. Y. A. Vlasov, X.-Z. Bo, J. C. Sturm, and D. J. Norris, On-chip natural assembly of silicon photonic bandgap crystals, *Nature.* **414**(6861), 289–293 (2001).

24. A. K. Sood, Structural ordering in colloidal suspensions, *Solid State Phys.* **45**, 1–73 (1991).

25. W. Poon, Colloids as big atoms, *Science.* **304**(5672), 830–831 (2004).

26. E. M. Terentjev, D. A. Weitz, et al., *W.C.K. Poon in The Oxford Handbook of Soft Condensed Matter.* Oxford University Press (2015).

27. S. C. Glotzer and M. J. Solomon, Anisotropy of building blocks and their assembly into complex structures, *Nature Mater.* **6**(8), 557–562 (2007).

28. S. Sacanna, W. Irvine, P. M. Chaikin, and D. J. Pine, Lock and key colloids, *Nature.* **464**(7288), 575–578 (2010).

29. P. M. Johnson, C. M. van Kats, and A. van Blaaderen, Synthesis of colloidal silica dumbbells, *Langmuir.* **21**(24), 11510–11517 (2005).

30. D. J. Kraft, W. S. Vlug, C. M. van Kats, A. van Blaaderen, A. Imhof, and W. K. Kegel, Self-assembly of colloids with liquid protrusions, *J. the Amer. Chem. Soc.* **131**(3), 1182–1186 (2008).

31. L. Hong, A. Cacciuto, E. Luijten, and S. Granick, Clusters of amphiphilic colloidal spheres, *Langmuir.* **24**(3), 621–625 (2008).

32. A. B. Pawar and I. Kretzschmar, Fabrication, assembly, and application of patchy particles, *Macromolecul. Rapid Commun.* **31**(2), 150–168 (2010).

33. L. Hong, A. Cacciuto, E. Luijten, and S. Granick, Clusters of charged janus spheres, *Nano Lett.* **6**(11), 2510–2514 (2006).

34. A. B. Pawar and I. Kretzschmar, Multifunctional patchy particles by glancing angle deposition, *Langmuir.* **25**(16), 9057–9063 (2009).

35. O. Cayre, V. N. Paunov, and O. D. Velev, Fabrication of asymmetrically coated colloid particles by microcontact printing techniques, *J. Mater. Chem.* **13**(10), 2445–2450 (2003).

36. F. Huo, A. K. Lytton-Jean, and C. A. Mirkin, Asymmetric functionalization of nanoparticles based on thermally addressable dna interconnects, *Advan. Mater.* **18**(17), 2304–2306 (2006).

37. P. N. Pusey and W. van Megan, Phase behaviour of concentrated suspensions of nearly hard colloidal spheres, *Nature.* **320**(6060), 340 (1986).

38. P. Pusey, E. Zaccarelli, C. Valeriani, E. Sanz, W. C. Poon, and M. E. Cates, Hard spheres: crystallization and glass formation, *Philo. Trans. Royal Society A: Math., Phys. Engin. Sci.* **367**(1909), 4993–5011 (2009).

39. W. W. Wood and J. Jacobson, Preliminary results from a recalculation of the monte carlo equation of state of hard spheres, *The J. Chem. Phys.* **27**(5), 1207–1208 (1957).

40. B. Alder and T. Wainwright, Phase transition for a hard sphere system, *The J. Chem. Phys.* **27**(5), 1208 (1957).

41. S. Asakura and F. Oosawa, On interaction between two bodies immersed in a solution of macromolecules, *J. Chem. Phys.* **22**(7), 1255–1256 (1954).

42. S. Sanyal, N. Easwar, S. Ramaswamy, and A. Sood, Phase separation in binary nearly-hard-sphere colloids: evidence for the depletion force, *EPL (Europhysics Letters).* **18**(2), 107 (1992).

43. S. M. Ilett, A. Orrock, W. Poon, and P. Pusey, Phase behavior of a model colloid-polymer mixture, *Phys. Rev. E.* **51**(2), 1344 (1995).

44. W. Poon, The physics of a model colloid–polymer mixture, *J. Phys.: Condensed Matter.* **14**(33), R859 (2002).

45. V. J. Anderson and H. N. Lekkerkerker, Insights into phase transition kinetics from colloid science, *Nature.* **416**(6883), 811–815 (2002).

46. C. Wilson et al., Gelation in colloid–polymer mixtures, *Faraday Discussions.* **101**, 65–76 (1995).

47. A. Dinsmore and D. Weitz, Direct imaging of three-dimensional structure and topology of colloidal gels, *J. Phys.: Condensed Matt..* **14**(33), 7581 (2002).

48. K. N. Pham, A. M. Puertas, J. Bergenholtz, S. U. Egelhaaf, A. Moussaïd, P. N. Pusey, A. B. Schofield, M. E. Cates, M. Fuchs, and W. C. K. Poon, Multiple glassy states in a simple model system, *Science.* **296**(5565), 104 (2002).

49. C. K. Mishra, A. Rangarajan, and R. Ganapathy, Two-step glass transition induced by attractive interactions in quasi-two-dimensional suspensions of ellipsoidal particles, *Phys. Rev. Lett.* **110**(18), 188301 (2013).

50. T. Wilson, *Confocal Microscopy*, Academic Press: London, etc. **426**, 1–64 (1990).

51. M. H. Chestnut, Confocal microscopy of colloids, *Current Opinion in Colloid & Interface Sci.* **2**(2), 158–161 (1997).

52. A. D. Dinsmore, E. R. Weeks, V. Prasad, A. C. Levitt, and D. A. Weitz, Three-dimensional confocal microscopy of colloids, *Appl. Optics.* **40**(24), 4152–4159 (2001).

53. V. Prasad, D. Semwogerere, and E. R. Weeks, Confocal microscopy of colloids, *J. Phys.: Condensed Matter.* **19**(11), 113102 (2007).

54. U. Gasser, E. R. Weeks, A. Schofield, P. Pusey, and D. Weitz, Real-space imaging of nucleation and growth in colloidal crystallization, *Science.* **292**(5515), 258–262 (2001).

55. M. E. Leunissen, C. G. Christova, A.-P. Hynninen, C. P. Royall, A. I. Campbell, A. Imhof, M. Dijkstra, R. Van Roij, and A. Van Blaaderen, Ionic colloidal crystals of oppositely charged particles, *Nature.* **437**(7056), 235–240 (2005).

56. Q. Chen, S. C. Bae, and S. Granick, Directed self-assembly of a colloidal kagome lattice, *Nature.* **469**(7330), 381–384 (2011).

57. N. Vogel, C. K. Weiss, and K. Landfester, From soft to hard: the generation of functional and complex colloidal monolayers for nanolithography, *Soft Matter.* **8**(15), 4044–4061 (2012).

58. Y. Yin, Y. Lu, B. Gates, and Y. Xia, Template-assisted self-assembly: a practical route to complex aggregates of monodispersed colloids with well-defined sizes, shapes, and structures, *J. Amer. Chem. Soc.* **123**(36), 8718–8729 (2001).

59. N. V. Dziomkina and G. J. Vancso, Colloidal crystal assembly on topologically patterned templates, *Soft Matter.* **1**(4), 265–279 (2005).

60. R. Ganapathy, M. R. Buckley, S. J. Gerbode, and I. Cohen, Direct measurements of island growth and step-edge barriers in colloidal epitaxy, *Science.* **327**(5964), 445–448 (2010).

61. J. R. Savage, S. F. Hopp, R. Ganapathy, S. J. Gerbode, A. Heuer, and I. Cohen, Entropy-driven crystal formation on highly strained substrates, *Proc. Nat. Acad. Sci.* **110**(23), 9301–9304 (2013).

62. C. K. Mishra, A. Sood, and R. Ganapathy, Site-specific colloidal crystal nucleation by template-enhanced particle transport, *Proc. Nat. Acad. Sci.* **113**(43), 12094–12098 (2016).

63. D. Kaya, N. Green, C. Maloney, and M. Islam, Normal modes and density of states of disordered colloidal solids, *Science.* **329**(5992), 656–658 (2010).

64. A. M. Alsayed, M. F. Islam, J. Zhang, P. J. Collings, and A. G. Yodh, Premelting at defects within bulk colloidal crystals, *Science.* **309**(5738), 1207–1210 (2005).

65. K. H. Nagamanasa, S. Gokhale, R. Ganapathy, and A. Sood, Confined glassy dynamics at grain boundaries in colloidal crystals, *Proc. Nat. Acad. Sci.* **108**(28), 11323–11326 (2011).

66. Y. Peng, F. Wang, Z. Wang, A. M. Alsayed, Z. Zhang, A. G. Yodh, and Y. Han, Two-step nucleation mechanism in solid–solid phase transitions, *Nature Mater.* **14**(1), 101–108 (2015).

67. T. Palberg, W. Mönch, J. Schwarz, and P. Leiderer, Grain size control in polycrystalline colloidal solids, *The J. Chem. Phys.* **102**(12), 5082–5087 (1995).

68. S. Gokhale, K. H. Nagamanasa, R. Ganapathy, and A. K. Sood, Grain growth and grain boundary dynamics in colloidal polycrystals, *Soft Matter.* **9**(29), 6634 (2013).

69. P. Schall, I. Cohen, D. A. Weitz, and F. Spaepen, Visualization of dislocation dynamics in colloidal crystals, *Science.* **305**(5692), 1944–1948 (2004).

70. P. Schall, I. Cohen, D. A. Weitz, and F. Spaepen, Visualizing dislocation nucleation by indenting colloidal crystals, *Nature.* **440**(7082), 319–323 (2006).

71. S. Suresh, Crystal deformation: Colloid model for atoms, *Nature Mater.* **5**(4), 253–254 (2006).

72. P. Schall, Laser diffraction microscopy, *Rep. Prog. Phys.* **72**(7), 076601 (2009).

73. A. Cotterell, *Dislocation and Plastic Flow in Crystals* (1953).

74. F. Frank, Report of the symposium on the plastic deformation of crystalline solids, *Carnegie Instit. Technol., Pittsburgh.* p. 150 (1950).

75. P. Schall, D. A. Weitz, and F. Spaepen, Structural rearrangements that govern flow in colloidal glasses, *Science.* **318**(5858), 1895–1899 (2007).

76. A. Ghosh, V. K. Chikkadi, P. Schall, J. Kurchan, and D. Bonn, Density of states of colloidal glasses, *Phys. Rev. Lett.* **104**(24), 248305 (2010).

77. K. Chen, W. G. Ellenbroek, Z. Zhang, D. T. Chen, P. J. Yunker, S. Henkes, C. Brito, O. Dauchot, W. Van Saarloos, A. J. Liu, et al., Low-frequency vibrations of soft colloidal glasses, *Phys. Rev. Lett.* **105**(2), 025501 (2010).

78. P. Yunker, Z. Zhang, K. B. Aptowicz, and A. G. Yodh, Irreversible rearrangements, correlated domains, and local structure in aging glasses, *Phys. Rev. Lett.* **103**(11), 115701 (2009).

79. K. H. Nagamanasa, S. Gokhale, A. Sood, and R. Ganapathy, Experimental signatures of a nonequilibrium phase transition governing the yielding of a soft glass, *Phys. Rev. E.* **89**(6), 062308 (2014).

80. C. Eisenmann, C. Kim, J. Mattsson, and D. A. Weitz, Shear melting of a colloidal glass, *Phys. Rev. Lett.* **104**(3), 035502 (2010).

81. P. Pusey and W. Van Megen, Observation of a glass transition in suspensions of spherical colloidal particles, *Phys. Rev. Lett.* **59**(18), 2083 (1987).

82. E. Leutheusser, Dynamical model of the liquid-glass transition, *Phys. Rev. A.* **29**(5), 2765 (1984).

83. U. Bengtzelius, W. Gotze, and A. Sjolander, Dynamics of supercooled liquids and the glass transition, *J. Phys. C: Solid State Phys.* **17**(33), 5915 (1984).

84. W. Götze, J. Hansen, D. Levesque, and J. Zinn-Justin. *Liquids, Freezing and The Glass Transition* (1991).

85. W. Gotze and L. Sjogren, Relaxation processes in supercooled liquids, *Rep. Prog. Phys.* **55**(3), 241 (1992).

86. W. Götze, *Complex Dynamics of Glass-Forming Liquids: A Mode-Coupling Theory: A Mode-Coupling Theory.* Vol. 143, Oxford University Press (2008).

87. J. Barrat, W. Gotze, and A. Latz, The liquid-glass transition of the hard-sphere system, *J. Phys.: Cond. Matter.* **1**(39), 7163 (1989).

88. W. Van Megen and P. Pusey, Dynamic light-scattering study of the glass transition in a colloidal suspension, *Phys. Rev. A.* **43**(10), 5429 (1991).

89. W. Götze and L. Sjögren, β relaxation at the glass transition of hard-spherical colloids, *Phys. Rev. A.* **43**(10), 5442 (1991).

90. M. D. Ediger, Spatially heterogeneous dynamics in supercooled liquids, *Ann. Rev. Phys. Chem.* **51**(1), 99–128 (2000).

91. R. Böhmer, Nanoscale heterogeneity if glass-forming liquids: experimental advances, *Current Opinion in Solid State and Materials Science.* **3**(4), 378–385 (1998).

92. H. Sillescu, Heterogeneity at the glass transition: a review, *J. Non-Crystalline Solids.* **243**(2), 81–108 (1999).

93. S. C. Glotzer, Spatially heterogeneous dynamics in liquids: insights from simulation, *J. Non-Crystalline Solids.* **274**(1), 342–355 (2000).

94. L. Berthier, G. Biroli, J.-P. Bouchaud, L. Cipelletti, and W. van Saarloos, *Dynamical Heterogeneities in Glasses, Colloids, and Granular Media.* Oxford University Press (2011).

95. W. K. Kegel and A. van Blaaderen, Direct observation of dynamical heterogeneities in colloidal hard-sphere suspensions, *Science.* **287**(5451), 290–293 (2000).

96. E. R. Weeks, J. C. Crocker, A. C. Levitt, A. Schofield, and D. A. Weitz, Three-dimensional direct imaging of structural relaxation near the colloidal glass transition, *Science.* **287**(5453), 627–631 (2000).

97. F. W. Starr, J. F. Douglas, and S. Sastry, The relationship of dynamical

heterogeneity to the adam-gibbs and random first-order transition theories of glass formation, *The J. Chem. Phys.* **138**(12), 12A541 (2013).

98. M. Letz, R. Schilling, and A. Latz, Ideal glass transitions for hard ellipsoids, *Phys. Rev. E.* **62**(4), 5173 (2000).

99. C. De Michele, R. Schilling, and F. Sciortino, Dynamics of uniaxial hard ellipsoids, *Phys. Rev. Lett.* **98**(26), 265702 (2007).

100. P. Pfleiderer, K. Milinkovic, and T. Schilling, Glassy dynamics in monodisperse hard ellipsoids, *EPL (Europhysics Letters).* **84**(1), 16003 (2008).

101. Z. Zheng, F. Wang, Y. Han, et al., Glass transitions in quasi-two-dimensional suspensions of colloidal ellipsoids, *Phys. Rev. Lett.* **107**(6), 065702 (2011).

102. Z. Zheng, R. Ni, F. Wang, M. Dijkstra, Y. Wang, and Y. Han, Structural signatures of dynamic heterogeneities in monolayers of colloidal ellipsoids, *Nature Commun.* **5** (2014).

103. J. P. Garrahan and D. Chandler, Geometrical explanation and scaling of dynamical heterogeneities in glass forming systems, *Phys. Rev. Lett.* **89**(3), 035704 (2002).

104. D. Chandler and J. P. Garrahan, Dynamics on the way to forming glass: Bubbles in space-time, *Ann. Rev. Phys. Chem.* **61**, 191–217 (2010).

105. F. Ritort and P. Sollich, Glassy dynamics of kinetically constrained models, *Adv. Phys.* **52**(4), 219–342 (2003).

106. A. S. Keys, L. O. Hedges, J. P. Garrahan, S. C. Glotzer, and D. Chandler, Excitations are localized and relaxation is hierarchical in glass-forming liquids, *Phys. Rev. X.* **1**(2), 021013 (2011).

107. M. Evans, Anomalous coarsening and glassy dynamics, *J. Phys.: Condensed Matter.* **14**(7), 1397 (2002).

108. S. Gokhale, K. H. Nagamanasa, R. Ganapathy, and A. Sood, Growing dynamical facilitation on approaching the random pinning colloidal glass transition, *Nature Commun.* **5** (2014).

109. C. K. Mishra, K. H. Nagamanasa, R. Ganapathy, A. Sood, and S. Gokhale, Dynamical facilitation governs glassy dynamics in suspensions of colloidal ellipsoids, *Proc. Nat. Acad. Sci.* **111**(43), 15362–15367 (2014).

110. T. Kirkpatrick and D. Thirumalai, Colloquium: Random first order transition theory concepts in biology and physics, *Rev. Mod. Phys.* **87**(1), 183 (2015).

111. T. Kirkpatrick and D. Thirumalai, p-spin-interaction spin-glass models: Connections with the structural glass problem, *Phys. Rev. B.* **36**(10), 5388 (1987).

112. W. Kauzmann, The nature of the glassy state and the behavior of liquids at low temperatures., *Chem. Rev.* **43**(2), 219–256 (1948).

113. C. Donati, J. F. Douglas, W. Kob, S. J. Plimpton, P. H. Poole, and S. C. Glotzer, Stringlike cooperative motion in a supercooled liquid, *Phys. Rev. Lett.* **80**(11), 2338 (1998).

114. J. D. Stevenson, J. Schmalian, and P. G. Wolynes, The shapes of cooperatively rearranging regions in glass-forming liquids, *Nature Phys.* **2**(4), 268–274 (2006).

115. J.-P. Bouchaud and G. Biroli, On the adam-gibbs-kirkpatrick-thirumalai-wolynes scenario for the viscosity increase in glasses, *The J. Chem. Phys.* **121**(15), 7347–7354 (2004).

116. L. Berthier and W. Kob, Static point-to-set correlations in glass-forming liquids, *Phys. Rev. E.* **85**(1), 011102 (2012).

117. W. Kob, S. Roldán-Vargas, and L. Berthier, Non-monotonic temperature evolution of dynamic correlations in glass-forming liquids, *Nature Phys.* **8** (2), 164–167 (2012).

118. K. H. Nagamanasa, S. Gokhale, A. Sood, and R. Ganapathy, Direct measurements of growing amorphous order and non-monotonic dynamic correlations in a colloidal glass-former, *Nature Phys.* (2015).

119. D. Ganapathi, K. H. Nagamanasa, A. Sood, and R. Ganapathy, Measurements of growing surface tension of amorphous–amorphous interfaces on approaching the colloidal glass transition, *Nature Commun.* **9**(1), 397 (2018).

120. B. Zhang and X. Cheng, Structures and dynamics of glass-forming colloidal liquids under spherical confinement, *Phys. Rev. Lett.* **116**, 098302 (Mar, 2016). doi: 10.1103/PhysRevLett.116.098302.

121. G. Brambilla, D. El Masri, M. Pierno, L. Berthier, L. Cipelletti, G. Petekidis, and A. B. Schofield, Probing the equilibrium dynamics of colloidal hard spheres above the mode-coupling glass transition, *Phys. Rev. Lett.* **102** (8), 085703 (2009).

122. J. E. Hallett, F. Turci, and C. P. Royall, Local structure in deeply supercooled liquids exhibits growing lengthscales and dynamical correlations, *Nature Commun.* **9**(1), 3272 (2018).

123. M. Vogel and S. C. Glotzer, Spatially heterogeneous dynamics and dynamic facilitation in a model of viscous silica, *Phys. Rev. Lett.* **92**(25), 255901 (2004).

124. S. Gokhale, R. Ganapathy, K. H. Nagamanasa, and A. Sood, Localized excitations and the morphology of cooperatively rearranging regions in a colloidal glass-forming liquid, *Phys. Rev. Lett.* **116**(6), 068305 (2016).

125. G. Tarjus, S. A. Kivelson, Z. Nussinov, and P. Viot, The frustration-based approach of supercooled liquids and the glass transition: a review and critical assessment, *J. Phys.: Condensed Matter.* **17**(50), R1143 (2005).

126. H. Tanaka, Two-order-parameter description of liquids. i. a general model of glass transition covering its strong to fragile limit, *The J. Chem. Phys.* **111**(7), 3163–3174 (1999).

Chapter 10

Linear Magnetoelectrics and Multiferroics

Premakumar Yanda and A. Sundaresan[*]

*Chemistry and Physics of Materials Unit and School of Advanced Materials,
Jawaharlal Nehru Centre for Advanced Scientific Research, Jakkur,
Bangalore 560 064, Karnataka, India*
[*]sundaresan@jncasr.ac.in

Coupling between magnetism and electricity in materials has been the subject of intense investigation in the recent years. Such a cross coupling in materials started with the symmetry restricted *linear magnetoelectric effect,* which involves induction of electric polarization that is proportional to applied magnetic field or vice versa. Later, this field has evolved into *multiferroics* in which magnetism and ferroelectricity coexist with a varying degree of their coupling depending on the origin of ferroelectricity. In this chapter, we first present a brief history and the basics of linear magnetoelectric effect, and then discuss some examples of recently discovered linear magnetoelectric materials. Later, we discuss the two classes (type-I and type-II) of single-phase multiferroics and introduce polar pyroelectric magnets as type-III multiferroics. Finally, we describe a dc bias current measurement that can differentiate intrinsic electric polarization from thermally induced free charge carriers.

1. Introduction

The behavior of electric and magnetic fields is well known from the time of Ørsted or even earlier and described by Maxwell's equations. But the coupling between electric and magnetic fields in solids was proposed by Pierre Curie in 1894.[1] After a long interval, Landau and Lifshitz[2] in 1959 pointed out two phenomena that could exist for certain classes of

224

magneto crystalline symmetry; one is piezomagnetism, which consists of linear coupling between a magnetic field in a solid and a deformation (analogous to piezoelectricity). The other is the linear coupling between magnetic and electric fields in a media, which would cause, for example, a magnetization proportional to an electric field.

Soon after these remarks, based on symmetry consideration, Dzyaloshinskii predicted that the antiferromagnetic Cr_2O_3 should exhibit such *linear magnetoelectric (ME) effect*.[3] Following this prediction, Astrov confirmed this effect experimentally in a single crystal of Cr_2O_3.[4] Since then a number of materials including $GaFeO_3$ were found to exhibit ME effect.[5] Later, the coexistence of magnetic and ferroelectric order was reported in magnetically diluted ferroelectric oxides such as $Pb(Fe_{1/2}Nb_{1/2})O_3$ and $Pb(Fe_{1/2}Ta_{1/2})O_3$.[6] A clear coexistence of weak ferromagnetism and ferroelectricity was shown in the nickel iodine boracite $Ni_3B_7O_{13}I$.[7] Following this, Hans Schmid[8] proposed a new word, *multiferroic* to describe materials having two or more primary ferroic properties in the same phase. However, the field of multiferroicity did not progress much until the end of the 20th century presumably because ferroelectricity and magnetism have been practiced independently.

The article *Why Are There So Few Magnetic Ferroelectrics* triggered both ferroelectric and magnetism communities towards the importance of studying magnetoelectrics and multiferroics.[9] Thus, the study of coupling between magnetism and electricity in materials has become one of the hottest topics of current interest in condensed matter physics not only because of their interesting chemistry and physics but also for their promising applications such as electric field control of magnetism in memory devices, four state logic devices and magnetic field sensors. Consequently, there has been a flurry of activity that has led to the discovery of various kinds of multiferroics with a different extent of coupling between the two order parameters depending on the origin of the electric polarization.

There are several general review articles, focused small reviews on multiferroics published in the special issue of Journal of Physics: Condensed Matter, and also a book containing a collection of articles with various aspects of multiferroics.[10–18] In this chapter, we have attempted to provide an overview on *linear magnetoelectrics* and *multiferroics* and made a clear demarcation between them. To have practical applications, the multiferroics materials should exhibit a strong coupling with a significantly large electric polarization at room temperature. Though the magnetism induced multiferroics (type-II) exhibit a strong coupling between magnetism and ferroelectricity, the unusual magnetic ordering that breaks the inversion symmetry occurs at low temperatures because these complex magnetic structures result from frustrated magnetic interactions. As an alternative route to enhance the magnetoelectric coupling at high temperatures, the polar pyroelectric magnets have been suggested. Since the electric polarization in spin-induced multiferroics is small ($\sim 10^{-2}$ $\mu C/cm^2$), pyroelectric measurements have been widely used to measure electric polarization. Most often, the pyroelectric current mimics thermally induced free charge carriers and thus leads to ambiguity in identifying a true ferroelectric transition. A new DC bias current method, which has been successful in differentiating these two origins, has been discussed at the end of this chapter.

2. Linear Magnetoelectrics

Materials that show electric polarization under magnetic field and magnetization by electric field, where they exhibit linear response to the applied fields are called linear magnetoelectrics and the phenomenon is simply known as *magnetoelectric effect*. It should be noted that these materials exhibit polarization at the magnetic ordering temperature only under applied magnetic fields and thereby differ from the multiferroics where a spontaneous electric polarization occurs at the magnetic ordering temperature or independently at high temperatures. The magnetoelectric effect, which is coupling between magnetic and electric degrees of freedom, is thus described by using the free energy equation;

$$P_i(E,H) = -\frac{\partial F}{\partial E} = P_i^S - \frac{1}{2}\varepsilon_0\varepsilon_{ij}E_j - \alpha_{ij}H_j - \frac{1}{2}\beta_{ijk}H_jH_k \qquad (1)$$

$$M_i(E,H) = -\frac{\partial F}{\partial H} = M_i^S - \frac{1}{2}\mu_0\mu_{ij}H_j - \alpha_{ij}E_j - \frac{1}{2}\gamma_{ijk}E_jE_k. \qquad (2)$$

Here, P_i^S and M_i^S are the ith component of the spontaneous electric polarization and spontaneous magnetization of the material. The second term in Eq. (1) (or (2)) is associated with the polarization (magnetization) contributed by the electric field (magnetic field) where ε and μ are electric and magnetic susceptibilities, respectively. The magnetoelectric coefficient (α_{ij}), is an axial tensor of rank 2, which couples the electric and magnetic order parameters. From Eqs. (2) and (3), polarization depends linearly on the applied magnetic field and magnetization depends linearly on the applied electric field.

2.1. *Symmetry considerations*

Symmetry plays an important role in deciding the magnetoelectric and multiferroic properties of the materials. Mainly, symmetry in magnetoelectrics or multiferroics refers to spatial inversion I and time reversal T. First, let us consider polarization $\vec{P}$ which is a polar vector, which changes sign under the spatial inversion symmetry and invariant under time reversal symmetry, in other words polarization is odd under I and even under T:

$$I\vec{P} = -\vec{P}$$
$$T\vec{P} = \vec{P} \, . \tag{3}$$

On the other hand, magnetization $\vec{M}$ is an axial or pseudo vector, changes sign under the time reversal symmetry and invariant under spatial inversion.

$$I\vec{M} = \vec{M}$$
$$T\vec{M} = -\vec{M} \, . \tag{4}$$

These properties together with the Eqs. (1) and (2) are very important for understanding the magnetoelectric properties. From Eqs. (3) and (4) with the above transformation rules, it is clear that the linear magnetoelectric effect exists in a system only if inversion and time reversal symmetries are simultaneously broken which leaves the magnetic structure invariant under the combined operation of IT. In other words, for a material to be linear magnetoelectric, at least the magnetic point group must break the inversion symmetry with respect to spin and thus allow the occurrence of E_iH_j – term in the free energy equation.

2.2. *Microscopic origin*

Symmetry considerations are very useful to predict whether a material can be magnetoelectric or not. However, it cannot tell us the magnitude of the magnetoelectric response. In order to determine the magnetoelectric coefficients α_{ij}, it is important to understand the microscopic mechanisms that induce the magnetoelectric effect. Various mechanisms such as single-ion anisotropy, symmetric and antisymmetric super-exchange, dipolar interactions, Zeeman energy have been suggested for different types of magnetic ordering.[19] Theoretical analyses of various microscopic origins suggest a weak magnetoelectric effect. It was shown that the magnetoelectric response is limited by the relation[20],

$$\alpha_{ij} \leq \sqrt{\chi_{ii}^{e}\chi_{jj}^{m}}$$

where χ^e and χ^m are the electric and magnetic susceptibilities. This equation suggests that the magnetoelectric effect can be large in ferroelectric and ferromagentic materials.

2.3. *Linear magnetoelectric materials*

Soon after the experimental demonstration of magnetoelectric effect in Cr_2O_3 many other materials such as $GaFeO_3$, $TbPO_4$, $TbCoO_3$, $LiTPO_4$ (T = Mn, Fe, Co, Ni), $RAlO_3$ (R = Tb, Dy & Gd) were reported to exhibit a linear magnetoelectric effect.[5,21–23] As of today, it is believed that nearly 100 compounds have been identified. Recently, linear magnetoelectric effect has been reported in $MnTiO_3$, $A_2M_4O_9$ (A = Nb & Ta, M = Mn, Fe & Co), $NdCrTiO_5$, Cr_2WO_6, etc.[24–30] In all these compounds, electric polarization appears at the magnetic ordering only under an applied external magnetic field and the magnitude of polarization varies linearly with the magnetic field. Spinel compounds with magnetic ions present at the A-site exhibit a plethora of magnetic phenomena, such as spin liquid, orbital liquid, or spin glass due to frustration pertinent to the A-site diamond lattice with a competing nearest (J_1) and next nearest neighbor (J_2) couplings. However, some of the spinels show a long range antiferromagnetic ordering at low temperatures. Herein, we discuss the observation of the linear magnetoelectric effect in a family of A-site antiferromagnetic spinels, $MnGa_2O_4$, $MnAl_2O_4$, Co_3O_4 and $CoAl_2O_4$.[31,32]

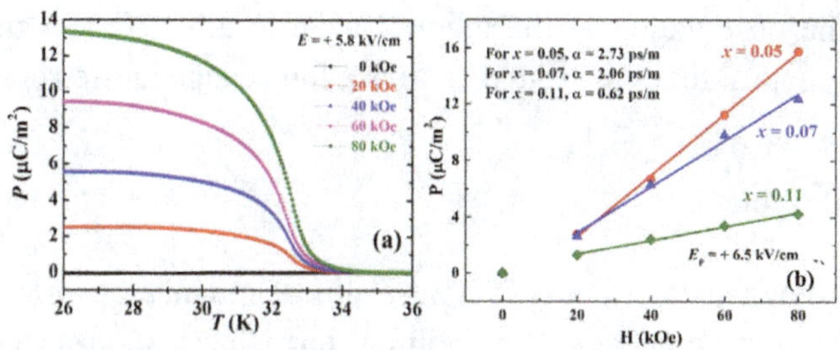

Fig. 1. Temperature and field dependent magnetoelectric polarization for (a) Co_3O_4 and (b) $CoAl_2O_4$. Adapted from Refs. 31 & 32.

All these compounds crystallize in a normal spinel structure with the space group $Fd\bar{3}m$ where the magnetic ions $Co^{2+}(Mn^{2+})$ occupy the tetrahedral (8a) site and the nonmagnetic $Co^{3+}(Ga^{3+})$ ions occupy the octahedral (16d) site. Here, we discuss briefly the compounds, Co_3O_4 and $MnGa_2O_4$, which undergo antiferromagnetic ordering at 30 and 32 K, respectively. Electric polarization appears at the magnetic ordering temperature in both these compounds only under an applied magnetic field as shown in Figure 1(a), demonstrating the linear magnetoelectric effect.[31] The calculated values of magnetoelectric coefficient α is 2.6 ps/m for Co_3O_4 and 0.17 ps/m for $MnGa_2O_4$ at 20 K, which are comparable to that of $NdCrTiO_5$. The magnetic space group of $MnGa_2O_4$ is $R\bar{3}'m'$, which breaks both inversion and time reversal symmetries and conform to the observed magnetoelectric effect. The observation of the linear magnetoelectric effect in the collinear antiferromagnetic spinels confirms the theoretical prediction that the single ion anisotropy of the magnetic ions Co^{2+}/Mn^{2+} located at the local noncentrosymmetric crystal environment (T_d) can cause a magnetoelectric effect.

The isostructural compound, $CoAl_2O_4$, represents an interesting case where the ratio (J_2/J_1) of the two competing interactions lie close to the critical region which separates the long range antiferromagnetic and the spiral spin liquid states. Unlike the other A-site magnetic spinels discussed above, the magnetic property of this compound is sensitive to the degree of antisite disorder. With increasing antisite disorder in $Co_{1-x}Al_x[Al_{2-x}Co_x]O_4$ (x = 0.05, 0.07, 0.11), the magnitude of magnetoelectric coefficient decreases and the sample with x = 0.14 exhibits a spin glass behavior without the magnetoelectric effect as shown in Figure 1(b).

Further, the observation of the magnetoelectric effect in low disordered materials suggest that the ground state is a long range antiferromagnet.[32]

3. Multiferroics

Since the magnetoelectric coefficient in linear magnetoelectric materials is small and the suggestion of combining ferroelectric and ferromagnetic materials with large electric and magnetic susceptibilities opened up a new route to enhance the magnetoelectric effect. Due to the limitation of having insulating ferromagnets, in general, the multiferroics are referred to materials that exhibit simultaneously spontaneous electric polarization and any kind of long-range magnetic ordering. To have a large magnetoelectric effect, it is important to have a strong coupling between the magnetic and electric order parameters. Most of the high-performance ferroelectric materials are perovskite oxides, for example $BaTiO_3$, where the transition metal ion in the highest oxidation state (Ti^{4+}) with d^0 electronic state is responsible for ferroelectric polarization. On the other hand, the magnetism requires partially filled d-shell. Thus, the mechanism of the second order Jahn-Teller distortion or covalency of transition metal ions with the surrounding oxygen ions associated with the classical ferroelectrics does not allow the incorporation of magnetism in such materials. However, the last two decades of research on multiferroics generated new routes to combine ferroelectricity and magnetism in the same phase. Based on the mechanism of generation of ferroelectricity, the known multiferroic materials have been classified into type–I and type–II multiferroics.[15,18] After discussing briefly these two types of multiferroics, including their symmetry requirements, we propose a new class of multiferroics based on polar pyroelectric magnets.

3.1. *Type-I multiferroics*

Type-I multiferroics are polar magnetic materials that exhibit an independent origin of ferroelectric and magnetic ordering at different temperatures. Since the polar distortion in these materials occurs at high temperatures, the spatial inversion symmetry (I) is already broken and the time reversal symmetry (T) is broken at the magnetic ordering temperature. Thus, the symmetry conditions are met easily in this class of multiferroics. However, the polar distortion in these materials is predominantly associated with the off-center distortion of nonmagnetic ions. Thus, these materials show a weak coupling between magnetism and ferroelectricity. Despite this weakness, there has been a lot of interest in combining ferroelectricity of various origins with magnetism in the recent years. Based on the mechanism of origin polar distortion, the type-I multiferroics are further classified; some of them are discussed below.

3.1.1. *Ferroelectricity due to lone pairs*

In the well-known perovskite oxide $PbTiO_3$, in addition to the covalency of d^0 transition metals with their surrounding oxygen ions, the 6s lone pair electrons of Pb^{2+} ion due to its stereochemical activity also contributes significantly to the ferroelectric polarization.[33] The latter mechanism allows combining the ferroelectricity and magnetism in $PbVO_3$ where the ferroelectricity arises from the off centering of Pb^{2+} ions and the magnetism is due to V^{4+} ions.[34] This approach has been applied to the family of perovskite compounds $BiMO_3$ (M =3d transition metal) to explore the possibility of lone pair effect associated with Bi^{3+} ions. Only $BiFeO_3$ has been the centre of attraction because of its high ferroelectric transition (T_C = 1103 K) with large ferroelectric polarization (60 $\mu C/cm^2$) at room temperature but the Fe^{3+} spins undergo spiral antiferromagnetic ordering at T_N = 643 K.[35] Except $BiFeO_3$, all other compounds in this family require high pressure and high temperature to stabilize the perovskite structure. However, not all compounds in this family are polar. The compounds with M = Cr and Ni crystallize in centrosymmetric structure. $BiCoO_3$ has a polar structure (*P4mm*) but there is no report on the ferroelectric properties probably due to difficulties in preparing an insulating sample. $BiMnO_3$ was reported to be ferromagnetic and ferroelectric with the monoclinic (*C2*) crystal structure. Later, it has been shown that the structure and physical properties are very sensitive to oxygen nonstoichiometry and yet there is no consensus on the polar nature of this compound.[36] With the extension of this approach to nonmagnetic *M* cations such as Al, Sc, Ga and In, polar structure (*R3c*) and ferroelectricity have been shown in BiAlO$_3$ with a remanent polarization of 12 μC/cm^2.[37] Recently, it has been shown that the combination of 1:1 mixture of the isostructural polar (*R3c*) compounds, BiFeO$_3$ and BiAlO$_3$ under high pressure and high temperature conditions resulted in the polar (*R3*) double perovskite with an ordering of isovalent cations (Fe^{3+}/Al^{3+}), as shown in Figure 2, where the butterfly piezoelectric loop confirms the ferroelectric behavior of this compound.[38]

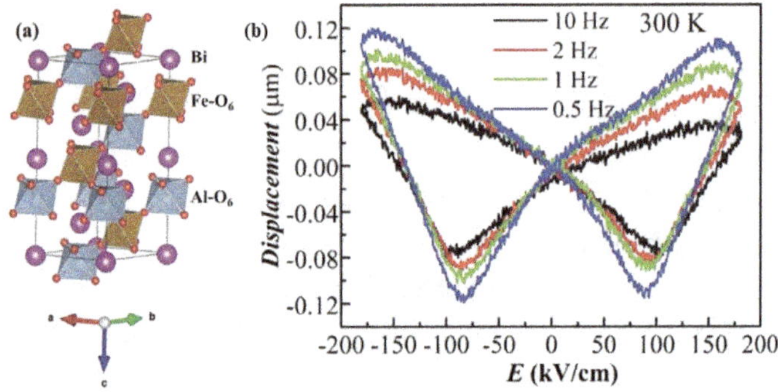

Fig. 2. (a) Crystal structure of Bi$_2$FeAlO$_6$. (b) Piezoelectric displacement loop in Bi$_2$FeAlO$_6$ obtained at 300 K with 180 kVcm^{-1} electric field at various frequencies. Adapted from Ref. 38.

3.1.2. *Ferroelectricity due to geometric frustration*

The manganites *R*MnO$_3$ with bigger *R*-ions (R = La-Dy) crystallize in a centrosymmetric orthorhombic (*Pnma*) perovskite structure. For smaller *R*-ions (R = Ho-Lu and Y), at an ambient pressure condition, these compounds crystallize in a polar hexagonal structure (*P6$_3$cm*). In YMnO$_3$, the ferroelectric transition occurs at T_C = 950 K and the Mn^{3+} ions undergo antiferromagnetic ordering at T_N = 77 K.[39] Unlike the perovskite oxides where the *proper* ferroelectricity comes from zone-center polar structural instability, the hexagonal manganites are known as *improper* ferroelectric where the polar distortion is a result of a complex lattice distortion. Structurally the hexagonal manganites are quite different from the perovskite, where the Mn^{3+} ions are five coordinated with oxygen in trigonal bipyramidal configuration. Consequently, we have different crystal field scheme where the Mn-3d levels are split into two lower energy doublets and a high-energy singlet. More importantly, the buckled MnO_5 polyhedra drive the displacement of Y-ions, which results in a net electric polarization. Since the ferroelectricity and magnetism have different origins the magnetoelectric coupling in these materials remains weak.

3.1.3. *Ferroelectricity due to charge ordering*

Certain type of charge ordering has been suggested to be a source of electric polarization. For example, the coexistence of site-centered and bond-centered ordering can induce ferroelectric polarization in $(Pr,Ca)MnO_3$ and $LuFe_2O_4$. However, the ferroelectricity in these compounds has not yet been confirmed.[40]

3.2. *Type-II multiferroics*

Unlike type-I multiferroics, type-II multiferroics have a centrosymmetric crystal structure and the inversion symmetry is broken due to magnetic ordering with certain type of spin structures where it induces a spontaneous electric polarization.[12,41] Since the polarization is the secondary effect of magnetic ordering, these materials are called type-II multiferroics with improper ferroelectricity, in which the electric polarization is two orders less than that in type-I multiferroics. As the polarization originates from magnetism, the coupling between ferroelectricity and magnetism is strong in these materials. We can classify the type-II multiferroics based on various magnetic structures that break the inversion symmetry when they are placed in a certain lattice structure. Here, we classify the known magnetic structures in type-II multiferroics into two types, namely, (i) noncollinear and incommensurate spiral spin structure and (ii) commensurate collinear spin structures. Several types of spiral spin structures have been reported in the literature and different microscopic mechanisms of inducing electric polarization have been suggested.[12,42–46] We discuss below the various spiral spin structures with specific examples, followed by the collinear spin structure.

3.2.1. *Spiral magnetic structures*

3.2.1.1. *Cycloidal magnetic structure*

The orthorhombic (*Pbnm*) perovskite TbMnO$_3$ is a well-known example for cycloidal-spin induced ferroelectric polarization, in which a strong coupling between ferroelectric polarization and magnetism has been demonstrated.[47] This compound undergoes a collinear sinusoidal antiferromagnetic ordering at $T_{N1} \sim 42$ K followed by a cycloidal spin ordering at $T_{N2} \sim 28$ K. The cycloidal spins lie in the *bc* plane with the modulation vector along the *b*-axis. The cycloidal spin ordering breaks the inversion symmetry and induces electric polarization along the *c*-axis ($P_c \sim 600\text{-}800$ μC/m^2), which is explained theoretically by the spin-current or the inverse Dzyaloshinskii-Moriya (DM) model.[43,45] According to this model, the electric polarization is governed by the equation,

$$P \propto A_0\, e_{ij} \times (S_i \times S_j)$$

where e_{ij} is the unit vector connecting the neighboring spins S_i and S_j, and A_0 is determined by the spin-orbit and spin-exchange interactions. The direction of polarization in the cycloidal-spin induced ferroelectricity can be controlled by applied magnetic fields. When the field is applied along the *b*-axis ($H\|$b), the polarization flops from P_c to P_a above a critical field due to the rotation of cycloidal-spins from *bc* to *ab* plane. This effect may be considered as a giant magnetoelectric effect. The flop of *bc* to *ab* cycloidal plane was also observed with change in temperature as in Eu$_{0.55}$Y$_{0.45}$MnO$_3$, where it undergoes a successive magnetic transition to a collinear spin density wave ($T_{N1} = 46$ K), then *bc*-cycloidal ($T_{N2} = 24$ K) and followed by *ab*-cycloidal at $T_{N3} = 22$ K.[48] Thus, the effects of the magnetic field and temperature on the magnetic ground state suggest that the energy of the two cycloidal states is nearly degenerate due to a small difference in the magnetic anisotropy. There are reports of *ab*-cycloidal induced polarization either in the ground state or under applied magnetic field in GdMnO$_3$.[49]

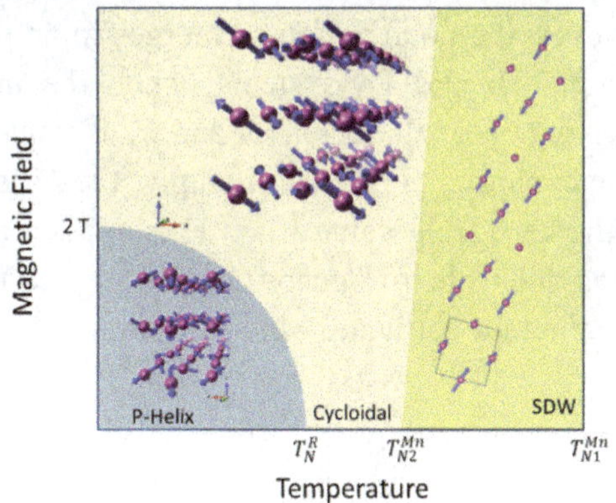

Fig. 3. Schematic phase diagram of various magnetic phases with temperature and magnetic field in Gd$_{0.5}$Dy$_{0.5}$MnO$_3$. Adapted from Ref. 50.

Interestingly though, the compound Gd$_{0.7}$Tb$_{0.7}$MnO$_3$ first undergoes the usual spin density wave transition but transforms to *ab*-cycloidal and finally to a nonpolar A-type antiferromagnetic phase.[51] The 4*f*-moments of Tb^{3+}- ions in TbMnO$_3$ order at low temperature ($T_N^{Tb} = 7$ K) but do not contribute to the ferroelectric polarization. On the other hand, Dy^{3+}-moments in DyMnO$_3$ are induced to order at the Mn-ordering temperature with the same wave vector as the Mn and hence contribute to the total ferroelectric polarization ($P_c \sim 2000$ µC/m^2).[52] However, upon independent commensurate ordering of Dy-moments the polarization reduces to the value of TbMnO$_3$. Applied magnetic fields recover the original polarization.

A recent study on a single crystal of mixed magnetic rare-earth system Gd$_{0.5}$Dy$_{0.5}$MnO$_3$, which can be considered equivalent to TbMnO$_3$ according to the average size of Gd and Dy ions, shows an unusual temperature and magnetic field behavior of polarization.[50] Similar to *R*MnO$_3$ systems (*R* = Tb and Dy), this compound undergoes a nonpolar spin density wave transition followed by the cycloidal polar phase as seen in Figure 3. Intriguingly, the cycloidal spins are located in a tilt plane (102) with respect to the crystallographic axes and thus induce polarization along both *a*- and *b*-directions with $P_a > P_c$ as shown in

Figure 4. The polarization is suppressed below the short-range ordering of *R*-ions (10 K) to a small but finite value due to the transformation of the cycloidal spiral into the helical spin structure with a cycloidal component. However, the polarization reemerges in an applied magnetic field. In contrast to Tb and Dy-systems, when the magnetic field is applied along *a*-axis ($H\|$a), P_a decreases and P_c is enhanced because of the rotation of the cycloidal plane. Surprisingly, the P_a is enhanced three times of that of the zero-field value when H is applied along the *b*-axis. This demonstrates the role of anisotropy of rare-earth ions and the coupling between the rare-earth and Mn-subsystems.

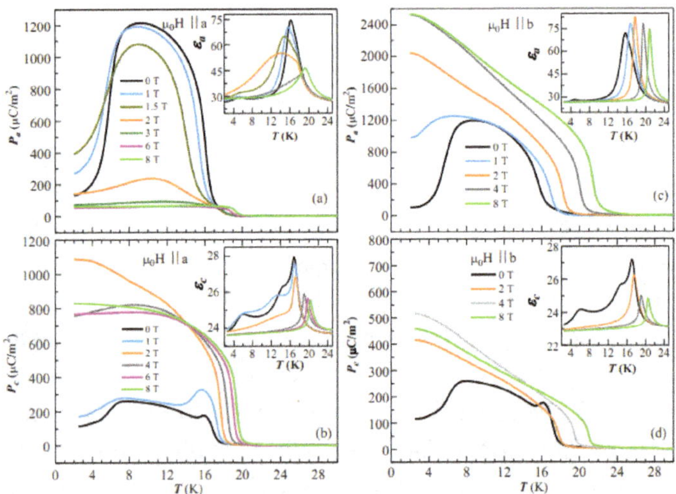

Fig. 4. (a)-(b) show the suppression and enhancement electric polarization of P_a and P_c when magnetic field is applied along *a* direction. (c)-(d) shows temperature-dependence of P_a and P_c under applied field along *b*-direction in $Gd_{0.5}Dy_{0.5}MnO_3$. Adapted from Ref. 50.

Many other type-II multiferroic compounds with cycloidal spin induced ferroelectric polarization have been reported. Due to page limitation, we briefly mention several materials exhibiting multiferroic properties arising from cycloidal spin structures. Some of the materials in which the cycloidal spin structure results from frustrated magnetic interactions are, $Ni_3V_2O_8$, $MnWO_4$, $LiCuVO_4$, $BaYFeO_4$ and CuO.[53–57]

3.2.1.2. *Conical magnetic structure*

The prototypical material that exhibit a conical spin structure is the normal cubic spinel ($Fd\bar{3}m$) $CoCr_2O_4$, which undergoes a ferrimagnetic transition at T_C = 93 K.[58] At a low temperature, it transforms to the incommensurate conical spin states at T_S = 26 K, which is a combination of uniform and transverse spiral spin states, where the spontaneous electric polarization occurs. At T ~ 15 K, it undergoes a lock-in transition below which the spin structure is commensurate. The mechanism of the inverse DM interaction or spin-current model has been successful in explaining the origin of the electric polarization in the conical spin structure as well. The other material that belongs to this category is the Y-type hexaferrite $Ba_2Mg_2Fe_{12}O_{22}$, which shows the longitudinal spin structure but under low magnetic field it transforms to a transverse conical structure and thus induces polarization.[59]

3.2.1.3. *Proper-screw spin structure*

Materials with proper-screw spin structures are most ubiquitous among the spiral magnets. In this structure, the spins rotate in the plane perpendicular to the propagation vector (k). The prototypical examples for this spin structure are the delafossite oxides AMO_2 (A = Cu, Ag and M = Cr, Fe) where the magnetic MO_2 units form a layered triangular lattice.[60,61] The spin current model mentioned in section 3.2.1.1 does not produce spontaneous electric polarization in this structure. An alternative model is the spin-dependent p-d hybridization from Ref. 47, which involves only a single magnetic site coupled to a ligand ion, was suggested to account for the observed electric polarization. In this model, the covalency between the transition metal and ligand is modulated depending on the direction of the local spin moment through relativistic spin orbit coupling. The typical example is the rhombohedral $CuFeO_2$ where a proper screw magnetic structure with a 120° spin rotation angle allows electric polarization along the bond direction.[61] The other compounds that exhibit proper-screw spin induced ferroelectric are MI_2 (M = Mn, Ni) with CdI_2 structure.[62] However, a recent symmetry analysis study has shown that the p-d hybridization cannot induce electric polarization if the magnetic ions are located at the local centrosymmetric site.[63] But this mechanism is applicable to noncentrosymmetric multiferroics such as the proper screw $Ba_2CoGe_2O_7$ with a tetragonal crystal structure ($P\bar{4}2_1m$) and the collinear antiferromagnet $RFe_3(BO_3)_4$ having a $R32$ structure.[64,65]

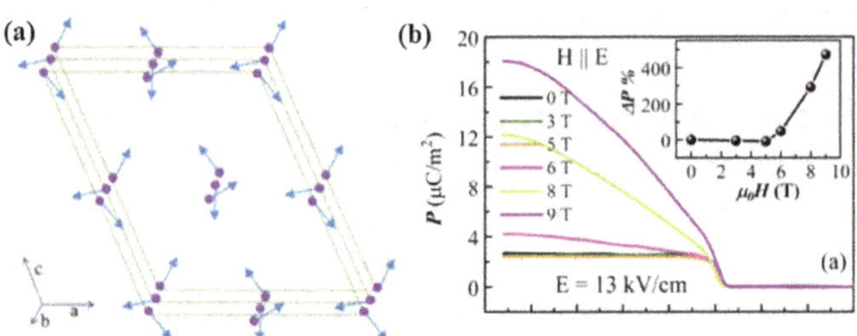

Fig. 5. (a) Magnetic structure of $MnSb_2S_4$. (b) Temperature evolution of electric polarization under various magnetic fields. Inset shows the magnetic field variation of electric polarization. Adapted from Ref. 66.

Recently, a spin driven ferroelectricity has been reported in monoclinic ($C2/m$) $MnSb_2S_4$ where the MnS_6 octahedra share their edges along the b-axis.[66] The chains are connected along the a-axis through a distorted square pyramid of Sb ions and form layers parallel to the c-axis. This compound undergoes an antiferromagnetic ordering at 25 K where a dielectric anomaly and a spontaneous electric polarization occur. A remarkable enhancement of polarization (~ 500%) is observed under the magnetic field (5 T) applied parallel to the electric field (Figure 5(b)). On the other hand, for $H \perp E$, the polarization does not change significantly, demonstrating the spin driven ferroelectricity. The neutron diffraction study has shown an incommensurate helical spin structure where the magnetic moments of manganese lie in the ac plane (Figure 5(a)). The angle between the adjacent spin is ~133° along the chains of MnS_6 octahedra (b-axis) while it is 66.85° along the a-axis and collinear along the c-axis. As the propagation vector lies along the b-axis, this magnetic structure can be considered as a proper screw type as in the case of delafossite except for the angle between adjacent spins. Since the Mn^{2+} ions in this structure are located in the centrosymmetric environment, which precludes single spin contributions to the magnetoelectric effect, although consideration of clusters of spins can induce electric polarization according to the *p-d* hybridization mechanism. Therefore, the magnetoelectric effect in this material was explained by the more general form of inverse Dzyaloshinskii-Moriya (DM) type interactions that allows polarization independent of whether the magnetic atoms are located in the centrosymmetric environment or not.

3.2.2. *Collinear spin structures*

Unlike the spiral multiferroics that involve relativistic spin orbit interactions in a specific crystal lattice, there are other multiferroics that exhibit a certain type of collinear spin structure which breaks the inversion symmetry and induces ferroelectricity. For example, the compound Ca_3CoMnO_6 with one-dimensional chains of alternating Co^{2+} and Mn^{4+} ions shows an antiferromagnetic ordering with up-up-down-down spin configuration, which breaks the inversion symmetry and thus induces ferroelectric polarization along the chain direction.[67] In the paramagnetic state, the crystal structure has an inversion centre which is broken below the magnetic ordering by the exchange-striction mechanism which shortens the distance between the ions of parallel spins and elongates the bonds between the antiparallel spins.[68]

A similar up-up-down-down spin configuration induced multiferroicity is reported in the orthorhombic $RMnO_3$ with smaller R-ions (R=Ho, Er, Tm, Yb and Lu) synthesized at high pressure.[41,44] In these materials, the magnetic ions are identical but the Mn^{3+}-O-Mn^{3+} bond direction is zig-zag. With the up-up-down-down spin configuration, it has been shown that the exchange striction causes a shift of oxygen ions perpendicular to the Mn-Mn bonds and thus induces polarization along the same direction. Other examples of multiferroics exhibiting exchange striction induced ferroelectric polarization are: RMn_2O_5 (R = Y, Tb, Ho, Er or Tm) and $RFeO_3$ (R = Gd, Tb and Dy).[69–71] In general, the magnitude of electric polarization emerging from the exchange striction mechanism is higher than that induced by the mechanism involving relativistic spin orbit interactions.

3.3. *Type-III multiferroics*

Here, we discuss a new class of polar magnets, where the polar distortion is driven by a chemical or charge ordering of cations and thus the polar structure is stabilized right at the formation temperature of these compounds. Consequently, some of these polar materials have been shown to be pyroelectric at all temperatures. Hence, they differ from type-I multiferroics where they undergo a ferroelectric (polar) - paraelectric (nonpolar) transition. However, a change in polarization (ΔP) occurs at the magnetic ordering temperatures in type-III multiferroics, which evidences magnetoelectric coupling but the extent of the magnetoelectric effect depends on the magnetic spin structure and therefore the microscopic mechanisms that couple the magnetism with electric polarization. Though this behavior is similar to that of type-II multiferroics, the inversion symmetry in the later is broken only at the magnetic ordering temperature by complex magnetic structures. For these reasons, we can classify the polar pyroelectric magnets as type-III multiferroics. This class of multiferroics is promising because they do not necessarily involve a complex magnetic structure arising from spin frustration and therefore one can combine appropriate magnetic ions with the polar structure to achieve a magnetoelectric effect at room temperature. Furthermore, because of the pyroelectric nature, this class of multiferroics may not require poling electric field to measure electric polarization. Here, we discuss some examples of pyroelectric magnets that can be classified as type-III multiferroics.

The first example to be listed in this class of pyroelectric magnets is the mineral kamiokite $Fe_2Mo_3O_8$ that crystallizes in the polar hexagonal structure ($P6_3mc$).[72] Another example could be the pyroelectric ferrimagnet $CaBaCo_4O_7$ crystallizing in the polar orthorhombic structure ($Pbn2_1$) where CoO_4 tetrahedra form alternating layers of triangular and Kagome lattices with Co^{2+}/Co^{3+} charge ordering in four different crystallographic sites.[73] The other known charge ordered polar magnets that may be classified under type-III multiferroics are, $GaFeO_3$, Ni_3TeO_6 and related compounds.[74,75] Several other polar magnets have been reported to be potential for magnetoelectric properties.[76]

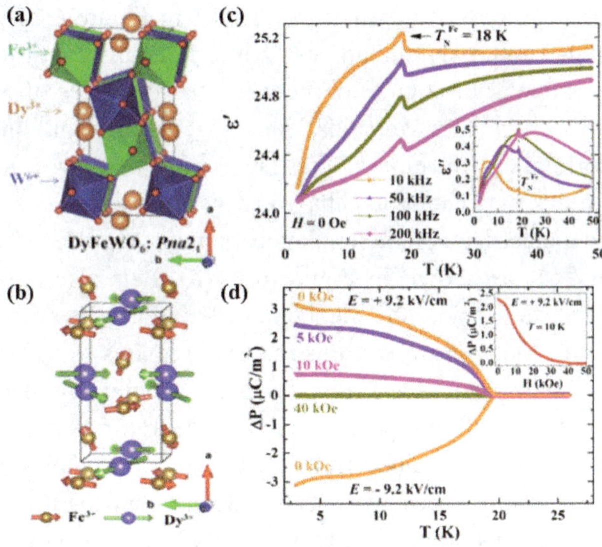

Fig. 6. (a)-(b) Crystal and magnetic structures of DyFeWO₆. (c) Temperature variation of the real part of dielectric constant $\varepsilon'(T)$ of DyFeWO₆ measured with different frequencies. The inset shows the corresponding imaginary part of the dielectric constant $\varepsilon''(T)$. (d) Temperature evolution of electric polarization $\Delta P(T)$ of DyFeWO₆ below $T_N^{Fe} = 18$K. The inset shows a magnetic-field-dependent change in electric polarization measured at 10 K. Adapted from Ref. 77.

Recently, we reported a new family of polar ($Pna2_1$) magnet RFeWO₆ (R = Eu, Tb, Dy and Y), which is an ordered derivative of centrosymmetric ($Pnma$) aeschynite structure.[77] In this structure, the Fe^{3+} and W^{6+} ions are ordered alternatively because of a significant difference in size and charge of these ions, which is responsible for stabilizing the polar structure. The FeO₆ and WO₆ octahedra form a dimer by sharing their edges and the dimers are linked by corner sharing of oxygen, as shown in Figure 6(a). The Fe-moments in these compounds undergo antiferromagnetic ordering at $T_N^{Fe} \sim$ 15-18 K. The magnetic R-ions show a different magnetic behavior. Europium remains paramagnetic down to 2 K and Tb moments order antiferromagnetically at 2.4 K. Magnetization, heat capacity and neutron diffraction measurements in DyFeWO₆ confirmed the antiferromagnetic ordering of Fe at T_N = 18 K with the propagation vector $\mathbf{k}$ = (0, ½, ½). Intriguingly, the Dy-moments are also found to be ordered at the Fe-ordering temperature with the same wave vector. The magnetic structure is given in Figure 6(b) where it is seen that both Fe and Dy sublattices form a noncollinear structure with nearly perpendicular configuration between the two sites of Fe and Dy. The magnetic structure remains unchanged although independent magnetic correlations develop among Dy-moments below 5 K, which is evident from the metamagnetic behavior.

A clear dielectric anomaly and appearance of electric polarization is observed at T_N^{Fe}, as shown in Figure 6(c)-(d). It is also clear that the electric polarization below the magnetic ordering can be switched by an external electric field. However, this material remains pyroelectric in the paramagnetic state. From the symmetry point of view, the paramagnetic space group $Pna2_1$ allows polarization of the form $P = (0,0,p_z)$. The magnetic space group obtained from the analysis of neutron data is C_ac and the point group $m1'$ allows electric polarization of the form $\mathbf{P}_m = (p_x,0,p_z)$. It requires further measurements on a single crystal to understand the magnetism induced electric polarization. The electric polarization disappears under magnetic fields (> 1 T) indicating a possible change of the magnetic structure. Similar dielectric anomaly and electric polarization were observed at the Neel temperature of Fe in the compounds, $RFeWO_6$ with $R =$ Eu, Tb and Y. However, the effect of the magnetic field on the suppression of polarization is different for different R-ions. The electric polarization in $YFeWO_6$ remains significantly high even at the field of 8 T, demonstrating the strong coupling of $4f$-$3d$ moments and the magnetoelectric effect.

4. DC Bias Technique

The standard P-E loop measurements are successful with the classical ferroelectrics and type-I multiferroics with large polarization. However, this method cannot be used in the case of type-II and type-III multiferroics because the electric polarization in these materials is of the order $\sim 10^{-2}$ $\mu C/cm^2$. In this case, we measure pyroelectric current, after cooling the sample in a poling electric field, across the ferroelectric transition and integrate the current to get the temperature dependence of the spontaneous electric polarization. In many materials, a symmetric peak in pyrocurrent is observed in the vicinity of magnetic transition due to the leakage current and thermally stimulated charge carriers, which does not indicate an intrinsic electric polarization. In the literature, there are many publications that associate such pyroelectric currents to spontaneous electric polarization. The shape of the pyrocurrent peaks is asymmetric for intrinsic electric polarization. In many cases, the heating rate dependent pyroelectric current is useful to differentiate the intrinsic polarization from the extrinsic effects discussed above. We have introduced a slightly different method, which we termed as a DC bias current measurement that can identify the intrinsic polarization.[78]

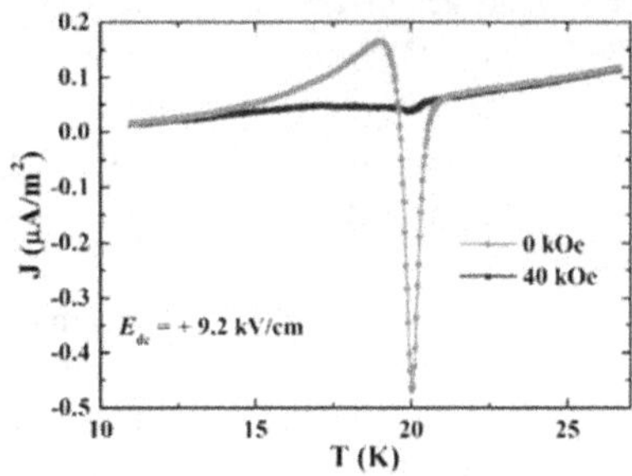

Fig. 7. Temperature dependence of a DC-biased current of $DyFeWO_6$. Adapted from Ref. 77.

In this method, we cool the sample to the lowest temperature without the poling field and apply an electric field, then measure the pyrocurrent while warming across the ferroelectric transition in the presence of the electric field. According to this protocol, we expect positive and negative peaks corresponding to polarization and depolarization currents, as shown in Figure 7, for $DyFeWO_6$ discussed in section 3.3.[77] In the case of non-ferroelectrics, we do not observe any feature at the magnetic transition but a monotonous increase of pyrocurrent appears for leaky dielectrics. Therefore, the DC bias method is very useful to identify intrinsic ferroelectricity induced by magnetism.

5. Summary

In this chapter, we have made an attempt to distinguish the linear magnetoelectric and multiferroic materials emphasizing the stringent symmetry requirements. The focus is made mainly on the recent developments in single-phase magnetoelectrics and multiferroics. The extensive work on these materials have led to the identification of polar pyroelectric magnets which although resemble type-II multiferroics, the magnetic structure and the mechanism of inducing change in polarization at the magnetic ordering are quite different and thus they are classified as type-III multiferroics.

References

1. Curie, P. On symmetry in physical phenomena, symmetry of an electric field and of a magnetic field. *J. Phys.* **3**, 401 (1894).
2. Landau, L. D., Bell, J. S., Kearsley, M. J., Pitaevskii, L. P., Lifshitz, E. M. and Sykes, J. B. *Electrodynamics of Continuous Media.* **8**, (Elsevier, 2013).
3. Dzyaloshinskii, I. E. On the magneto-electrical effects in antiferromagnets. *Sov. Phys. JETP* **10**, 628–629 (1960).
4. Astrov, D. N. The magnetoelectric effect in antiferromagnetics. *Sov. Phys. JETP* **11**, 708–709 (1960).
5. Rado, G. T. Observation and possible mechanisms of magnetoelectric effects in a ferromagnet. *Phys. Rev. Lett.* **13**, 335 (1964).
6. Smolenskiĭ, G. A. and Chupis, I. E. Ferroelectromagnets. *Sov. Phys. Uspekhi* **25**, 475 (1982).
7. Ascher, E., Rieder, H., Schmid, H. and Stössel, H. Some Properties of Ferromagnetoelectric Nickel-Iodine Boracite, $Ni_3B_7O_{13}I$. *J. Appl. Phys.* **37**, 1404–1405 (1966).
8. Schmid, H. Multi-ferroic magnetoelectrics. *Ferroelectrics* **162**, 317–338 (1994).
9. Hill, N. A. Why are there so few magnetic ferroelectrics? *J. Phys. Chem. B* **104**, 6694–6709 (2000).
10. Fiebig, M. Revival of the magnetoelectric effect. *J. Phys. D. Appl. Phys.* **38**, R123 (2005).
11. Eerenstein, W., Mathur, N. D. and Scott, J. F. Multiferroic and magnetoelectric materials. *Nature* **442**, 759–765 (2006).
12. Cheong, S.-W. and Mostovoy, M. Multiferroics: a magnetic twist for ferroelectricity. *Nat. Mater.* **6**, 13–20 (2007).
13. Tokura, Y. Multiferroics-toward strong coupling between magnetization and polarization in a solid. *J. Magn. Magn. Mater.* **310**, 1145–1150 (2007).
14. Wang, K. F., Liu, J. M. and Ren, Z. F. Multiferroicity: the coupling between magnetic and polarization orders. *Adv. Phys.* **58**, 321–448 (2009).
15. Khomskii, D. I. Classifying multiferroics: mechanisms and effects. *Physics (College. Park. Md).* **2**, 20 (2009).

16. Rao, C. N. R., Sundaresan, A. and Saha, R. Multiferroic and magnetoelectric oxides: the emerging scenario. *J. Phys. Chem. Lett.* **3**, 2237–2246 (2012).

17. Khomskii, D. I. Ferroelectrics, magnetoelectrics, and multiferroics. in *Transition Metal Compounds* pp. 269–309 (Cambridge University Press, 2014).

18. Wang, J. *Multiferroic Materials: Properties, Techniques, and Applications.* (CRC Press, 2016).

19. Gehring, G. A. On the microscopic theory of the magnetoelectric effect. *Ferroelectrics* **161**, 275–285 (1994).

20. Brown Jr, W. F., Hornreich, R. M. and Shtrikman, S. Upper bound on the magnetoelectric susceptibility. *Phys. Rev.* **168**, 574 (1968).

21. Rado, G. T., Ferrari, J. M. and Maisch, W. G. Magnetoelectric susceptibility and magnetic symmetry of magnetoelectrically annealed $TbPO_4$. *Phys. Rev. B* **29**, 4041 (1984).

22. Mercier, M. M. Mercier, J. Gareyte, and EF Bertaut, *CR Seances Acad. Sci., Ser. B* **264**, 979 (1967).

23. Hornreich, R. M. The magnetoelectric effect: some likely candidates. *Solid State Commun.* **7**, 1081–1085 (1969).

24. Mufti, N., Blake, G. R., Mostovoy, M., Riyadi, S., Nugroho, A. A. and Palstra, T. T. M. Magnetoelectric coupling in $MnTiO_3$. *Phys. Rev. B - Condens. Matter Mater. Phys.* **83**, 104416 (2011).

25. Fischer, E., Gorodetsky, G. and Hornreich, R. M. A new family of magnetoelectric materials: $A_2M_4O_9$(A = Ta, Nb; M = Mn, Co). *Solid State Commun.* **10**, 1127–1132 (1972).

26. Fang, Y., Zhou, W. P., Yan, S. M., Bai, R., Qian, Z. H., Xu, Q. Y., Wang, D. H. and Du, Y. W. Magnetic-field-induced dielectric anomaly and electric polarization in $Mn_4Nb_2O_9$. *J. Appl. Phys.* **117**, 17B712 (2015).

27. Maignan, A. and Martin, C. $Fe_4Nb_2O_9$: A magnetoelectric antiferromagnet. *Phys. Rev. B* **97**, 161106 (2018).

28. Fang, Y., Song, Y. Q., Zhou, W. P., Zhao, R., Tang, R. J., Yang, H., Lv, L. Y., Yang, S. G., Wang, D. H. and Du, Y. W. Large magnetoelectric coupling in $Co_4Nb_2O_9$. *Sci. Rep.* **4**, 3860 (2014).

29. Hwang, J., Choi, E. S., Zhou, H. D., Lu, J. and Schlottmann, P. Magnetoelectric effect in $NdCrTiO_5$. *Phys. Rev. B* **85**, 24415 (2012).

30. Fang, Y., Wang, L. Y., Song, Y. Q., Tang, T., Wang, D. H. and Du, Y. W. Manipulation of magnetic field on dielectric constant and electric polarization in Cr_2WO_6. *Appl. Phys. Lett.* **104**, 132908 (2014).

31. Saha, R., Ghara, S., Suard, E., Jang, D. H., Kim, K. H., Ter-Oganessian, N. V and Sundaresan, A. Magnetoelectric effect in simple collinear antiferromagnetic spinels. *Phys. Rev. B* **94**, 14428 (2016).

32. Ghara, S., Ter-Oganessian, N. V. and Sundaresan, A. Linear magnetoelectric effect as a signature of long-range collinear antiferromagnetic ordering in the frustrated spinel $CoAl_2O_4$. *Phys. Rev. B* **95**, 094404 (2017).

33. Cohen, R. E. Origin of ferroelectricity in perovskite oxides. *Nature* **358**, 136 (1992).

34. Shpanchenko, R. V, Chernaya, V. V, Tsirlin, A. A., Chizhov, P. S., Sklovsky, D. E., Antipov, E. V, Khlybov, E. P., Pomjakushin, V., Balagurov, A. M. and Medvedeva, J. E. Synthesis, structure, and properties of new perovskite $PbVO_3$. *Chem. Mater.* **16**, 3267–3273 (2004).

35. Wang, J., Neaton, J. B., Zheng, H., Nagarajan, V., Ogale, S. B., Liu, B., Viehland, D., Vaithyanathan, V., Schlom, D. G. and Waghmare, U. V. Epitaxial BiFeO3 multiferroic thin film heterostructures. *Science (80-.)*. **299**, 1719–1722 (2003).

36. Sundaresan, A., Mangalam, R. V. K., Iyo, A., Tanaka, Y. and Rao, C. N. R. Crucial role of oxygen stoichiometry in determining the structure and properties of $BiMnO_3$. *J. Mater. Chem.* **18**, 2191–2193 (2008).

37. Mangalam, R. V. K., Bhat, S. V, Iyo, A., Tanaka, Y., Sundaresan, A. and Rao, C. N. R. Dielectric properties, thermal decomposition and related aspects of $BiAlO_3$. *Solid State Commun.* **146**, 435–437 (2008).

38. De, C., Arévalo-López, Á. M., Orlandi, F., Manuel, P., Attfield, J. P. and Sundaresan, A. Isovalent cation ordering in the polar rhombohedral perovskite Bi_2FeAlO_6. *Angew. Chemie Int. Ed.* **57**, 16099–16103 (2018).

39. Van Aken, B. B., Palstra, T. T. M., Filippetti, A. and Spaldin, N. A. The origin of ferroelectricity in magnetoelectric $YMnO_3$. *Nat. Mater.* **3**, 164–170 (2004).

40. Van den Brink, J. and Khomskii, D. I. Multiferroicity due to charge ordering. *J. Phys. Condens. Matter* **20**, 434217 (2008).

41. Tokura, Y., Seki, S. and Nagaosa, N. Multiferroics of spin origin. *Rep. Prog. Phys.* **77**, 076501 (2014).

42. Tokura, Y. and Seki, S. Multiferroics with spiral spin orders. *Adv. Mater.* **22**, 1554–1565 (2010).

43. Katsura, H., Nagaosa, N. and Balatsky, A. V. Spin current and magnetoelectric effect in noncollinear magnets. *Phys. Rev. Lett.* **95**, 57205 (2005).

44. Sergienko, I. A., Şen, C. and Dagotto, E. Ferroelectricity in the magnetic e-phase of orthorhombic perovskites. *Phys. Rev. Lett.* **97**, 227204 (2006).

45. Mostovoy, M. Ferroelectricity in spiral magnets. *Phys. Rev. Lett.* **96**, 1–4 (2006).

46. Arima, T. Ferroelectricity induced by proper-screw type magnetic order. *J. Phys. Soc. Japan* **76**, 73702 (2007).

47. Kimura, T., Goto, T., Shintani, H., Ishizaka, K., Arima, T. and Tokura, Y. Magnetic control of ferroelectric polarization. *Nature* **426**, 55–58 (2003).

48. De, C., Bag, R., Singh, S., Orlandi, F., Manuel, P., Langridge, S., Sanyal, M. K., Rao, C. N. R., Mostovoy, M. and Sundaresan, A. Highly tunable magnetic spirals and electric polarization in $Gd_{0.5}Dy_{0.5}MnO_3$. *Phys. Rev. Mater.* **3**, 44401 (2019).

49. Murakawa, H., Onose, Y., Kagawa, F., Ishiwata, S., Kaneko, Y. and Tokura, Y. Rotation of an electric polarization vector by rotating magnetic field in cycloidal magnet $Eu_{0.55}Y_{0.45}MnO_3$. *Phys. Rev. Lett.* **101**, 197207 (2008).

50. Noda, K., Nakamura, S., Nagayama, J. and Kuwahara, H. Magnetic field and external-pressure effect on ferroelectricity in manganites: comparison between $GdMnO_3$ and $TbMnO_3$. *J. Appl. Phys.* **97**, 10C103 (2005).

51. Yamasaki, Y., Sagayama, H., Abe, N., Arima, T., Sasai, K., Matsuura, M., Hirota, K., Okuyama, D., Noda, Y. and Tokura, Y. Cycloidal spin order in the a-axis polarized ferroelectric phase of orthorhombic perovskite manganite. *Phys. Rev. Lett.* **101**, 97204 (2008).

52. Goto, T., Kimura, T., Lawes, G., Ramirez, A. P. and Tokura, Y. Ferroelectricity and giant magnetocapacitance in perovskite rare-earth manganites. *Phys. Rev. Lett.* **92**, 257201 (2004).

53. Lawes, G., Harris, A. B., Kimura, T., Rogado, N., Cava, R. J., Aharony, A., Entin-Wohlman, O., Yildirim, T., Kenzelmann, M., Broholm, C. and Ramirez, A. P. Magnetically driven ferroelectric order in $Ni_3V_2O_8$. *Phys. Rev. Lett.* **95**, 87205 (2005).

54. Khomskii, O. H., Hollmann, N., Klassen, I., Jodlauk, S., Bohaty, L., Becker, P., Mydosh, J. A., Lorenz, T., and Khomskii, D. I. New multiferroic material: $MnWO_4$. *J. Phys. Condens. Matter* **18**, L471 (2006).

55. Naito, Y., Sato, K., Yasui, Y., Kobayashi, Y., Kobayashi, Y. and Sato, M. Ferroelectric transition induced by the incommensurate magnetic ordering in $LiCuVO_4$. *J. Phys. Soc. Japan* **76**, 23708 (2007).

56. Cong, J.-Z., Shen, S.-P., Chai, Y.-S., Yan, L.-Q., Shang, D.-S., Wang, S.-G. and Sun, Y. Spin-driven multiferroics in $BaYFeO_4$. *J. Appl. Phys.* **117**, 174102 (2015).

57. Kimura, T., Sekio, Y., Nakamura, H., Siegrist, T. and Ramirez, A. P. Cupric oxide as an induced-multiferroic with high-T_C. *Nat. Mater.* **7**, 291–294 (2008).

58. Yamasaki, Y., Miyasaka, S., Kaneko, Y., He, J. P., Arima, T. and Tokura, Y. Magnetic reversal of the ferroelectric polarization in a multiferroic spinel oxide. *Phys. Rev. Lett.* **96**, 207204 (2006).

59. Kimura, T. Magnetoelectric hexaferrites. *Annu. Rev. Condens. Matter Phys.* **3**, 93–110 (2012).

60. Seki, S., Onose, Y. and Tokura, Y. Spin-driven ferroelectricity in triangular lattice antiferromagnets $ACrO_2$ (A = Cu, Ag, Li, or Na). *Phys. Rev. Lett.* **101**, 067204 (2008).

61. Kimura, T., Lashley, J. C. and Ramirez, A. P. Inversion-symmetry breaking in the noncollinear magnetic phase of the triangular-lattice antiferromagnet $CuFe\,O_2$. *Phys. Rev. B - Condens. Matter Mater. Phys.* **73**, 220401(R) (2006).

62. Kurumaji, T., Seki, S., Ishiwata, S., Murakawa, H., Tokunaga, Y., Kaneko, Y. and Tokura, Y. Magnetic-field induced competition of two multiferroic orders in a triangular-lattice helimagnet MnI_2. *Phys. Rev. Lett.* **106**, 167206 (2011).

63. Matsumoto, M., Chimata, K. and Koga, M. Symmetry analysis of spin-dependent electric dipole and its application to magnetoelectric effects. *J. Phys. Soc. Japan* **86**, 34704 (2017).

64. Hutanu, V., Sazonov, A., Murakawa, H., Tokura, Y., Náfrádi, B. and Chernyshov, D. Symmetry and structure of multiferroic $Ba_2CoGe_2O_7$. *Phys. Rev. B - Condens.*

Matter Mater. Phys. **84**, 212101 (2011).

65.   Zvezdin, A. K., Krotov, S. S., Kadomtseva, A. M., Vorob'ev, G. P., Popov, Y. F.,
      Pyatakov, A. P., Bezmaternykh, L. N. and Popova, E. A. Magnetoelectric effects in
      gadolinium iron borate $GdFe_3(BO_3)_4$. *J. Exp. Theor. Phys. Lett.* **81**, 272–276
      (2005).

66.   De, C., Ter-Oganessian, N. V and Sundaresan, A. Spin-driven ferroelectricity and
      large magnetoelectric effect in monoclinic $MnSb_2S_4$. *Phys. Rev. B* **98**, 174430
      (2018).

67.   Choi, Y. J., Yi, H. T., Lee, S., Huang, Q., Kiryukhin, V. and Cheong, S.-W.
      Ferroelectricity in an Ising chain magnet. *Phys. Rev. Lett.* **100**, 47601 (2008).

68.   Wu, H., Burnus, T., Hu, Z., Martin, C., Maignan, A., Cezar, J. C., Tanaka, A.,
      Brookes, N. B., Khomskii, D. I. and Tjeng, L. H. Ising magnetism and
      ferroelectricity in Ca_3CoMnO_6. *Phys. Rev. Lett.* **102**, 026404 (2009).

69.   Hur, N., Park, S., Sharma, P. A., Ahn, J. S., Guha, S. and Cheong, S.-W. Electric
      polarization reversal and memory in a multiferroic material induced by magnetic
      fields. *Nature* **429**, 392–395 (2004).

70.   Tokunaga, Y., Furukawa, N., Sakai, H., Taguchi, Y., Arima, T. and Tokura, Y.
      Composite domain walls in a multiferroic perovskite ferrite. *Nat. Mater.* **8**, 558
      (2009).

71.   Tokunaga, Y., Iguchi, S., Arima, T. and Tokura, Y. Magnetic-field-induced
      ferroelectric state in $DyFeO_3$. *Phys. Rev. Lett.* **101**, 097205 (2008).

72.   Wang, Y., Pascut, G. L., Gao, B., Tyson, T. A., Haule, K., Kiryukhin, V. and
      Cheong, S.-W. Unveiling hidden ferrimagnetism and giant magnetoelectricity in
      polar magnet $Fe_2Mo_3O_8$. *Sci. Rep.* **5**, 12268 (2015).

73.   Johnson, R. D., Cao, K., Giustino, F. and Radaelli, P. G. $CaBaCo_4O_7$: A
      ferrimagnetic pyroelectric. *Phys. Rev. B* **90**, 45129 (2014).

74.   Saha, R., Shireen, A., Shirodkar, S. N., Waghmare, U. V, Sundaresan, A. and Rao,
      C. N. R. Multiferroic and magnetoelectric nature of $GaFeO_3$, $AlFeO_3$ and related
      oxides. *Solid State Commun.* **152**, 1964–1968 (2012).

75.   Oh, Y. S., Artyukhin, S., Yang, J. J., Zapf, V., Kim, J. W., Vanderbilt, D. and
      Cheong, S. W. Non-hysteretic colossal magnetoelectricity in a collinear
      antiferromagnet. *Nat. Commun.* **5**, 1–7 (2014).

76.   Cai, G. H., Greenblatt, M. and Li, M. R. Polar magnets in double corundum oxides.
      Chem. Mater. **29**, 5447–5457 (2017).

77.   Ghara, S., Suard, E., Fauth, F., Tran, T. T., Halasyamani, P. S., Iyo, A., Rodríguez-
      carvajal, J. and Sundaresan, A. Ordered aeschynite-type polar magnets $RFeWO_6$ (R
      = Dy, Eu, Tb, and Y): A new family of type-II multiferroics. *Phys. Rev. B* **95**,
      224416 (2017).

78.   De, C., Ghara, S. and Sundaresan, A. Effect of internal electric field on ferroelectric
      polarization in multiferroic $TbMnO_3$. *Solid State Commun.* **205**, 61–65 (2015).

Chapter 11

Phase Transitions in Materials

Divya Chalapathi, Priyanka Jain and Chandrabhas Narayana[*]

*School of Advanced Materials and Chemistry and Physics of Materials Unit,
Jawaharlal Nehru Centre for Advanced Scientific Research, Jakkur,
Bangalore 560 064, Karnataka, India*
[*]cbhas@jncasr.ac.in

Matter exhibits several phases as a function of pressure, temperature, composition, etc. which are important in understanding nature and our surroundings. These perturbations provide a detailed picture of the materials by correlating their physics and chemistry. From this knowledge, it is possible to predict, synthesize and characterize new materials. This study of material synthesis and their properties are significant for targeted applications. Further, by classifying these materials based on their phase transitions, a much easier comprehension for researchers is developed. In view of this, the chapter provides basic knowledge and a starting point for the students and researchers to understand the phenomenon of phase transitions in materials.

1. Introduction

Phase transitions in materials are an important phenomenon in the understanding of the physics and chemistry of materials and thus have an important role in materials science. In our day-to-day life activities, we come across many phases of materials and their properties. Understanding these materials provide the reasons for their existence and can lead to various new applications. Water is one of the classic examples used in any textbook to explain its phase transitions. The fact that water molecules can form a tetrahedral structure when they transform into a solid provides us many interesting facts about ice. The tetrahedral nature of water molecules with differential bonds is the reason ice exhibits 14 phases in the solid phase. The understanding of the solid phase of ice is important in its discovery on other planets in the universe. It is pertinent to ask if ice is a unique case of a solid to exhibit multiple phases in the solid state. Interestingly, the recent studies on NH_4F show that ice is not the only one showing a rich phase diagram in the solid phase.[1] From the studies on NH_4F, it is now clear that tetrahedral molecules with differential bonding will indeed produce ice-like phase diagram.[1]

Another interesting material which has been considered a "holy grail of condensed matter" is solid hydrogen. In 1935, Wigner and Huntington predicted that solid hydrogen under pressure would transform into a metal. Many researchers have worked tirelessly in reaching this elusive state of matter in solid hydrogen. It was shown that solid hydrogen remains molecular up to 342 GPa at room temperature, which made the theoreticians relook at the metallic phase of solid hydrogen.[2] Many attempts made till now have not conclusively proven the presence of metallic hydrogen phase. It is significant to understand other planets like Jupiter. It is adequately clear from these two examples that phase transitions in materials must be understood from both a theoretical and an experimental perspective. Together they unravel the properties exhibited by materials for fundamental understanding and applications.

This chapter aims to provide the classification of different phase transitions, their basics and interpret them using selected examples. Broadly, this classification is subject to perturbations caused by external parameters affecting the structural, magnetic and electronic properties of materials. There is an attempt to cover most of the possible phase transitions found in materials with an emphasis on the current trends in material science.

2. Structural Phase Transitions

During a structural phase transition, the overall structure of the materials changes, involving the movement of the atoms, changes in symmetry, volume, state of the matter, etc. These may result in changes in other properties of the material but are considered a consequence of the structural phase transitions. There are many types of structural transitions in materials, and these will be discussed in the subsequent sections.

2.1. *Solid-solid phase transition*

Thermodynamically, during a phase transition in a solid, the free energy remains constant, while quantities like volume, entropy and heat capacity change discontinuously. A transition is classified as first-order, second-order or higher depending upon the discontinuous behavior of the free energy derivative. The thermodynamics of first-order transitions is given by the following Clausius-Clapeyron equation;

$$\frac{\partial p}{\partial T} = \frac{\Delta S}{\Delta V} = \frac{\Delta H}{T \Delta V} \tag{1}$$

$$\frac{\partial p}{\partial T} = \frac{\Delta C_p}{V T \Delta \alpha}. \tag{2}$$

As seen from Eq. 1, if discontinuous changes in entropy and volume occur the transition is first order. While for a second-order transition, these quantities will have a nearly zero value and there will be a discontinuous change in the heat capacity from Eq. 2. Landau, in 1937, provided the basis of second-order transition. The free energy (F) was defined as a function of the order parameter (Q) and gave the following expression;

$$F = F_0 + A < Q >^2 + B < Q >^4 + higher\ order \tag{3}$$

where F_o contains all the degrees of freedom, $<Q> = Q\ (T)$ is the order parameter and is one-dimensional, $A = \alpha\ (T - T_c)$, B and α are independent, near critical temperature, T_c.

Transitions in a solid can occur in two ways; one by its reorganization into a new lattice; as in the case of graphite forming diamond. These are referred to as *reconstructive* transitions and are hard to achieve. The other form is when a regular lattice is slightly *distorted* without breaking any linkages. These displacements in lattice positions can be caused by atoms, molecular units or ordering of the atoms in equivalent positions.

Reconstructive transitions usually involve phases with no symmetry relations. The phase transition is very abrupt having no thermodynamic order parameter. It rather involves a mechanism in which the intermediate structure must be having a symmetry whose space group is a subgroup of the space groups of the two end phases. This intermediate state can be transformed from one phase to another according to changes in the kinetic order parameter.[3,4] Such transitions are most commonly observed in elemental crystals, alloys and minerals like the fcc→hcp transitions that occur in Co, Pb, Fe, ZnS and so on.[5] Many binary systems, like II-VI semiconductors, at ambient conditions, have a coordination number of 4 and crystallize in wurtzite ($F\bar{4}3m$) structure. But, at high pressure, they stabilize to crystallize in a cubic-NaCl ($Fm\bar{3}m$) structure having a coordination number of 6.

We focus our discussion on the distortive structural phase transition. A distortive phase transition as a function of temperature is fundamentally described by two phases, above and below the T_c. The high-temperature phase is known as the prototype phase, while the low-temperature phase is described by the order parameter, Q (T) which exists below T_c and has a group symmetry which is the subset of the symmetry of the prototype phase. This was also the basic assumption on which Landau's equation for free energy was expressed as mentioned in Eq. 3. Based on the physical nature of order-parameter, structural transitions can be distinguished between displacive and order-disorder transitions.

In the displacive type, there is a collective displacement of atoms, ions, etc. of the structure with respect to the average positions of the entities occupied at equilibrium, above T_c. This can be illustrated by the famous $BaTiO_3$ which undergoes a cubic to tetragonal transition at 126°C.[6] We consider the barium ions as a reference of a unit cell. All titanium ions shift slightly upwards along one of the cubic axes whereas two kinds of oxygen atoms i.e. oxygen atoms in non-equivalent positions, shift downwards. As a result of these displacements in ions, the symmetry of the cubic system is lowered from ($m\bar{3}m$) to ($4mmm$) at T_c. A mode of finite frequency exists below T_c and freezes out on approaching it. It is obtained from solving;

$$\frac{\partial^2 F}{\partial^2 Q} = m\omega_s^2(T) \tag{4}$$

where ω_s is termed as the "soft-mode". Based on the Landau theory and anharmonic interactions between the phonons, Cochran *et al.* gave the following expression;[7]

$$\omega_s^2 = K(T - T_c) \, . \tag{5}$$

These modes can be probed directly by Raman or neutron scattering. Conclusive proof for the soft-mode theory was given by experiments performed on the zone boundary mode of $SrTiO_3$. $SrTiO_3$ has a perovskite structure which is undistorted in its high-temperature phase. It has a unit cell with Sr atoms occupying (0, 0, 0) positions of the cube, Ti atoms at (½, ½, ½) and O atoms at (½, 0, ½) (0, ½, ½) and (½, ½, 0).[8] A spontaneous lattice distortion was shown by x-ray diffraction studies as a function of temperature. A cubic to tetragonal distortion at 110K with a modification of c/a=1.0005 was reported.[9] Since there was no volume change; the phase transition was predicted to be second-order in nature. It was explained that there was a slight distortion along three possible [100] directions of tetragonal symmetry. However, the subdomains were small and became large only near the transition point. Another line splitting occurred between 65K and 35K, which was attributed to structural transition but was later proved to be classical to quantum paraelectric phase transition by Müller et al.[10,11] In 1968, Fleury et al. modelled the Raman spectrum of $SrTiO_3$ below 110 K and highlighted essential features of soft phonon at the corner (R point) of the cubic [111] Brillouin zone near the transition point.[12] They also suggested that the angle of rotation of the oxygen octahedron increases on lowering the temperature, which gives rise to a new IR-active mode below T_c. The lattice-dynamical characteristic of this phase transition was experimentally validated using inelastic neutron scattering experiments.[13] The phonon dispersion relations of the [111] zone boundary at 120 K revealed two sets of dispersion curves as shown in Figure 1(a) and the Γ_{25} mode at R point was temperature-dependent. Then, the Γ_{25} phonon mode was studied as a function of temperature. As seen clearly, the

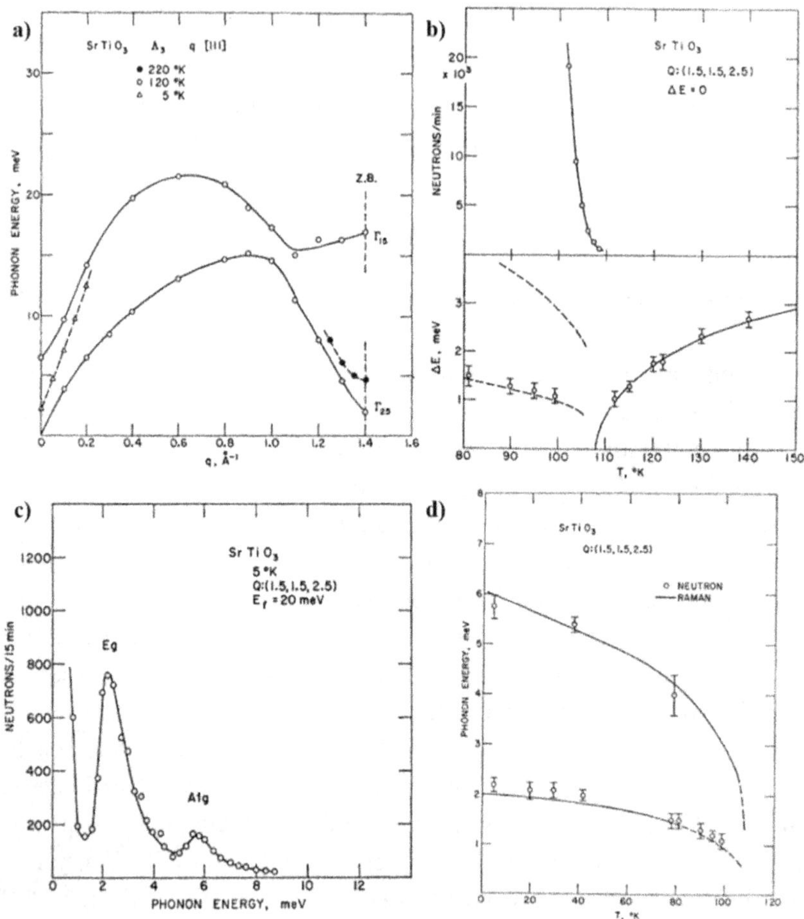

Fig. 1. (a) Phonon dispersion energy of lowest transverse branches with [111] zone boundary. (b) Temperature dependence of Γ_{25} mode at the R point. (c) Neutron profiles of the lowest phonons at the R point which become the zone-centre. (d) Temperature dependence of the phonons in tetragonal phase. Adapted from Ref. [13].

frequency of this function approaches zero at the transition temperature (Figure 1(b)). This mode which is triply degenerate in the cubic symmetry splits into two modes depending upon the polarization vector in the tetragonal phase. These modes are assigned as E_g and A_{1g} by Fleury *et al.* as observed Raman lines which becomes the zone center below T_c. The neutron (experimental) and Raman (modelled) data are in good agreement (Figure 1(c) & 1(d)).[12] Hence, indeed $SrTiO_3$ undergoes a cubic to tetragonal transition due to condensation of Γ_{25} at the [111] zone boundary.

In an order-disorder transition, the probability of certain atoms occupying the lattice site differs. In the high symmetry phase, they have equal probability while below T_c, occupancy varies at different sites. The structure of the prototype phase is nothing but a statistical average of the "degenerate" structures which are quite stable below T_c. In KH_2PO_4 which is an insulating crystal, the protons H^+ can occupy two different sites near the PO_4^- tetrahedron. Now below T_c, one of these sites is favored over the other.[5]

Distortive transitions can be ferroic or non-ferroic in nature.[14] Ferroic materials are those which involve a change in point symmetry along with a possibility of a change in the translational symmetry upon phase transition. Further, these ferroic materials may or may not undergo a change in the crystal structure causing ferroelasticity and have a possibility of introducing spontaneous polarization in the molecule, making them ferroelectric upon structural phase transition. The classification is described in the following Flowchart 1. The non-ferroic structural transitions have the same point group symmetry upon transition and the translational symmetry is broken at T_c.

The systems discussed above were commensurate systems, i.e. contained soft-modes having wave-vectors at the center or zone-boundaries in a Brillouin zone. But, in certain systems, there also exists a phase in which the periodic lattice cannot be defined as a simple multiple as that in the high-symmetry phase (or the prototype phase). These incommensurate phases are stable only in a certain temperature range, beyond which the more stable commensurate structure dominates. During an incommensurate modulation, the soft modes phase is unknown with a broken continuous symmetry. Also, the symmetry of the soft mode must be even as it considers wave-vectors with both positive and negative signs. K_2SeO_4, which is an insulator, undergoes a transition of second-order from Pnam to an incommensurate phase followed by a ferroelectric (Pna2$_1$, commensurate) phase.[8] Gen *et al.* studied an incommensurate transition in K_2SeO_4, wherein a clear soft mode appears and it softens around 130K over some fraction of the Brillouin zone.

Such phase transitions also occur in other ferroelectrics like $NaNO_2$, $SC(NH_2)_2$ and in many one-dimensional systems like KCP $(K_2Pt(CN)_4Br_{0.3}.3.2D_2O)$, $Hg_{3-\delta}AsF_6$ and so on.[15]

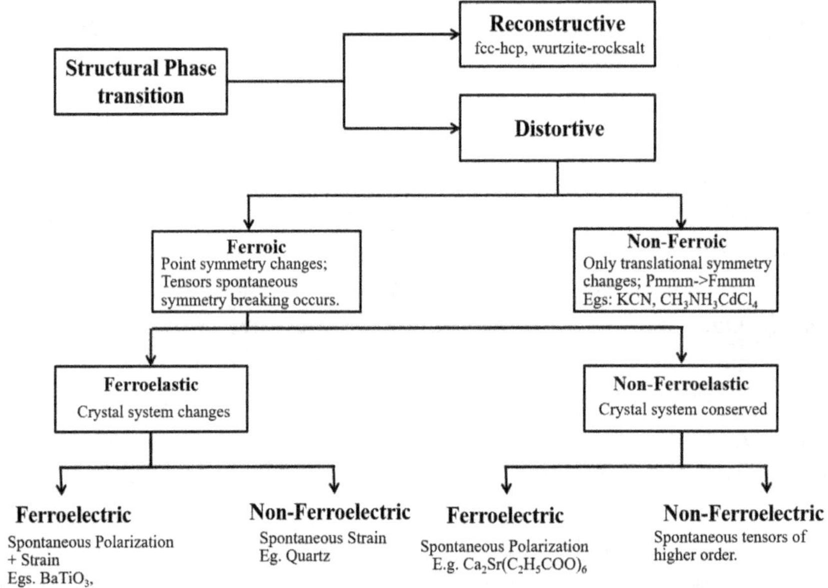

Flowchart. 1. Classification of structural phase transitions.

2.2. *Solid-liquid phase transitions*

All solid-liquid phase transitions are first order in nature due to the associated discontinuity in entropy corresponding to latent heat. Thermodynamically, the equilibrium between liquid and solid is represented by the coexistence line in the (p, T)–plane corresponding to $\mu_s = \mu_l$. (μ is the chemical potential). Microscopically, the crystallization of a liquid is accompanied by symmetry breaking. The onset of transition is strongly influenced by the geometry of atoms (or molecules) and their interatomic interactions. In some cases, delays in the transition can occur, causing the system to go into a metastable equilibrium. For example, liquids can supercool (without solidifying) and solids can superheat (without melting).

Melting is a phase transition which involves a discontinuity in the first derivative of the free energy (dF/dT) at the melting point. The free energy (F = E − TS) is not the same for the liquid and solid phases of a pure element as represented in Figure 2(a). The two curves cross at the melting temperature, Tm. At equilibrium, the material follows the solid curve at low temperatures and then follows the liquid curve above T_m.[16] Differential Thermal Analysis (DTA), is a standard technique that can determine the latent heat and thus the phase change associated with it. Here, the sample and a reference are heated simultaneously under similar conditions in one furnace. The differential temperature change is plotted against temperature for both heating and cooling cycles to give us information about the phase change. The general response of DTA during cooling and heating for a series of alloys, in a eutectic system where the solid solubility is limited, is given schematically in Figure 2(b) and 2(c).[16]

A eutectic system is a homogenous mixture of substances that melts or solidifies at a single temperature that is lower than the melting point than either one of these. P.R Subramanian *et al.* evaluated the phase diagram for Ag-Cu system. The melting point of Ag was observed at 961.78°C and that of Cu was found 1084.62 °C, while that of 39% Cu, Ag-Cu eutectic mixture was measured at 779.1 °C.[17]

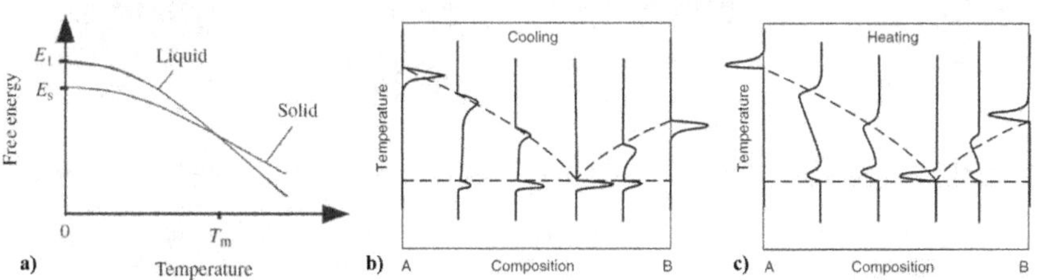

Fig. 2. (a) Free energy plots for liquid and solid phases of a pure element. (b) and (c) Schematic DTA response on cooling and heating of pure A and B and four other alloys superimposed on a simple Eutectic system. Adapted from Ref. [16].

Other direct techniques include Differential Scanning Calorimetry (DSC), Heat Flux DSC, Power Compensating DSC, etc and Indirect methods for detecting melting include X-ray diffraction (XRD), Infrared (IR) Microscopy in layered double hydroxide/poly(ethylene terephthalate) nanocomposites[18]; Raman Spectroscopy in long-chain n-alkanes[19] and UV absorption spectroscopy.

2.3. *Liquid crystals*

Partially ordered liquid phases are termed as Liquid Crystals (LCs). A crystal melts when the thermal energy creating disorder exceeds the intermolecular interaction energy (ensures cohesion). When the molecules are rod-shaped, then positional and orientational order exists. If the thermal energy is enough to disrupt the positional order alone, the system occurs in a mesomorphic state called the liquid crystal phase. The orientational ordering of liquid crystals is non-polar in nature. Based on the order parameter involved in a phase transition, there are many types of liquid crystals including Nematics, Smectics, and Discotics as seen in Figure 3.[20]

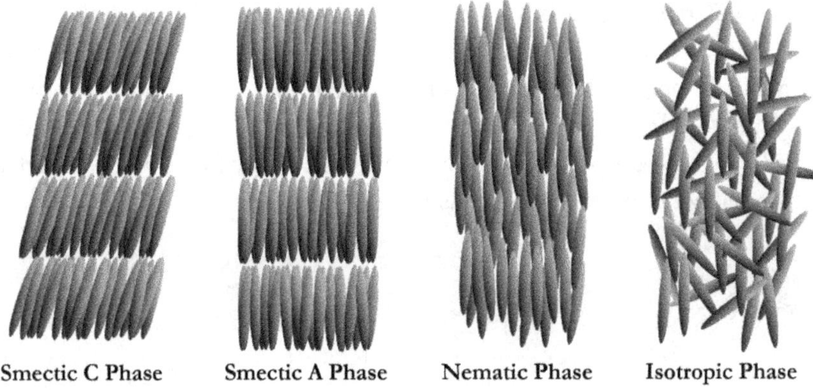

Fig. 3. Schematic representations of various liquid crystal phases.

In a phase diagram, the LC phase (or mesomorphs) is intermediate to the stability regions of liquid and a three-dimensional crystal. The LC phase has a rich polymorphism. Within a small temperature range, slight changes in temperature give rise to many LC phases. All the phase transitions in LCs are either second-order or weak first-order transitions. There are three major groups of LC phase transitions. The first group corresponds to second-order orientational phase transitions occurring due to rotational symmetry breaking. The second group is crystallization phase transitions that are associated with the formation of partial/ complete translational ordering. The third group is isotropic liquid–uniaxial nematic phase transitions.[21]

Upon cooling the system, a point is reached where the continuous rotational symmetry breaks down, but the translational symmetry is maintained. Thus, many rod-shaped/ ellipsoid molecules are observed pointing in one direction. They are still free to move but are fixed in one direction. It is termed as the nematic phase with an alignment direction $\hat{n}$.

Such systems have cylindrical symmetry, i.e the system remains the same with rotation about $\hat{z}$ and invariant under $\hat{z} \rightarrow -\hat{z}$. On further cooling of the system, the oriented ellipsoids of the system become layered with $\hat{n} = \hat{z}$, termed as the smectic A LC phase.

Light Scattering, Calorimetry, and X-ray techniques can be used to study the phase transitions in Liquid crystals. High-Resolution X-ray diffraction was used to study the smectic A-smectic C in 4-n-pentyl-phenylthiol-4'-n-octyloxybenzoate in a large magnetic field.[22]

2.4. *Glass transition*

Certain materials can become liquids with high viscosity (η in the range ~10^{13} Pa s) when melted. When cooled below the melting point, instead of solidifying immediately, they remain in a super cooled phase. Because of the increase in the viscosity of the system, it freezes in the form of a "glass", a non-crystalline solid phase. The topology of glasses can be defined as a continuous random network.

There are two different time constants which control this transition: τ_1 – the time required for crystallization of a given volume of the liquid phase. This value decreases when a system is supercooled, but with a further decrease in temperature, it increases.
τ_2 – the internal relaxation time/ structural relaxation time. This is the time required for the molecules in the material to rearrange to find equilibrium. This is proportional to viscosity.

Thus, glass formation is known to be a result of a dramatic increase in the relaxation time of the liquid, such that it acts as a solid in a timescale of any practicable scientific experiment.

Thermodynamically a glass is out of equilibrium, or partially in a state of metastable equilibrium. This is because the vibrational energy in glass, whose temperature is uniform, is equilibrated and the thermal conduction ensures local redistribution of heat. This formation of a glassy phase is not accompanied by any latent heat. Viscosity plays a crucial role in its behavior and is correlated with the internal relaxation time and hence is a measure of the duration of shear resistance.[23]

There are many proposed models to explain the glass transition. According to the entropy model, the increase in relaxation time near a glass transition is due to a decrease in the number of available conformations. It assumes that all molecules reorient under influence of their neighbors thus defining a cooperatively rearranging region. There are at least two different possible configurations for the same. As the temperature decreases, the minimum volume V of these regions grows and the activation energy for transitions between systems increases. If S_{conf} is the system's entropy, $\Delta E \sim V \sim 1/S_{conf}$.

For these models to be true, the cooperatively rearranging regions would be so small, in order of a few molecules that it would be difficult to observe experimentally. There are many other models including the free-volume model, energy model, mode-coupling models, frustration-based model, elastic model, harmonic models, models of local expansion, shoveling model, percolation model, etc.

The glass transition of amorphous SiO_2 (AS) was studied using calorimetry. AS was heated in a Pt-Rh 15% crucible, filled with 3-4 g of aluminosilicate powder, which was then dropped into an ice calorimeter to obtain the values of relative enthalpies.[24] A plot of mean heat capacities C_m (J/mol K) vs T (K) was plotted.[24] Two breaks in the slope were observed, one around 1600K, 1607 ±10K and at 1480K, thus corresponding to glass transition temperatures.[24]

2.5. *Sol-Gel transition*

A gel exists in a solid phase with a homogenous disorder. It deforms under weak pressures and hence is a soft system. Formation of a "gel" is a two-step process. Initially, a "sol" is formed, when the solvent (usually liquid) is mixed with a compound (usually a polymeric material). A sol to gel transition occurs under certain conditions of temperature/pH/ concentration.

According to P H Hermans, a gel can be characterized by,[25]

- A homogenous phase with 2 components (solid solute dissolved in a liquid solvent).
- The dissolved/ dispersed constituent and the solvent occupy the entire volume of the gel phase, i.e they are interconnected in the form of a 3-d network.
- It behaves like a solid under the effect of microscopic forces.

Gels can be divided into two categories–Physical and Chemical gels. Physical gels are held by weak van der Waals/ Hydrogen bonds, hence are thermo-reversible and have a long-range order. Some of them include gelatin, casein, agarose, pectin, etc. Chemical gels are held by strong covalent bonds and have localized bond zones, thus making them irreversible in nature. Some of these include polyacrylamide, polystyrene, vulcanized rubber and silica gel.[23]

The gelation temperature, the temperature at which the formation of a network of solute molecules is observed is not unique for a unique system (fixed concentration of solute in a solvent). This gelation is a function of the thermal history of the material, thus thermal hysteresis is observed. Hence, a state of thermal in-equilibrium exists.

DSC and DTA are two common techniques used to determine the formation of a gel. For example, Gelatin, a protein of animal origin, has a considerable absence of internal order and has a random configuration of polypeptide chains in aqueous solutions. It is soluble in water and forms thermo-reversible gels, in which it forms a 3D network with zones of intermolecular microcrystalline junctions.[23]

DSC measurements reveal a glass transition initially followed by an endotherm representing a helix-coil transition which is of first-order and kinetic in nature. This happens, because heating the sample beyond the glass transition provokes disruption of the crystalline region and the gelatin behaves as a molecularly dispersed molten plastic.

3. Magnetic Phase Transitions

The origin of magnetism lies in the orbital and spin motions of electrons and how much the electrons interact with one another. There are different kinds of magnetic materials, based on the arrangements of the spins in the lattice, namely paramagnetic, diamagnetic, ferromagnetic, ferrimagnetic and antiferromagnetic materials. Under varying external parameters, these materials undergo magnetic phase transitions, including Curie transition (ferromagnetic to paramagnetic, when heated over Curie temperature T_C), Neel transition (antiferromagnetic to paramagnetic, when heated over Neel temperature, T_N), frustration and spin-glass transition.

Ferromagnetism can be explained using the Ising Model.[26] As seen in Figure 4(a), in an Ising lattice, with local magnetic moments, the energy to flip a single spin is determined by the degree of magnetic alignment of the neighboring spins. In this model, the energy of ↑-spin depends on the orientation of its z first nearest neighbors and similarly for ↓-spin,

$$e_\uparrow = -J(n_\uparrow - n_\downarrow) \quad \& \quad e_\downarrow = -J(n_\downarrow - n_\uparrow) = J(n_\uparrow - n_\downarrow) \tag{6}$$

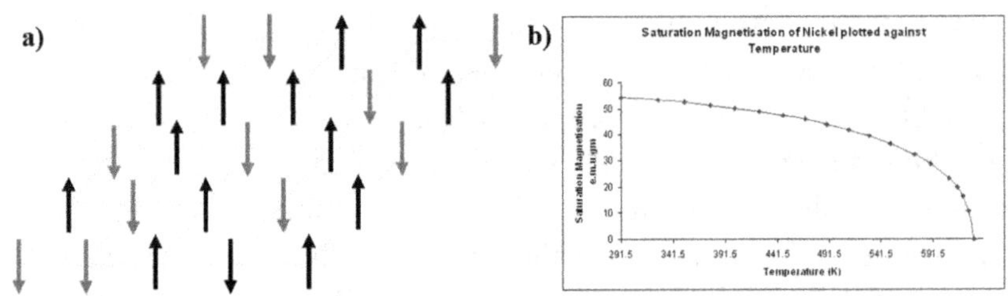

Fig. 4. (a) Schematic of a configuration of a 2D Ising model on a square lattice. (b) Variation of saturation magnetization with temperature for Ni. Adapted from Ref. [27].

where $n_\uparrow$ is the number of nearest neighbors of ↑-spin and vice-versa. Since it is difficult to precisely obtain values of $n_\uparrow$ and $n_\downarrow$, they are approximated as their average overall sites. Let the magnetic moment of the spin be μ and the external applied magnetic field, B, the favorable energy for ↑-spin is -μB and the unfavorable energy for ↓-spin is μB. Thus,

$$e_\uparrow = -Jz\left(\frac{N_\uparrow}{N} - \frac{N_\downarrow}{N}\right) - \mu B \quad \& \quad e_\downarrow = +Jz\left(\frac{N_\uparrow}{N} - \frac{N_\downarrow}{N}\right) + \mu B \ . \tag{7}$$

Here, the order parameter, L will be equivalent to $\left(\frac{N_\uparrow - N_\downarrow}{N}\right)$, where it takes values from -1<L<1. Thus,

$$e_\uparrow = -JzL - \mu B \quad \& \quad e_\downarrow = +JzL + \mu B \ . \tag{8}$$

At a finite temperature, the relative probabilities of the $\uparrow$-spin and $\downarrow$-spin are their Boltzmann factors,

$$\frac{N_\uparrow}{N_\downarrow} = \frac{e^{-e_\uparrow/k_B T}}{e^{-e_\downarrow/k_B T}} = \frac{e^{(JzL+\mu B)/k_B T}}{e^{-(JzL+\mu B)/k_B T}} = e^{2(JzL+\mu B)/k_B T} \tag{9}$$

$$L = \tanh\left(\frac{JzL+\mu B}{k_B T}\right) \quad \& \quad N_\uparrow + N_\downarrow = \frac{1+l}{1-L} = e^{2(JzL+\mu B)/k_B T} \ . \tag{10}$$

At the critical temperature, L will be infinitesimally small, and the above equation can be linearized to,

$$2L = \frac{2(JzL+\mu B)}{k_B T} \ . \tag{11}$$

When the external applied magnetic field, B = 0, then,

$$T_C = \frac{Jz}{k_B} \ . \tag{12}$$

Thus, the critical condition occurs when the thermal energy, $k_B T$ equals the exchange energy over the first nearest neighbor shell, zJ. Thus, below the Curie temperature, the magnetic order develops cooperatively, and the materials have a bulk magnetic moment. The Curie temperature can be determined from a plot of saturation magnetization versus temperature as seen in Figure 4(b).[27]

The measurement of a magnetic phase transition is made using a vibrating sample magnetometer or a SQUID (Superconducting Quantum Interference Device), where the magnetic moment of the sample is determined in the presence of an external magnetic field, H_{app}. As magnetic field lines are always closed loops, every material of a specific aspect ratio in the presence of H_{app} produces a demagnetizing field, $H_D =$

DM. Thus, the net field felt by the sample is, $H = H_{app} - DM$. Each magnetic state has its own characteristic field and temperature dependence. Plots of magnetization versus applied field can be made. Paramagnets would show a straight line with slope ~1. Ferromagnets and ferrimagnets have a hysteresis loop. Antiferromagnets show a linear field dependence. A high T_c SQUID measurement of a Mn_3O_4 film revealed a magnetic phase transition in the range of 30K-50K.[28]

4. Transport Phase Transitions

Pressure, temperature, and doping can change the electronic band structures, especially when the material is near any critical point. The change in the electronic structure of the material could also be influenced by the electronic spins which may help in hopping between different orbital levels or formation of Cooper pairs etc. Such changes cause transport property changes and leading to metal-insulator transitions, quantum phase transition, and superconductivity. The following section details various such phenomena seen in materials.

4.1. *Metal-insulator phase transition*

The two major directions of investigations in the field of metal-insulator transitions (MIT) have been (a) interaction-driven Mott-Hubbard transitions and (b) disorder driven Anderson transitions. In recent years, the two directions have converged in the studies of interacting, disordered systems and the concept of many-body localization has emerged. We will elaborate on the two directions and subsequently on the convergence. A paradigm for investigating strongly correlated electron systems is the single-orbital Hubbard model (HM), comprising a kinetic energy term, and a double occupancy prohibiting Coulomb repulsion term. The HM has been the standard model for exploring interaction-driven MIT and almost every method present in the repertoire has been employed. These methods include dynamical mean-field theory and quantum cluster theories combined with quantum Monte-Carlo (QMC), numerical renormalization group, exact diagonalization, density matrix renormalization group, iterated perturbation theory, local moment approach; determinantal QMC, slave particle theories, etc. A consensus has emerged that the paramagnetic Mott-Hubbard transition in the Hubbard model is first-order in nature, accompanied by hysteresis, and a finite temperature critical point. Incorporating symmetry breaking leads to a very rich phase diagram in the doping-interaction-temperature space, that in two dimensions, appears to mimic very closely the experimentally found phase diagram of high-temperature superconducting cuprates.

A very famous compound undergoing metal to insulator transitions is V_2O_3. The phase diagram involves a high temperature paramagnetic metallic (PM) and a paramagnetic insulating (PI) phase, along with a low temperature antiferromagnetic insulating (AFI) phase. The crystal structure in the disordered (or high temperature) phase is rhombohedral (corundum), while in the ordered phase is monoclinic. The two metal to insulator transitions occurring in V_2O_3, i.e PM to AFI at low temperatures and PM to PI at high temperatures are Mott Transitions.

According to band theory, metals and insulators are defined using their electronic interactions with the ion field into account, while in the case of Mott insulator, electron-electron interactions are also considered. As a result, along with the lattice spacing, d, (from ion field), the Bohr radius a_B for

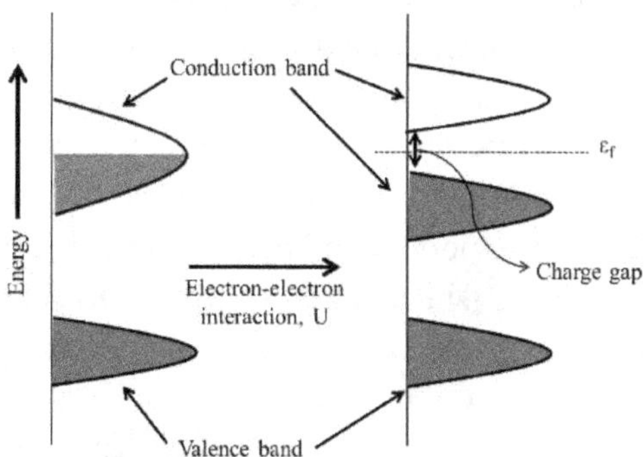

Fig. 5. Hubbard model of metal-insulator transition.

Coulomb interaction is also included. If $d \gg a_B$ then the materials are an insulator and vice versa. Hence, there will be a critical value d_0 where this transition occurs. Mott-Hubbard model is one of the important models which emphasized electron-electron interaction. In the HM model, two Hamiltonians were incorporated, one with onsite interaction of electrons, U, and another with hopping energy (or tunneling energy) of the electron, t. Assuming that Mott insulator phase will emerge at half-filling, the valence band remains unchanged upon including the electron interaction U, whereas the conduction band split into two and opened a charge gap as shown in Figure 5. Thus, the corresponding state became an insulator. This is how the insulating behavior of NiO was proved when the band theory predicted it to be metallic.[29] There are two types of HM for the metal-insulator transitions: the filling control transition and the bandwidth control transition. The former type is changing the electron concentration or the chemical potential, which can be experimentally achieved by dopants. While the bandwidth control transition occurs by varying the hopping energy of electron or the bandwidth.

4.2. *Topological phase transitions*

An electronic topological transition (ETT) has been proposed by Lifshitz in 1960[30] where the size and shape of the Fermi surface can be tuned with external parameters such as the pressure, temperature, magnetic field or composition. During such a transition, a void can be created or destroyed, or the neck connecting two parts of the Fermi surface can be opened as shown in Figure 6(a). This can influence the electron density near the boundary and the electron dynamics acquire many unique features which lead to anomalies of the electron characteristics of the material. According to Lifshitz, ETT can cause no discontinuity in the volume (first derivative of the Gibbs free energy), but an alteration in the second derivative of the Gibbs free energy, like the compressibility, could take place.[31] The ambient 3D topological insulators such as Bi_2Se_3, Bi_2Te_3, and Sb_2Te_3 have been proved to show ETT under a hydrostatic pressure.[32]

Topological Insulators (TI) are materials having an insulating electronic bandgap in their bulk, but exhibit conducting gapless surface states due to strong spin-orbit coupling (SOC).[33]

The energy bands of TIs are characterized by the order parameter Z_2 (i.e. topological invariant quantity) which is calculated from geometric properties of electronic states as a function of the Bloch vector.[34]

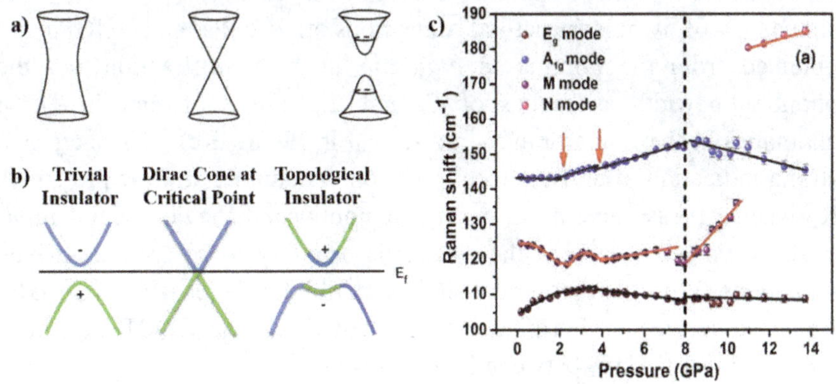

Fig. 6. (a) Schematic diagram of fermi surface neck disruption in ETT. (b) Schematic representation of TQPT process. (c) Pressure versus Raman shift of phonon modes (A_{1g}, M,E_g and N) of TiTe$_2$. Adapted from Ref. [41].

This quantity helps to differentiate between the topologically trivial ($Z_2 = 0$) and nontrivial ($Z_2 = 1$) states and hence assists to identify a topological quantum phase transition (TQPT) in any system. There are some SOC narrow bandgap materials which are trivial insulators at ambient conditions, but they can behave as nontrivial topological insulators under the application of strain. This transition from trivial insulator to topological insulator is known as the topological quantum phase transitions (TQPT). Theoretically, TQPT is characterized by band inversion (i.e. crossing of conduction and valence bands) with the closing of bandgap at a time-reversal invariant **k** point along with a parity change resulting in the change of Z_2 topological invariant as shown in Figure 6 (b).[35] TQPT under strain has been observed in SOC materials by both chemical and physical methods. For instance, chemical doping in TiBi(S$_{1-x}$Se$_x$)$_2$[36,37] and Pb$_{1-x}$Sn$_x$Se[38] systems causes TQPT. Similarly, hydrostatic pressure-induced TQPT has been observed in several systems such as BiTeI, BiTeBr, and Sb$_2$Se$_3$. [34,39,40]

There are only a few materials in which TQPT and ETT have been experimentally observed, using either a direct method like ARPES or indirect methods. Usually, a combination of indirect methods like Raman and IR spectroscopy, X-ray diffraction, transport measurement, etc. along with the first principle calculations has been used for the detection of topological transitions. For example, in case of 1T-TiTe$_2$,[41] the signatures of two isostructural transitions at ~2GPa and ~4GPa are obtained from the minima in c/a ratio in X-ray diffraction and the phonon linewidth anomalies of E_g and A_{1g} modes of Raman. These anomalies in the Raman modes are shown in Figure 6(c). This acts as a strong indication of unusual electron-phonon coupling at these pressures. Resistance measurements also present nonlinear behavior over similar pressure ranges proposing the electronic origin of these pressure-driven transitions. With the theoretical studies exhibiting the possibility of ETT from the electronic Fermi surface calculations, the TQPT has been identified in 1T-TiTe$_2$ between 2GPa and 4Gpa.

4.3. *Superconductivity*

In 1911, Onnes *et al.* discovered a steep decrease in the resistance of Hg at 4.2 K, below which it goes to a superconducting state as shown in Figure 7.[42] As the term suggests, superconductors become different from a "normal" conductor as the resistivity goes to zero; $\rho \approx 0$. Any material is expected to have a finite resistivity even at an absolute zero due to the presence of defects and phonons in a system. But for some materials (mostly having high resistivity at ambient conditions), at a certain critical temperature, T_c, undergoes a transition to a superconducting state. In 1933, Walter Meissner discovered that superconductors are diamagnetic and expel magnetic field from it (known as the Meissner effect). This becomes an important factor to distinguish them from normal conductors. But, they switched to normal conductors at a very low critical magnetic field, H_c. Later, superconductors were classified as Type-I and Type-II based on these critical magnetic fields. Most pure elements like Sn, Hg, Nb, etc. are Type-I superconductors wherein, the transition to a normal conducting state is rapid at a fairly low H_c. Whereas, the Type-II superconductors are generally compounds like Nb$_3$Ge where there are two critical fields H_{c1} and H_{c2}. At the lower critical field H_{c1}, the external magnetic field penetrates only to a certain extent whilst the material maintains its superconducting property, which vanishes completely at H_{c2}.

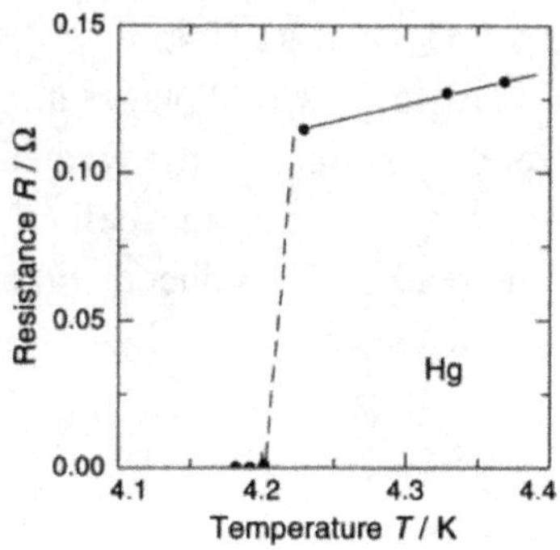

Fig. 7. Superconductivity transition in Hg at 4.2K. Adapted from Ref. [42].

In 1957, Bardeen, Cooper, and Schrieffer gave a theory (BCS theory) explaining the mechanism of superconductivity.[43] It assumes that superconductivity arises when the Cooper pair interaction dominates the Coulomb interaction. A Cooper pair is essentially a weak electron-electron bound pair mediated by phonon interaction. We do not discuss any detailed quantum mathematical calculations but qualitatively, the interaction of electrons with the cationic lattice will have a net positive charge in the vicinity of the attraction site. An electron with opposite momentum and spin is attracted to this force and acts as the paired electron. Even though this theory gives a good understanding of the cause of superconductivity, it does not predict materials undergoing such a phase transition. The BCS theory mainly works well with weakly coupled superconductors but not for high T_c superconductors.

Until 1986, most discovered superconductors had a low T_c, of about 23.2K in Nb and 38K in compounds of form $Ba\text{-}PbBi\text{-}O_3$.[44] Only, in early 1987, compound Y-Ba-Cu-O (YBCO) was discovered to have a T_c of 93K, which meant liquid nitrogen (boiling point- 77K) could be used instead of liquid He which is much more expensive, less abundant and has more complex cryogenic system than liquid nitrogen. These compounds are known as High-T_c superconductors. The superconductor

YBCO is a defect perovskite of the composition $YBa_2Cu_3O_{7-x}$ with x varying from 0 to 1 in small fractions.[44] From the neutron diffraction studies, by varying the oxygen concentration, it was shown that oxygen atoms and its vacancies are significant in superconductivity. Only $YBa_2Cu_3O_7$ and $YBa_2Cu_3O_{0.68}$ were found to be orthorhombic and superconducting mainly due to the Cu-O planes along the b-axis. Oxygen atoms occupy four different positions in the former compound where the O(4) forms chains along b-axis in the unit cell. It was observed that as the oxygen vacancies increased, T_c reduced indicating a disruption in conduction pathways by vacancies.

4.4. *Quantum phase transitions*

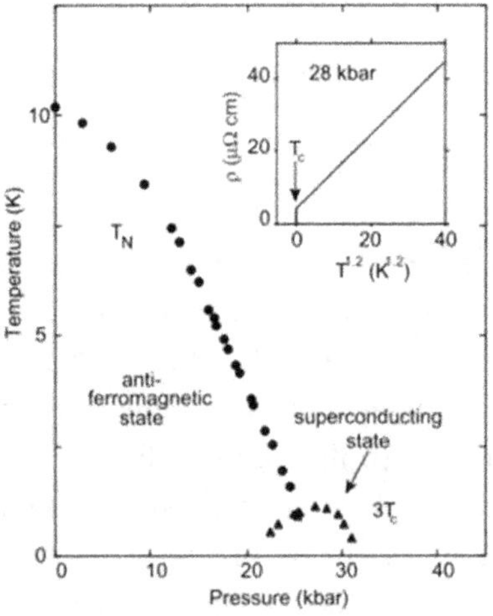

Fig. 8. Temperature–pressure phase diagram of high-purity single-crystal $CePd_2Si_2$. Adapted from Ref. [46].

Exotic states are achieved for an electron at zero temperature as a function of pressure, magnetic field, composition and so on. These transitions around zero temperature are known as quantum phase transitions. They have many competing interactions around the quantum critical point and minute changes or fluctuations can largely affect the phase diagram as compared to thermal fluctuations.[45]

Such an exotic state was observed in a heavy-fermion system, $CePd_2Si_2$ for a single crystal at the edge of magnetic order.[46] This system undergoes an anti-ferromagnetic ordering at low temperature, which is suppressed by the application of pressure. However, a superconducting state was observed just before the ordering suppresses completely at the quantum phase transition as shown in Figure 8.

Here, the competing state is argued to be of the spin fluctuations associated with the antiferromagnetic state, but magnetism is sufficiently weakened. Thereby, the superconducting state is a result of these bound magnetic spin-spin interacting charge carriers.

5. Summary

Phase transitions in materials can be broadly classified into structural, magnetic and transport phase transition. Structural phase transitions are induced by the changes in the atomic environment while magnetic and transport is induced by the electronic environment, either due to the spin or charge. At a given critical point, certain cases can have multiple phase transitions, like structural and transport, or magnetic and transport. The above classification of various phase transition encompasses a good number of the observed phase transitions in nature and could even provide insight into the understanding of the new materials being discovered. The theoretical and experimental explanation provided here would be a good beginning for the students and researchers embarking on material science research.

References

1. Bellin, C., Mafety, A., Narayana, C., Giura, P., Rousse, G., Itié, J.-P., Polian, A., Saitta, A. M. & Shukla, A. Disorder-order phase transition at high pressure in ammonium fluoride. *Phys. Rev. B* **96**, 094110 (2017).
2. Narayana, C., Luo, H., Orloff, J. & Ruoff, A. L. Solid hydrogen at 342 GPa: no evidence for an alkali metal *Nature* **393**, 46 (1998).
3. Tolédano, P. & Dmitriev, V. Domains and reconstructive phase transitions in crystals, quasicrystals and complex fluids. *Ferroelectrics* **191**, 85-107 (1997).
4. Capillas, C., Perez-Mato, J. & Aroyo, M. Maximal symmetry transition paths for reconstructive phase transitions. *J. Phys. Condens. Matter* **19**, 275203 (2007).

5. Tolédano, J. C. & Tolédano, P. *The Landau Theory of Phase Transitions: Application to Structural, Incommensurate, Magnetic and Liquid Crystal Systems.* (1987).

6. von Hippel, A. Ferroelectricity, Domain structure, and phase transitions of barium titanate. *Rev. Mod. Phys.* **22**, 221-237 (1950).

7. Müller, K. A. & Thomas, H. *Structural Phase Transitions I.* (Springer Berlin Heidelberg, 2012).

8. A. Cowley, R. & M. Shapiro, S. Structural phase transitions. *J. Phys. Soc. Jpn* **75**, 111001 (2006).

9. Lytle, F. W. X-ray diffractometry of low-temperature phase transformations in strontium titanate. *J. Appl. Phys.* **35**, 2212-2215 (1964).

10. Fang, X. Phase transitions in strontium titanate. *Conf. Proc.* (2013).

11. Müller, K. A. & Burkard, H. $SrTiO_3$: An intrinsic quantum paraelectric below 4 K. *Phys. Rev. B* **19**, 3593 (1979).

12. Fleury, P., Scott, J. & Worlock, J. Soft phonon modes and the 110 K phase transition in $SrTiO_3$. *Phys. Rev. Lett.* **21**, 16 (1968).

13. Shirane, G. & Yamada, Y. Lattice-dynamical study of the 110K phase transition in $SrTiO_3$. *Phys. Rev.* **177**, 858-863 (1969).

14. Tagantsev, A. K., Cross, L. E. & Fousek, J. in *Domains in Ferroic Crystals and Thin Films* (eds Alexander K. Tagantsev, L. Eric Cross, & Jan Fousek) 11-107 (Springer New York, 2010).

15. Ishibashi, Y. Incommensurate phase transitions in ferroelectrics. *Ferroelectrics* **24**, 119-126 (1980).

16. Boettinger, W. J., Kattner, U. R., Moon, K.-W. & Perepezko, J. H. in *Methods for Phase Diagram Determination,* pp.151-221 (Elsevier, 2007).

17. Subramanian, P. R. & Perepezko, J. H. The Ag-Cu (silver-copper) system. *J. Phase Equilibria* **14**, 62-75 (1993).

18. Lee, W. D., Im, S. S., Lim, H.-M. & Kim, K.-J. Preparation and properties of layered double hydroxide/poly (ethylene terephthalate) nanocomposites by direct melt compounding. *Polymer* **47**, 1364-1371 (2006).

19. Corsetti, S., Rabl, T., McGloin, D. & Kiefer, J. Intermediate phases during solid to liquid transitions in long-chain n-alkanes. *Phys. Chem. Chem. Phys.* **19**, 13941-13950 (2017).

20. Gray, G. W. *Molecular Structure and the Properties of Liquid Crystals.* (Academic press, 1962).

21. Dierking, I. & Al-Zangana, S. Lyotropic Liquid crystal phases from anisotropic nanomaterials. *Nanomaterials* **7**, 305 (2017).

22. Safinya, C., Kaplan, M., Als-Nielsen, J., Birgeneau, R., Davidov, D., Litster, J., Johnson, D. & Neubert, M. High-resolution x-ray study of a smectic-A—smectic-C phase transition. *Phys. Rev. B* **21**, 4149 (1980).

23. Papon, P., Leblond, J. & Meijer, P. H. *Physics of Phase Transitions.* (Springer, 2002).

24. Richet, P. & Bottinga, Y. Glass transitions and thermodynamic properties of amorphous SiO_2, $NaAlSinO_{2n+2}$ and $KAlSi_3O_8$. *Geochimica et Cosmochimica Acta* **48**, 453-470 (1984).

25. Hermans, P. Gels. *Colloid Science* **2**, 483-651 (1949).

26. Fultz, B. *Phase Transitions in Materials.* (Cambridge University Press, 2014).

27. Weiss, P. & Forrer, R. Magnetisation of Nickel and the magnetocaloric effect. *Ann. Phys.* **5**, 153 (1926).

28. Hanson, M., Ivanov, Z., Johansson, C., Kislinski, Y. & Larsson, P. A magnetic phase transition studied with high-Tc SQUIDs. *J. Magn. Magn. Mater* **177**, 519-520 (1998).

29. Meng, Q. *Metal Insulator Transition.* (2010).

30. Lifshitz, I. Anomalies of electron characteristics of a metal in the high pressure region. *Sov. Phys. JETP* **11**, 1130-1135 (1960).

31. Polian, A., Gauthier, M., Souza, S. M., Trichês, D. M., de Lima, J. C. & Grandi, T. A. Two-dimensional pressure-induced electronic topological transition in Bi_2Te_3. *Phys. Rev. B* **83**, 113106 (2011).

32. Manjón, F., Vilaplana, R., Gomis, O., Pérez-González, E., Santamaría-Pérez, D., Marín-Borrás, V., Segura, A., González, J., Rodríguez-Hernández, P. & Muñoz, A. High-pressure studies of topological insulators Bi_2Se_3, Bi_2Te_3, and Sb_2Te_3. *Phys. Status Solidi (b)* **250**, 669-676 (2013).

33. Hasan, M. Z. & Kane, C. L. Colloquium: topological insulators. *Rev. Mod. Phys.* **82**, 3045 (2010).

34. Bera, A., Pal, K., Muthu, D., Sen, S., Guptasarma, P., Waghmare, U. V. & Sood, A. Sharp Raman anomalies and broken adiabaticity at a pressure induced transition from band to topological insulator in Sb_2Se_3. *Phys. Rev. Lett.* **110**, 107401 (2013).

35. Fu, L., Kane, C. L. & Mele, E. J. Topological insulators in three dimensions. *Phys. Rev. Lett.* **98**, 106803 (2007).

36. Xu, S.-Y., Xia, Y., Wray, L., Jia, S., Meier, F., Dil, J., Osterwalder, J., Slomski, B., Bansil, A. & Lin, H. Topological phase transition and texture inversion in a tunable topological insulator. *Science* **332**, 560-564 (2011).

37. Sato, T., Segawa, K., Kosaka, K., Souma, S., Nakayama, K., Eto, K., Minami, T., Ando, Y. & Takahashi, T. Unexpected mass acquisition of Dirac fermions at the quantum phase transition of a topological insulator. *Nat. Phys.* **7**, 840 (2011).

38. Dziawa, P., Kowalski, B., Dybko, K., Buczko, R., Szczerbakow, A., Szot, M., Łusakowska, E., Balasubramanian, T., Wojek, B. M. & Berntsen, M. Topological crystalline insulator states in $Pb_{1-x}Sn_xSe$. *Nat. Mater.* **11**, 1023 (2012).

39. Xi, X., Ma, C., Liu, Z., Chen, Z., Ku, W., Berger, H., Martin, C., Tanner, D. & Carr, G. Signatures of a pressure-induced topological quantum phase transition in BiTeI. *Phys. Rev. Lett.* **111**, 155701 (2013).

40. Ohmura, A., Higuchi, Y., Ochiai, T., Kanou, M., Ishikawa, F., Nakano, S., Nakayama, A., Yamada, Y. & Sasagawa, T. Pressure-induced topological phase transition in the polar semiconductor BiTeBr. *Phys. Rev. B* **95**, 125203 (2017).

41. Rajaji, V., Dutta, U., Sreeparvathy, P., Sarma, S. C., Sorb, Y., Joseph, B., Sahoo, S., Peter, S. C., Kanchana, V. & Narayana, C. Structural, vibrational, and electrical properties of 1T–TiTe$_2$ under hydrostatic pressure: Experiments and theory. *Phys. Rev. B* **97**, 085107 (2018).

42. Khachan, J. & Bosi, S. Superconductivity. *Conf. Proc.* (2008).

43. Hott, R., Kleiner, R., Wolf, T. & Zwicknagl, G. Review on Superconducting Materials. *arXive-prints* (2013). <https://ui.adsabs.harvard.edu/abs/2013arXiv1306.0429H>.

44. Lundy, D. R., Swartzendruber, L. J. & Bennett, L. H. A brief review of recent superconductivity research at NIST. *J Res Natl Inst Stand Technol* **94**, 147-178 (1989).

45. Transitions in focus. *Nat. Phys.* **4**, 157 (2008).

46. Mathur, N., Grosche, F., Julian, S., Walker, I., Freye, D., Haselwimmer, R. & Lonzarich, G. Magnetically mediated superconductivity in heavy fermion compounds. *Nature* **394**, 39 (1998).

Chapter 12

Advances in Electrode Materials for Sodium-ion Batteries

Vinita Ahuja and Premkumar Senguttuvan[*]

New Chemistry Unit, International Centre for Materials Science and School of Advanced Materials, Jawaharlal Nehru Centre for Advanced Scientific Research, Jakkur, Bangalore 560064, Karnataka, India
[*]*prem@jncasr.ac.in*

The pursuit of alternative battery chemistries based on the low cost and sustainable materials to replace the existing lithium ion batteries has opened new avenues. From this view point, Na-ion batteries are attractive due to the inexpensive and earth abundant sodium precursors. Thanks to the "know-how" knowledge gained from LIBs, the development of electrode and electrolyte materials for sodium ion batteries has been accelerated in the recent years. In this review, we have highlighted the recent progress on the Na-ion electrode materials such as layered oxides, polyanionic compounds, hard carbon and alloys.

1. Introduction

In response to our ever-increasing energy demand and CO_2 levels, several nations are committed towards a paradigm shift from conventional to renewable energy sources. Given the intermittent nature of renewable sources and their geological locations, the energy storage technologies play a vital role to distribute the energy from the grids to the customer utilities. The present large scale electrical storage technologies store energy in different forms such as electrical (capacitor and supercapacitor), kinetic (fly wheels), potential (pumped hydro and compressed air) and chemical (batteries) and correspondingly their response time varies from a few seconds to several hours. Among them batteries have ubiquitous advantages such as high round trip efficiency,

275

low maintenance, flexible energy and power characteristics.[1,2] Different classes of battery technologies for various kinds of applications were developed during the last century and some of them were successfully implemented in the grid application. A large installation of lead-acid batteries (10 MW/40 MWh) was built in Chino, CA in 1988. Despite its largest market share and low cost, the limited cycle life, high maintenance and toxicity remain as major challenges for their grid application. Na/S technology has also been demonstrated for the grid application, however, its high temperature operation and safety related to molten sodium and sulfur electrodes impede their wide spread application. Modern Li-ion batteries (LIBs) have already captured the portable electronics and electric vehicles market and at present, have been considered for grid application. The major concerns about the successful implementation of LIBs in the grid storage are not only the cost and availability and accessibility of lithium and cobalt resources but also safety, wide operational temperature and long cycle life.

In the present scenario, scientists have started to consider other possibilities of developing new battery chemistries for the grid storage application with a special emphasis on cost and abundance of the materials. Chemically similar sodium ion batteries (SIBs) are appealing from the view point of both raw material abundance as well as the cost in comparison with the existing LIBs. The operation principle of SIBs is analogous to that of LIBs. The SIB consists of two insertion-based electrodes (positive and negative electrodes) which are separated by an electrolyte (i.e. generally solution containing sodium salts dissolved in aprotic solvents). During the discharge, sodium ions stored in the negative electrodes get released and travel towards the positive electrode through the electrolyte, whereas the electrons liberated at the negative electrode pass through the external circuit to complete the electrochemical reaction. Upon charging, the above-mentioned processes reverse.

2. A Brief History

Historically, the development of electrode materials for modern rechargeable batteries had originated from the progress made in solid state ionic conductors. Despite the fundamental research on the ionic conductors that had started in 1839 by Faraday, the discovery of β-Al_2O_3 opened new perspectives for the applications.[3] Its structure is made of two spinel blocks which are separated by a mirror plane containing one oxygen and one vacancy. This plane contains many vacancies through which the Na^+ ions diffuse. Ford Company successfully used this electrolyte in high temperature Na/S batteries for the electric vehicle application. Another important milestone in the development of solid ionic conductors was the discovery of Na super ionic conductor (NASICON)-$Na_{1+x}Zr_2(PO_4)_{3-x}(SiO_4)_x$ by Goodenough.[4] It was reported to exhibit higher ionic conductivities (closer to that of β-Al_2O_3) at 200 °C. Simultaneously, several research groups focused on the reversible intercalation of lithium ions in various host structures in aprotic electrolytes and Whittingham demonstrated the reversible operation of Li/TiS_2 cells.[5] The first concept of "rocking chair" battery was proposed by Armand in which lithium ions move back and forth between two intercalation electrodes.[6] In the late 70s', Delmas initiated the electrochemical sodium (de)-intercalation studies on Na_xMO_2 layered oxides.[7] In parallel, Goodenough reported the reversible (de)-intercalation of lithium ions into $LiCoO_2$ cathode and Yoshino and coworkers from Asahi Kasei demonstrated the lithium ion battery comprising of low temperature coke and the modified version of $LiCoO_2$ as anode and cathode respectively.[8] Soon, Sony Corporation commercialized the lithium ion batteries in 1991. After that, most of the research efforts focused on the development of Li-ion materials, whilst the Na-ion materials received less attention. It was in the late 2000's, when the researcher started to seek alternative battery chemistries based on low cost and abundant raw materials, that Na-ion chemistry emerged as the front runner.

3. Design Consideration

A battery is an energy storage device which is made of several electrochemical cells connected in series or parallel to increase the output voltage and current. The electrochemical cell consists of two electrodes, the anode and cathode, which are separated by the electrolyte. During

                    V. Ahuja & P. Senguttuvan

the (primary) battery operation, electrochemical reactions take place at the anode and the cathode (oxidation and reduction respectively), while the electrolytes allow the passage of ions. If the corresponding electrochemical reactions could be reversed by the application of an external power, the battery can be classified as secondary or rechargeable battery. The total charge stored or the capacity Q (Ah g^{-1} or Ah L^{-1}), energy density (Wh kg^{-1} or Wh L^{-1}) and Coulombic efficiency (C.E. in %) of the battery are given in Eqs. (1), (2) and (3) respectively, wherein *I*, *V* and *t* represent discharge/charge current, voltage and time respectively.

$$Q = \int_0^{\Delta t} I dt \tag{1}$$

$$Energy\ density = \int_0^{\Delta t} IV(t)dt \tag{2}$$

$$C.E = 100 \times \frac{Q_{disch}}{Q_{ch}} \tag{3}$$

Thus, the energy density of the battery could be varied by carefully tuning the voltage and the capacity of the individual cell. These two parameters are strongly related to the thermodynamics of electrode and electrolyte materials.[9] Figure 1 shows the open circuit energy diagram of the electrode and electrolyte in terms of relative electron energies. The

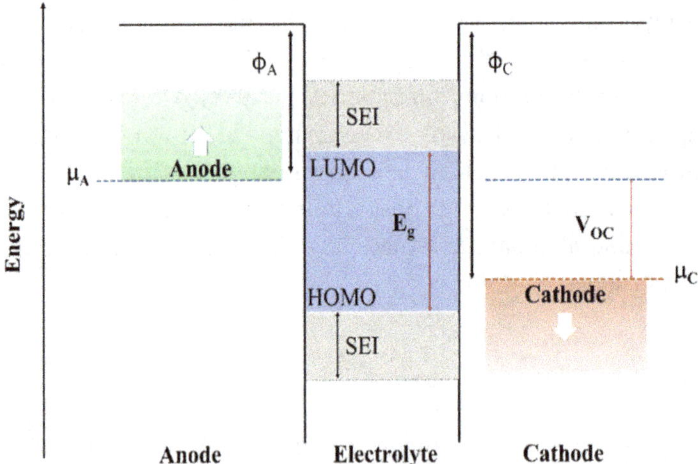

Fig. 1. Schematic open-circuit energy diagram. Adapted with permission from J.B. Goodenough & Y. Kim, *Chem. Mater.* **22**, 587-60 (2010). Copyright 2009 American Chemical Society.

anode is the electron source (reductant), the cathode is the electron sink (oxidant) and the energy gap E_g between the lowest occupied molecular orbital (LUMO) and the highest unoccupied molecular orbital (HOMO) is the electrolyte window. The open circuit voltage (V_{oc}) of the cell is determined by

$$V_{oc} = \mu_A \text{-} \mu_C / e \tag{4}$$

where μ_A and μ_c are the electrochemical potentials of the anode and the cathode respectively and e is the magnitude of the electron charge. If μ_A and μ_c lie above and below the LUMO and the HOMO, the reduction and oxidation of electrolyte occurs at the anode and cathode respectively. When a passivation barrier is created at the electrode and electrolyte interphase (so called solid electrolyte interphase (SEI)), the exchange of electrons between the anode/cathode and electrolyte is prevented, thus suppressing the above-mentioned parasitic reactions and increasing the cell voltage. On the other hand, the total charges stored or the capacity of the electrode materials is governed by the number of electrons exchanged in the corresponding electrochemical reaction (Δx) and the molar mass of the electrode materials (M):

$$Capacity = \frac{26.8 \times \Delta x}{M} . \tag{5}$$

Moving away from lithium to sodium-based systems, first the output voltages of SIBs reduce due to the higher reducing potential of Na^+/Na^0 couple (-2.71 V vs. standard hydrogen electrode (SHE)) in comparison with its lithium counterpart (-3.04 V vs. SHE). Second, the storage capacities of Na-ion electrodes are also decreased due to the higher molar mass of Na-ion electrodes. Overall, the SIBs have lower energy densities in comparison with the LIBs. Nevertheless, energy and power densities are not the critical figures of merit for their application in the grid storage, rather cost, cycle life, round-trip efficiency and safety are the key factors to be considered.

4. Electrode Materials for SIBs

The renaissance of SIB chemistry has resulted in the exploration of a large variety of electrode materials in the last decade (Figure 2).[10, 11] On the cathode side, several layered oxides, polyanionic, Prussian blue type

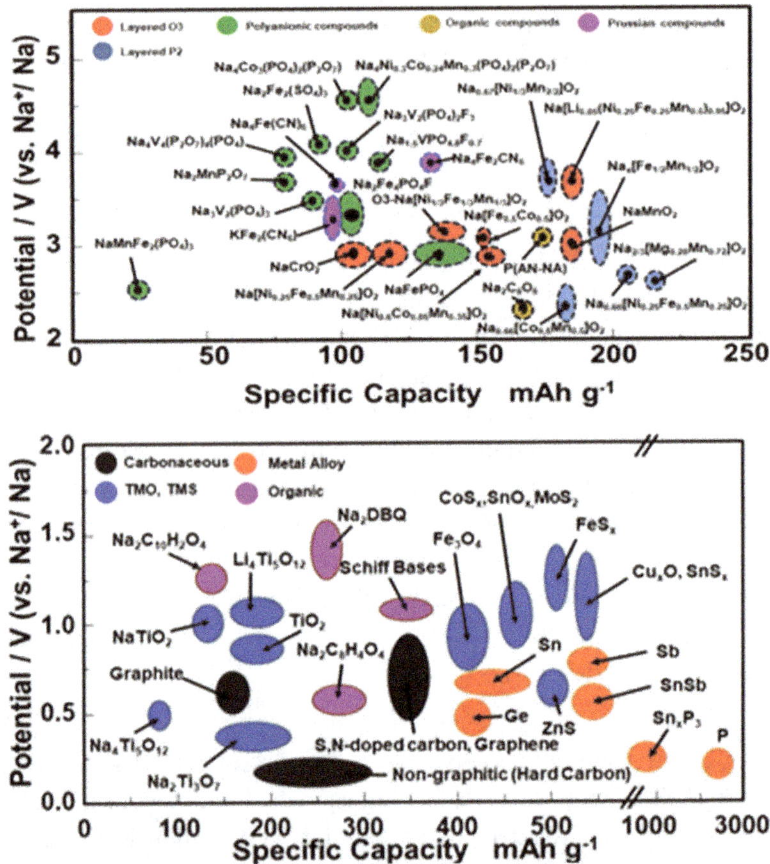

Fig. 2. Current state of art of the Na-ions electrode materials.

electrodes exhibit high voltages and moderate capacities. On the other hand, different types of carbonaceous materials, transition metal oxides and alloys as negative electrodes show stellar electrochemical performances. Such a vast exploration of electrode materials has been made possible based on the "know-how" knowledge gained from Li-ion chemistry. In this chapter, we will focus on selected candidates of electrode materials such as layered oxides and phosphates on the cathode section and carbonaceous materials, titanium oxides and alloys on the anode section.

4.1. *Positive electrode materials*

4.1.1. *Layered oxides*

Amid the all known cathode candidates, Na_xMO_2 layered oxides ($0 < x \leq 1$; M = Ti-Cu; Nb &Mo) are the most explored and promising family of cathode materials for SIBs because of their high capacities, appropriate operating potentials, as well as the convenient synthesis routes.[12] The interesting fact is that all the 3d transition metal ions from Ti to Cu are electrochemically active in the Na-ion layered structures. A typical layered oxide Na_xMO_2 structure consists of alternating layers of edge shared MO_6 octahedra and Na^+ ions (Figure 3). Further, Delmas[13] classified the layer oxides classified into different groups (O3, P2, P3, etc.), in accordance with the coordination environment of Na^+ ions and the number of oxide layers in a single unit cell. The symbols P" or "O" represent the prismatic or the octahedral coordination sites and the "2" or "3" suggests the number of transition metal layers in the single unit cell. The crystal structures of P2, P3 and O3 phases are depicted in Figure 3. P2-type Na_xTMO_2 (space group of $P6_3/mmc$) consists of two kinds of MO_2 layers (AB and BA) with the Na^+ ions in two different types of trigonal prismatic sites: Na_f contacts the two MO_6 octahedra of the adjacent slabs along its face, whereas Na_e shares the six surrounding TMO_6 octahedra along its edges. $O3$-Na_xMO_2 (space group: $R3m$) is composed of three kinds of AB, CA, and BC MO_2 layers in the unit cell, and all the Na^+ ions are accommodated in the "octahedral" (O) sites between MO_2 layers.

In terms of electrochemical performances, O3 type cathodes exhibit high capacities due to the presence of a large amount of sodium ions whereas the P2 type structures show high rate capabilities due to the high diffusivity of sodium ions. O3 phase oxide cathodes undergo phase transformations from O3 structure as O3 $\leftrightarrow$ O'3 $\leftrightarrow$ P3 $\leftrightarrow$ P'3. In comparison with O3-type structures, P2 phases undergo limited number of phase transitions upon electrochemical cycling, thus exhibiting better cyclability. However, the P2 phase transforms to the O2 phase at a high voltage oxidation due to the gliding of some MO_2 sheets which lead to rapid capacity decay.

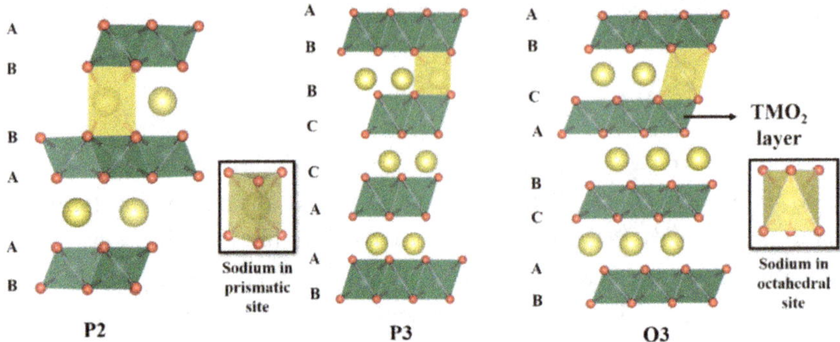

Fig. 3. Crystal structures of P2, P3 and O3 type layered Na$_x$MO$_2$ oxides.

The electrochemical sodium (de)-intercalation properties of several layered oxides have been extensively reported in recent reviews [12, 14] and herein, we will highlight the signatures of single and some of the mixed metal transition layered oxides. Starting from the early transition metal-based cathodes, O3-NaScO$_2$ was reported to be electrochemically inactive due to the empty 3d orbital. O3-NaTiO$_2$ delivered a reversible capacity of ~150 mAh g^{-1} with an average voltage of ~1.0 V vs. Na$^+$/Na^0 in the voltage range of 1.6-0.6 V vs. Na$^+$/Na^0.[15] At a high voltage window, the irreversible phase change occurs presumably due to the migration of titanium ions from the slabs to the interslab space. Interestingly, vanadium based layered oxide crystalizes in both O3 and P2 type structures. Whilst both of them reversibly exchanged 0.5 moles of sodium ions in the voltage range of 2.5-1.4 V vs. Na$^+$/Na^0 with the multistep voltage profiles, the P2 phase exhibited lower polarization.[16, 17] Despite the fact that the Li-analogue of layered chromium oxide was found to be electrochemically inactive,[18] O3-NaCrO$_2$ was reported for its electrochemical activity in Na cells.[19] During electrochemical sodium (de)-intercalation, it reversibly transforms into O'3 and P'3 phases up to x=0.5. Beyond this limit, an irreversible structural transformation results due to the migration of chromium ions into the interslab space.[20, 21] The manganese analogue of O3 phase crystallizes in the monoclinic structure (O'3) due to the Jahn-Teller distortion of the Mn^{3+} ions.[22] It delivered a reversible capacity of 185 mAh g^{-1} (i.e. ~0.8 moles of sodium ions) in the voltage range of 3.8-2.0 V vs. Na$^+$/Na^0. In contrast to the electrochemically inactive $LiFeO_2$, surprisingly $NaFeO_2$ exhibited reversible capacities of ~120 $mAhg^{-1}$ at an average voltage of 3.3 V vs. Na^+/Na^0 due to the redox activity of Fe^{3+}/Fe^{2+}.[23] Delmas first investigated the electrochemical (de)-intercalation of sodium ions into P2 and O3-$NaCoO_2$. Upon sodium extraction, the O3 cathode transforms into the O'3 phase followed by the formation of the P3 phase whereas the P2 cathode preserves its structure throughout the electrochemical cycling.[24] Recently, Delmas *et al.*[25] revisited the P2 cathode and identified various single and bi-phasic domains due to the Na^+/vacancy ordering at different sodium concentrations during the electrochemical cycling which arises from repulsive interactions between Na-Na and Na-Co (figure 4). O3-$NaNiO_2$ cathode is much interested because of the high potential of $Ni^{2+}/Ni^{3+}/Ni^{4+}$ couples and delivered reversible capacities of ~100 mAh g^{-1} in the voltage range of 3.75-1.25 V vs. Na^+/Na^0 with the multistep voltage profiles.[26,27]

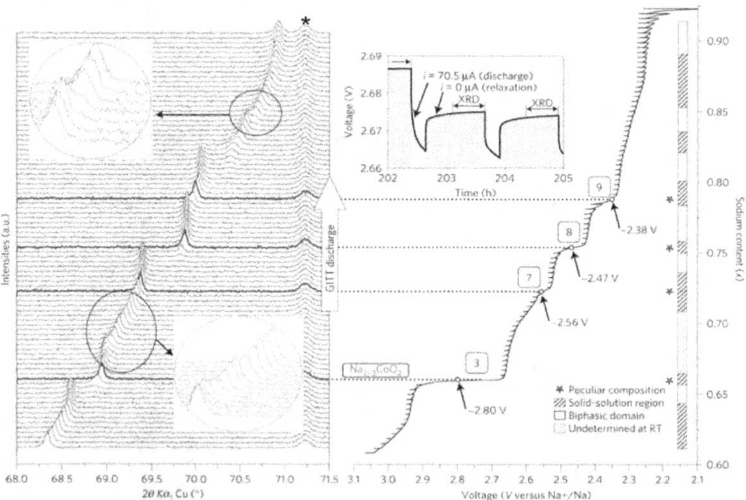

Fig. 4. *In-situ* XRD patterns during sodium ion intercalation in P2-Na_xCoO_2. Adapted with permission from R. Berthelot, D. Carlier & C. Delmas, *Nat. Mater.* **10**, 74-80 (2011). Copyright 2011 Nature Publishing Group.

Recently, several research groups have extensively focused on the development of mixed transition metal based layered oxide cathodes in order to improve the intercalation voltage, structural stability and capacity and suppress the phase transitions. Among them, Ni/Mn-based binary metal oxides containing Ni^{2+} and Mn^{4+} are the most widely studied electrode materials mainly due to two factors: (1) the overall battery performance could make full use of the $Ni^{2+}/Ni^{3+}/Ni^{4+}$ redox reaction which has a relatively high operating voltage; (2) Ni^{2+} and Mn^{4+} are the Jahn–Teller inactive centers, which provides better structural stability of electrodes during cycling. In early 2000, Dahn *et al.*[28] first reported a P2–$Na_{2/3}Ni_{1/3}Mn_{2/3}O_2$ cathode with the relatively high theoretical capacity (173 mAh g^{-1}), and an average operating voltage greater than 3.7 V vs. $Na^+/Na^{0.}$ However, the prime shortcoming was its poor cycling due to detrimental P2-O2 transformation by gliding of transition metal layers upon sodium ion removal at high voltages. To overcome this problem, doping with other divalent cations was attempted in the P2– $Na_{2/3}Mn_{1-x}M_xO_2$ (M=Cu^{2+} and Mg^{2+}) which resulted in the smoothing of the charge/ discharge voltage profiles and the formation of OP4 intergrowth structure at higher voltages (Figure 5).[29,30] From the view

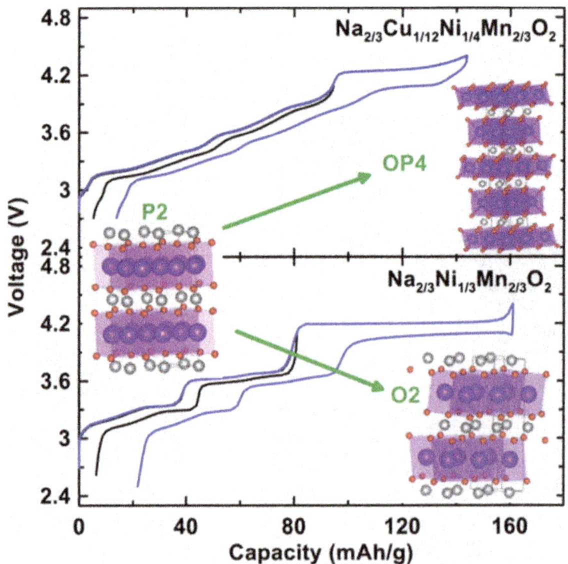

Fig. 5. Voltage vs. capacity profiles of $Na_{2/3}Cu_xNi_{1/3-x}Mn_{2/3}O_2$ cathodes cycled against sodium metal in between 4.1 to 2.5 V vs. Na^+/Na^0. Adapted with permission from L. Zheng, J. Li and M. N. Obrovac, *Chem. Mater.* **29**, 1623, (2017). Copyright 2017 American Chemical Society.

of low cost and earth abundant raw materials, the P2-$Na_{2/3}Fe_{2/3}Mn_{2/3}O_2$ cathode received much attention with a high reversible capacity of ~190 mAh g^{-1} and an average voltage of 2.75 V vs. Na^+/Na^0.[31]

4.1.2. *Polyanionic compounds*

Polyanionic compounds exhibit numerous advantages over oxides and other classes of cathode materials.[32] First, they possess rich crystal chemistry in terms of chemical composition and structural diversity. Second, they exhibit high structural, chemical and thermal stabilities. Third and most importantly, they show higher intercalation voltages for a given redox couple $M^{(n-1)+}/M^{n+}$ in comparison with oxides due to an inductive effect.

Olivine-$LiFePO_4$ is the most studied polyanionic cathode in LIBs.[33] Its three-dimensional crystal structure is built by the corner sharing FeO_6 which is connected to the PO_4 edges (Figure 6(a)). It delivers high reversible capacities of ~ 170 mAh g^{-1} with an average insertion voltage of 3.45 V vs. Li^+/Li^0. On the other hand, the most thermodynamically favored sodium ion bearing phase is maricite-$NaFePO_4$.[34] The maricite phase has edge shared FeO_6 units which share corners with PO_4 units (Figure 6(b)) and there is no Na^+ ion channel for diffusion, thus making it electrochemically inactive. When the chemically sodiated olivine-$NaFePO_4$ was heated above 480 °C, it undergoes an irreversible phase transformation to maricite.[35] It delivers reversible capacities of ~120 mAh g^{-1} in the sodium cell with an average insertion voltage of 2.8 V vs. Na^+/Na^0. Casas-Cabanas et al.[36] found the huge volume mismatch (~17.58 %) between $FePO_4$ and $NaFePO_4$ end members and attributed the formation of the intermediate $Na_{0.7}FePO_4$ to the Na^+ ions/vacancy ordering.

The incorporation of fluorine anions in the phosphate network is expected to enhance the intercalation voltage, which was first demonstrated by Nazar's group.[37] They have prepared a two dimensional layered Na_2FePO_4F cathode and tested it in lithium cells. Further,

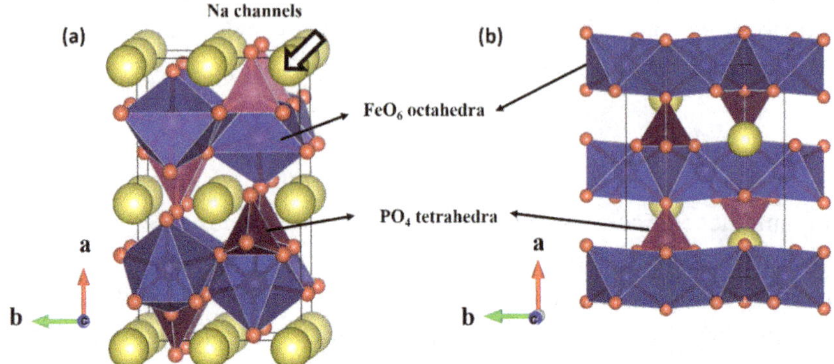

Fig. 6. The crystal structures of (a) Olivine-NaFePO$_4$ and (b) Maricite-NaFePO$_4$.

Tarascon's group evaluated the electrochemical performance of the same cathode in sodium cells and found two well-defined voltage plateaus at ~3.1 and 2.9 V vs. Na$^+$/Na^0 with a reversible capacity of ~110 mAh g^{-1}.[38] Replacing iron with manganese has resulted in the formation of a three-dimensional Na$_2$MnPO$_4$F structure wherein Mn$_2$F$_2$O$_8$ chains are linked by tetrahedra PO$_4$. This Mn-based fluorophosphate cathode showed higher intercalation voltage and reversible capacities of 100 mAh g^{-1}.

In 2003 Barker et al.[39] first reported the synthesis and electrochemical performance of NaVPO$_4$F in hybrid ion cells. It showed two voltage plateaus and a reversible capacity of ~110 mAh g^{-1} in the sodium cell. Very recently, Masquelier et al.[40] revisited the Na$_3$V$_2$(PO$_4$)$_2$F$_3$ phase using the synchrotron radiation diffraction measurement and found that this compound crystallizes in the orthorhombic *Amam* space group. Its crystal structure is built by the corner sharing of V$_2$O$_8$F$_3$ bioctahedral units with the PO$_4$ tetrahedra and contains large tunnels along [1 1 0] and [1 $\bar{1}$ 0] direction wherein the sodium ions occupy (Figure 7(a)). Prof. Rojo's group have extensively studied this class of materials and found that the intercalation mechanism includes the formation of the solid solution as well as the two-phase behavior.[41, 42]

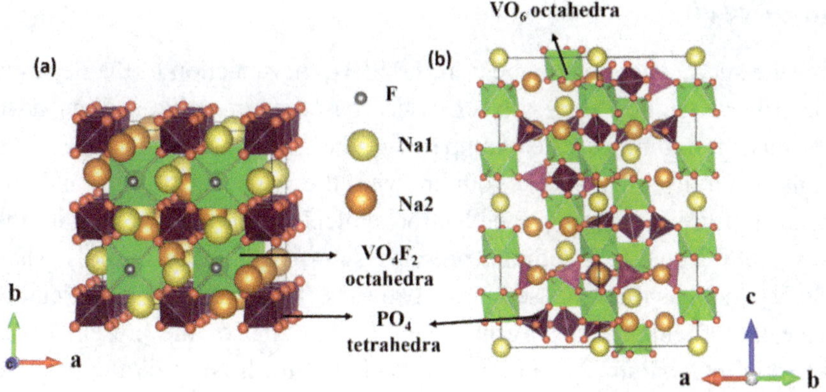

Fig. 7. The crystal structures of (a) $Na_3V_2(PO_4)_2F_3$ and (b) $Na_3V_2(PO_4)_3$.

NASICON (Na super ionic conductor) family of compounds are known for their high sodium conductivity as well as structural stability and have received much attention for their application as Na-ion electrodes.[43] The NASICON crystal structure is built by the so called "lantern units" consisting of two MO_6 octahedra and three PO_4 tetrahedra stacked along the c-axis as shown in Figure 7(b). In this three dimensional framework, sodium ions occupy two different crystallographic sites, i.e. Na(1) and Na(2). Among the NASICON compounds studied so far, $Na_3V_2(PO_4)_3$ has shown a stellar electrochemical performance, exhibiting two voltage plateaus at 3.4 and 1.6 V vs. Na^+/Na^0 which were ascribed to the operation of V^{4+}/V^{3+} and V^{3+}/V^{2+} respectively.[44] The corresponding capacities were found to be ~120 and 60 mAh g^{-1} respectively. Several attempts were made to improve the electrochemical performance of the $Na_3V_2(PO_4)_3$ cathode through carbon coating and cation doping in the place of vanadium ions.[45-47] On other hand, Goodenough et al.[48] attempted to replace toxic and high cost vanadium by manganese cation and obtained the fully sodiated NASICON-$Na_4VMn(PO_4)_3$ structure. When the electrode cycled in between 3.8 and 2.8 V, a two-step voltage profile with an average intercalation voltage of ~3.5 V vs. Na^+/Na^0 was noticed with a reversible capacity of ~100 mAh g^{-1}. Upon increasing the upper cut-off voltage to attain higher amount of capacities, severe structural degradation was observed with the *in-situ* XRD experiments.[49]

4.2. *Negative electrode materials*

For the successful development of the SIBs, the selection of the negative electrode is significantly vital. Devices based on metallic sodium have various shortcomings when compared to their lithium metal anodes. The high reactivity of metallic sodium with the organic solvents and the dendrite formation during the electrochemical cycling paired up with the low melting point of sodium, presents significant safety hazards when the Na anode is directly used in the batteries. Therefore, it is beneficial to have the "rocking chair" format, similarly to that of the present LIBs, wherein the sodium ions are cycled back and forth between the positive and negative electrodes. Likewise, in the Li-ion chemistry, NIB anodes have been categorized in three groups, i.e. intercalation, alloying and conversion-based electrodes.[50] The first is built on the insertion reaction which has been demonstrated on carbonaceous and transition metal oxide materials. The second is based on the alloying reaction where the compounds of group 14 and 15 are the promising candidates of this class. The third is the conversion reaction wherein the transition metal-based compounds (oxides, nitrides, fluorides) are reduced to nano-sized transition metals and sodium compounds such as Na_2O and Na_2S.

4.2.1 *Carbonaceous materials*

Till date, graphite has been widely used as a negative electrode material in LIBs because of its high gravimetric capacity of 372 mAh g^{-1}. Upon electrochemical reduction, Li^+ ions are inserted in the van der Waals gap between graphene layers forming Li-graphite intercalation compounds (GIC) and LiC_6 which are formed towards the end of the discharge.[51] Nevertheless, the graphite is electrochemically less active in the Na cells due to its bigger ionic size and small interplanar distance between graphitic layers.[52] In 2014, the reversible sodium (de)-intercalation into graphite was reported using diglyme as a solvent, wherein the solvent co-intercalation assisted the insertion of sodium ions in between graphene layers.[53]

On other hand, amorphous carbonaceous materials have also been investigated for their electrochemical sodium (de)-intercalation properties ever since the seminal report by Doeff.[54] The amorphous carbonaceous materials could be categorized into two groups, i.e. soft carbon and hard carbon. In soft carbon, the crosslinking between the sp^2 graphene layers is weak, these layers are mobile during the heat treatment (>1500 °C) which leads to the formation of graphitic carbon. On the contrary, the layers in hard carbon are immobile due to strong crosslinking, thus it keeps the disordered structure. Amorphous carbons are made by the pyrolysis of organic precursor in the temperature range of 500 °C to 2000 °C.

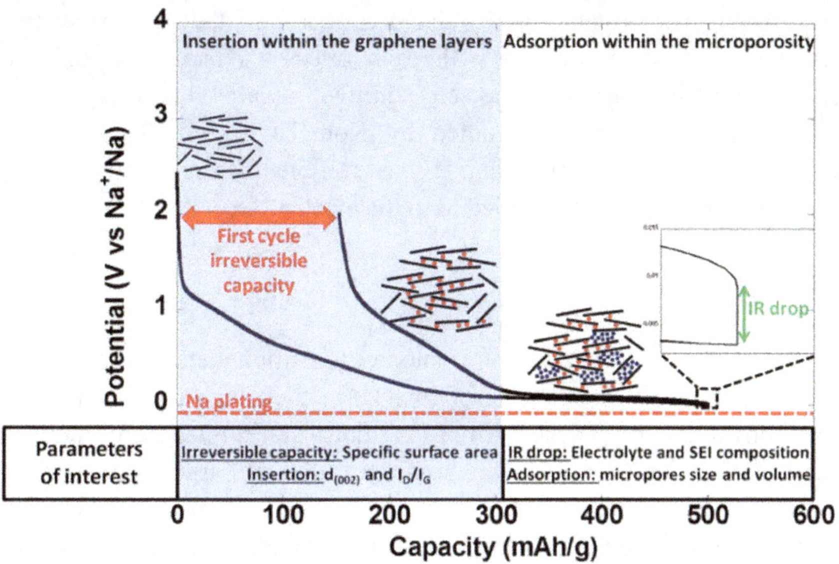

Fig. 8. Potential vs. capacity profiles of hard carbon anode. Adapted with permission from E. Irisarri, A. Ponrouch, M. R. Palacin, *J. Electrochem. Soc.* **162**, A2476 (2015). Copyright 2015 Electrochemical Society.

The reversible electrochemical insertion of sodium ions into hard carbon was first reported by Dahn in 2000.[55] They synthesized hard carbon through pyrolysis of sugar at 1000 °C. The voltage-capacity profiles of the hard carbon anode consist of two distinct regions as shown in Figure 8: A sloping voltage profile from 1.2 V to 0.1 V vs. Na$^+$/Na^0 followed by a low voltage intercalation plateau at ~ 0.1 V vs. Na$^+$/Na^0. The first was attributed to the insertion of sodium ions in between the layers whereas the second was correlated to the adsorption of sodium ions in the micropores.[56] On the contrary, recent reports ascribed the low voltage plateau to the insertion of sodium ions into the interlayers whereas the slopping voltage profiles were connected to pseudo adsorption at the edges of the crystallites based on the *in-situ* XRD and Raman experiments. Recently, Rojo *et al.*,[57] have summarized the complexity of the underlying electrochemical process on hard carbon to the sensitivity of different experimental probes.

Further, the cyclability of hard carbon anode in various electrolytes was studied by Komaba and Palacin groups and their corresponding findings were in agreement with each other.[58,59] The effect of FEC (Fluoroethylene carbonate) as an additive towards the hard carbon cycling performance was studied by Komaba *et al.*[60] The capacity degradation of hard carbon anode was sufficiently suppressed when a small amount of FEC was added in to the electrolyte.

4.2.2   Titanium oxides

In comparison with carbonaceous anodes, transition metal oxides exhibit higher volumetric energy density, thanks to their high density. Despite the conventional wisdom, which predicts the limitation of sodium intercalation into TiO_2 crystal structures due to larger ionic radius, various reports demonstrated the reversible sodium insertion into different polymorphs of TiO_2.[61,62] Lithium titanate $Li_4Ti_5O_{12}$-one of the most studied spinel compounds in lithium ion batteries, intercalates three moles of lithium ions at 1.5 V vs. Li^+/Li^0 with negligible volume changes. When it was tested in the Na-cells, the same anode showed the intercalation behaviour at ~ 0.7 V vs. Na^+/Na^0 with a reversible capacity of 155 mAh g^{-1}.[63] During the first discharge, this anode undergoes a phase transition to produce a mixture of $Li_7Ti_5O_{12}$ and $Na_6LiTi_5O_{12}$, which further reversibly exchanges sodium ions during the following cycles. Among the ternary sodium titatnates, $Na_2Ti_3O_7$ has manifested much attention due to its attractive capacity (170 mAh g^{-1}) and low voltage (~0.3 V vs. Na^+/Na^0).[64] Upon the sodium intercalation, it undergoes a two-phase transformation to produce $Na_4Ti_3O_7$ as confirmed by the *in-situ* XRD. However, its cycling stability and Coulombic efficiency remain as major challenges to successfully implement this material into practical cells. Recently, lithium substituted layered sodium titanate, $P2-Na_{0.66}[Li_{0.22}Ti_{0.78}]O_2$ has been reported to exhibit high reversible capacities of ~116 mAh g^{-1} with a cycle life of more than 1200 cycles.[65]

4.2.3. *Alloys*

At present, alloy materials are considered to be the most attractive and potential candidates as anode materials for SIBs due to their higher volumetric energy densities. In principle, alloys uptake large number of sodium ions when compared to the intercalation-based anodes, thus they have a higher gravimetric capacity. However, such an enhancement is paired with other shortcomings. Due to the presence of a large number of sodium ions, the electrode undergoes huge volume changes upon cycling which leads to pulverization of particles and loss of electronic contact upon cycling, thus the capacity fades rapidly.

Recently, various research groups demonstrated about the stellar cycling performances of the Sb anode. It showed high reversible capacities of about 500-600 mAh g^{-1} with an excellent capacity retention over hundreds of cycles. Monoconduit *et al.*[66] probed the electrochemical alloying reaction mechanism of Sb with sodium using the *in-situ* XRD. During the first discharge, crystalline Sb first transforms into amorphous Na_xSb followed by the formation of cubic and hexagonal Na_3Sb mixture. Upon the subsequent charge, crystalline Na_3Sb is transformed into amorphous Sb. The better cyclability of Sb anode in Na cells in comparison with Li cells is attributed to the formation of amorphous Na_xSb which acts as a buffer for volume changes as well as for the use of the FEC additive (Figure 9). Apart from the bulk Sb anode, the nano-sized Sb too has shown a high capacity with a better cycle life.[67] Binary alloy compounds such as M_xSb_y [68] have been exploited to improve electrochemical performance where M is an electrochemically inactive component, which acts as buffering matrix to reduce the volume changes and prevent aggregation during the repeated cycling. An alternative approach to suppress the volume expansion is the utilization of alloy-carbon composites. Here the role of carbon is to provide an electrical conduction path and mechanical strength to the alloy materials. This strategy was demonstrated by Qian *et al.* to obtain high electrochemical capacities of 300 mAh g^{-1} [69], when the Sb alloy was mechanically ball milled with conductive carbon. Several attempts were made to fabricate the Sb/MWCNT and Sb/C fiber composites for the Na-ion battery application which are summarized in the ref. 70.

Tin based anodes are attractive due to their higher theoretical capacity of 847mAh g^{-1} (based on the formation of $Na_{15}Sn_4$ phase). During the first discharge, they undergoe a series of phase transformations ($NaSn_5$ to $NaSn$ to Na_9Sn_4 to $Na_{15}Sn_4$).[71] These results are in agreement with the theoretical calculations using the density functional theory. The final crystalline phase of $Na_{15}Sn_4$ is resulted from the expansion of Sn particles by 420%. The microstructural evolution and phase transformation of tin nanoparticles was investigated by using *in-situ* transmission microscopy.[72] This study showed the formation of the amorphous $NaSn_2$ phase at the first step followed by the formation of the amorphous Na_9Sn_4 and Na_3Sn and finally crystalline $Na_{15}Sn_4$ phases. The extreme volume expansion (~420% based on the formation of $Na_{15}Sn_4$) and the contraction of the tin anode during the sodium uptake and release remain as the major obstacle for its successful implementation in the practical SIBs. Various strategies including Sn/C composites and intermetallic compounds were attempted in the recent times. The $Sn_{0.9}Cu_{0.1}$ system showed reversible capacities of 420 mAh g^{-1} with a 97% of cycle retention after 100 cycles.[73] The reason for its great reversible capacity is attributed to the suppression of aggregation among the nanoparticles owing to the presence of Cu. The SnSb/C composite anode prepared via high-energy ballmilling showed a high initial capacity (541 mAh g^{-1}) whilst 80% of it was retained after 50 cycles.[74]

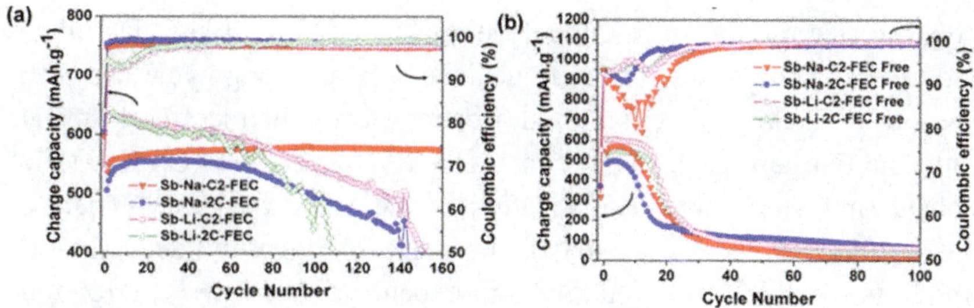

Fig. 9. Comparison of the performances of the bulk Sb anode in Li and Na cells with and without the FEC additive. Adapted with permission from A. Darwiche, C. Marino, M. T. Sougrati, B. Fraisse, L. Stievano and L. Monconduit, *J. Am. Chem. Soc.* **134**, 20805–20811, (2012). Copyright 2012 American Chemical Society.

Phosphorous exhibits a high theoretical gravimetric capacity (2596 mAh g^{-1}) compared to any other known sodium anode materials. It electrochemically alloys with sodium to form Na_3P. Nonetheless, Na_3P reacts spontaneously with water to form phosphine (PH_3), which limits its practical application. Kim *et al.* reported high reversible capacity of 1890 mAh g^{-1} for the red phosphorous/carbon composite electrode with a negligible capacity fading for the first 30 cycles.[75] On the other hand, black phosphorous, which is the least reactive allotrope, also showed stellar reversible capacities with an excellent capacity retention.[76]

5. Summary and Outlook

In summary, we briefly discussed the recent progress on the selected Na-ion electrodes in this report. Despite their large reversible capacities, layered oxides still suffer from air instability and multiple phase transformations with poor cyclability.[77] On the other hand, phosphate electrodes demonstrate high voltages and stable cycling with moderate capacities. On the anode side, carbonaceous materials show stellar long-standing cycling performances. However, the low voltage insertion of carbon anodes raises serious concerns over safety issues for their practical application. In this aspect, transition metal oxides appear to be good, yet their capacity and cyclability need to be improved. Despite their large gravimetric capacities, alloy anodes still suffer from key issues such as low first cycle Coulombic efficiency and cyclability. Apart from the electrode materials, the development of the suitable electrolytes and binders are also important to achieve safe and long cycle life SIBs. Recently Palacin's group reported a better electrochemical performance with the Na-ion electrolyte consisting of $NaPF_6$ in EC:PC:DMC.[78] Considering, the practical application in the SIBs, the development of water-soluble binders are vital in the place of polyvinylidene fluoride which is soluble in the toxic and expensive N-methyl pyrrolidone solvent. Binders such as sodium carboxymethyl cellulose, poly (acrlic acid) and sodium alginate showed better adhesion of electrode particles and a stable SEI formation.[79] Further, the studies dedicated on the development of full Na-ion cells are vital to hasten the commercialization of SIBs. Despite its known low energy density, recent reports call for an improvement on the rate performance, cycle life and safety of SIBs.[80,81]

References

1. Z. Yang, J. Zhang, M. C.W. Kinter-Meyer, X. Lu, D. Choi, J. P. Lemmon and J. Liu, *Chem. Rev.* **111**, 3577–3613, (2011).
2. B. Dunn, H. Kamath and J. M. Tarascon, *Science*, **334**, 928, (2011).
3. Y. Yao and J. Kummer, *Inorg. Nucl. Chem* **29**, 2453, (1967).
4. J. B. Goodenough, H. Y.-P. Hong and J. A. Kafalas, *Mater. Res. Bull.* **11**, 203, (1976).
5. M. S. Whittingham, *Chem. Rev.* **104** (10), 4271–4302, (2004).
6. Armand M. *Intercalation Electrodes*. (Plenum Press, New York, 1980).
7. C. Delmas, *Adv Energy Mater.* **8**, 170, (2018).
8. G. E. Blomgren, *J. Electrochem. Soc.* **164**(1), A5019–A5025, (2017).
9. J. B. Goodenough and K. S. Park, *J. Am. Chem. Soc.* **135**(4), 1167–1176, (2013).
10. N. Yabuuchi, K. Kubota, M. Dahbi and S. Komaba, *Chem. Rev.* **114**, 11636, (2014).
11. J. Y. Hwang, S. T. Myung and Y. K. Sun, *Chem. Soc. Rev.* **46**(12), 3529–3614, (2017).
12. K. Kubota, S. Kumarkura, Y. Yoda, K. Kuroki and S. Komaba, *Adv. Energy Mater.* **8**, 1703415, (2018).
13. C. Delmas, C. Fouassier, P. Hagenmuller, *P. Physica B*, **99**, 81, (1980).
14. M. H. Han, E. Gonzalo, G. Singh and T. Rojo, *Energy Environ. Sci.* **8**, 81–102, (2015).
15. D. Wu, X. Li, B. Xu, N. Twu, L. Liu and G. Ceder, *Energy Environ. Sci.* **8**, 195, (2015).

16. C. Didier, M. Guignard, C. Denage, O. Szajwaj, S. Ito, I. Saadoune, J. Darriet and C. Delmas, *Electrochem. Solid-State Lett.* **14**, A75, (2011).

17. M. Guignard, C. Didier, J. Darriet, P. Bordet, E. Elkaim and C. Delmas, *Nat. Mater.* **12**, 74, (2013).

18. S. Komaba, C. Takei, T. Nakayama, A. Ogata and N. Yabuuchi, *Electrochem. Commun.* **12**, 355, (2010).

19. J. J. Braconnier, C. Delmas and P. Hagenmuller, *Mater. Res. Bull.* **17**, 993, (1982).

20. K. Kubota, I. Ikeuchi, T. Nakayama, C. Takei, N. Yabuuchi, H. Shiiba, M. Nakayama and S. Komaba, *J. Phys. Chem. C.* **119**, 166, (2015).

21. S. H. Bo, X. Li, A. J. Toumar and G. Ceder, *Chem. Mater.* **28**, 1419, (2016).

22. X. H. Ma, H. L. Chen and G. Ceder, *J. Electrochem. Soc.* **158**, A1307, (2011).

23. N. Yabuuchi, H. Yoshida, S. Komaba, *Electrochemistry.* **80**, 716, (2012).

24. J.-J. Braconnier, C. Delmas, C. Fouassier and P. Hagenmuller, *Mater. Res. Bull.* **15**, 1797, (1980).

25. R. Berthelot, D. Carlier and C. Delmas, *Nat. Mater.* **10**, 74, (2011).

26. J. J. Braconnier, C. Delmas and P. Hagenmuller, *Mater. Res. Bull.* **17**, 993, (1982).

27. P. Vassilaras, X. H. Ma, X. Li and G. Ceder, *J. Electrochem. Soc.* **160**, A207, (2013).

28. Z. H. Lu and J. R. Dahn, *J. Electrochem. Soc.* **148**, A1225, (2001).

29. G. Singh, N. Tapia-Ruiz, J. M. Lopez del Amo, U. Maitra, J. W. Somerville, A. R. Armstrong, J. Martinez de Ilarduya, T. Rojo, and P. G. Bruce, *Chem. Mater.* **28**, 5087, (2016)

30. L. Zheng, J. Li and M. N. Obrovac, *Chem. Mater.* **29**, 1623, (2017).

31. N. Yabuuchi, M. Kajiyama, J. Iwatate, H. Nishikawa, S. Hitomi, R. Okuyama, R. Usui, Y. Yamada and S. Komaba, *Nat. Mater.* **11**, 512, (2012).

32. C. Masquelier and L. Croguennec, *Chem. Rev.* **113**(8), 6552–6591, (2013).

33. A. K. Padhi, K. S. Nanjundaswamy, J. B. Goodenough, *J. Electrochem. Soc.* **144**, 1188, (1997).

34. K. Zaghib, J. Trottier, P. Hovington, F. Brochu, A. Guerfi, A. Mauger, C. M. Julien, *J. Power Sources.* **196**, 9612, (2011).

35. P. Moreau, D. Guyomard, J. Gaubicher, F. Boucher, *Chem. Mater.* **22**, 4126, (2010).

36. C. Cabanas, V. V. Roddatis, D. Saurel, P. Kubiak, J. Carretero-Gonzalez, V. Palomares, Pl. Serras and T. Rojo, *J. Mater. Chem.* **22**, 17421–17423, (2012).

37. B. L. Ellis, W. R. M. Makahnouk, Y. Makimura, K. Toghill and L. F. Nazar, *Nat. Mater.* **6**, 749–753, (2007).

38. N. Rechan, J.-N. Chotard, L. Dupont, K. Djellab, M. Armand and J. M. Tarascon, *J. Electrochem. Soc.* **156**, A993–A999, (2009).

39. J. Barker, M. Y. Saidi and J. L. Swoyer, *Electrochem. Solid- State Lett.* **6**, Al–A4, (2003).

40. M. Bianchini, N. Brisset, F. Fauth, F. Weill, E. Elkaim, E. Uard, C. Masquelier and L. Croguennec, *Chem. Mater.*, **26**, 4238–4247, (2014).

41. P. Serras, V. Palomares, J. Alonso, N. Sharma, J. M. López del Amo, P. Kubiak, M. L. Fdez-Gubieda and T. Rojo, *Chem. Mater.* **25**, 4917, (2013).

42. N. Sharma, P. Serras, V. Palomares, H. E. A. Brand, J. Alonso, P. Kubiak, M. L. Fdez-Gubieda and T. Rojo, *Chem. Mater.* **26,** 3391, (2014).

43. S. Chen, C. Wu, L. Shen, C. Zhu, Y. Huang, K. Xi, J. Maier and Y. Yu, *Adv. Mater.* **29**(48), 1700431, (2017).

44. S. Y. Lim, H. Kim, R. A. Shakoor, Y. Jung and J. W. Choi, *J. Electrochem. Soc.* **159,** A1393, (2012).

45. A. Inoishi, Y. Yoshioka, L. Zhao, A. Kitajou and S. Okada, *ChemElectroChem,* **4**(11), 2755–2759, (2017).

46. W. Shen, H. Li, Z. Guo, Z. Li, Q. Xu, H. Liu and Y. Wang, *RSC Adv.* **6** (75), 71581–71588, (2016).

47. F. Lalère, V. Seznec, M. Courty, R. David, J. N. Chotard and C. Masquelier, *J. Mater. Chem. A,* **3**(31), 16198–16205, (2015).

48. W. Zhou, L. Xue, X. Lü, H. Gao, Y. Li, S. Xin, G. Fu, Z. Cui, Y. Zhu and J. B. Goodenough, *Nano Lett.* **16**(12), 7836–7841, (2016).

49. F. Chen, V. M. Kovrugin, R. David, O. Mentré, F. Fauth, J. N. Chotard and C. A. Masquelier, *Small Methods,* 1800218, (2018).

50. M. Dahbi, N. Yabuuchi, K. Kubota, K. Tokwa and S. Komaba, *Phys. Chem. Chem. Phys.* **16**, 15007–15028, (2014).

51. T. Ohzuku, Y. Iwakoshi and K. Sawai, *J. Electrochem. Soc.* **140**, 2490–2498, (1993).

52. P. Ge and M. Fouletier, *Solid State Ionics,* **28–30**, 1172–1175, (1988).

53. B. Jache and P. Adelhelm, *Angew. Chem., Int. Ed.* **53**, 10169–10173, (2014).

54. M. M. Doeff, Y. P. Ma, S. J. Visco and L. C. Dejonghe, *J. Electrochem. Soc.* **140**, L169–L170, (1993).

55. D. A. Stevens and J. R. Dahn, *J. Electrochem. Soc.* **147**, 4428–4431, (2000).

56. H. Hou, X. Qui, W. Wei, Y. Zhang and X. Ji, *Adv. Energy Mater.* **7**, 1602898, (2017).

57. D. Saurel, B. Orayech, B. Xiao, D. Carriazo, X. Li and T. Rojo, *Adv. Energy Mater.* **8**, 1703268, (2018).

58. S. Komaba, W. Murata, T. Ishikawa, N. Yabuuchi, T. Ozeki, T. Nakayama, A. Ogata, K. Gotoh and K. Fujiwara, *Adv. Funct. Mater.* **21**, 3859–3867, (2011).

59. E. Irisarri, A. Ponrouch, M. R. Palacin, *J. Electrochem. Soc.* **162**(14), A2476–A2482, (2015).

60. S. Komaba, T. Ishikawa, N. Yabuuchi, W. Murata, A. Ito and Y. Ohsawa, *ACS Appl. Mater. Interfaces,* **3**, 4165–4168, (2011).

61. S. Lunell, A. Stashans, L. Ojamäe, H. Lindström and A. Hagfeldt, *J. Am. Chem. Soc.* **119**, 7374–7380, (1997).

62. J. P. Huang, D. D. Yuan, H. Z. Zhang, Y. L. Cao, G. R. Li, H. X. Yang and X. P. Gao, *RSC Adv.* **3**, 12593–12597, (2013)

63. Y. Sun, L. Zhao, H. Pan, X. Lu, L. Gu, Y.-S. Hu, H. Li, M. Armand, Y. Ikuhara, L. Chen and X. Huang, *Nat. Commun.* **4**, 1870–1880, (2013).

64. P. Senguttuvan, G. Rousse, V. Seznec, J. M. Tarascon and M. R. Palacin, *Chem. Mater.* **23**, 4109–411, (2011).

65. Wang, Y. S.; Yu, X. Q.; Xu, S. Y.; Bai, J. M.; Xiao, R. J.; Hu, Y. S.; Li, H.; Yang, X. Q.; Chen, L. Q.; Huang, X. J. *Nat. Commun.* **4**, 2365, (2013).

66. A. Darwiche, C. Marino, M. T. Sougrati, B. Fraisse, L. Stievano and L. Monconduit, *J. Am. Chem. Soc.* **134**, 20805–20811, (2012).

67. M. He, K. Kravchyk, M. Walter and M. V. Kovalenko, *Nano Lett.* **14**, 1255–1262, (2014).

68. L. Baggetto, M. Marszewski, J. G_rka, M. Jaroniec and G. M. Veith, *J. Power Sources.* **243**, 699–705, (2013)

69. J. Qian, Y. Chen, L. Wu, Y. Cao, X. Ai and H. Yang, *Chem. Commun.* **48**, 7070–7072, (2012).

70. Y. Kim, K Ha, S. M. Oh and K. T. Lee, *Chem. Eur. J.* **20**(38), 11980–11992, (2014).

71. L. D. Ellis, T. D. Hatchard and M. N. Obrovac, *J. Electrochem. Soc.* **159**, A1801–A1805, (2012).

72. J. W. Wang, X. H. Liu, S. X. Mao and J. Y. Huang, *Nano Lett.* **12**, 5897–5902, (2012).

73. Y. M. Lin, P. R. Abel, A. Gupta, J. B. Goodenough, A. Heller and C. B. Mullins, *ACS Appl. Mater. Interf.* **5**, 8273–8277, (2013).

74. L. F. Xiao, Y. L. Cao, J. Xiao, W. Wang, L. Kovarik, Z. M. Nie and J. Liu, *Chem. Commun.* **48**, 3321–3323, (2012).

75. Y. Kim, Y. Park, A. Choi, N.-S. Choi, J. Kim, J. Lee, J. H. Ryu, S. M. Oh and K. T. Lee, *Adv. Mater.* **25**, 3045–3049, (2013).

76. T. Ramireddy, T. Xing, M. M. Rahman, Y. Chen, Q. Dutercq, D. Gunzelmann and A. M. Glushenkov, *J. Mater. Chem. A* **3**, 5572–5584, (2015).

77. P. -F. Wang, Y. You, Y. -X. Yin and Y. -G. Guo, *Adv. Energy Mater.* **8**, 1701912, (2018).

78. A. Ponrouch, D. Monti, A. Boschin, B. Steen, P. Johansson and M. R. Palacin, *J. Mater. Chem. A,* **3**, 22, (2015).

79. C. Bommier and X. Ji, *Small* **14**, 1703576, (2018).

80. J. Deng, W.-B. Luo, S.-L. Chou, H.-K. Liu and S.-X. Dou, *Adv. Energy Mater.* **8**, 1701428, (2018).

81. A. Bauer, J. Song, S. Vail, W. Pan, J. Barker and Y. Lu, *Adv. Energy Mater.* **8**, 1702869, (2018).

Chapter 13

Supercapacitors Based on Graphene, Borocarbonitrides and Molybdenum Sulphides

C. N. R. Rao[*] and K. Gopalakrishnan

Chemistry and Physics of Materials Unit, New Chemistry Unit, International Centre for Materials Science, Sheikh Saqr Laboratory and School of Advanced Materials, Jawaharlal Nehru Centre for Advanced Scientific Research, Jakkur, Bangalore, 560064, India
[*]*cnrrao@jncasr.ac.in*

Two-dimensional layered materials such as graphene, borocarbonitrides and molybdenum sulphides have shown interesting electrochemical supercapacitor properties. The unique structure, surface properties and the electrical conductivity of these materials contribute to the excellent performance as well as good cyclic stability.

1. Introduction

Supercapacitors are electrochemical energy storage devices that store electrical charge electrostatically by polarizing an electrolytic solution.[1] They can store a large quantity of energy when compared to that of capacitors but lesser than most of the batteries. Supercapacitors have various applications in mobile devices and automobiles due to their excellent reversible charge storage process and long cyclic stability. Supercapacitors are of two types, namely the electrical double-layer capacitor and the pseudocapacitor. The electrical double capacitor stores the charge at the electrode/electrolyte interface when the electrodes are polarised, forming a double layer and the process is non-faradaic. The most used electrical double capacitor electrodes are carbon based materials due to their electrical conductivity, high surface area and unique porosities. Materials like activated carbon, carbon nanotubes, graphene, borocarbonitrides and transition metal chalcogenides (TMDs) have been used as supercapacitor electrodes.[2] In the case of pseudocapacitors, the charge between the electrode and electrolyte are stored faradaically. Metal oxides like RuO_2, IrO_2 and conducting polymers like polyaniline and polypyrrole are most studied materials for pseudocapacitors. However, these materials undergo a structural degradation during continuous charge/discharge process. To sort this out, pseudocapacitve materials have been made composites with EDLC materials.

In this chapter, we discuss the synthesis and supercapacitor applications of graphene, borocarbonitrides and molybdenum sulphide. It is noteworthy that all these materials possess outstanding electrochemical properties and exhibit an excellent supercapacitor performance.

2. Two-dimensional Layered Material Based Supercapacitors

2.1. *Graphene*

Graphene, the mother of all graphitic carbons and the Nobel material, has shown outstanding chemical and physical properties such as ballistic transport, robustness, high thermal conductivity and optical transmittance.[3] These properties of graphene are unique when compared to other allotropes of carbon like fullerenes and carbon nanotubes. The electronic properties of graphene change with respect to the number of layers, where a single-layer graphene shows a metallic behaviour and a few-layer graphene has a tiny band-gap. Graphene can be synthesized in several ways either by physical or by chemical methods.[3] A few of the methods of synthesis are mentioned below.

Geim et al.[5] have successfully exfoliated the single-layer of graphene by micromechanical cleavage of highly oriented pyrolytic graphite (HOPG). Graphene produced from this method is of high quality but the quantity obtained is less. A large quantity of high quality few-layered graphene (2-3 layers) can be produced using the arc-discharge method without the use of a catalyst.[6] In addition, the single-layer or few-layer graphene can be synthesized by the chemical vapour deposition (CVD) method using transition metal (Cu, Ni) sheets as a catalyst.[7] Besides using the physical exfoliation method, graphite can be exfoliated using chemical reagents like N-methylpyrrolidone or dimethylformamide yielding mostly single-layer graphene sheets.[8]

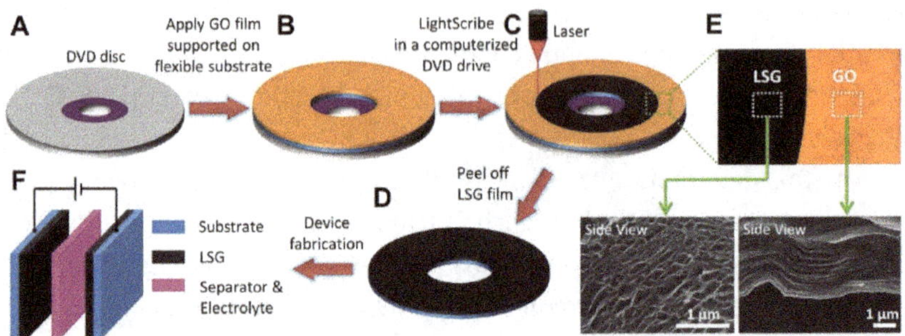

Scheme 1. Schematic illustration and fabrication of laser-scribed graphene-based electrochemical capacitors. (a)-(d) GO film supported on a flexible substrate is placed on top of a LightScribe-enabled DVD media disc, and a computer image is then laser-irradiated on the GO film in a computerized LightScribe DVD drive. (e) As shown in the photograph, the GO film changes from a golden brown color to black as it has reduced to a laser-scribed graphene. The low-power infrared laser changes the stacked GO sheets immediately into well-exfoliated few-layered LSG film, as shown in the cross-sectional SEM images. (f) A symmetric EC is constructed from two identical LSG electrodes, ion-porous separator, and electrolyte. Reproduced with permission from Ref. 4.

Graphene has been utilized as a supercapacitor electrode because of its unique electrical conductivity, high surface area, chemical stability and low-cost production.[3] Theoretical calculations predict that the surface area of graphene can range up to 2675 m^2 g^{-1} and can exhibit a specific capacitance of 550 F g^{-1}. Rao et al.[9] have shown graphene based supercapacitor electrodes made from exfoliated graphene oxide (EG) and nanodiamond (DG). The specific capacitance of EG and EG are 117 and 35 F g^{-1} respectively in 1M H_2SO_4 and in an organic electrolyte, the values of EG and DG are 75 F g^{-1} and 40 F g^{-1} and the energy densities are 31.9 and 17.0 Wh kg^{-1} respectively. Graphene reduced from graphene oxide has shown a specific capacitance of 135 and 99 F g^{-1} in aqueous and organic electrolyte[10] respectively whereas the microwave exfoliated graphene oxide shows a surface area of 463 m^2 g^{-1} with a specific capacitance as high as 191 F g^{-1} in an aqueous electrolyte.[11] Liu et al.

have fabricated high performance graphene supercapacitors with an energy density in the order of 10^{-3} Wh cm^{-3}. Bi-layer graphene shows up to 80 µF cm^{-2}, whereas the few-layer graphene shows a much better specific capacitance of 394 µFcm^{-2}.[12] Reduced graphene oxide prepared at a low temperature has a mesoporous surface and reaches up to 284 F g^{-1} with an energy density of 131 Wh kg^{-1}.[13] The 3D self-supported graphene synthesized using sugar blowing technique possesses a high specific surface area of 1005 m^2 g^{-1} and the capacitance is 250 F g^{-1} at 1 A g^{-1} and at higher current densities the capacitance decreases to 130 F g^{-1}.[14] Downard et al.[15] have chemically linked aryl spacer groups to graphene using aryldiazonium salts and evaluated their performance. When compared to the unmodified graphene, the modified graphene show an excellent double layer capacitance even after 20000 cycles of charge/discharge. This determines that these chemically linked spacer functional groups stop graphene aggregation and increase the performance.

Microsupercapacitors based on laser-induced graphene made on commercial polyimide sheets show an areal capacitance of >9 mF cm^{-2} at 0.02 mA cm^{-2}.[16] Laser scribed graphene films show an excellent electronic conductivity of 1738 S m^{-1} and the surface of 1520 m^2 g^{-1}.[4] The electrochemical supercapacitors based on these electrodes show an areal capacitance of 4.04 mF cm^2 in 1.0 M H$_2$SO$_4$. Scheme 1 shows the fabrication of these supercapacitor electrodes and the scanning electron microscopy images. Graphene hydrogels withstand a maximum stress of 0.9 mPa and have a surface area of 614 m^2 g^{-1} and exhibit an aerial capacitance of 33.8 mF cm^{-2} at 1 mA cm^{-2}.[17] The hydrogel retains 97.8 % of capacitance even after 4000 cycles of charge/discharge and shows an energy density of 2.73 µWh cm^{-3}. The carbon aerogel/graphene composite shows a better performance (300 F g^{-1}) when compared to that of activated carbon (189 F g^{-1}).[18] Ultrathin printable graphene supercapacitors made from exfoliated graphene have a volumetric capacitance of 348 F cm^{-3} and are stable at a high scan rate of 2000 V s^{-1}. Three-dimensional freestanding graphene exhibits a capacitance of 266 µF in an aqueous electrolyte.[19] In ionic liquids, the capacitance increases up to 636 µF with an energy density of 40.94 Wh kg^{-1}. Electrolyte dependent supercapacitor measurements by Liu et al.[20] show that the performance of supercapacitors can vary depending on the ionic sizes of the electrolyte ions.

2.2. *Graphene composites with conducting polymers*

Composites based on conducting polymers like polyaniline (PANI) and polypyrrole could lead to a high capacitance due to the pseudocapacitve nature of the polymers. Composites of graphene and polyaniline have shown a specific capacitance of 210 F g^{-1} at 0.3 A g^{-1}.[21] The conductivity of the composite film containing 44% reduced graphene is around 5.5 x 10^2 S m^{-1} which is about 10 times higher than that of polyaniline nanofibers. Flexible graphene/polyaniline nanofiber composite films show specific capacitances ranging from 400-425 F g^{-1}.[22] Polyaniline coated curved-graphene based supercapacitors have shown excellent capacitive behaviours in the redox electrolyte when compared to the one measure in a non-redox electrolyte, where the capacitance has increased from 36 to 92%.[23] The increase in capacitance is mainly from the pseudocapacitance from polyaniline and the electric double layer capacitance from the graphene sheet. In addition to PANI, polypyrrole based graphene composites increases the charge accumulation in supercapacitors, where the capacitance is around 277.8 F g^{-1}.[24] Asymmetric supercapacitors based on graphene supported iron oxide nanosheets show a high capacitance of 720 F g^{-1} and provides a high energy density of 140 Wh kg^{-1} with a reasonable cyclic stability.[22] Nafion/graphene modified screen printed electrodes show a capacitance of 169.5 F g^{-1} at 1.3 μA.[25] The addition of Nafion alters the orientation of graphene, increases the accessibility of the electrolyte and edge plane like sites.

2.3. *Nitrogen doped graphene*

Although, graphene has a very high theoretical specific capacitance of 540 F g^{-1}, but experimentally only half the capacitance is achieved and the energy density is low. Energy densities of graphene can be increased by making composites with metal oxides or conducting polymers as seen in the earlier section. However, it remains a big challenge to improve the capacitive nature of graphene without losing its EDLC behaviour and its long cyclic stability. Doping graphene with heteroatoms (B, N, S) can effectively increase the capacitive properties and the energy density of graphene, because doping affects the electronic structure which increases the attraction of electrolyte ions in the solution towards the electrode.[26] In general, there are four different nitrogen species doped in the graphene lattice.[27] The large hindrance from the sp^2-hybridized carbons make doping difficult into the graphene lattice. Thus, quaternary nitrogen is a difficult form and leads to less doping levels whereas the pyridinic nitrogen and pyrrolic nitrogen forms are much easier and are proven to be more electrochemically active. Thus, regulating the concentrations of pyridinic and pyrrolic nitrogen in the graphene lattice is highly vital to increase the specific capacitance of graphene. The redox properties of pyrrolic and pyridinic nitrogen increases the pseudocapacitance whereas the quaternary nitrogen plays role in the EDL capacitance. In addition, quaternary-type nitrogen and the pyridinic-nitrogen bound to oxygen increases the wettability between the electrode and the electrolyte and increases the final capacitance. Furthermore, these nitrogens in graphene exhibit different electronic properties and structures. Importantly, the quaternary-type nitrogen increases the electrical conductivity and helps in the fast charge/discharge process. This highlights the role of different nitrogen species in the electrochemical performance and it is important to synthesize selective nitrogen contents in the graphene lattice.

Several synthetic methods have been employed to synthesize nitrogen-doped graphene. Thus, Rao et al.[28] have synthesized the nitrogen-doped few-layer graphene (~3-4 layers) containing 1-1.4 at.% of nitrogen using arc-discharge method in the presence of H_2, He and NH_3/pyridine vapours. Heating graphene oxide with NH_3 at different temperatures (300 to 1100 °C) yields up to 5 at.% of nitrogen and the doping concentration depends on the carboxyl/hydroxyl functionalities and also the heating temperature. GO treated with melamine/urea at varying mass proportions yields nitrogen content ranging from 5-10 at.%.[29,30] Under hydrothermal/solvothermal conditions the doping level can reach up to 13 at.% by treating CCl_4 with lithium nitride.[31]

304   *C. N. R. Rao & K. Gopalakrishnan*

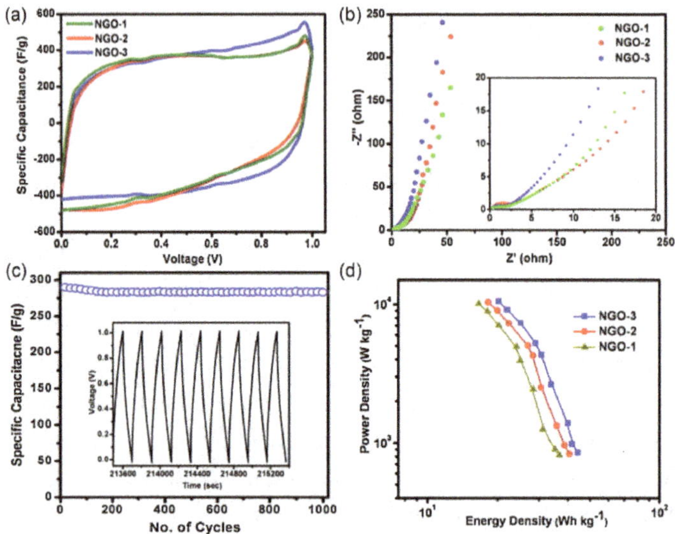

Fig. 1. (a) Cyclic voltammograms of NGOs at a scan rate of 20 mV/s. (b) Nyquist curves for NGO electrodes. (c) Specific capacitance versus the cycle number of NGO measured at a current density of 0.5 A/g within an operational window of 0.0-1 V (the inset shows the charge/discharge curves of the last few cycles for NGO). (d) Ragone plots of NGO based supercapacitors. Reproduced with permission from Ref. 32.

Thermal treatment of RGO in the presence of ammonia yields around 2 at.% of nitrogen and the surface area (630.6 m^2 g^{-1}) as well conductivity (~1000–3000 S m^{-1}) are found to be higher as compared to that of the un-doped RGO.[33] The specific capacitance of doped graphene is 145 F g^{-1} at 0.5 A g^{-1} which is found to be superior to that of a high surface activated carbon and the un-doped reduced graphene oxide. N-doped graphene synthesized by hydrothermal treatment of GO and urea shows 326 F g^{-1} at 0.2 A g^{-1}.[34] The ammonia released during the decomposition of urea reacts with the carboxyl groups of graphene oxide and dopes the nitrogen in the graphene lattice forming an N-doped graphene. When treating graphene oxide with hexamethylenetetramine it yields 8.62 at.% of nitrogen containing different nitrogen species and shows a capacitance of 161 F g^{-1} at 0.5 A g^{-1} and a good cyclic stability.[35] In addition, graphene oxide treated with cynamide yields crumbled N-doped graphene sheets with a specific capacitance 248 F g^{-1} at 5 mVs^{-1} and high pore volume (3.42 cm^3 g^{-1}).[36] Nitrogen-doped graphene hydrogel synthesized by using an organic amine shows a specific capacitance of 185 F g^{-1}.[37] The capacitance of the hydrogels can be tuned by choosing the right organic amine.

RGO heated with urea at different temperatures yields 2 to 8 wt. % range of nitrogen, the specific capacitance is 126 F g^{-1} at 10 mV s^{-1} in 6M KOH while in the organic electrolyte the specific capacitance value is 258 F g^{-1} at 5 mV s^{-1}.[30] It is noteworthy that the capacitance increases with increase in the nitrogen concentration. When treating GO and urea under microwave conditions the supercapacitor performance increases with a value of 440-461 F g^{-1}.[32] In a typical synthesis, graphene oxide was finely ground with different proportions of urea and treated under the microwave for 30s. The nitrogen content varies from 14-18 wt.% with different weight ratios of graphene oxide and urea. The cyclic voltammogram of NGO is shown in Figure 1(a). The equivalent series resistances are in 0.35, 0.36 and 0.33 Ω for NGO-1, NGO-2 and NGO-3 respectively as seen from the Nyquist plots (Fig. 1(b)). Figure 1(c) shows the cyclic stability of the NGO. These electrodes show an excellent capacitance retention of 97% after 1000 cycles at 0.5 A/g. The maximum energy density obtained is 44.4 W h kg^{-1} and the power densities are from 852–10 524 W kg^{-1} (Figure 1(d)) and shows an excellent cycling stability (last few cycles of charge/dishcharge is shown in inset of Fig. 1(c)). Furthermore, XPS reveals three different nitrogen species present, in which, pyrrolic and pyridinic nitrogens play an important role in increasing the capacitance whereas the graphitic nitrogen increases the conductivity. Thus, the graphene containing pyrrolic nitrogen shows a specific capacitance of 194 F g^{-1}.[38] DFT calculations predict that the binding energy of electrolyte ions and the pyrrolic nitrogen are in the proper level, which increases the capacitance of the doped graphene. The microwave treatment of graphene oxide in ethylene glycol and ammonia solution gives up to an energy density of 42.8 W h kg^{-1} and shows good long term stability.[39] Solid-state microwave irradiation and heat treatment of graphene oxide under NH$_3$ shows a capacitance of 291 F g^{-1} at 1 A g^{-1}.[40] The cycle life, leakage current and impedance of N-doped graphene based supercapacitors can be optimized by heating the material at different temperatures.[41]

N-doped graphene aerogels show very good electrical conductivity and wettability and also shows a very good capacitive behaviour (223 F g^{-1} at 0.2 A g^{-1}).[42] The photoreduction of graphene oxide in the presence of ammonia for 15 minutes yields around 6 at.% of nitrogen content and shows 247.1 F g^{-1}.[43] Under low temperatures, Xin et al[44] have prepared nitrogen doped graphene with 13.44 at.% nitrogen and a capacitance of 218 F g^{-1} at 1 A g^{-1}. It also shows an excellent cyclic stability and retains ~95% of the initial capacitance even after 1000 cycles at 20 A g^{-1}. Crumbled nitrogen doped graphene possesses gravimetric and volumetric capacitances of 128 F g^{-1} and 98 F cm^{-3}, respectively in an ionic liquid medium.[45]

Thermal treatment of nitrogen-containing resin gives covalently linked 3-D graphene sheets.[46] These graphene sheets possess excellent chemical stability and robustness. The specific capacitance is about 509 F g^{-1} at 1 A g^{-1} and the energy density and power density are 40.7 Wh kg^{-1} and 0.3 kW kg^{-1} respectively. Since, the electrical conductivity of the N-doped graphene is far superior than the un-doped graphene, the supercapacitor electrodes can be fabricated without any carbon additives. A binder free 3-D nitrogen doped graphene sheet based supercapacitor electrodes shows a volumetric capacitance of 437.5 F cm^{-3} (334.0 F g^{-1}) at 0.5 A g^{-1}.[47] 3-D N-doped graphene layers synthesized by a gas foaming method shows a surface of 1196 m^2 g^{-1}.[47] The specific capacitance and the energy density reaches up to 335 F g^{-1} and 58.1 Wh kg^{-1} respectively. Activation of 3-D N-doped graphene by synchronous graphitization-activation-doping combination method shows a surface area of 1815 m^2 g^{-1} and the specific capacitance is of 383.2 F g^{-1}.[48] 3-D nitrogen-doped graphene hydrogels synthesized by hydrothermal method from graphene oxide and 1,4-butane diamine exhibit a high specific capacitance of 268.8 Fg^{-1} at 0.3 A g^{-1}.[49] It shows a good cycle stability and a slight increase in the initial specific capacitance after 10000 cycles. This is due to the fact that the pores get open after long charge/discharge cycles and thus the capacitance is increased. 3D N-doped mesoporous graphene synthesized by amidation of graphene oxide shows a considerable supercapacitor performance.[49] The amidation process not only avoids graphene sheet agglomeration but also dopes nitrogen in the graphene lattice. N-doped 3D graphene with

15.8 at.% of nitrogen content shows a high electrical conductivity (3.33 S cm^{-1}) and a large surface area (583 m^2 g^{-1}).[50] This material shows up to 380, 332, and 245 F g^{-1} in alkaline, acidic and neutral electrolytes respectively. Nitrogen doped graphene with aniline additive show 5.7 times better capacitance (2.02 F m^{-1} at 5 mV s^{-1}) than without the additive.[51] Nitrogen-doped holey graphene shows a high volumetric capacitance of 439 F cm^{-3}.[52]

Thermal KOH activation of nitrogen doped graphene increases the capacitance by 5 times where the capacitance reaches up to 132 F g^{-1} at 0.1 A g^{-1}.[53], whereas, the H_3PO_4 activation results in a phosphorus and nitrogen-doped graphene.[54] The resultant product shows a 244 F g^{-1} at 0.1 A g^{-1} with an excellent stability. The reacting GO with 3,4-diaminopyridine in the presence of an acid introduces a large amount of pyridinic nitrogen in the graphene lattice.[55] The nitrogen content can be maintained by varying the amount of the nitrogen source and the maximum specific capacitance achieved is 214 F g^{-1} at 0.1 A g^{-1}. Hydrothermal reaction at weak basic conditions (ammonium phosphate $((NH_4)_2HPO4)$) generates large amount of oxygen functionalities and also produces pyridine/pyrrole nitrogens.[55] The residual oxygen functionalities (12.8 at.%) in the nitrogen doped graphene increases the pseudocapacitance and the cyclic stability. Polypyrrole-modified GO treated with methane plasma results in GO-polypyrrole-CH_4 and the supercapacitors made from this N-doped graphene shows an initial specific capacitance of ~312 F g^{-1} at 0.1 A g^{-1} with a capacitance retention of ~100% after 1000 charge/discharge cycles at 10.0 A g^{-1}.[56]

2.4. Boron doped graphene

In addition to the nitrogen-doped graphene, boron-doped graphenes have shown extraordinary electrochemical supercapacitor properties. Boron doped graphene can be made by arc-discharge using boron stuffed graphite rods or in the presence of diborane yields up to 3 at.% of boron doping.[28] Thermal annealing of graphene with boron oxide dopes 3.2% of boron into the graphene lattice[57] while the porous B-doped graphene made from the Fried-ice method shows 281 F g^{-1} with an excellent stability of more than 4000 cycles.[58] Graphene oxide treated with boric acid at 900 °C yields a boron-doped graphene containing 4.7 at.% and exhibits a specific capacitance of 172.5 F g^{-1} at 0.5 A g^{-1} which is about 80% higher than that of a pristine graphene.[59] This is due to the formation of BC_2O/BCO_2 during the reaction process, which helps to increase the wettability of the material and the capacitance. The large scale synthesis of a chemically modified boron doped graphene can be achieved by treating graphene oxide in borane-tetrahydrofuran adduct under reflux conditions.[60] The final product will have a specific capacitance of 200 F g^{-1} with an excellent stability up to 4500 cycles. The high content of boron doping (6.04 ± 1.44 at. %) can be achieved by heating graphene oxide and B_2O_3 at 1000 °C.[61] The supercapacitor electrodes made from this material shows an EDLC behaviour with a specific capacitance of 448 F g^{-1} without using any additives. A boron-doped porous graphene made using a laser induction process shows an areal capacitance of 16.5 mF cm^{-2}.[62] In addition, these devices show a good cyclability and mechanical flexibility. The dielectric barrier discharge plasma technology based synthesis of B-RGO (1.4 at.% of B) shows a significant increase in the capacitance (446.24 F g^{-1} at 0.5 A g^{-1}) when compared to the reduced graphene oxide.[63] The increase in the specific capacitance of the B-doped graphene can be attributed to its enhancement in the conductivity and to the specific surface when compared to that of an un-doped graphene.

2.5. *Borocarbonitrides*

Borocarbonitrides, $B_xC_yN_z$ are nanosheets comprising of graphene and BN domains, probably along with B-C-N rings.[64,65] $B_xC_yN_z$ contain hexagonal networks of B-C, B-N, C-N, and C-C bonds but no B-B or N-N bonds. The electronic properties of BCN can be altered by changing the combination of graphene and boron nitride domains. Atomic sheets of BCN can be synthesized under CVD conditions in the presence of CH_4 and ammonia-borane at 1000 °C[66] while under low pressure CVD conditions, BCN forms at 1050 °C with the same precursors.[67] Gas phase synthesis using methane with ammonia and BCl_3 yields BCN[68] whereas $BC_{1.6}N$ is formed when BBr_3, NH_3 and ethylene are used.[69] The solid-state reaction of the activated carbon, boric acid and urea form few-layer borocarbonitrides with well-defined compositions.[64]

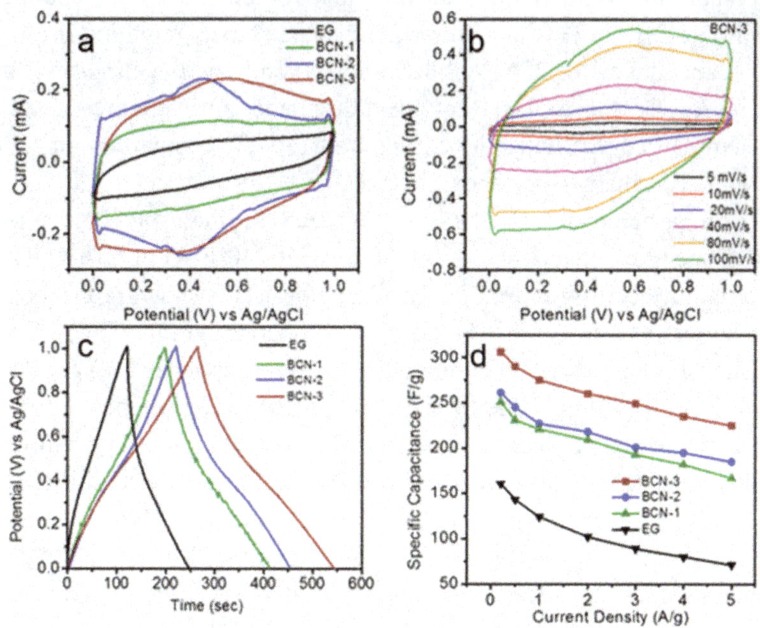

Fig. 2. Supercapacitor performance of BCN samples. (a) CV curves at 40 mV/s, (b) CV curves of BCN-3, (c) Galvanostatic C–D curves (at 1 A/g). (d) Specific capacitance vs current density. Reproduced with permission from Ref. 70.

The BCN synthesized from this method has a high surface area (1500-2000 m^2 g^{-1}) and exciting gas adsorption properties. When the few-layer graphene is used one can achieve BCN containing 5–7 layers.[71] Similarly, when graphene oxide is used with B_2O_3 and ammonia at a different temperature it results in BCN with varying B-N contents.[72] The solid-state supercapacitors based on the boron and nitrogen doped graphene shows a specific capacitance of 62 F g^{-1} with a good energy density of 8.65 W h kg^{-1}.[73] The reduction and simultaneous doping of boron and nitrogen in graphene oxide can be done by the solvothermal treatment with the ammonia borane.[60] The specific capacitance reached up to 130 F g^{-1} at 1 A g^{-1}. Annealing anionic graphene oxide nanosheets with the cationic poly-L-Lysine and H_3BO_3 forms a microporous B,N-doped graphene.[74] The micro-supercapacitors based on this electrode exhibits a large volumetric capacitance of ~488 F cm^{-3} and an excellent rate capability. BCN with a surface area of 802.35 m^2 g^{-1} and boron content of 2.19 at.% show very good capacitance retention from the initial capacitance of 254 F g^{-1} at 0.25 A g^{-1} and good cycling stability.[75] The reason is due to fact that the boron and nitrogen present in the graphene lattice facilitates the charge transfer between carbon atoms and thus increases the performance.

Borocarbonitrides (BC$_{4.5}$N) synthesized by the solid-state method show a specific capacitance of 178 F g^{-1} and 240 F g^{-1} in 6M KOH and the organic electrolyte. While, BCN synthesised using graphene show even better performance.[70,76] The cyclic voltammograms (CV) of BCN exhibits large curves when compared to that of graphene (EG) as show in Figure 2 (a),(b). This is due to the high contents of nitrogen and boron in the graphene lattice. The CV curves almost rectangular indicating the excellent charge storing ability of BCN electrodes. In addition, these electrodes show very good cyclic stability of over 1000 cycles. The charge- discharge curves look nearly symmetrical indicating the resemblance of an ideal capacitor. The highest specific capacitance achieved is about 306 F g^{-1} at 0.2 A g^{-1} for BCN-3 (Fig. 2(c)). Figure 2(d) shows that the capacitance of BCN decrease with the increase in the applied current. Covalently linked BN$_{1-x}$G$_x$ composites have shown a specific capacitance of 238 F g^{-1} (BN$_{0.25}$G$_{0.75}$) at 0.3 A g^{-1}. The rise in the performance is due to the microporous networks created during the cross-linking. When BCN-BCN is covalently cross-linked the specific capacitance increases to 261 F g^{-1} at 5 mV s^{-1}.[77] The enhancement in the capacitance is attributed to the high surface area and the pores generated on cross-linking the individual layers of BCN. BN/reduced graphene oxide composites prepared by the physical mixture through the hydrothermal reaction show 824 F g^{-1}.[78]

2.6. Molybdenum sulphide

Unlike graphene, MoS$_2$ has a band-gap of 1.23 eV.[79] MoS$_2$ shows excellent electrochemical properties due to electron-electron correlations among Mo atoms. A simplest method of making MoS_2 will be the exfoliation of bulk MoS_2.[80,81]

Flower-like MoS_2 synthesized using the hydrothermal method shows a specific capacitance of 122 F g^{-1} at 1 A g^{-1} whereas at 2 mV s^{-1} it is 114 F g^{-1}.[82] A one-pot synthesis of MoS_2 using ammonium heptamolybdate and thiourea forms randomly oriented MoS_2 layers.[83] These layers exhibit 106 F g^{-1} at 5 mV s^{-1} and a stability of 93.8% after 1000 cycles. The hydrothermal reaction in the presence of sodium molybdate and thioacetamide yields mesoporous MoS_2 with a specific capacitance of 376 and 403 F g^{-1}.[84] The porous nature of MoS_2 provides a good surface area and also increases the mobility of ions and increases the capacitance. Hybrids of 1T and 2H MoS_2 show a specific capacitance of 366.9 F g^{-1}.[85] MoS_2 intercalated with H^+, Li^+, Na^+ and K^+ ions shows a volumetric capacitance of 400 to 700 F cm^{-3} (Fig. 3(a)).[86] The CV curves obtained from K_2SO_4 and KCl electrolytes are almost similar (Fig. 3(b)). This reveals that the anion radii has no role and the cation is the one, which intercalates in-between the layers. The CV curves measured at different scan rates are shown in Fig. 3(c). The electrodes show good capacitance even at a higher scan rate of 200 mV s^{-1}. Figure 3(d) shows the plot of specific capacitance versus scan rates. This high observation of high specific capacitance is due to the large anionic polarizability of S^{2-} ion. Figure 3(e) shows the constant current charge/discharge measurements measured in the Na_2SO_4 electrolyte. The curves look similar to that of an ideal capacitor and they also show very good cyclic stability of more than 5000 cycles at 2 A g^{-1}, retaining about 93% of the capacitance (Fig. 2(f)). In addition, they show a high voltage supercapacitor performance as well as good coulombic efficiency.

MoS_2 shows a comparable supercapacitance performance when compared to that of a pristine graphene. Like graphene, MoS_2 shows a double layer capacitance but at slow scan rates, it exhibits a faradaic behaviour. In addition, MoS_2 possesses a lower electronic conductivity and thus making composites with graphene or conducting polymers will increase the overall capacitance of MoS_2 and increase the electronic conductivity. Huang et al.[87] have shown a specific capacitance of 243 F g^{-1} at 1 A g^{-1} with an energy density of 73.5 Wh kg^{-1} at a power density of 19.8 kW kg^{-1}. After charge/discharging for 1000 cycles the composite retains 92.3% of its capacitance. MoS_2 grown on graphene oxide by the microwave method shows a specific capacitance of 128 F g^{-1}.[88]

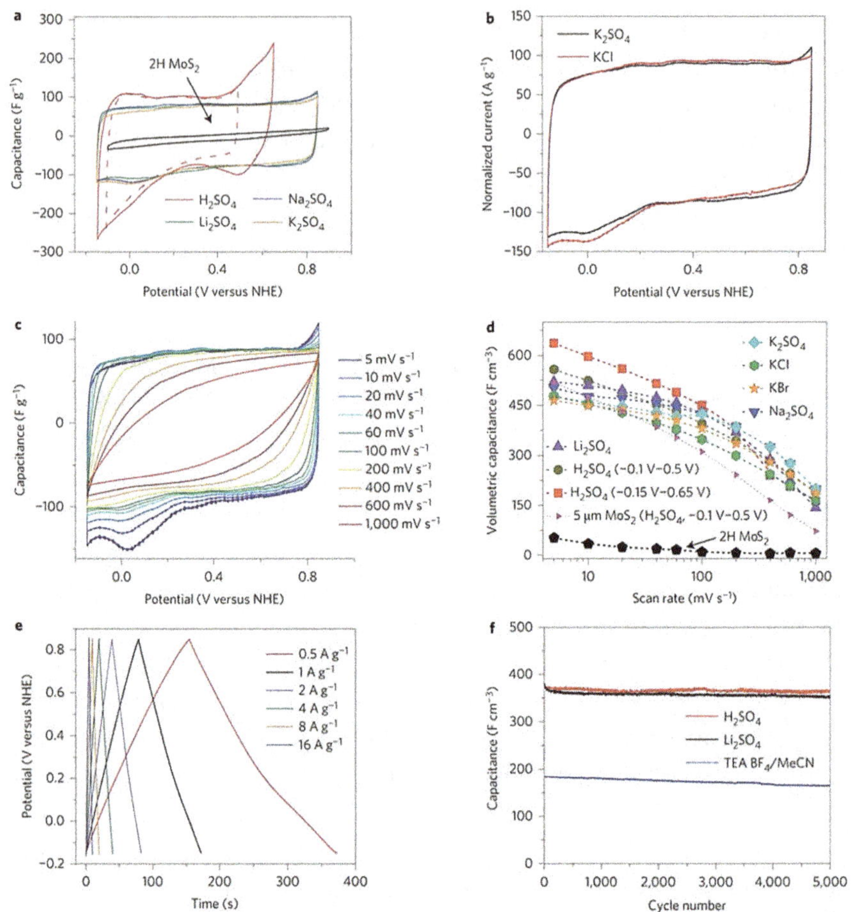

Fig. 3. (a) CVs of the 1T phase MoS_2 nanosheet paper in 0.5 M sulphate-based electrolyte solutions at scan rates of 20 mV s^{-1}. (b) Comparison of the CV curves of 1T MoS_2 in 0.5 M K_2SO_4 and 1 M KCl. (c) CVs of 1T phase MoS_2 electrodes in 0.5 M Na_2SO_4 from 5 mV s^{-1} -1,000 mV s^{-1}. The capacitance remains virtually constant up to the scan rate of 100 mV s^{-1}. (d) Evolution of the volumetric capacitance of the 1T phase MoS_2 electrodes with the scan rate for different electrolytes and 1-μm- and 5-μm-thick films. The concentration of the cations in the electrolyte solutions were fixed at 1 M. (e) Galvanostatic cycles from 0.5 A g^{-1}-16 A g^{-1} in K_2SO_4. (f) Capacitance retention after 5,000 cycles in 0.5 M Li_2SO_4, H_2SO_4 and 1 M $TEABF_4$ in acetonitrile. Reproduced with permission from Ref. 86.

When the concentration of MoS_2 is increased, the capacitance of the composite increases up to 148 F g^{-1}. It is found that MoS_2 forms a covalent bond with a reduced graphene oxide (Mo- O- C) and maximum energy density obtained is 63 W h kg^{-1}. Porous MoS_2 embedded on carbon derived from sucrose shows a capacitance of 589 F g^{-1}.[89] Coin-cell supercapacitors based on MoS2/graphene membranes show an areal capacitance of 11 mF cm^{-2} at 5 mV s^{-1}.[90] It is observed that the overall capacitance of these membranes increases after continuous charge/discharge cycles. Composites of MoS_2 (a combination of 1T and 2H phase) with graphene forms an aerogel and exhibits a specific capacitance of 416 F g^{-1}, showing an excellent cyclic stability of over 50000 cycles.[91] The improvement in the electrochemical performance can be attributed towards the interlinked graphene network, which promotes the electrolyte ion diffusion as well as the overall electronic conductivity.

3D tubular MoS_2/PANI composites prepared by the co-polymerization of aniline on MoS_2 shows 552 F g^{-1}.[92] When measured in a two-electrode configuration it is around 124 F g^{-1}. In-situ polymerization of PANI on MoS_2 layers show 575 F g^{-1} at 1 A g^{-1}.[93] They also show good cyclic stability with the initial capacitance loss of 2% after 500 cycles. Large scale synthesis of polyaniline and MoS_2 needles show 853 F g^{-1} at 1 A g^{-1}.[94] Hierarchical microflowers of MoS_2/polyaniline exhibit a specific capacitance of 245.5 F g^{-1} at 10 A g^{-1}.[95] The increase in the capacitance is due to the synergistic effect between MoS_2 and PANI.

References

1. B. E. Conway, Transition from "supercapacitor" to "battery" behavior in electrochemical energy storage. *J. Electrochem. Soc.* **138**, 1539-1548 (1991).
2. C. N. R. Rao, K. Gopalakrishnan, U. Maitra, Comparative study of potential applications of graphene, MoS2, and other two-dimensional materials in energy devices, sensors, and related areas. *ACS Appl. Mater. Interfaces* **7**, 7809-7832 (2015).
3. C. N. R. Rao, A. K. Sood, K. S. Subrahmanyam, A. Govindaraj, Graphene: the new two-dimensional nanomaterial. *Angew. Chem. Int. Ed.* **48**, 7752-7777 (2009).

4. M. F. El-Kady, V. Strong, S. Dubin, R. B. Kaner, Laser scribing of high-performance and flexible graphene-based electrochemical capacitors. *Science* **335**, 1326-1330 (2012).

5. K. S. Novoselov, A. K. Geim, S. V. Morozov, D. Jiang, M. I. Katsnelson, I. V. Grigorieva, S. V. Dubonos, A. A. Firsov, Two-dimensional gas of massless Dirac fermions in graphene. *Nature* **438**, 197 (2005).

6. K. S. Subrahmanyam, L. S. Panchakarla, A. Govindaraj, C. N. R. Rao, Simple method of preparing graphene flakes by an arc-discharge method. *J. Phys. Chem. C* **113**, 4257-4259 (2009).

7. Y. Zhang, L. Zhang, C. Zhou, Review of chemical vapor deposition of graphene and related applications. *Acc. Chem. Res.* **46**, 2329-2339 (2013).

8. C.-J. Shih, S. Lin, M. S. Strano, D. Blankschtein, Understanding the stabilization of liquid-phase-exfoliated graphene in polar solvents: molecular dynamics simulations and kinetic theory of colloid aggregation. *J. Am. Chem. Soc.* **132**, 14638-14648 (2010).

9. S. R. C. Vivekchand, C. S. Rout, K. S. Subrahmanyam, A. Govindaraj, C. N. R. Rao, Graphene-based electrochemical supercapacitors. *J. Chem. Sci.* **120**, 9-13 (2008).

10. M. D. Stoller, S. Park, Y. Zhu, J. An, R. S. Ruoff, Graphene-based ultracapacitors. *Nano Lett.* **8**, 3498-3502 (2008).

11. Y. Zhu, S. Murali, M. D. Stoller, A. Velamakanni, R. D. Piner, R. S. Ruoff, Microwave assisted exfoliation and reduction of graphite oxide for ultracapacitors. *Carbon* **48**, 2118-2122 (2010).

12. J. J. Yoo, K. Balakrishnan, J. Huang, V. Meunier, B. G. Sumpter, A. Srivastava, M. Conway, A. L. Mohana Reddy, J. Yu, R. Vajtai, P. M. Ajayan, Ultrathin planar graphene supercapacitors. *Nano Lett.* **11**, 1423-1427 (2011).

13. H. Yang, S. Kannappan, A. S. Pandian, J.-H. Jang, Y. S. Lee, W. Lu, Nanoporous graphene materials by low-temperature vacuum-assisted thermal process for electrochemical energy storage. *J Power Sources* **284**, 146-153 (2015).

14. X. Wang, Y. Zhang, C. Zhi, X. Wang, D. Tang, Y. Xu, Q. Weng, X. Jiang, M. Mitome, D. Golberg,Y. Bando, Three-dimensional strutted graphene grown by substrate-free sugar blowing for high-power-density supercapacitors. *Nature Commun.* **4**, 2905 (2013).

15. A. K. Farquhar, P. A. Brooksby, A. J. Downard, Controlled spacing of few-layer graphene sheets using molecular spacers: capacitance that scales with sheet number. *ACS App. Nano Materials* **1**, 1420-1429 (2018).

16. Z. Peng, J. Lin, R. Ye, E. L. G. Samuel, J. M. Tour, Flexible and stackable laser-induced graphene supercapacitors. *ACS Appl. Mater. Interfaces* **7**, 3414-3419 (2015).

17. U. N. Maiti, J. Lim, K. E. Lee, W. J. Lee, S. O. Kim, Three-dimensional shape engineered, interfacial gelation of reduced graphene oxide for high rate, large capacity supercapacitors. *Adv. Mater.* **26**, 615-619 (2014).

18. Y. J. Lee, G.-P. Kim, Y. Bang, J. Yi, J. G. Seo, I. K. Song, Activated carbon aerogel containing graphene as electrode material for supercapacitor. *Mater. Res. Bull.* **50**, 240-245 (2014).

19. M. P. Down, C. E. Banks, Freestanding three-dimensional graphene macroporous supercapacitor. *ACS App. Energy Materials* **1**, 891-899 (2018).

20. W.-w. Liu, X.-b. Yan, J.-w. Lang, J.-b. Pu, Q.-j. Xue, Supercapacitors based on graphene nanosheets using different non-aqueous electrolytes. *New J. Chem.* **37**, 2186-2195 (2013).

21. Q. Wu, Y. Xu, Z. Yao, A. Liu, G. Shi, Supercapacitors based on flexible graphene/polyaniline nanofiber composite films. *ACS Nano* **4**, 1963-1970 (2010).

22. C. Long, T. Wei, J. Yan, L. Jiang, Z. Fan, Supercapacitors based on graphene-supported iron nanosheets as negative electrode materials. *ACS Nano* **7**, 11325-11332 (2013).

23. W. Chen, R. B. Rakhi, H. N. Alshareef, Capacitance enhancement of polyaniline coated curved-graphene supercapacitors in a redox-active electrolyte. *Nanoscale* **5**, 4134-4138 (2013).

24. H. P. de Oliveira, S. A. Sydlik, T. M. Swager, Supercapacitors from free-standing polypyrrole/graphene nanocomposites. *J. Phys. Chem. C* **117**, 10270-10276 (2013).

25. D. A. C. Brownson, C. E. Banks, Fabricating graphene supercapacitors: highlighting the impact of surfactants and moieties. *Chem. Commun.* **48**, 1425-1427 (2012).

26. C. N. R. Rao, K. Gopalakrishnan, A. Govindaraj, Synthesis, properties and applications of graphene doped with boron, nitrogen and other elements. *Nano Today* **9**, 324-343 (2014).

27. R. Lv, Q. Li, A. R. Botello-Méndez, T. Hayashi, B. Wang, A. Berkdemir, Q. Hao, A. L. Elías, R. Cruz-Silva, H. R. Gutiérrez, Y. A. Kim, H. Muramatsu, J. Zhu, M. Endo, H. Terrones, J.-C. Charlier, M. Pan, M. Terrones, Nitrogen-doped graphene: beyond single substitution and enhanced molecular sensing. *Sci. Rep.* **2**, 586 (2012).

28. L. S. Panchakarla, K. S. Subrahmanyam, S. K. Saha, A. Govindaraj, H. R. Krishnamurthy, U. V. Waghmare, C. N. R. Rao, Synthesis, Structure, and properties of boron- and nitrogen-doped graphene. *Adv. Mater.* **21**, 4726-4730 (2009).

29. Z. Mou, X. Chen, Y. Du, X. Wang, P. Yang, S. Wang,, Forming mechanism of nitrogen doped graphene prepared by thermal solid-state reaction of graphite oxide and urea. *Appl. Surf. Sci.* **258**, 1704-1710 (2011).

30. K. Gopalakrishnan, K. Moses, A. Govindaraj, C. N. R. Rao, Supercapacitors based on nitrogen-doped reduced graphene oxide and borocarbonitrides. *Solid State Commun.* **175-176**, 43-50 (2013).

31. D. Deng, X. Pan, L. Yu, Y. Cui, Y. Jiang, J. Qi, W.-X. Li, Q. Fu, X. Ma, Q. Xue, G. Sun, X. Bao, Toward N-doped graphene via solvothermal synthesis. *Chem. Mater.* **23**, 1188-1193 (2011).

32. K. Gopalakrishnan, A. Govindaraj, C. N. R. Rao, Extraordinary supercapacitor performance of heavily nitrogenated graphene oxide obtained by microwave synthesis. *J. Mater. Chem. A* **1**, 7563-7565 (2013).

33. Y. Qiu, X. Zhang, S. Yang, High performance supercapacitors based on highly conductive nitrogen-doped graphene sheets. *Phys. Chem. Chem. Phys.* **13**, 12554-12558 (2011).

34. L. Sun, L. Wang, C. Tian, T. Tan, Y. Xie, K. Shi, M. Li, H. Fu, Nitrogen-doped graphene with high nitrogen level via a one-step hydrothermal reaction of graphene oxide with urea for superior capacitive energy storage. *RSC Adv.* **2**, 4498-4506 (2012).

35. J. W. Lee, J. M. Ko, J.-D. Kim, Hydrothermal preparation of nitrogen-doped graphene sheets via hexamethylenetetramine for application as supercapacitor electrodes. *Electrochim Acta* **85**, 459-466 (2012).

36. Z. Wen, X. Wang, S. Mao, Z. Bo, H. Kim, S. Cui, G. Lu, X. Feng, J. Chen, Crumpled nitrogen-doped graphene nanosheets with ultrahigh pore volume for high-performance supercapacitor. *Adv. Mater.* **24**, 5610-5616 (2012).

37. P. Chen, J.-J. Yang, S.-S. Li, Z. Wang, T.-Y. Xiao, Y.-H. Qian, S.-H. Yu, Hydrothermal synthesis of macroscopic nitrogen-doped graphene hydrogels for ultrafast supercapacitor. *Nano Energy* **2**, 249-256 (2013).

38. F. M. Hassan, V. Chabot, J. Li, B. K. Kim, L. Ricardez-Sandoval and A. Yu,, Pyrrolic-structure enriched nitrogen doped graphene for highly efficient next generation supercapacitors. *J. Mater. Chem. A* **1**, 2904-2912 (2013).

39. F. N. I. Sari, J.-M. Ting, One step microwaved-assisted hydrothermal synthesis of nitrogen doped graphene for high performance of supercapacitor. *App. Surf. Sci.* **355**, 419-428 (2015).

40. H.-C. Youn, S.-M. Bak, M.-S. Kim, C. Jaye, D. A. Fischer, C.-W. Lee, X.-Q. Yang, K. C. Roh, K.-B. Kim, High-surface-area nitrogen-doped reduced graphene oxide for electric double-layer capacitors. *ChemSusChem* **8**, 1875-1884 (2015).

41. W. Li, H.-Y. Lü, X.-L. Wu, H. Guan, Y.-Y. Wang, F. Wan, G. Wang, L.-Q. Yan, H.-M. Xie, R.-S. Wang, Electrochemical performance improvement of N-doped graphene as electrode materials for supercapacitors by optimizing the functional groups. *RSC Adv.* **5**, 12583-12591 (2015).

42. Z.-Y. Sui, Y.-N. Meng, P.-W. Xiao, Z.-Q. Zhao, Z.-X. Wei, B.-H. Han, Nitrogen-doped graphene aerogels as efficient supercapacitor electrodes and gas adsorbents. *ACS Appl. Mater. Interfaces* **7**, 1431-1438 (2015).

43. H. Huang, G. Luo, L. Xu, C. Lei, Y. Tang, S. Tang, Y. Du, NH3 assisted photoreduction and N-doping of graphene oxide for high performance electrode materials in supercapacitors. *Nanoscale* **7**, 2060-2068 (2015).

44. H. Xin, D. He, Y. Wang, W. Zhao, X. Du, Low-temperature preparation of macroscopic nitrogen-doped graphene hydrogel for high-performance ultrafast supercapacitors. *RSC Adv.* **5**, 8044-8049 (2015).

45. J. Wang, B. Ding, Y. Xu, L. Shen, H. Dou, X. Zhang, Crumpled nitrogen-doped graphene for supercapacitors with high gravimetric and volumetric performances. *ACS Appl. Mater. Interfaces* **7**, 22284-22291 (2015).

46. Y. Qin, J. Yuan, J. Li, D. Chen, Y. Kong, F. Chu, Y. Tao, M. Liu, Crosslinking graphene oxide into robust 3D porous N-doped graphene. *Adv. Mater.* **27**, 5171-5175 (2015).

47. X. Zhao, H. Dong, Y. Xiao, H. Hu, Y. Cai, Y. Liang, L. Sun, Y. Liu, M. Zheng, Three-dimensional Nitrogen-doped graphene as binder-free electrode materials for supercapacitors with high volumetric capacitance and the synergistic effect between nitrogen configuration and supercapacitive performance. *Electrochimica Acta* **218**, 32-40 (2016).

48. J. Hao, D. Shu, S. Guo, A. Gao, C. He, Y. Zhong, Y. Liao, Y. Huang, J. Zhong, Preparation of three-dimensional nitrogen-doped graphene layers by gas foaming method and its electrochemical capactive behavior. *Electrochim Acta* **193**, 293-301 (2016).

49. Y. Zhang, G. Wen, P. Gao, S. Bi, X. Tang, D. Wang, High-performance supercapacitor of macroscopic graphene hydrogels by partial reduction and nitrogen doping of graphene oxide. *Electrochim Acta* **221**, 167-176 (2016).

50. W. Zhang, C. Xu, C. Ma, G. Li, Y. Wang, K. Zhang, F. Li, C. Liu, H.-M. Cheng, Y. Du, N. Tang, W. Ren, Nitrogen-superdoped 3D graphene networks for high-performance supercapacitors. *Adv. Mater.* **29**, 1701677 (2017).

51. K. V. Sankar, R. K. Selvan, R. H. Vignesh, Y. S. Lee, Nitrogen-doped reduced graphene oxide and aniline based redox additive electrolyte for a flexible supercapacitor. *RSC Adv.* **6**, 67898-67909 (2016).

52. Y. Zhang, L. Ji, W. Li, Z. Zhang, L. Lu, L. Zhou, J. Liu, Y. Chen, L. Liu, W. Chen, Y. Zhang, Highly defective graphite for scalable synthesis of nitrogen doped holey graphene with high volumetric capacitance. *J. Power Sources* **334**, 104-111 (2016).

53. B. Zheng, T.-W. Chen, F.-N. Xiao, W.-J. Bao, X.-H. Xia, KOH-activated nitrogen-doped graphene by means of thermal annealing for supercapacitor. *J Solid State Electrochem.* **17**, 1809-1814 (2013).

54. P. Wang, H. He, X. Xu, Y. Jin, Significantly enhancing supercapacitive performance of nitrogen-doped graphene nanosheet electrodes by phosphoric acid activation. *ACS Appl. Mater. Interfaces* **6**, 1563-1568 (2014).

55. M. S. Lee, H.-J. Choi, J.-B. Baek, D. W. Chang, Simple solution-based synthesis of pyridinic-rich nitrogen-doped graphene nanoplatelets for supercapacitors. *App. Energy* **195**, 1071-1078 (2017).

56. K. Wang, M. Xu, Y. Gu, Z. Gu, J. Liu, Q. H. Fan, Low-temperature plasma exfoliated n-doped graphene for symmetrical electrode supercapacitors. *Nano Energy* **31**, 486-494 (2017).

57. Z.-H. Sheng, H.-L. Gao, W.-J. Bao, F.-B. Wang, X.-H. Xia, Synthesis of boron doped graphene for oxygen reduction reaction in fuel cells. *J. Mater. Chem.* **22**, 390-395 (2012).

58. Z. Zuo, Z. Jiang, A. Manthiram, Porous B-doped graphene inspired by Fried-Ice for supercapacitors and metal-free catalysts. *J. Mater. Chem. A* **1**, 13476-13483 (2013).

59. L. Niu, Z. Li, W. Hong, J. Sun, Z. Wang, L. Ma, J. Wang, S. Yang, Pyrolytic synthesis of boron-doped graphene and its application as electrode material for supercapacitors. *Electrochimica Acta* **108**, 666-673 (2013).

60. V. H. Pham, S. H. Hur, E. J. Kim, B. S. Kim, J. S. Chung, Highly efficient reduction of graphene oxide using ammonia borane. *Chem. Commun.* **49**, 6665-6667 (2013).

61. D.-Y. Yeom, W. Jeon, N. D. K. Tu, S. Y. Yeo, S.-S. Lee, B. J. Sung, H. Chang, J. A. Lim, H. Kim, High-concentration boron doping of graphene nanoplatelets by simple thermal annealing and their supercapacitive properties. *Sci. Rep.* **5**, 9817 (2015).

62. Z. Peng, R. Ye, J. A. Mann, D. Zakhidov, Y. Li, P. R. Smalley, J. Lin, J. M. Tour, Flexible boron-doped laser-induced graphene microsupercapacitors. *ACS Nano* **9**, 5868-5875 (2015).

63. S. Li, Z. Wang, H. Jiang, L. Zhang, J. Ren, M. Zheng, L. Dong, L. Sun, Plasma-induced highly efficient synthesis of boron doped reduced graphene oxide for supercapacitors. *Chem. Commun.* **52**, 10988-10991 (2016).

64. N. Kumar, K. Moses, K. Pramoda, S. N. Shirodkar, A. K. Mishra, U. V. Waghmare, A. Sundaresan, C. N. R. Rao, Borocarbonitrides, BxCyNz. *J. Mater. Chem. A* **1**, 5806-5821 (2013).

65. C. N. R. Rao, K. Gopalakrishnan, Borocarbonitrides, $B_xC_yN_z$: synthesis, characterization, and properties with potential applications. *ACS Appl. Mater. Interf.* **9**, 19478-19494 (2017).

66. L. Ci, L. Song, C. Jin, D. Jariwala, D. Wu, Y. Li, A. Srivastava, Z. F. Wang, K. Storr, L. Balicas, F. Liu and P. M. Ajayan, Atomic layers of hybridized boron nitride and graphene domains. *Nature Mater.* **9**, 430 (2010).

67. C.-K. Chang, S. Kataria, C.-C. Kuo, A. Ganguly, B.-Y. Wang, J.-Y. Hwang, K.-J. Huang, W.-H. Yang, S.-B. Wang, C.-H. Chuang, M. Chen, C.-I. Huang, W.-F. Pong, K.-J. Song, S.-J. Chang, J.-H. Guo, Y. Tai, M. Tsujimoto, S. Isoda, C.-W. Chen, L.-C. Chen, K.-H. Chen, Band gap engineering of chemical vapor deposited graphene by in situ BN doping. *ACS Nano* **7**, 1333-1341 (2013).

68. R. B. Kaner, J. Kouvetakis, C. E. Warble, M. L. Sattler, N. Bartlett, Boron-carbon-nitrogen materials of graphite-like structure. *Mater. Res. Bull.* **22**, 399-404 (1987).

69. N. Kumar, K. Raidongia, A. K. Mishra, U. V. Waghmare, A. Sundaresan, C. N. R. Rao, Synthetic approaches to borocarbonitrides, BC_xN $_{(x=1-2)}$. *J. Solid State Chem.* **184**, 2902-2908 (2011).

70. M. B. Sreedhara, K. Gopalakrishnan, B. Bharath, R. Kumar, G. U. Kulkarni, C. N. R. Rao, Properties of nanosheets of 2D-borocarbonitrides related to energy devices, transistors and other areas. *Chem. Phys. Lett.* **657**, 124-130 (2016).

71. Y. Kang, Z. Chu, D. Zhang, G. Li, Z. Jiang, H. Cheng, X. Li, Incorporate boron and nitrogen into graphene to make BCN hybrid nanosheets with enhanced microwave absorbing properties. *Carbon* **61**, 200-208 (2013).

72. T.-W. Lin, C.-Y. Su, X.-Q. Zhang, W. Zhang, Y.-H. Lee, C.-W. Chu, H.-Y. Lin, M.-T. Chang, F.-R. Chen, L.-J. Li, Converting graphene oxide monolayers into boron carbonitride nanosheets by substitutional doping. *Small* **8**, 1384-1391 (2012).

73. Z.-S. Wu, A. Winter, L. Chen, Y. Sun, A. Turchanin, X. Feng, K. Müllen, Three-dimensional nitrogen and boron co-doped graphene for high-performance all-solid-state supercapacitors. *Adv. Mater.* **24**, 5130-5135 (2012).

74. Z.-S. Wu, K. Parvez, A. Winter, H. Vieker, X. Liu, S. Han, A. Turchanin, X. Feng, K. Müllen, Layer-by-layer assembled heteroatom-doped graphene films with ultrahigh volumetric capacitance and rate capability for micro-supercapacitors. *Adv. Mater.* **26**, 4552-4558 (2014).

75. Z. Chen, L. Hou, Y. Cao, Y. Tang, Y. Li, Gram-scale production of B, N co-doped graphene-like carbon for high performance supercapacitor electrodes. *App. Surf. Sci.* **435**, 937-944 (2018).

76. M. Barua, M. B. Sreedhara, K. Pramoda, C. N. R. Rao, Quantification of surface functionalities on graphene, boron nitride and borocarbonitrides by fluorescence labeling. *Chem. Phys. Lett.* **683**, 459-466 (2017).

77. N. K. Singh, K. Pramoda, K. Gopalakrishnan, C. N. R. Rao, Synthesis, characterization, surface properties and energy device characterstics of 2D borocarbonitrides, $(BN)_xC_{1-x}$, covalently cross-linked with sheets of other 2D materials. *RSC Adv.* **8**, 17237-17253 (2018).

78. S. Saha, M. Jana, P. Khanra, P. Samanta, H. Koo, N. C. Murmu, T. Kuila, Band gap engineering of boron nitride by graphene and its application as positive electrode material in asymmetric supercapacitor device. *ACS Appl. Mater. Interfaces,* **7**, 14211-14222 (2015).

79. C. N. R. Rao, U. Maitra, U. V. Waghmare, Extraordinary attributes of 2-dimensional MoS_2 nanosheets. *Chem. Phys. Lett.* **609**, 172-183 (2014).

80. H. S. S. Ramakrishna Matte, A. Gomathi, A. K. Manna, D. J. Late, R. Datta, S. K. Pati, C. N. R. Rao, MoS_2 and WS_2 Analogues of graphene. *Angew. Chem. Int. Ed.* **49**, 4059-4062 (2010).

81. U. Maitra, U. Gupta, M. De, R. Datta, A. Govindaraj and C. N. R. Rao, Highly effective visible-light-induced H2 generation by single-layer 1T-MoS_2 and a nanocomposite of few-layer 2H-MoS_2 with heavily nitrogenated graphene. *Angew. Chem. Int. Ed.* **52**, 13057-13061 (2013).

82. X. Zhou, B. Xu, Z. Lin, D. Shu, L. Ma, Hydrothermal Synthesis of flower-like MoS2 nanospheres for electrochemical supercapacitors. *J. Nanosci. Nanotech.* **14**, 7250-7254 (2014).

83.	K. Krishnamoorthy, G. K. Veerasubramani, S. Radhakrishnan, S. J. Kim, Supercapacitive properties of hydrothermally synthesized sphere like MoS_2 nanostructures. *Mater. Res. Bull.* **50**, 499-502 (2014).

84.	A. Ramadoss, T. Kim, G.-S. Kim, S. J. Kim, Enhanced activity of a hydrothermally synthesized mesoporous MoS_2 nanostructure for high performance supercapacitor applications. *New J. Chem.* **38**, 2379-2385 (2014).

85.	L. Jiang, S. Zhang, S. A. Kulinich, X. Song, J. Zhu, X. Wang, H. Zeng, Optimizing hybridization of 1T and 2H phases in MoS_2 monolayers to improve capacitances of supercapacitors. *Mater. Res. Lett.* **3**, 177-183 (2015).

86.	M. Acerce, D. Voiry, M. Chhowalla, Metallic 1T phase MoS_2 nanosheets as supercapacitor electrode materials. *Nature Nanotech.* **10**, 313 (2015).

87.	K.-J. Huang, L. Wang, Y.-J. Liu, Y.-M. Liu, H.-B. Wang, T. Gan, L.-L. Wang, Layered MoS_2-graphene composites for supercapacitor applications with enhanced capacitive performance. *Int. J. Hydro. Energy* **38**, 14027-14034 (2013).

88.	E. G. da Silveira Firmiano, A. C. Rabelo, C. J. Dalmaschio, A. N. Pinheiro, E. C. Pereira, W. H. Schreiner, E. R. Leite, Supercapacitor electrodes obtained by directly bonding 2d MoS_2 on reduced graphene oxide. *Adv. Energy Mater.* **4**, 1301380 (2014).

89.	H. Ji, C. Liu, T. Wang, J. Chen, Z. Mao, J. Zhao, W. Hou, G. Yang, Porous hybrid composites of few-layer MoS_2 nanosheets embedded in a carbon matrix with an excellent supercapacitor electrode performance. *Small* **11**, 6480-6490 (2015).

90.	M. A. Bissett, I. A. Kinloch, R. A. W. Dryfe, Characterization of MoS_2–Graphene composites for high-performance coin cell supercapacitors. *ACS Appl. Mater. Interfaces* **7**, 17388-17398 (2015).

91.	A. Gigot, M. Fontana, M. Serrapede, M. Castellino, S. Bianco, M. Armandi, B. Bonelli, C. F. Pirri, E. Tresso, P. Rivolo, Mixed 1T–2H phase MoS2/reduced graphene oxide as active electrode for enhanced supercapacitive performance. *ACS Appl. Mater. Interfaces* **8**, 32842-32852 (2016).

92.	L. Ren, G. Zhang, Z. Yan, L. Kang, H. Xu, F. Shi, Z. Lei, Z.-H. Liu, Three-dimensional tubular MoS2/PANI hybrid electrode for high rate performance supercapacitor. *ACS Appl. Mater. Interfaces* **7**, 28294-28302 (2015).

93.	K.-J. Huang, L. Wang, Y.-J. Liu, H.-B. Wang, Y.-M. Liu, L.-L. Wang, Synthesis of polyaniline/2-dimensional graphene analog MoS_2 composites for high-performance supercapacitor. *Electrochimica Acta* **109**, 587-594 (2013).

94.	J. Zhu, W. Sun, D. Yang, Y. Zhang, H. H. Hoon, H. Zhang, Q. Yan, Multifunctional architectures constructing of PANI nanoneedle arrays on MoS_2 thin nanosheets for high-energy supercapacitors. *Small* **11**, 4123-4129 (2015).

95.	J. Chao, J. Deng, W. Zhou, J. Liu, R. Hu, L. Yang, M. Zhu, O. G. Schmidt, Hierarchical nanoflowers assembled from MoS2/polyaniline sandwiched nanosheets for high-performance supercapacitors. *Electrochimica Acta* **243**, 98-104 (2017).

Chapter 14

Photovoltaics: Materials and Devices

K.S. Narayan[*]

*School of Advanced Materials, Jawaharlal Nehru Centre for Advanced
Scientific Research, Jakkur, Bangalore 560064, Karnataka, India*
[*]*narayan@jncasr.ac.in*

A summary of the different types of solar cells is presented. Essential
concepts, material features and the device architecture for photovoltaic
applications are discussed. Recent innovations and developments in the
solution-processed solar cells are emphasized. A roadmap for the near
future and prospects for different technology are presented.

1. Introduction

A green future demands a clean renewable source of energy. India is
expected to contribute a major fraction of the world economic growth in
the coming decades, with the demand and consumption of energy
surging towards high levels. The current portion of the renewable energy,
as per the installed capacity is about 21%, of which solar power
contributes about 32%. With the rapid growth in the capacity addition on
the back of a strong government push, the renewable sector now
accounts for more than one-fifth of the country's overall installed power
capacity, a significant achievement in the clean energy push. The
national solar mission launched by the Government of India envisions a
landmark target of 100 GW solar capacity to be installed by 2022. The
state of Karnataka has achieved an installed capacity exceeding 5 GW
currently and is aggressively implementing several projects to create new
records.[1] With this level of large-scale implementation, the market

321

economics has begun to indicate that the photovoltaic (PV) option can be less expensive than thermal power, due to a variety of reasons including the quality of coal, the logistics of transporting coal and regulatory structures. This is an encouraging sign for the entire world at large, indicating that a viable shift to non-carbon emitting sources of fuel is possible.

The next level of innovation in this sector is to utilize green materials and green manufacturing processes. This poses a set of scientific challenges since there is an inherent requirement of placing such systems in an outdoor harsh environment which needs to deliver over a minimum of twenty years. The research challenge of coming up with new materials and processes and simultaneously controlling stability, degradation, ageing and failure is a critical need of the hour. The present chapter explores the global progress in this direction emphasizing key breakthroughs. Since silicon technology has been a focus of commercial PV systems[2], a brief primer of widely prevalent concepts are covered initially, and in the process, it also sets a benchmark for other materials and technology. The selected concepts contextualize the advent of new designer materials and architecture.

The range of novel solar cell materials which include organic small molecules, polymers, quantum dots and hybrid perovskites are also discussed in the chapter. A common thread in these recent materials is the availability of low temperature, solution-processing methods justifying greener routes for PV. Each of these systems has their unique processing-structure-property correlations and needs to be well understood. One of the justifications to pursue new materials for PV technology is the possibility of tandem combination with the widely prevalent silicon panels.The author's interest in studying the impact of defects in solution processed PVs will form one of the core discussions in the chapter. Another growing need which requires fundamental approaches are issues related to stability and ageing. A general outlook of PV in the near future and the directions of the research field is presented.

2. Some Important Concepts in PV

The central operation of solar cells relies on converting incident solar radiation and using local mechanisms for free carrier generation, separation and transport to the electrodes. These processes may coexist in the same region of the cell or there may be selective regions or components for the different processes in the cell. Eventually, the electrical transport to the contacts must compete against recombination, and the distribution of the electric field inside the solar cell has a profound impact on the charge carrier collection. A classification of the different solar cell technologies introduced by Kirchartz, Bisqueret et al.[3] (Figure 1) in terms of the ratio between the thickness of the active layer and the space charge (depletion layer) width is useful and widely applicable.

2.1. *The photoactive layer of the solar cell: The p-n junction*

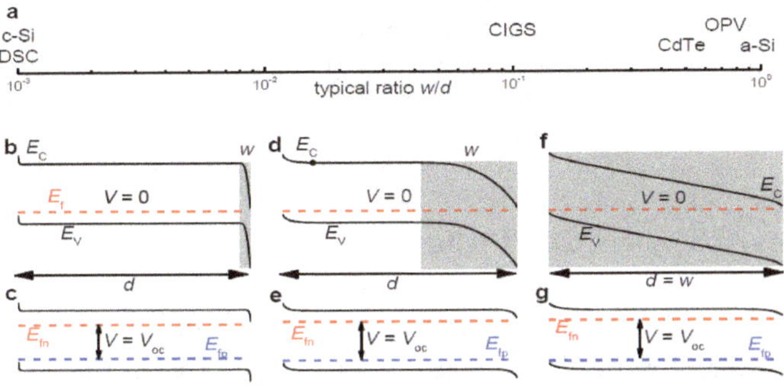

Fig. 1. (a) The typical ratio of the space charge region width w and thickness d for different type of solar cells. (b), (c) Solar cell band diagrams for small w/d, (d), (e) for intermediate w/d and (f), (g) depict for $w = d$ from Ref. 3 (adapted with permission).

When a p-doped and n-doped semiconductor (Si) forms an interface, mobile carriers diffuse across and reach an electronic equilibrium to form a p-n junction. Upon illumination, the electron-hole pairs generated in the semiconductor can be separated by the built-in field to realize a photocurrent across a load resistor connected externally. The difference in the electrochemical potential of the electrons between the p- and n-type regions separates the photogenerated carriers. The carriers in the vicinity of the junction diffuse to the p–n junction where they are also separated. The photogenerated carriers then diffuse to the electrodes to give rise to a current accompanied by a finite loss due to the free carrier recombination or recombination via trap-sites.

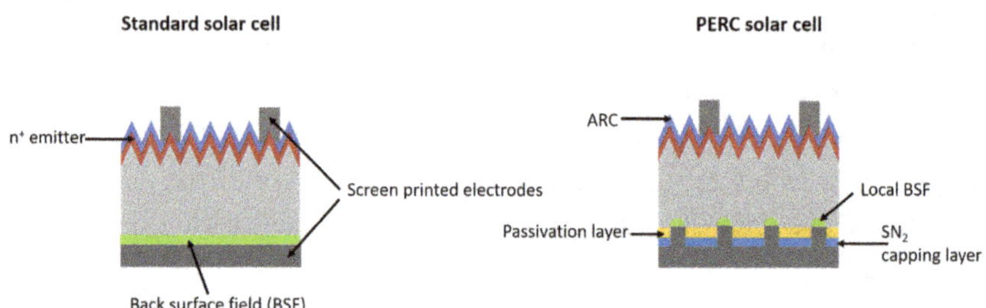

Fig. 2. Typical current density and output power P versus voltage V characteristics of solar cells. The circles represent the open circuit voltage V_{oc}, the short-circuit current density J_{sc} and P_{max}. The open circuit voltage (V_{oc}) is the voltage at which the net current becomes zero, which is equivalent to an open circuit. The short circuit current density (J_{sc}) is the current density through the cell when no external bias is applied ($V = 0$), Fill Factor $= ((V_{max}J_{max})/((V_{oc}J_{sc})$, where V_{max} and I_{max} correspond to P_{max}. Power conversion efficiency $(\eta) = P_{out}/P_{in} = (FF)(V_{oc}J_{sc})/P_{in}$

3. Types of Si Solar Cell Technologies

Compared to monocrystalline silicon, the surface of the multi-crystalline silicon wafer is more difficult to be passivated due to the existence of grain boundaries and the various grain crystallization orientations, which typically results in about 0.5% conversion efficiency loss. The crystallization defects in the bulk of multi-crystalline silicon, such as grain boundaries and metalimpurity contamination, generate carrier recombination centres and hence degrade η further. It should be noted that there is a considerable difference between the performances achieved at an academic-laboratory setting as compared to large-scale production in a manufacturing assembly. The criteria for a viable manufacturing process of large-scale PVs have a completely different set of criteria and parameters as compared to isolated small area cells.[4-6]

Screen printing technology mainstreams the manufacturing processes for traditionalsilicon-based solar cells. Till recently, solar cells were typically manufactured applying silicon wafers and a screen-printed full-area aluminum (Al) layer contacting the complete rear wafer surface. The local Al doping of the rear wafer surface, (aluminum back-surface field (Al-BSF)) marginally reduces the recombination of photo-generated charge carriers at the Al rear contact. Additionally, the Al layer increases the IR absorption, η of conventional industrial silicon solar cells with a full-area Al-BSF which was limited to around 20%. A cell concept that overcomes these limitations introduced in the University of New South Wales "passivated emitter and rear cell" (PERC) was utilized to improve this limit.[7-8] This first lab-type PERC solar cell applied a silicon dioxide layer at the rear silicon wafer surface and the evaporated Al rear layer only locally contacted the rear silicon surface, hence minimizing the recombination of charge carriers at the rear surface and increasing the internal rear reflectivity and absorption of infrared light. The process of lithography is then introduced to pattern the local point contact openings in the silicon dioxide passivation layer.

In the case of the bifacial solar cells, the presence of a textured surface on both sides facilitates a more complete absorption of incident light. The bifacial solar cell also has an additional surface passivation (enables increased current collection) on its rear surface. Additionally, instead of a full metallic back surface, the bifacial cell has a finger grid at the rear side. All these factors enhance power performance from the module. The HIT (heterojunction with an intrinsic thin layer) solar cell is composed of a thin monocrystalline silicon wafer surrounded by ultra-thin amorphous silicon layers. The achievement of high conversion efficiency for monocrystalline silicon solar cell was realized with the heterojunction back contact structure, a fusion of back contact structure utilized in Sharp Electronics Inc.'s high-efficient solar module BLACKSOLAR and heterojunction structure, forming an amorphous silicon film on the surface of the monocrystalline silicon substrate. The η

of Panasonic Co. Ltd., $\approx 25.6\,\%$, is close to the theoretical limit, and it uses an n-type monocrystalline silicon heterojunction and an intrinsic thin-layer interdigitated back contact (HIT-IBC) structure.[10] A steady improvement in the standard p-n junction PV manufacturing process appears to justify Si as the mainstay choice for PVs. This has led to the development and implementation of PERC, HIT and bifacial solar cell technology platforms. The availability of these panels in the market depends on the cost to benefit factors, and it is established that the standard solar cell technology that has paved the way to these alternate designs, constitutes viable platforms to manufacture high power and high η solar panels.

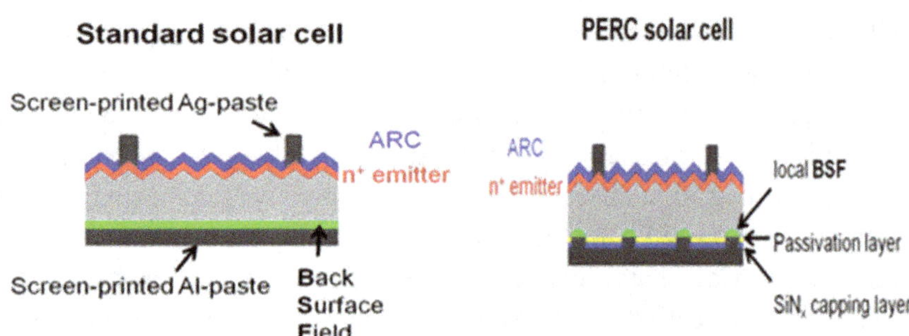

Fig. 3. Standard solar cell and PERC solar cell device schematic.

4. Non Silicon-based PVs

The next-generation devices are a class of cells that evolved to be recognized as thin film solar cells comprising of a few to several layers of different materials. Amorphous silicon-based cells can be included here since it is possible to produce them on a flexible substrate with the expenditure of a relatively less energy than crystalline Si solar cells. Other representative examples include cadmium telluride (CdTe) cells or multi-component copper indium sulfide or copper indium gallium sulfide/selenide (CIGS) cells.[2] These cells are also single junction devices and are generally less complex in processing compared to first-generation crystalline solar cells. CIGS-based PVs have made rapid strides and have demonstrated advantages such as steel foil substrates and building integrated features. Despite promising efficiencies, these materials have not been able to penetrate into the commercial scene. However, the possibility of thin-flexible cells combined with a high performance is the driving factor for these cells. Heterojunction devices have an inherent advantage over homojunction devices. Many PV materials can be only p-type or only n-type doped. Since heterojunctions allow us to forego the requirement of ambipolar materials, many promising PV materials can be investigated to produce optimal cells. Also, a high-bandgap window layer reduces the cell's series resistance, so a window material can be made highly conductive. In principle, its thickness can be also be increased without reducing the light transmittance. As a result, light-generated carriers can easily flow laterally in the window layer to reach an electrical contact.

5. Shockley–Quessier(S–Q) Limit

The S–Q efficiency limit is taken as a universal measure to gauge maximum thermodynamically possible efficiencies of different solar cell technologies. The fact is that for many cell types it presents only the first step in efficiency analysis, because, as already noted, the S–Q analysis is for an ideal model system. In 1961, Shockley and Queisser arrived at an estimate of maximum theoretical efficiency in a homojunction solar cell based on a detailed balance formalism. The premise is that the absorption of a photon creates an electron-hole pair which can either separate to give free charges or recombine. The electron-hole pair can recombine either radiatively emitting another photon or non-radiatively as heat. The radiative recombination leads to a photon emission which can either be reabsorbed or can exit the cell. This is a loss process and is an unavoidable consequence of detailed balance. On the other hand, non-radiative recombination can be reduced by using high-quality trap-free materials. Radiative recombination turns out to be an unavoidable consequence of this theory.[11] Since $V_{oc} \propto E_g$ (band gap) and $I_{sc} \propto 1/E_g$ there is a tradeoff between the current and the open circuit voltage. Considering a single junction with an impedance matching, the theoretical limit for the maximum μ turns out to be 30% for a band gap of 1.28 eV at $T_c = 300$ K. Hence, there is a need to implement strategies like tandem or multi-junction cells, multiple excitonic systems and photonic crystal based solar cells to beat the 30% limit. The main limitation of such cells is the thermalization of hot carriers generated by photons with energies greater than the bandgap energy and the non-absorption of photons with an energysmaller than the energy gap.

6. Overcoming the S–Q Limit

Multi-junction or tandem solar cells incorporate two or more cells that have complementary absorption spectra and can absorb light in the complete solar spectrum. There have been successful attempts by many companies and laboratories that surpassed the 40% mark. Themes that have emerged recently aim at overcoming the limit by engineering materials or implementing photonic designs. Some of the strategies are mentioned in the following section.

6.1. *Tandem-junction and multijunction cells*

Tandem-junction cell architectures present a path toward higher module efficiencies over single-junction designs due to the ability to split the solar spectrum into multiple bands that can be more efficiently converted by separate devices. This enables surpassing the limiting efficiency of a single-junction. As module costs drop, the balance-of-systems costs dominate the cost of PV installations, and gains in efficiency become a more powerful lever to influence the overall system costs. In case of a dual junction tandem cell, two complementary cells are stacked with a large and a small band gap. The two sub-cells in the tandem device may be connected in series or in parallel by varying the interconnecting scheme. The series connection is the most widely adopted one where the individual sub-cells are interconnected electrically and optically. However, it should be mentioned that the overall power generated is not additive due to the current losses. The choice of the interconnecting layer plays a crucial role. The function of the interconnecting layer is to ensure the alignment of the quasi-Fermi level of electrons in the acceptor of the front cell with the quasi-Fermi level of holes in the donor of the rear cell (or vice versa in an inverted architecture). In other words, the intermediate layer should allow the recombination of holes coming from one sub-cell with electrons coming from the other, and the open circuit voltage of the tandem solar cell will be the sum of the two sub-cells.

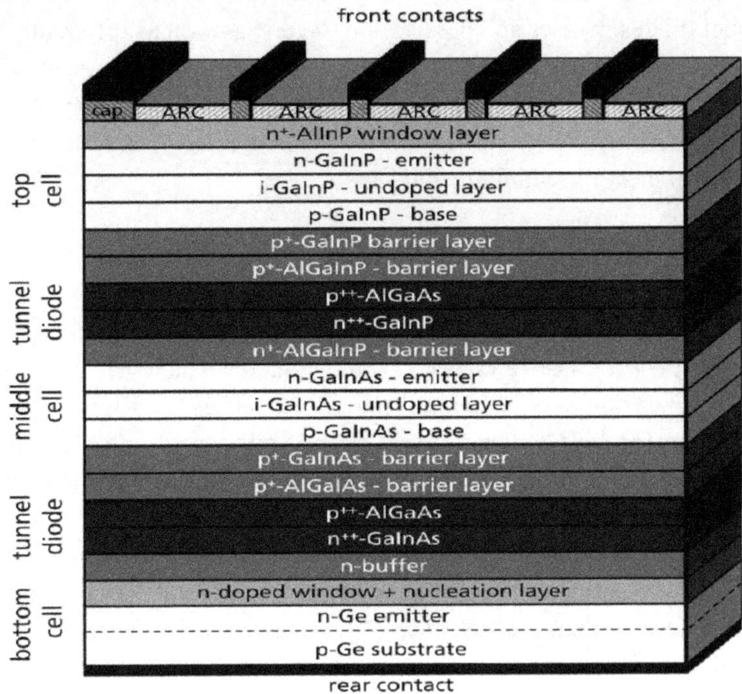

Fig. 4. Lattice matched triple junction $Ga_{0.5}In_{0.5}P/Ga_{0.99}In_{0.01}As/Ge$ device (adapted with permission from Ref. 12).

The highest efficiencies of any photovoltaic technology, so far, have been reached with solar cells made of III–V compound semiconductors. They are particularly suitable for multi-junction solar cells, in which the solar cells with different bandgaps are used to absorb distinct parts of the solar spectrum. III–V multi-junction solar cells are also particularly suitable for the need in space due to their radiation hardness, i.e., the high end-of-life efficiency, small temperature coefficients, high reliability and the combination of high voltage and low currents. The above figure depicts a triple junction where the Voc from each sub-cell adds up in this case. A milestone in multijunction concept is the demonstration of MOVPE (metal–organic vapour phase epitaxy) grown $Ga_{0.5}In_{0.5}P/Ga_{0.99}In_{0.01}As/Ge$ triple-junction, consisting of lattice-matched layers (Figure 4).[12,13] The internal architecture of the monolithically grown solar cell is complex. Appropriate combination and sequence of tunnel diodes, barrier and passivation layers, as well as differently doped layers, are needed as shown in the accompanying shows an internal structure. These devices have achieved a record efficiency of 41.6% (AM1.5d, 364 suns).[12, 13] Currently, tandem solar cells hold the record for the highest η scaling the 44% mark.

6.2. *Multiple exciton generations*

The concept of multiple exciton generations (MEG) has been extensively studied over the last few years. Semiconductor nanocrystals or quantum dots (QDs) possess tunable optical properties and serve as ideal platforms for MEG[14.] The absorption of a photon of a sufficiently high energy by a QD, results in an electron–hole (e–h) pair exciton formation. Excitons in confined regions of space interact with the Coulomb potential, in contrast to bulk semiconductors, where excitons exist only if k_BT is lessthan the exciton binding energy. If the incident photon has an energy greater than the band-gap (E_g) of the material, it can be absorbed by higher energy states to produce a "hot exciton" with excess energy, $\Delta E_{ex} = h\nu - E_g$. If this excess energy is converted into kinetic energy and not lost as heat, under the condition $\Delta E_{ex} > E_g$ ($h\nu > 2E_g$), additional e–h pairs can be generated for a single incident photon via impact ionization, where additional energy of the photon can excite another e–h pair. In bulk semiconductors, the likelihood of hot carriers undergoing impact ionization is reduced due to a) a limited density of final states and b) rapid hot carrier cooling via the phonon emission, suppressing multiple e–h pairs even for $h\nu > (4-5)\ E_g$. Hot-carrier relaxation can be controlled in QDs to make the MEG process efficient. Recent reports demonstrated that an external quantum η of over 100%, implying a realization of MEG in these systems.

Singlet exciton fission: Singlet fission is another process to realize MEG where a singlet exciton leads to the formation of two triplet excitons, which can be further dissociated to produce charge carriers.[15,16] This process is analogous to internal conversion or a radiation-less transition between two electronic states of equal multiplicity. This transition can be a very fast process and if the coupling is favorable, it can occur on a picosecond time scale and can out-compete fluorescence. The implications of such a process are useful in the field of photovoltaic technology. The quantitative analysis for incorporating singlet fission sensitizer results in a theoretical limit of 50% as compared to the Shockley–Queisser limit of about 33%. Hence the process of generation of two triplets resulting from the absorption of a sufficiently energetic single photon can improve the device performance drastically.

Photonic crystals: Another method to overcome the S–Q limit is by using photonic crystals to capture emitted light and re-direct it back to cause further excitations. Placing a photonic crystal (PC) on top of the solar cell modifies the absorption and emission. If the photonic crystal is selected with a band-gap that extends from the semiconductor band-gap energy E_g to the photonic band-gap E_{pc}, photons in this range of energies will be reflected and cannot reach the cell. Similarly, photons are emitted as a result of radiative recombination with energy greater than E_g. Furthermore, only photons with energy greater than E_{pc} can escape the cell. Photons with intermediate energies in between E_g and E_{pc} will be reflected by the PC and will be absorbed and re-emitted continuously–a process called photon recycling at the open circuit. This process leads to a high concentration of separated charge carriers and an increased open circuit voltage.[17]

6.3. *Solar concentrators*

3rd-Gen PVs aim at a better utilization of the complete solar spectrum in order to get the maximum efficiency for a particular system. Some examples include intermediate band-gap cells, quantum dot based concentrators and up/down-converters. Conversion of the incident solar spectrum into a precisely matched spectrum of the absorber can significantly increase the efficiency of solar cells. One such example is the down conversion and the technology is referred to as luminescent solar concentrators (LSC). An LSC essentially consists of a highly transparent plastic matrix, in which luminescent materials like organic dye molecules are dispersed. Upon illumination, these dyes absorb and re-emit at a red-shifted wavelength, often with very high quantum efficiency. Since the refractive index of the plastic is higher than its surroundings, total internal reflection ensures that much of this light reaches the solar cell placed at the sides of the plastic matrix. LSCs can capture light that is direct or diffuse and concentrate them sufficiently, eliminating the need for solar tracking, thereby minimizing costs. Furthermore, they can be applied to any solar cell technology-from silicon to organic materials. Quantum dots have been a promising alternative to dyes with superior stability offering tunability and can be easily embedded in plastics.[18-20]

The implementation of a micro-concentrator in PVs, by reducing the cell size to the sub-mm range can result in a widegain, but at the expense of introducing some manufacturing challenges linked to the tiny size of the cells and the increased number of units to manipulate.[21]

Losses in Solar Cells: In general, solar cells do not operate at the theoretical limits of efficiency. This is due to the presence of several losses of photons and charge carriers. There are intrinsic loss mechanisms apart from a host of other extrinsic losses like Fresnel's reflection and non-ideal back reflectors. Intrinsic losses cause a reduction in current or voltage, modifying current-voltage characteristics. These losses include: i) Carnot loss—there is always an energy penalty incurred when the thermal energy from the Sun is converted to electrical work leading to a voltage drop. ii) Boltzmann loss—caused by increased entropy as a result of a mismatch between absorption and emission angles. This loss mechanism can be described by Boltzmann's equation and can be minimized for devices with restricted emission angles. iii) Thermalization loss—occurs as a result of the strong interaction of excited charge carriers and lattice phonons. This loss can be minimized in multi-junction or hot carrier solar cells. iv) Emission loss—caused by charge recombination leading to emission, leading to the lowering of current v) Band gap loss—caused by the limited band gap of the device.

Photons with energies lower than the band gap are not absorbed leading to a loss in the current.

7. Solution-Processed Solar Cells

A key factor in the choice of technology to fabricate solar cells should be ideally based on the total energy-input which goes into the making of the energy conversion device. The energy-payback should not ideally go beyond a year or two. The assessment of this energy-assessment cycle is quite complex since all contributions need to be included right from glass manufacturing, semiconductor crystals in a fabrication plant, metal contacts and numerous other sources. Solution-processed solar cells based on organic photovoltaics, quantum dots, hybrid perovskites fall under this class of manufacturing technology which ensure energy recovery within a fraction of its life-cycle of usage.

8. Organic Solar Cells and Bulk Heterojunction Concept

Organic solar cell designs involving a bare, thin semiconductor layer with electrodes having different work functions are not efficient. In a typical organic/polymer semiconductor film, the optical absorption is excitonicin nature resulting in the formation of neutral quasiparticles representing a bound singlet state of electrons and holes. Due to the lack of crystallinity and the disordered nature of these systems, excitons have a lifetime in the range of nanoseconds and diffusion lengths in the range 10–20 nm[22] [23]. In organic systems with low dielectric constants and strong electron correlations, excitons typically exhibit a relatively high binding energy as compared to the kT and are of the order 0.3–1 eV[23] [24]. The limited region of photoactivity in the vicinity of the electrode-interface is not sufficient. Further, modification by introducing a complementary layer to form a Donor – Acceptor bilayer[25] is also not sufficient to generate reasonable η due to the limited exciton diffusion length. This issue was resolved by the introduction of the concept of bulk heterojunction (BHJ), a nanostructured bi-continuous interpenetrating network.[26,27] The active layer is then the entire film consisting of a binary mixture of donor and acceptor species. The generation of photo-carriers in a BHJ involves: Incident photons of energy higher than the bandgap of the donor (acceptor) molecule are absorbed to create excitons. The exciton generated inside the donor (acceptor) domain subsequently diffuses to the donor: acceptor interface, where it is quenched by an electron (hole) transfer to the acceptor (donor) molecule. This electron (hole) transfer process does not necessarily generate dissociated free charge carriers. At this point, the electron and hole are located on different materials but can be bound by the Coulomb attraction (~0.1 – 0.5 eV) forming an interfacial bound electron–hole pair which is more commonly referred as charge transfer (CT) states. The dissociation of these bound electron-hole pairs into free charge carriers is the most important step in the overall process. The free charges are then transported through the bulk and finally extracted at the device electrodes. This interpenetrating bi-continuous network of donor/ acceptor molecules enhances the interfacial area available for exciton dissociation. If the inter-mixed D/A microphase lengths are in the range of 10–20 nm, then most of the excitons reach the D/A interface to dissociate into free charges. Hence the efficiency of the charge generation in these bulk heterojunction films is governed by the crystallinity and phase-separation property of the individual components.[28] Reports during the early stages of BHJ devices showed promising results in which the donor polymers were blended with fullerene derivatives acceptors. The donor polymer poly-[3-hexylthiophene] (P3HT) and acceptor molecule [6, 6]-phenyl C61-butyric acid methyl ester ($PC_{60}BM$) has been a model BHJ system for decades. Of late, D-A based BHJs have risen to efficiencies exceeding 15%. The success of the C60 acceptors is both due to its electron acceptor characteristics and its ability to form BHJ nano-morphology with donor polymers. However, the relatively low absorption of fullerenes in the visible region along with a limited tunability of energy levels posed a limitation. In this regard, the development of new electron acceptor materials was of importance to overcome these limitations.

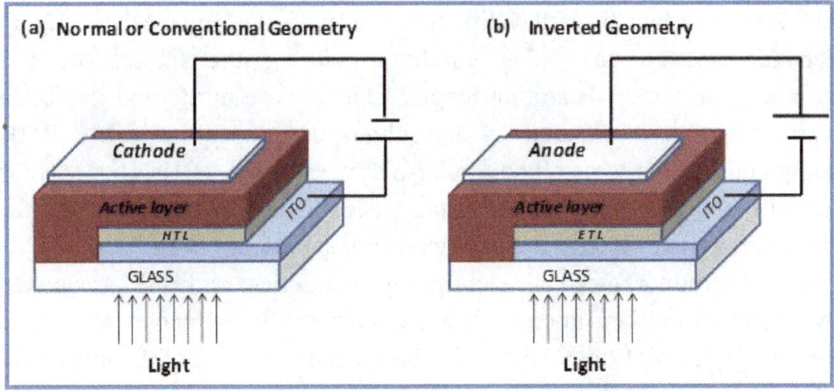

Fig. 5. Normal and inverted geometry of organic solar cells. In the normal geometry, light is illuminated from anode side, whereas in the inverted geometry light is illuminated from cathode side. Work functions of transparent conducting oxides are modified using additional buffer layers. Generally, ZnO, TiO2 used to selectively extract electrons (ETL), and PEDOT:PSS, MoOx were used to selectively extract holes (HTL).

In parallel to the improvements insingle small-area cells, the wide-spread development of large area modules using roll-to-roll fabrication techniques were also made. To address the stability issue, the inverted geometry was proposed as an alternative to the conventional device architecture. In the normal or conventional architecture (Figure 5), low work function metals such as Ca or Al are used as an electron collecting top electrode, which is prone to get oxidized in the presence of moisture and oxygen. In addition, the hole transporting buffer layer PEDOT:PSS is inherently hygroscopic and acidic which accelerates the degradation of devices.[29] To overcome this problem, the inverted geometry of the device was implemented (see Figure 5).

8.1. *Photophysics of OSCs*

The singlet excitons are spatially more localized on the backbone of the conjugation. Electronic transitions are accompanied with the local relaxation of the neighboring molecular structure, due to the strong electron-phonon coupling. Hence BHJ films having an adequate donor–acceptor interfacial area along with optimum domain sizes for efficient diffusion of excitons and charges, offer a guideline to select D-A pairs.

The size and density of the different phases in the organic BHJ highly depends on the processing conditions during the fabrication. For instance, the phase separation length scales, phase-purity and demixing features are quite different in amorphous donor polymer based BHJs compared to the crystalline donor-polymer based BHJs.The Marcus model then can be used to provide a basic framework to understand the key intermediate steps in the photocarrier generation.

The importance of identifying energetics and the associated dynamics of the key intermediate CT state has been recognized. Since the energy level of the CT state lies below the band gap of the individual components in the blend, it is expected to show a red-shifted band in both the absorption and emission spectra.[30]

Time-resolved transient absorption spectroscopic measurements are generally employed to study the charge generation and recombination dynamics in the donor-acceptor systems. Transient absorption spectroscopy is a pump-probe technique, in which the pump laser pulse excites the sample from the ground state to the excited state and the probe pulse monitors the excited state dynamics. The delay between the pump and probe pulse assists in monitoring the dynamics of the excited species. Along with radiative species like singlet excitons, the non-radiative excited species like polarons and triplet excitons can also be monitored using this technique. The key challenge of this technique is to resolve and assign all the bands observed in the transient spectra. Although quasi-steady state spectra are utilized in assigning basic bands, the combination of the quantum chemical calculation and the global fitting of transient spectra has proved to be more useful in identifying the photophysical processes leading to a charge carrier generation.

The presence of the bound CT state and formation of free charge carriers in ultrafast timescales is a highly debated topic. Considering the energy diagram of the photocharge generation, the CT state possesses excess thermal energy initially due to the difference in energy between the donor singlet exciton and the CT state. The dissociation of this hot-CT state occurs through two possible pathways: (1) formation of free charges from the hot-CT state or (2) thermal relaxation of the CT state to its ground state followed by the dissociation into free charges. It has been found that the charge extraction is independent of incident photon energies (even at sub-gap regime). This has led to an interpretation that the relaxed CT states can produce a similar quantity of charges as that of the pristine singlet excitation in organic BHJs.

8.2. *Recombination losses*

The magnitude of the difference $E_{CT} - eV_{OC}$ is a measure of the energy lost to the charge recombination and the consequent decrease in quasi-Fermi level splitting. $E_{CT} - eV_{OC}$ can then be divided into radiative and non-radiative recombination losses. A detailed balance analysis shows that all solar cells must lose some voltage to radiative recombination. A certain amount of energy loss due to the radiative recombination is therefore intrinsic and sets the upper limit for the V_{OC}. The process of recombination of charge carriers (electron and hole pair) via the radiative relaxation process of the Coulombically bound electron–hole pair (CT state) or the free carriers across the D/A interface generated is the geminate recombination, which follows monomolecular kinetics. The non-geminate recombination involves the recombination of free charges which have their origin in different excitons and are of three predominant types: trap-assisted recombination, bimolecular recombination and Auger recombination. The important efficiency limiting recombination in organic BHJ devices is bimolecular recombination.[31,32] If n and p represent the electron and the hole charge density respectively and n_i is the intrinsic carrier concentration, the bimolecular recombination rate k_R is proportional to the carrier concentration $\sim k_L(np - n_i^2)$. The charge carriers in these materials follow Langevin kinetics, where the recombination process is proportional to the diffusion of the two charge carriers (n and p), hence k_R is directly proportional to the charge carrier μ. $k_L \sim q/\varepsilon\ (\mu_e + \mu_h)$ and $k_R = q/\varepsilon\ (\mu_e + \mu_h)(np - n_i^2)$. It should be noted that a higher μ does not directly imply increased recombination rates. (by introducing the Langevin reduction factor) In addition to the bimolecular recombination, the monomolecular trap-assisted recombination also decides the photocurrent generation efficiency in organic BHJ devices. In the trap-assisted recombination, a single carrier gets localized first by an energetic trap followed by the recombination of the opposite charge with the localized charge. In this case, the recombination rate depends on the density of traps and the rate of trapping and detrapping of mobile charges from the localized traps. The Shockley-Read-Hall recombination[33], where the rate of trap-assisted recombination is:

$$R_{SRH} = C_e C_h N_{tr} \frac{(np - n_i^2)}{[C_e(n + n_1)C_h(p - p_1)]}$$

where C_e represents the probability per unit time that an electron in the conduction band will be captured by an empty trap. Correspondingly, C_h represents the probability per unit time that a hole will be captured by the electron in the trap. N_{tr} indicates the density of traps, ni denotes the intrinsic carrier concentration. The trap-assisted recombination in organic semiconductors is decided by the diffusion of free charges towards the trapped charges.[33]. In spite of the presence of traps in organic BHJ devices, the majority of high efficiency solar cells exhibit a near unity quantum yield indicating that the devices are not exclusively limited by the trap-assisted recombination. However, for low efficient devices, the trap-assisted recombination seems to be an important limiting factor for charge transport.

8.3. *BHJs: donors and acceptor molecules*

The prospect of non-fullerene acceptor materials as an alternative to replace fullerene derivatives has altered the course of the BHJ development.[34] These molecules are synthesized from diversified, low-cost routes with significant chemical, thermal and photostability. The optically active acceptor, unlike fullerenes, offer additional spectral tunability and with a choice of the appropriate donor can cover a wide solar spectral range.[34,35] In essence, a facile functionalization of these molecules can tailor the charge generation and transport features. The design rules of conjugated polymers are also valuable for the development of novel NFAs. Many approaches can be employed to tailor the optical and electrochemical properties such as adjusting the conjugation lengths to tune the spectral response, using fluorination to alter the frontier energy levels, and tuning the extent of the HOMO–LUMO overlap to modify the extinction coefficients. As described in previous sections, there has been evidence suggesting the traditionally believed >0.2 eV excess energy might not be a strict restriction for an efficient free carrier generation.

Table 1. Electronic levels and band gap of acceptor and donor molecules.

Material	HOMO (eV)	LUMO (eV)	Band gap (E_g) (eV)
$PC_{60}BM$	- 6.1	- 4.3	~1.8
$PC_{70}BM$	- 6.0	- 4.3	~1.7
Bis-PCBM	- 6.1	- 4.1	~2.0
ICBA	- 5.9	- 4.0	~1.9
Twisted Perylene	- 6.1	- 4.1	~2.0
Planar Perylene	- 6.0	- 4.1	~2.1
ITIC	- 5.5	-3.8	~1.7
BT-CIC	-5.5	- 3.6	~1.9

Material	HOMO (eV)	LUMO (eV)	Band gap (eV)
P3HT	- 5.1	- 3.0	~2.1
PBTTT	- 5.15	-3.05	~2.1
MEHPPV	- 5.02	- 2.7	~2.3
PBDTTT-CT	- 5.1	- 3.5	~1.6
PTB7	- 5.15	- 3.5	~1.65
PCPDTBT	- 5.1	- 3.6	~1.5
PBDB-T	- 5.3	- 3.5	~1.8
PffBTOT-2OD	-5.3	- 3.7	~1.6
DTS(FBTTh2)2	- 5.1	- 3.3	~1.8
BTID-2F	- 5.2	- 3.4	~1.8

In comparison to fullerene-based organic cells where the newly designed polymers must meet the energy level requirement set by PCBM, the capability of synergistically adjusting the energy levels for both donor and acceptor clearly shows the design flexibility of materials for non-fullerene organic solar cells. Based on the steady developments and improvements in all aspects, the road ahead for organic solar cells indeed appears promising.

8.4. *Hybrid organic-inorganic perovskite solar cells*

This decade has witnessed the emergence of HOIP materials. These class of materials exhibit superior photoelectric properties and a relatively simpler processing. Within just a span of seven years of research, solar cells with the hybrid perovskite-based absorbing material fabricated by a simple low-temperature solution processing technique which started from a 4% efficiency are now capable of delivering efficiencies above 23%. The perovskite structure typically implies a structurewith the stoichiometry ABX_3 where A and B represent cations and X is an anion. The ideal perovskite structure exists in a simple cubic crystal form with the corner sharing BX_6 octahedral network. In hybrid organic-inorganic perovskites (HOIPs), 'A' is an organic cation which is monovalent. According to Goldschmidt tolerance factor, the formation or non-formation of a perfect cubic crystal is governed by the size restrictions of the ions. The synthesis of HOIPs is carried out by various methods such as combining a metal salt such as PbI_2 with an organic halide salt like MA iodide in one step and spin-coating of the solution of both the salts. Co-evaporation and a two-step process of first forming the metal salt film followed by the exposure of the organic halide are other common synthesizing techniques.

8.4.1. *Properties of HOIP: Reasons for unprecedented interest*

These materials can be easily solution proceed into thin films and have a low enthalpy of formation. In solar cells involving direct bandgap semiconductors, charge transport is efficient as light absorption can proceed without the phonon's assistance. In PV devices, the carrier generation profile depends on the absorption coefficient which is $\sim 10^5$ cm^{-1} in HOIPs and hence the absorber layer can be significantly thinner than the carrier diffusion length. Nearly 100% of the absorbed light is captured within a 300 nm thick layer. The sharp and clear absorption onset is expressed in the small Urbach tail and energy and is 15 meV which is comparable to 11 meV in Si and 8 meV in GaAs, implying a low density of sub-bandgap states near the band edge.[39]

HOIPs show many notable properties like the easy tunability of the band gap, geometrical structure and dimensionality tuning and a high PL quantum yield. Charge carriers in HOIPs have a low effective mass of the order of ~ 0.10–0.15 m_0, resembling values for Si and GaAs.[39] The measured diffusion coefficients are 0.05–0.2 cm^2 s^{-1} and $\mu \sim$ (1-30 cm^2 V^{-1} s^{-1}). In single crystals, the μ exceeding 100 V^{-1} s^{-1} are observed, however, it is limited by the scattering processes in HOIP materials. Long carrier lifetimes ranging from 100 ns to >1 µs at 1 sun are responsible for the efficient carrier collection.[40] In spite of the modest μ, the long carrier lifetimes result in long diffusion length-scales which are substantially higher than the absorption depth and contributes to the efficient carrier collection. HOIPs potentially provide an ideal companion for silicon in a tandem system.[41]

Although efficiencies of hybrid perovskite solar cells have rapidly reached high values, these HOIPs suffer from the serious drawback of fast degradation without encapsulation. The degradation of perovskite solar cells has been attributed to multiple external factors including moisture, oxygen and UV light along with temperature variations. Ion migration, electromigration and interfacial reactions are some of the internal intrinsic factors which appear to play a critical role in the degradation. Irreversible reactions occurring at the interface of the hole transfer layer and the perovskite layer have also been cited as a factor for the deterioration of the devices. Metal migration from the electrode into the hole transporting layer is also a mechanism for lowering the performance. The flow of components due to the thermal evaporation during the deposition and illumination of light while operation, and the migration induced from the electric fields which results in electro-migration and hysteresis effect are other contributing factors. Ion migration is especially sensitive to the concentration of mobile vacancies. These features which appear to be the sources for instability in these systems also may be linked to the dependence of the I(V) response on the direction (forward or reverse) of scan and/or speed of the bias scan results in a hysteresis.[42]

In earlier versions of the devices (2011), an un-encapsulated device PCE could get reduced by 80% within just 10 minutes because the perovskite quantum dots tend to dissolve gradually into the redox electrolyte. The decomposition of perovskite to PbI_2 in the presence of moisture is another major challenge although improvements in the stability and degradation rates are being vigorously pursued to ensure the commercial feasibility of these hybrid perovskite solar cells. For a technology to be feasible in the real-world application, consistent performance under real conditions for durations over twenty years is imperative. Some of the recent developments include major strides in overcoming the issues stated above.

An increased efficiency and stability has been recently demonstrated in FA/Cs perovskite as a photoactive layer in a single-junction solar cell under a simulated concentrated solar irradiance. The PCE of 21.1% under 1 Sun illumination increased to 23.6% under 14 Suns. FA/Cs devices with a relatively stable interfacial layer sustained 90% of its efficiency after 150h of ageing at 10 Suns of concentrated light.[43] An interesting approach to improve the stability was recently demonstrated in a fluorous organic cation-an organic spacer in a 2D perovskite, mixed with a 3D perovskite which resulted in the vertical segregation of the hydrophobic low dimensional perovskite on top of 3D. This self-assembled hydrophobic fluorous low dimensional perovskite layer reduces water-induced degradation. Apart from stability, the 2D layer improves the interfacial contact with the buffer layers withan enhanced PCE from 18% to 20%.[44] An enhanced operational stability was recently achieved using a multifunctional molecular modulation on the morphology and electronic properties of perovskite films of the thermally stable $FA_{0.9}Cs_{0.1}PbI_3$, which resulted in a PCE of 20.9% without the use of antisolvents. The molecularly engineered bifunctional molecular modulator, SN, combines multiple functions by simultaneously passivating the surface defects while displaying a structure-directing function through an interaction with the perovskite with improved film quality.[45] Fluorene-terminated hole-transporting material with a high glass transition temperature ($\sim 433K$) and tuned energy level (Homo = -5.27 eV, better matching with VB of perovskite) as a replacement to the conventional spiro-OMeTAD appears promising. A device with this HTM shows an improved PCE of 20.9% and maintained 95% of the initial performance for 500 hours.[46] Perovskite (($FA_{0.95}PbI_{2.95})_{0.85}(MAPbBr_3)_{0.15}$) solar cells of the inverted planar structure with $\sim 21\%$ efficiency were achieved by reducing the NR recombination through a solution-processed secondary growth (SSG) technique. In SSG, the perovskite film was kept for a secondary growth with the assistance of the guanidinium bromide. By this means, a wider band gap top layer and a higher degree of the n-type layer was possible, which reduced the NR recombination losses and increased V_{oc} by 100 mV.[47] Mesoporous titanium dioxide (mp-TiO2) as the electron transporting layer is the best performing PSCs, however, with stability issues under illumination. Lanthanum (La)–doped BaSnO3 (LBSO) has been shown to be a better choice from this point of view. A superoxide colloidal solution route for preparing a Lanthanum (La)–doped BaSnO3 (LBSO) electrode under very mild conditions (below 300°C) was found to improve the efficiencyto 21.2% achieved with LBSO as an electron transport layer as compared to the mp-TiO2 based device having a 19.7% efficiency.[48]

9. Need for a Reliable Monitoring Tool

From the above discussions, the solution processed active layers (BHJ, perovskite, QDs) based solar cells need to be exhaustively studied for stability and a reliable long-term performance. The new architectures of Si panels (PERC and HIT) with additional components and a new process flow also need to be evaluated for long-term performance. Typical approaches for long-term studies involve the periodic assessment of the solar cells using the standard I-V testing, impedance measurements and imaging at a different resolution to identify regions of failure and degradation. An accelerated testing under controlled environment of humidity and light soaking are used in laboratories to predict long-term performance. The numerous parameters which are involved in degradation make the problem complex. The variations in the outdoor conditions especially in semi-urban, tropical locations which results in thermal cycling, periodic rainfall, particulate matter and exposure to chemicals in various forms; all leading to deterioration of the solar panels makes it difficult to arrive at a realistic model. The symptoms of the degradation appear in the form of encapsulant discoloration, delamination, the appearance of hot spots and IC discoloration, fractured cells, diode/J-box and glass breakage, backsheet insulation compromise, potential induced degradation PID and permanent soiling. The author's laboratory at JNCASR has pioneered two evaluation methods to ascertain and quantify the processes involved in the degradation.

Noise measurements can be a very effective tool in gauging the degradation profile and understanding the carrier transport mechanisms. Noise spectroscopy is non-destructive, an in-situ technique which has been used for reliability studies in semiconductor devices and is used to study quantum wires, graphene, FETs.

The instantaneous value of conductance, in any semiconductor device, fluctuates around a mean value. A random noisy signal is in general, depicted by their spectral characteristics/*spectral density* or correlated function and its *amplitude* characteristics, for instance, is the probability density function.

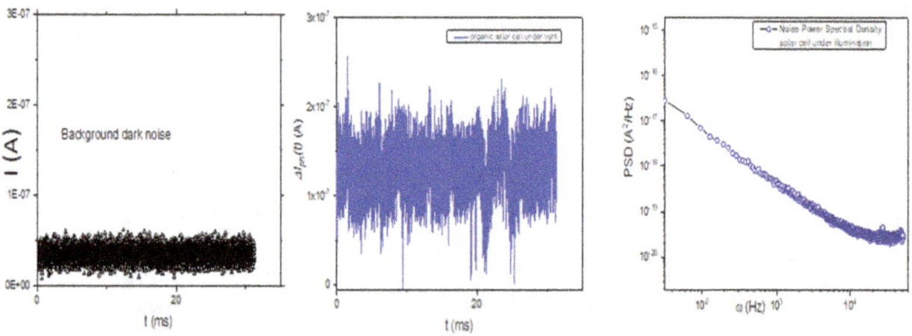

Fig. 6. Noise studies—time-series of photocurrent and power spectrum of a typical BHJ device from the author's laboratory.

A typical analysis of the noise characteristics for an electrical system is based on Hooge's formalism, and has two primary sources of fluctuations, mobility and the number of carriers. According to Hooge's empirical model the power spectral density is given by,

$$S_I = \frac{AI^\beta}{f^\gamma} A = \frac{\alpha_H}{N}.$$

This can then be expressed as $ln(S_I) = -\gamma \, ln(f) + ln(A) + \beta \, ln(I)$, where, S_I is the current power spectral density, γ is the frequency exponent, α_H is the Hooge's parameter and N is the total number of charge carriers. Recent studies from our laboratory of noise in photocurrent from OPV and HOPVs indicate its utility to study defects and degradation mechanism. In combination with a model framework, the defect density can be estimated using these measurements. In principle, noise measurements carried out in the early stages of the solar-cell decay, can predict the long-term performance trajectory of the cell based on the evolution of the defect density and energetics. [49]

The light beam induced current imaging (LBIC) scanning is a microscopic tool for identifying the local inhomogeneity in the performance of light to current conversion devices including solar cells, photodetectors and image sensors. The technique utilizes a tightly focused light beam as a probe and the response of the device in the form of current output (raw/amplified voltage output) is used to generate a map.Since the J_{sc} variation is directly related to the performance of the solar cell and LBIC can be considered as a power-map, it is possible to accurately monitor the signal from the localized illuminated spot with the rest of the region of the panel in the dark. Since the measurements are carried on real devices, the information contains all possible features of the basic device physics, which includes generation, transport and collection. An underperforming region will be represented by a lower signal level than average and the origin of the reduced signal can be associated with factors like inefficient generation (resulting from lower quantum yield), higher series resistance (resulting from broken or defective layers) and contact losses. Each of these losses can then be further identified using the spectral response and the transient features of the output signal. [50]

10. Near Future Outlook and Prospects

Solar PV research will continue to be the dominant theme. Large-scale implementation of renewables as a replacement for fossil fuel energy technology will be the driving factor. It should be noted that in comparison to the scale of the nuclear and plasma fusion projects, solar PVs, receive a considerably less investment in R&D. This trend in the policy hopefully will get modified.

In the near future, Si will continue as the mainstay PV technology. But, one can expect other material choices to open up with added mechanical and performance features. The economics of processing and large-scale scaling of these alternative PVs will, however, dictate the adoption of this technology. Hybrid solar cells will also emerge as interesting options. Rapid strides in material development will lead to an optimum guided choice. A combination of materials and range of phenomena should lead to an exciting range of research frontiers. Integration with storage technology will be an important criterion for the futuristic PV applications. A bright future lies ahead for PVs in general.

Acknowledgments

I acknowledge all my collaborators, Ph.D. scholars, and post-docs for the various input and contributions in this important research area. Specifically, I thank Ganesh, Abdul, Anshuman, Prashant, Swathi, Apoorva, Sukanya and Ravi for help with this chapter.

References

1. https://mercomindia.com/product/india-solar-project-tracker
2. D. S. Ginley, R. Collins and D. Cahen, Direct solar energy conversion with photovoltaic devices, *Fundamentals of Materials for Energy and Environmental Sustainability*, Cambridge Univ. Press (2012).
3. T. Kirchartz, J. Bisquert, I. Mora-Sero and G. Garcia-Belmonte, Classification of solar cells according to mechanisms of charge separation and charge collection, *Phys. Chem. Chem. Phys.*, **17**(6), 4007-4014 (2015).
4. M. A. Green, *Solar Cells: Operating Principles, Technology, and System Applications*. Prentice-Hall, Englewood Cliffs, N.J (1998).
5. Y. Lee, C. Park, N. Balaji, Y. J. Lee and V. A. Dao, High-efficiency silicon solar cells: a review, *Israel J. Chem.* **15**(10), 1050-1063 (2015).
6. D. Gupta, M. Bag and K. S. Narayan, Area dependent efficiency in organic solar cells, *Appl. Phys. Lett.* **93**, 163301-3 (2008).
7. T. Dullweber and J. Schmidt, Industrial silicon solar cells — a review *IEEE J. Photovol.* **6**(5), 1366-1381 (2016).
8. M. A. Green, The passivated emitter and rear cell (PERC): from conception to mass production, *Solar Energy Mater. Solar Cells*, **143**, 190-197 (2015).
9. http://www.sharp-world.com/corporate/news/180327.html)
10. https://news.panasonic.com/global/press/data/2014/04/en140410-4/en140410-4.html)

11. W. Shockley and H. J. Queisser, Detailed balance limit of efficiency of p-n junction solar cells. *J. Appl. Phys.* **32**(3), 510-519 (1961).

12. A. J. Nozik, G. Conibeer and M. C. Beard, *Advanced Concepts in Photovoltaics* Edited by, RSC Energy and Environment Series No. 11 (2014), *Multi-Junction Solar Cells*, Simon Philipps* and Andreas W. Bett.

13. D. Aiken, E. Dons and S. Je, Lattice-matched solar cells with 40% average efficiency in pilot production and a roadmap to 50%, *IEEE J. Photovoltaics*, **3**(1), 542-547 (2013).

14. M. C. Beard and R. J. Ellingson, multiple exciton generation in semiconductor nanocrystals: Toward efficient solar energy conversion. *Laser & Photon. Rev.* **2**(5), 377-399 (2008).

15. P. J. Jadhav, A. Mohanty, J. Sussman, J. Lee and M. A. Baldo, Singlet exciton fission in nanostructured organic solar cells. *Nano Lett.* **11**(4), 1495-1498 (2011).

16. B. Ehrler, M. W. B. Wilson, A. Rao, R. H. Friend and N. C. Greenham, Singlet exciton fission-sensitized infrared quantum dot solar cells. *Nano Lett.* **12**(2), 1053-1057 (2012).

17. J. N. Munday, The effect of photonic bandgap materials on the Shockley–Queisser limit. *J. Appl. Phys.* **112**(6), 064501-6 (2012).

18. T. Trupke, M. A. Green and P. Wurfel, Improving solar cell efficiencies by down-conversion of high-energy photons. *J. Appl. Phys.* **92**(3), 1668-1674 (2002).

19. V. Sholin, J. D. Olson and S. A. Carter, Semiconducting polymers and quantum dots in luminescent solar concentrators for solar energy harvesting. *J. Appl. Phys.* **101**(12), 123114-9 (2007).

20. A. J. Das and K. S. Narayan, Retention of power conversion efficiency–from small area to large area polymer solar cells, *Adv. Mater.* **25**(15), 2193-2199 (2013).

21. O. Fidaner, F. A. Suarez, M. Wiemer, V. A. Sabnis, T. Asano, A. Itou, D. Inoue, N. Hayashi, H. Arase, A. Matsushita, *and* T. Nakagawa High efficiency micro solar cells integrated with lensarray, *Appl. Phys. Lett. V*ol. **104**(10), 103902-5 (2014).

22. P. Peumans, V. Bulović and S. R. Forrest, Efficient photon harvesting at high optical intensities in ultrathin organic double-heterostructure photovoltaic diodes, *App. Phys. Lett.* **76**(19), 2650-3 (2000).

23. D. Joshi, R. Shivanna and K. S. Narayan, Organic photovoltaics: key photophysical, device and design aspects, *J. Mod. Opt.* **61**(21), 1703-1713(2014).

24. S. R. Forrest, The path to ubiquitous and low-cost organic electronic appliances on plastic, *Nature*, **428**(9686), 911-918 (2004).

25. C. W. Tang, Two-layer organic photovoltaic cell, *Applied Physics Letters,* **48**(2), 183-185 (1986).

26. G. Yu, J. Gao, J. C. Hummelen, F. Wudl and A. J. Heeger, Polymer Photovoltaic Cells: Enhanced Efficiencies via a Network of Internal Donor-Acceptor Hetero junctions, *Science*, **270**(5243), 1789-1791 (1995).

27. J. J. M. Halls, C. A. Walsh, N. C. Greenham, E. A. Marseglia, R. H. Friend, S. C. Moratti and A. B. Holmes, Efficient photodiodes from interpenetrating polymer networks, *Nature*, **376**(6540), 498-500 (1995).

28. D. Gupta, N. S. Vidhyadhiraja and K. S. Narayan, Transport of photogenerated charge carriers in polymer semiconductors, *proc. IEEE*, **97**(9), 1558-1569 (2009).

29. C. Waldauf, M. Morana, P. Denk, P. Schilinsky, K. Coakley, S. A. Choulisa, and C. J. Brabec, Highly efficient inverted organic photovoltaics using solution based titanium oxide as electron selective contact *Appl. Phys. Lett.* **89**, 233517-3 (2006).

30. T. M. Clarke and J. R. Durrant, Charge Photogeneration in Organic Solar Cells. *Chem. Rev.* **110**(11), 6736-6767 (2010).

31. L. J. A. Koster, M. Kemerink, M. M. Wienk, K. Maturová and R. A. J. Janssen, Quantifying Bimolecular Recombination Losses in Organic Bulk Heterojunction Solar Cells, *Adv. Mater.* **23**(14), 1670-1674 (2011).

32. C. M. Proctor, M. Kuik and Thuc-Quyen Nguyen, Charge carrier recombination in organic solar cells. *Progr. in Polymer Sci.* **38**(12), 1941-1960 (2013).

33. W. Shockley and W. T. Read, Statistics of the recombinations of holes and electrons. *Phys. Rev.* **87**(5), 835-842 (1952).

34. G. Zhang, J. Zhao, P. C. Y. Chow, K. Jiang, J. Zhang, Z. Zhu, J. Zhang, F. Huang and He Yan, Nonfullerene Acceptor Molecules for Bulk Heterojunction Organic Solar Cells, *Chem. Rev.* **118** (7), 3447-3507 (2018).

35. S. Rajaram, R. Shivanna, S. K. Kandappa andK. S. NarayanNonplanar perylene diimides as potential alternatives to fullerenes in organic solar cells. The *J. Phys. Chem. Lett.* **3** (17), 2405-2408 (2012).

36. V. M. Goldschmidt, Originalaufsätze Und Berichte Reine Und Technisch Angewandte Chemie Und Physikali sche Chemie, *Naturwissenschaften*, **14**(21), 477-485 (1926).

37. W. S. Yang, J. H. Noh, N. J. Jeon, Y. C. Kim, S. Ryu, J. Seo and S. I. Seok, High-performance photovoltaic perovskite layers fabricated through intramolecular exchange, *Science*, **348**(6240), 1234-1237 (2015).

38. A. Sadhanala, F. Deschler, T. H. Thomas, S. N. E. Dutton, K. C. Goedel, F. C. Hanusch, M. L. Lai, U. Steiner, T. Bein and P. Docampo, Preparation of Single-Phase Films of $CH_3NH_3Pb(I_{1-x}Br_x)_3$ with Sharp Optical Band Edges, The *J. Phys. Chem. Lett.* **5**, 2501-2505 (2014).

39. H. J. Snaith, Present status and future prospects of perovskite photovoltaics, *Nature Mater.* **17**, 372-376 (2018).

40. A. J. Neukirch, G. Gupta, J. J. Crochet, M. Chhowalla, S. Tretiak, M. A. Alam, H. Wang and A. D. Mohite, High-efficiency solution-processed perovskite solar cells with millimeter-scale grains, *Science*, **347**(6221), 522-525 (2015).

41. T. Leijtens, K. A. Bush, R. Prasanna and M. D. McGehee, Opportunities and challenges for tandem solar cells using metal halide perovskite semiconductors, *Nature Energy*, **3**, 828-838 (2018).

42. A. Singh, P. K. Nayak, S. Banerjee, Z. Wang, J. T. Wang, H. J Snaith, K. S. Narayan, Insights into the Microscopic and Degradation Processes in Hybrid Perovskite Solar Cells Using Noise Spectroscopy, *Solar RRL*, **2**(1), 1700173-1700179 (2018).

43. Z. Wang, Q. Lin, B. Wenger, M. Greyson Christoforo, Y. H. Lin, M. T. Klug, M. B. Johnston, L. M. Herz and H. J. Snaith, High irradiance performance of metal halide perovskites for concentrator photovoltaics. *Nature Energy*, **3**(10) 855-861 (2018).

44. K. T. Cho, Y. Zhang, S. Orlandi, M. Cavazzini, I. Zimmermann, A. Lesch, N. Tabet, G. Pozzi, G. Grancini **and** M. K. Nazeeruddin**,** Water-repellent low-dimensional fluorous perovskite as interfacial coating for 20% efficient solar cells. *Nano Lett.* **18**(9), 5467-5474 (2018).

45. D. Bi, X. Li, J. V. Milić, D. J. Kubicki, N. Pellet, J. Luo, T. LaGrange, P.Mettraux, L. Emsley, S. M. Zakeeruddin and M.Grätzel. Multifunctional molecular modulators for perovskite solar cells with over 20% efficiency and high operational stability. *Nature Commun.* **9**(1), 4482-4492 (2018).

46. N. J. Jeon, H. Na, E. H. Jung, T.-Y. Yang, Y. G. Lee, G. Kim, H. W. Shin, S. Seok, J. Lee and J. Seo. A fluorene-terminated hole-transporting material for highly efficient and stable perovskite solar cells. *Nature Energy*, **3**(8), 682-689 (2018).

47. D. Luo, W. Yang, Z. Wang, A.Sadhanala, Q. Hu, R.Su, R.Shivanna, G. F. Trindade, J. F. Watts, Z. Xu, T. Liu, K. Chen, F. Ye, P. Wu, L. Zhao, J. Wu, Y. Tu, Y. Zhang, X. Yang, W. Zhang, R. H. Friend, Q. Gong, H. J. Snaith and R. Zhu, Enhanced photovoltage for inverted planar heterojunction perovskite solar cells. *Science*, **360**.6396 : 1442-1446 (2018).

48. S. S. Shin, E. J. Yeom, W. S. Yang, S. Hur, M. G. Kim, J. Im, J. Seo, J. H. Noh and S. Seok, Colloidally prepared La-doped BaSnO3 electrodes for efficient, photostable perovskite solar cells, *Science*, **356**(6334), 167-171 (2017).

49. M. Bag, N. S. Vidhyadhiraja and K. S. Narayan, Fluctuations in Photocurrent of Bulk Heterojunction Polymer Solar Cells – A Valuable Tool to Understand Microscopic and Degradation Processes, *Appl. Phys. Lett.*, **101**(4), 043903-043906 (2012).

50. S. Mukhopadhyay, A. J. Das and K. S. Narayan, Perspective Article "High resolution photocurrent imaging of bulk heterojunction solar cells. *The J. Phys. Chem. Lett.* **4**(1), 161-169 (2013).

Chapter 15

Thermoelectric Energy Conversion

Manisha Samanta, Moinak Dutta and Kanishka Biswas[*]

New Chemistry Unit, International Centre for Materials Science and School of Advanced Materials, Jawaharlal Nehru Centre for Advanced Scientific Research Jakkur, Bangalore 560064, Karnataka, India
[*]*kanishka@jncasr.ac.in*

Thermoelectric materials, being capable of direct conversion of waste-heat into electricity without any moving parts, can play a prime role in the future energy generation, utilization, and management. However, the practical implementation of thermoelectric technology is limited by the low thermoelectric conversion efficiency of the solid materials, due to the conflicting interdependency of various thermoelectric parameters. This chapter describes the state-of the-art strategies in designing high-performance thermoelectric materials which are believed to enable the implementation of thermoelectric-technology for large-scale power generation applications. We proceed to discuss different ways of maximizing the thermoelectric figure of merit (zT) through enhancement of the Seebeck coefficient by modulating the electronic structure and reduction of the lattice thermal conductivity by composition and nano/micro-nanostructure designs. This book chapter also highlights the recent advances in finding new solids with intrinsically low thermal conductivity facilitating the decoupling of electron-phonon transport and tailoring the chemical bonding. Finally, the main challenges and future potential strategies for further improvement of the TE performance have been discussed at the end of the chapter.

1. Introduction

The ever increasing energy demands of the growing growing global population compels scientists to look for alternative sustainable energy resources to strike the right balance between the energy production and consumption. Now, with more than two-thirds of the worldwide utilized energy being lost as waste heat, it is of paramount economic and environmental benefit to capture this untapped waste heat and convert it into useful energy. Thermoelectric (TE) materials, being capable of direct and reversible conversion of heat into electricity, are thought to be a key solution for future global energy management.[1-4] However, the massive implementation of thermoelectric technology in commercial applications is restricted by the low thermo-electric conversion efficiency (η) of the thermoelectric materials. The main challenge to promote the thermoelectric power generation technology into broader mass-market applications is to increase the low η of current TE materials, given by the following equation,[1-4]

$$\eta = \frac{T_H - T_C}{T_H} \frac{\sqrt{1+zT}-1}{\sqrt{1+zT}+\frac{T_C}{T_H}} \tag{1}$$

where T_H is the temperature of the heat-source, T_C is the temperatures of the sink. zT which is known as *dimensionless figure of merit* of a material, is the main factor in determining the efficiency of the materials and can be expressed as,

$$zT = \sigma S^2 T / \kappa_{\text{total}} \tag{2}$$

where σ is the electrical conductivity, S is the Seebeck coefficient, T is the absolute temperature and κ_{total} is the total thermal conductivity of the material of interest. The term σS^2 is called the power factor. A large power factor means that the material will generate a high output power i.e. large voltage and a high current upon providing the temperature gradient. κ_{total} of a material is expressed as the summation of the electronic (κ_{el}), lattice (κ_{lat}) thermal conductivity and the bipolar thermal conductivity (κ_{b}), $\kappa_{\text{total}} = \kappa_{\text{el}} + \kappa_{\text{lat}} + \kappa_{\text{b}}$. Thus, a good thermoelectric material should simultaneously possess a large Seebeck coefficient (property of semiconductor), high electrical conductivity (property of metal) and poor thermal conductivity (property of glass). It is tricky to combine all these interdependent parameters in a single material, thus enabling its high thermoelectric performance.

Thermoelectric performance of a material can be improved by increasing the power factor and/or decreasing the thermal conductivity by careful control over the rational band structure engineering and microstructure designs, respectively. Maximization of the power factor is often acquired through the enhancement of the Seebeck coefficient with a minimal effect on the electrical conductivity of the materials. The two most widely used approaches for the Seebeck coefficient enhancement are i) the creation of the resonance level near the Fermi level and ii) band convergence.[1, 2] In addition to that, the carrier concentration optimization by doping/alloying[5–7] and carrier energy filtering[8, 9] are also well known strategies to improve the power factor of the materials. The realization of the reduced lattice thermal conductivity can be obtained by solid-solution alloying.[5–7, 10, 11] More recently, the concept of nanostructuring leads to a significantly reduced lattice thermal conductivity in the materials.[12–16]

Bismuth telluride[15,16] and lead chalcogenides[12–14] are the traditional thermoelectric materials which remain the top performing ones in their respective working temperature regions. However, the relatively high toxicity of lead in lead chalcogenides restricts their use in massive applications. As an alternative to lead chalcogenides based thermoelectrics, germanium chalcogenides[17,18] and tin chalcogenides[19,20] appear as the most promising ones. Materials having an *intrinsically* low κ_{lat} offer independent control of the electronic transport properties and hence have gained the attention of the thermoelectric community.[21] Thus, the search for new compounds with intriguing chemical and physical properties is a necessary activity to achieve a high thermoelectric performance in addition to optimizing the known systems and will be covered in details in the upcoming sections of this chapter.

2. Electronic Structure Modulation: Enhancement of Seebeck Coefficient

Thermoelectric power factor ($S^2\sigma$) is the product of the square of the Seebeck coefficient (S) and electrical conductivity (σ). For the typical degenerate semiconductors, Seebeck coefficient, electrical conductivity and power factor can be expressed as per the following equations,[17,22]

$$S = \frac{8\pi^2 k_B^2}{3eh^2} m^* T \left[\frac{\pi}{3n}\right]^{2/3} \tag{3}$$

$$\sigma = ne.\mu \tag{4}$$

$$S^2\sigma \alpha \frac{N_V C_l}{m_I^* E^2_{def}} T \tag{5}$$

where k_B - the Boltzmann constant, e - the electron charge, h - the Planck constant, m^* - the effective mass of carriers, n - the carrier concentration, μ - the carriers mobility, the C_l - the average longitudinal elastic moduli, m_I^* - the inertial effective mass, N_V - the total valley degeneracy and E_{def} - the deformation potential coefficient. From the above equations, it is quite evident that S has an inverse relation with n, whereas σ is directly proportional to n. Thus the reduction in the carrier concentration by doping or alloying will enhance the Seebeck coefficient, but simultaneously will decrease the electrical conductivity.[23] Thus optimization of the carrier concentration may not be always a helpful strategy to boost the power factor the materials. Interestingly, it is observed from equation (5) that power factor ($S^2\sigma$), is strongly correlated with valley degeneracy (N_V) and inertial effective mass (m_I^*) assuming the dominance of the acoustic phonon scattering mechanism. Thus, the modulation of the band structure with the help of appropriate dopants would be a crucial route to enhance the power factor, essentially through the enhancement of the Seebeck coefficient values with a minimal effect on the mobility of the charge carriers in this process. Till date, the most helpful strategy to enhance the Seebeck coefficient is distortion of the electronic structure near Fermi level via valence and/or conduction sub-bands convergence or formation of the resonant state at and around the Fermi level (E_F) which will be discussed in the following sections.

2.1. *Valence band convergence*

Valence band convergence is one of convenient ways to increase the effective band degeneracy, N_V and hence the Seebeck coefficient by converging different bands with the energy difference of a few $k_B T$ in the Brillouin zone (Fig. 1(a) and (b)). For example, we can consider IV-VI rock-salt tellurides which possess two valence bands, light hole and heavy hole valence band. Both the experimental and theoretical investigations confirm the presence of two valence bands in the electronic structure of lead chalcogenides: one at the L point (light hole valence band) with $N_V = 4$ and the other at the Σ point (heavy hole valence band) of the Brillouin zone with a very high $N_V = 12$ and the energy separation between them ($\Delta E_{L-\Sigma}$) are ~0.15 eV for PbTe[25-27] and ~0.24 eV PbSe.[28] Thus the access of both L- and Σ- bands for charge transport can be obtained by reducing the energy separation between them, which can be achieved by increasing the temperature or/and introduction of proper doping/alloying. Alloying of PbTe/PbSe with wide band gap semiconductor (for example, alloying of MgTe[25]/CdTe[26]/MnTe[27] with PbTe; SrTe in PbSe[28]) is proven to open up the principle band gap and decreases the $\Delta E_{L-\Sigma}$ of PbTe, thereby facilitating the valence band convergence and the enhancement Seebeck coefficient.

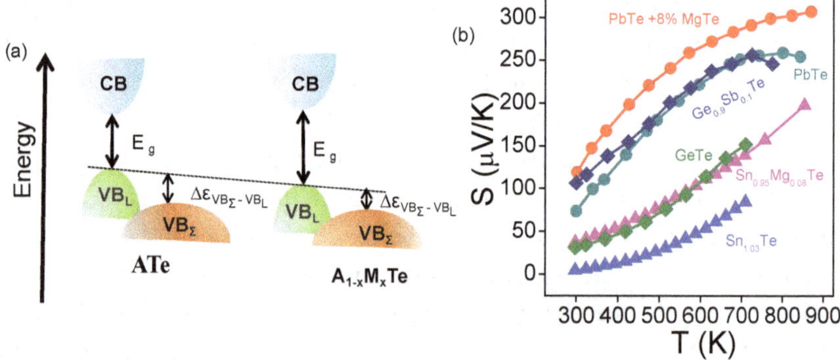

Fig. 1. (a) Schematic energy diagram of the electronic structure of cubic ATe (A = Ge/Sn/Pb) and A$_{1-x}$M$_x$Te near the Fermi level (E_F), where M denotes the dopants which facilitate valence band-convergence. (b) Typical examples of Seebeck coefficient (S) enhancement due to valence band convergence in GeTe, SnTe and PbTe.[5, 24, 25]

Such a band convergence strategy has also been widely employed in *p*-type SnTe, whose electronic structure is analogous to that of lead chalcogenides. Pristine SnTe has a $\Delta E_{L-\Sigma}$ of ~0.35 eV (at 300 K) which is comparatively higher than that of PbTe and PbSe, and acts as a bottleneck to avail its Σ valence band even at higher temperatures, hence the poor Seebeck coefficient value of SnTe (~ 16 μV/K at 300 K which reaches to ~ 90 μV/K at 715 K).[19, 20] Cd, Mg and Hg are found to be good dopants in terms of decreasing $\Delta E_{L-\Sigma}$ in SnTe, resulting in enhanced S value (see Fig. 1(b)).[24,29,30]

Similarly, the valence band convergence has also been realized in GeTe via. Pb, Sb and Mn doping in GeTe.[31-33] A notable conduction band convergence was established to be an effective approach in enhancing the Seebeck coefficients of the *n*-type Mg_2Si by alloying with Sn at the Si site.[34]

2.2. *Slight symmetry reduction*

Detailed electronic structure calculations demonstrate that rhombohedral distortion of GeTe along the [111] crystallographic direction (along the L point of BZ) splits up the 4 L pockets of cubic GeTe (*Fm-3m*) into 3 L + 1 Z pockets in rhombohedral GeTe (*R3m*) and 12 Σ pockets of cubic GeTe into 6 Σ + 6 η pockets in rhombohedral GeTe (Fig. 2(a)).[35,36] Thus manipulation of the degree of the rhombohedral distortion (also referred as slight symmetry reduction) can result in the band convergence of L and Σ bands. In a recent report by Li *et al.*[36], it has been shown that a slight symmetry reduction of cubic GeTe towards the rhombohedral symmetry by Pb and Bi doping causes an effective valence band (L and Σ bands) convergence, resulting in a high zT of ~2.4 in rhombohedral phase of Pb and Bi codoped GeTe at 600 K (Fig. 2(b)).

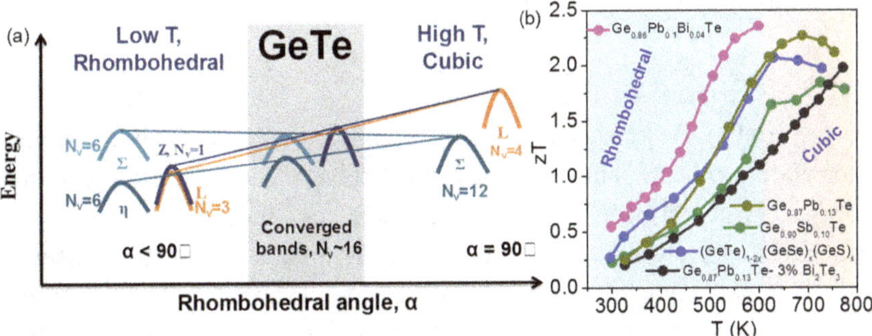

Fig. 2. (a) Schematic of evolution of valence bands of GeTe from rhombohedral phase to the cubic phase depending on the extent of rhombohedral distortion (α).[35] (b) Temperature dependent thermoelectric figure of merit (zT) of rhombohedral GeTe ($Ge_{0.86}Pb_{0.10}Bi_{0.04}Te$) with other high performance of GeTe based samples.[35] Fig. 2(a) and 2(b) reproduced with permission from Ref. 35. © 2018, Elsevier.

2.3. *Resonance level*

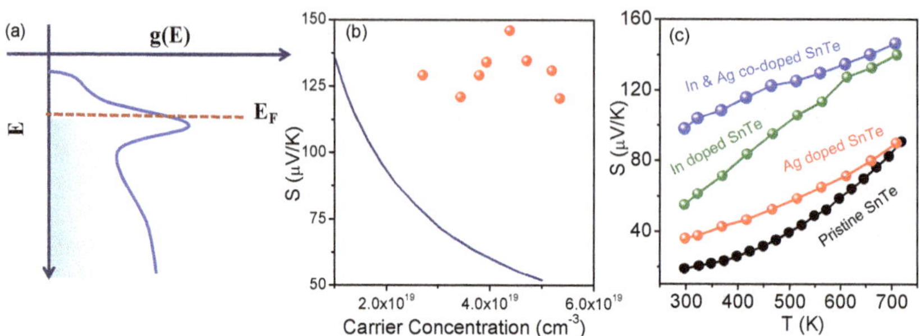

Fig. 3. (a) Schematic diagram of resonance level in the valence band, (b) Comparison of the experimentally observed room temperature *S vs. n* data for various Tl-doped PbTe samples with theoretical Pisarenko plot indicates the presence of resonant states in Tl doped PbTe.[37] (c) Illustration of synergistic effect of valence band convergence and resonance level leads to a significant improvement of Seebeck values in Ag, In co-doped SnTe samples.[38]

Favourable electronic interaction between the host (parent semiconductor) and the guest (dilute impurity) may introduce resonance states near E_F which cause excess DOS and improved band effective mass (m_b^*) near E_F and thereby enhance the Seebeck coefficient of the system (Fig. 3(a)). A fundamental relation between the local change in the density of states and Seebeck coefficient is provided by the famous Mott-equation, which is as follows,[2,38]

$$S = \frac{\pi^2 k_B^2}{3e} T . \frac{d[ln(\sigma(E))]}{dE} \Big| E = E_F \tag{6}$$

where, $\sigma(E)$ is energy-dependent electrical conductivity, strongly interrelated to DOS. Thus, an enhanced Seebeck coefficient can originate from a large $d[ln(\sigma(E))]/dE$ which can be induced by resonance impurities as aforementioned. Joseph P. Heremans[37] et al. first observed that Tl creates resonance levels near the valence band of the PbTe and as a consequence, 2% Tl doping in PbTe gives rise to an enhanced Seebeck coefficient of~ 100 µV/K at 300K and maximum zT ~ 1.5 at 773 K. This fact is further supported by Pisarenko plot where it is clearly seen that *S vs. n* data of the Tl doped PbTe Sample lie far above the Pisarenko Plot (theoretical S vs. n data) (Fig. 3(b)). Similarly, In doping in SnTe and

GeTe are known to create resonant level near the valence band edge.[10,39,40]

2.4. *Synergistic approach*

The introduction of the resonance level near Fermi Level is one of the efficient strategies to enhance the S value at room temperature by increasing the local DOS effective mass (m_b^*), however the effect of the resonance level becomes less prominent with increasing temperature because of the diminished resonant scattering. Interestingly, the effect band convergence becomes effective at an elevated temperature where the energy offset between two valence bands decrease, thereby increasing the valley degeneracy (N_V) and an effective enhancement of S value. Thus it is believed that the introduction of the synergistic effect of resonant levels and band convergence in a material via co-doping can enable us to achieve an enhanced Seeebeck coefficient and a high thermoelectric performance of the materials over a broad temperature range.[17,19] This concept has been successfully realized in Ag and In co-doped SnTe samples where In acts as resonance impurity to improve the S value at room temperature and Ag doping facilitates valence band convergence at high temperature (Fig. 3(c)).[38] This synergistic approach has also been proven to be influential for enhancing the thermoelectric efficiency of GeTe via Sb and In co-doping which resulted in a significantly high zT_{max} of ~2.3 at 680 K and zT_{avg} of ~1.6 in 300-723 K temperature range.[41]

3. Thermal Conductivity Minimization

An avant-garde approach to high performance thermoelectrics is by reducing the thermal conductivity of the solids.[42] In general, for solids, κ_{total} mainly constitutes of three parts, *viz.*, the electronic thermal conductivity (κ_{el}), the lattice thermal conductivity (κ_{lat}), and the bipolar thermal conductivity (κ_b). With the aid of Wiedemann–Franz–Lorenz law, $\kappa_{el} = L\sigma T$, for the electronic thermal conductivity, zT can be re-written as

$$zT = \frac{S^2}{L}\left[\frac{1}{1+\frac{\kappa_{lat}+\kappa_b}{\kappa_{el}}}\right].$$

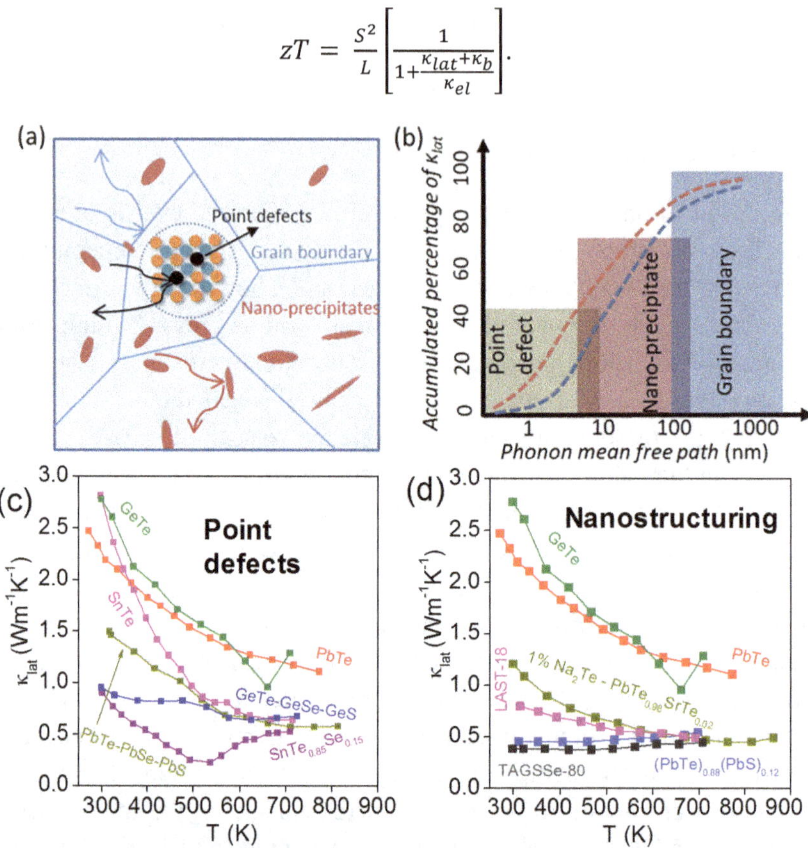

Fig. 4. (a) Schematic diagram describes all-scale hierarchical architecture to scatter phonon[42], (b) accumulated contributions to reduce lattice thermal conductivity with respect to phonon mean free path in PbTe.[42] Typical examples where the suppression of κ_{lat} was made possible using (c) solid-solution alloying[6, 10, 43] and (d) nanostructuring.[12–14, 44-46] Fig. 4(a) and 4(b) reproduced with the permission from Ref 42. © 2018, Nature Publishing Group.

Here, L is the Lorenz number, and the values mostly lie in the range of 1.6 to 2.5 × 10^{-8} V^2 K^{-2} for degenerate semiconducors. Materials exhibiting low κ_{lat}, and low κ_b is paramount in order to achieve high TE performance, while a high κ_{el} augurs well as it corresponds to high power factor.[2] Contributions from κ_b comes only at elevated temperatures and in very narrow-gap semiconductors, being almost negligible in room temperature conditions. Thus, exploring new strategies to maximize the suppression of κ_{lat} is crucial for a better TE performance. For bulk materials, $\kappa_{lat} = 0.33C_v v_g\,\tau$ provides proper description to the lattice thermal conductivity. Thus, to minimize the κ_{lat}, one has to engineer materials which decrease the specific heat (C_V), the group velocity (v_g) or the phonon relaxation time (τ). Typically, in thermoelectrics, the phonon relaxation time is the most focused variable to tweak and achieve the desirable κ_{lat} via extrinsically introducing 0D point defects, 1D dislocations or 2D grain boundaries or fine precipitates (Fig. 4(a)). Each of these defects enhances the phonon-scattering process and decreases the relaxation time (τ) and thereby decreasing the κ_{lat}.[42,47] Each of the aforementioned processes has their own frequency (ω) dependence (Fig. 4(b)). For example, the 0D point defects scatters the high frequency phonons ($\tau_{PD} \sim \omega^{-4}$); 1D dislocation scatters the mid frequency phonons ($\tau_{DC} \sim \omega^{-3}$ for dislocation cores and $\tau_{DS} \sim \omega^{-1}$ for dislocation strains); 2D interface scattering originating from grain boundaries or precipitates are effective for the low frequency phonons ($\tau_{inter} \sim \omega^{0}$).[42] The Umklapp process[48] which is ubiquitous has a relaxation time, $\tau_U \sim \omega^{-2}$, thus being effective in scattering the phonons of all frequencies. Callaway devised a phenomenological model considering all the contributions arising from the microstructural effects on phonon scattering at various length scales.[49] The model which is given as,[49,50]

$$\kappa_{lat} = \frac{k_B}{2\pi^2 v_g}\left(\frac{k_B T}{\hbar}\right)^3 \int_0^{\theta_D/T} \tau_C(x)\frac{x^4 e^4}{(e^x-1)^2}\,dx \tag{7}$$

acts as a guide to quantitatively access the contributions arising from each micro-structural effect. k_B in equation (7) corresponds to Boltzmann's constant; $\hbar$, T and τ_C denotes Plank's constant, absolute temperature and total relaxation time respectively. τ_C corresponds to the individual relaxation time via the relation $\tau_C^{-1} = \tau_U^{-1} + \tau_{PD}^{-1} + \tau_{DS}^{-1} + \tau_{DC}^{-1} + \tau_{inter}^{-1} + \dots$, where τ_U, τ_{PD}, τ_{DS}, τ_{DC} and τ_{inter} corresponds to relaxation times arising from the contributions of Umpklapp scattering, point defects, dislocation strain, dislocation cores and interface scattering respectively.

Apart from the extrinsic approaches to reduce the thermal conductivity, rational unearthing of materials with intrinsically low lattice thermal conductivity is an intriguing and efficient prospect. Since electrons and phonons propagate within the same sublattice, suppressing the phonon transport also handicaps the electron mobility. Thus materials with innate κ_{lat} offer an independent control to achieve high TE performances without having to compromise on the electrical mobility which is beneficial in maintaining high power factor.[21]

In this section, we have touched upon several strategies to reduce the κ_{lat} of the materials using extrinsic approaches such as alloying and nano-structuring as well as the intrinsic approaches which focuses on the enhanced level of anharmonicity arising from lone-pairs of electrons, rattler atoms or presence of soft phonon modes etc.

3.1. *Extrinsic approaches*

Extrinsic approaches refer to manually incorporating foreign elements to reduce the phonon relaxation time. External doping can vaguely result in two possibilities. The first type compromises of a single phase and is generally defined by a cluster of nano-sized particles or grains. We generally term them as solid-solution alloying. The second type comprises of a minor second phase embedded in a bulk matrix, generally in the form of nano-precipitates. Here, the second phase is generally of nanoscale size while the bulk matrix need not be. This second type is generally known as nano-structuring.

3.1.1. *Solid-solutions: Point defect phonon scattering*

A well-known pathway to minimize the lattice thermal conductivity is via introducing lattice imperfections in the form of point defects. Klemens[51] and Callaway[49] developed a thermal conductivity model to evaluate the degree of reduction caused due to mass fluctuations and chemical strain which are generally induced by point defects (Fig. 4(c)). This degree of reduction is depicted using the scattering parameter (Γ) which is given as[52]

$$\Gamma = x(1-x)\left[\left(\frac{\Delta M}{M}\right)^2 + \varepsilon\left(\frac{a_{disorder}-a_{pure}}{a_{pure}}\right)^2\right] \qquad (8)$$

where, a stands for doping fraction, $a_{disorder}$ and a_{pure} correspond to lattice constants of disordered and pure alloys respectively. $\Delta M/M$ is the rate of the change of atomic mass and ε corresponds to the elastic properties which are related to the adjusting parameter. One can clearly conclude from Eq. (8) that to minimize κ_{lat}, one had to maximise Γ. For that we need to either have a higher doping concentration (x); dopant having a contrasting mass difference compared to the host element; or a noteworthy lattice mismatch of the disordered phase with respect to the host phase.

PbTe one of the most promising TE materials for mid temperature range shows a high κ_{lat} of $\sim$ 2.2 $Wm^{-1}K^{-1}$ at room temperature.[53] It has been shown that 25% Se alloying in PbTe dramatically reduces the κ_{lat} to 1.25 $Wm^{-1}K^{-1}$ at 300 K.[53] Although most of the notable binary alloying is governed by enthalpy of the system, Kanatzidis and his co-workers argued that ternary alloying of $(PbTe)_{1-x-y}(PbSe)_x(PbS)_y$ is driven by the configurational entropy.[43] A low κ_{lat} of 0.5 $Wm^{-1}K^{-1}$ has been achieved at 800 K for 2% Na doped $(PbTe)_{0.90}(PbSe)_{0.05}(PbS)_{0.05}$ (Fig. 4(c)). On a similar note, sintered samples of 10% Sb doped $(GeTe)_{0.9}(GeSe)_{0.05}(GeS)_{0.05}$ shows a significant reduction in κ_{lat} (0.7 $Wm^{-1}K^{-1}$ at 728 K), compared to the pristine GeTe sample (Fig. 4(c)).[6] This huge reduction in κ_{lat} is due to the presence of multiple point scattering centres. The contrasting mass difference between the characteristic anions as well as antimony causes mass fluctuations ($\Delta M/M$), the size difference leads to a lattice mismatch and the Spark Plasma Sintering (SPS) enhances the grain boundary scattering. The consequence of all the above-mentioned effects enhances the phonon scattering processes and subsequently decreases the κ_{lat}. Solid solution alloying is although one of the most popular methods to suppress the lattice thermal conductivity of the system, and it affects only the high frequency phonons. To curb the propagation of mid and low frequency phonons, nanostructuring is employed as a popular tool by the TE community.

3.1.2. Nanostructuring

An innovative way to inhibit the transport of phonons having mid and longer wavelength is via introducing nano-scaled defects into the matix. An effective scattering of mid and low frequency phonons would be possible only if the nano-scaled defects are distributed uniformly and are of similar size to these phonons, typically up to to dozens of nanometres.[1,42,54] Quite a few approaches have been undertaken to achieve nanoscale in homogeneity viz. external addition of guest phase via chemical of mechanical mixing,[55] in-situ precipitation of second phase via kinetically or thermodynamically driven processes.[15,54-56] In-situ approach is the widely used due to even dispersion of the nano-precipitates which are also favourable for charge transport.[13,14]

Since the well-known TE materials such as PbTe, SnTe and GeTe have a wide array of phonons with mean free paths of 1–100 nm, nanostructuring is proved to be an effective pathway to reduce the κ_{lat} of these compounds (Fig. 4(d)).[17,57] Replacing Pb in PbTe with two aliovalent atoms (i.e., Ag and Sb) to form $AgPb_mSbTe_{m+2}$ (LAST-m) shows a substantial decrease in the κ_{lat} (Fig. 4(d)).[12] Particularly LAST-18 shows a significant reduction in the κ_{lat} (0.5 $Wm^{-1}K^{-1}$ at 700 K) as compared to PbTe. It has been argued that the formation of nanoprecipitates in the matrix takes place during the cooling period via nucleation and growth events. The nano-inclusions exhibit coherent of incoherent interfaces within the matrix.$(PbTe)_{0.88}(PbS)_{0.12}$ and 1.25% Sb doped PbTe, also shows a very low κ_{lat} due to the formation of nanostructures (Fig. 4(d)).[45-46] Formation of strained endotaxial nanostructures in PbTe- SrTe shows a huge reduction of κ_{lat} (~ 0.45 $Wm^{-1}K^{-1}$ at 823 K) (Fig. 4(d)).[13] Ge analogue of LAST-m, $(GeTe)_x(AgSbTe_2)_{100-x}$ (TAGS-x) systems also exhibit low κ_{lat} (~ 0.8 $Wm^{-1}K^{-1}$) due to the presence of minor second phases like Ag_8GeTe_6, Ag_6GeTe_5, $Ag_2Ge_{8.5}SbTe_{10}$ in TAGS-85 and Ag_3GeTe_2 in TAGS-80.[58] $(GeTe)_x(AgSbSe_2)_{100-x}$ (TAGSSe-x) has currently shown a promising ultralow κ_{lat} of ~ 0.4 $Wm^{-1}K^{-1}$ (Fig. 4(d)) within the measured temperature range (300–700 K) due to Ag_2Te nanodots and lies close to the theoretical κ_{min} of GeTe.[44] Bi_2Te_3 a champion material for low temperature thermoelectric applications, shows a significant drop in the κ_{lat} when alloyed with Sb to form $Bi_{0.5}Sb_{1.5}Te_3$. The ball milled, and hot-pressed samples show low κ_{lat} of 0.6 $Wm^{-1}K^{-1}$ due to 2–10 nm Sb rich nanodots formation.[15]

An intriguing and alternate approach to minimize κ_{lat} is to curb all the phonon propagation frequencies in the form of all-scale hierarchical architectures. Devising a thermoelectric material which combines the point defect, nanostructuring and mesoscale scattering will cover the whole frequency region of the phonon propagation. SPS samples of 2% Na doped PbTe-SrTe (4 mol%) reaches beyond the conventional alloy or nano scattering processes.[14] The system is comprised of all-scale hierarchical structures, and thus simultaneously hinders the phonon propagation from all relevant frequencies, resulting in the κ_{lat} value of $\sim$ 0.4 Wm^{-1}K^{-1} at 873 K. Till date it remains one of the most promising approach to extrinsically reduce the κ_{lat}.

3.2. *Intrinsically low thermal conductivity: role of chemical bonding*

The large lowering κ_{lat} through all scale hierarchical architectures amounts to a relatively poorer engineering at an extrinsic level and is generally difficult to achieve without a trade-off between electron transport. A careful selection is required of the scattering centres to ensure phonon propagation is suppressed while the mobility remains unaffected. In contrast, an intrinsic approach to obtain low κ_{lat} is seen as a fascinating route to overcome such interdependent issues (Fig. 5(a)).[21] In the past several years, great success has been achieved in seeking and understanding the fundamentals behind the cause new intrinsically low κ_{lat} materials, for instance, Zintl compounds,[60,64,65] Cu_2Se,[66] AgCuX (X= S, Se, Te),[61,62,67] BiSe,[7] $AgCrSe_2$,[68] I-V-VI$_2$ chalcogenides,[63,69,70] SnSe,[71,72] MgAgSb[73] etc. The intrinsically low κ_{lat} of all the above-mentioned TE materials is based on the bonding environment among the constituent atoms and varies from system to system. Here, we will be discussing the various reasons that lead to the intrinsically low κ_{lat} in these groups of TE materials. Above the Debye temperature, if phonon–phonon Umklapp scattering predominates the phonon transport, then the κ_{lat} is expressed as,[69]

$$\kappa_{lat} = A\frac{\bar{M}\theta_D^3\delta}{\gamma^2 N^{2/3}T} \qquad (9)$$

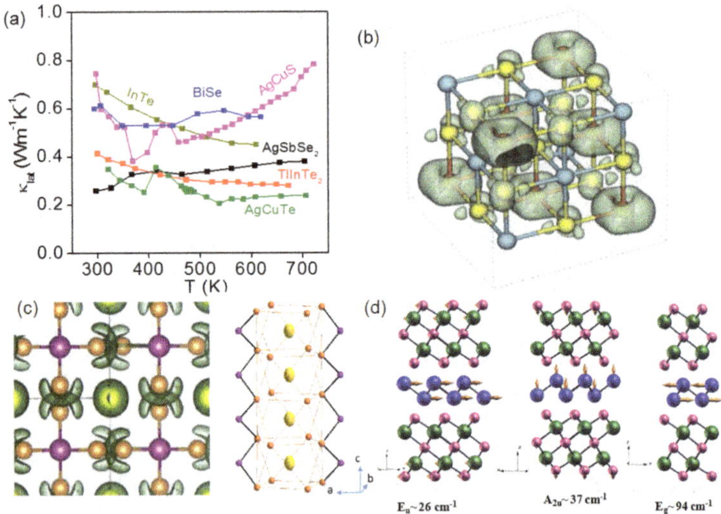

Fig. 5. (a) Typical examples of compounds showing intrinsically low κ_{lat} such as BiSe,[7] InTe,[59] TlInTe₂,[60] AgCuS,[61] AgCuTe,[62] AgSbSe₂[63]. (b) Schematic showing the presence of lone pairs on cations inducing lattice anharmonicity; (c) electron localization function diagram of TlInTe₂ (left). The yellow spheres represent Tl and show no strong bonding interactions while violet and orange represent In and Se atoms respectively, showing strong covalent interaction. Right side of the figure shows the ADP of Tl being anisotropically directed in z-direction indicating its rattling nature towards c-axis. (d) Localized vibrations of Bi-bilayer in BiSe with their corresponding eigen vectors. Fig. 5(b) reproduced with permission from Ref. 70. © 2013, The Royal Society of Chemistry. Fig. 5(c) reproduced with permission from Ref. 60. © 2017, American Chemical Society. Fig. 5(d) reproduced with permission from Ref. 7. © 2018, American Chemical Society.

where a corresponds to collection of physical constants, $\overline{M}$ is the average mass of the atoms in the crystal; θ_D and δ^3 corresponds to Debye temperature volume per atom respectively. N is the number of atoms in the primitive unit cell, and γ is Grüneisen parameter which indicates towards the anharmonicity of materials. The formula predicts that for a material possessing intrinsically low κ_{lat}, must accompany with it certain structural features *viz.* complex cell, low Debye temperature, strong anharmonicity and heavy constituent elements.[69,47] The presence of a large Grüneisen parameter indicates strong lattice anharmonicity, which leads to strong phonon–phonon interactions, and hence-forth possessing intrinsically low κ_{lat} in several materials. Compounds such as I-V-VI$_2$ chalcogenides[69,74,75] (I = Cu/Ag/Alkali metals, V = Sb/Bi and VI = chalcogens) show a high degree of lattice anharmonicity due to the presence of ns^2 lone pair of electrons on group V elements (Fig. 5(b)) which lowers their κ_{lat} For a material to possess strong anharmonicity, a large nonlinear dependence of the restoring force on the atomic displacement is essential. This usually depends directly on the atoms and their strength of bonding with the nearest neighbours. Usually atoms having large coordination numbers (i.e., atoms with many neighbours) show greater degree of anharmonicity. AgSbSe$_2$ which attains a rock-salt structure exhibits high Gruneisen value of 3.7 and a resultant low $\kappa_{lat}<$ 0.4 W/mK at 300 K.[74] The high Gruneisen value and corresponding low κ_{lat} can be ascribed to the cation site disorder and presence of ns^2 lone pairs of electrons on Sb which induces strong lattice anharmonicity. Similar cases of low κ_{lat} arising due to anharmonicity and cation disorder are also found in AgSbTe$_2$.[76,77]

Alternatively rattling of guest atoms in hollow cages of Skutterudites[78,79] and Clathrates[80] are also known to decrease κ_{lat} in compounds. Zintl compounds like TlInTe$_2$[60], InTe[59] and Tl$_3$VSe$_4$[65] also have cations rattlers which scatters the acoustic phonons thereby decreasing the κ_{lat} of the compounds. The characteristic feature in all these compounds lies in their structure. A case in point, TlInTe$_2$ crystallizes in tetragonal (I4/mmm) structure and comprises of bonding hierarchy.[60] In^{+3} covalently bonds with Te^{2-} forming anionic chains of $(InTe_2)_n^{-n}$ while being electro-statically interlocked with the Tl$^+$ chains. The presence of the bonding hierarchy and the rattling motion of Tl$^+$ enhances the phonon scattering process (Fig. 5(c)) and ultimately lowers the κ_{lat} of these Zintl compounds.

Cation disordered compounds such as Cu$_{2-\delta}$X (X= S, Se)[66,81] and their derivatives AgCuX (X= S, Se, Te) [61,62,67] also show a low κ_{lat} due to their "phonon glass electron crystal" (PGEC) properties at high temperatures. The characteristic feature about them is the co-existence of rigid anion sublattice along with dynamic cation sublattice. The liquid like movements of the cations lead to thermal damping while the crystalline sublattice of the anions promotes facile electron transfer, thus an effective decoupling of phonons and electrons takes place which is paramount for high efficient TE properties. Similarly, $AgCrSe_2$ also shows similar liquid like thermal damping effects where dynamical movements of Ag suppress the transverse acoustic (TA) modes and thereby possessing intrinsically low κ_{lat}.[68]

BiSe, a weak topological insulator, exhibits ultralow κ_{lat} of ~ 0.6 W/mK (at 300 K) where as Bi_2Se_3, strong TI shows a comparatively high κ_{lat} of ~ 1.8 W/mK at 300 K, inspite of the fact that both the compound belongs to the same layered homologous family $(Bi_2)_m(Bi_2Se_3)_n$.[7] The crystal structure of BiSe differs from that of Bi_2Se_3 by a bismuth bilayer (Bi_2) which is sandwiched between two Bi_2Se_3 quintuple [Se-Bi-Se-Bi-Se] layers. Theoretical calculations of phonon dispersion and experimentally measured low-temperature heat capacity recognize that localized vibrations of Bi-bilayer produce soft-optical phonons (Fig. 5(d)) which strongly couple with the heat carrying acoustic phonons, inhibit their propagation, resulting in ultralow κ_{lat} in BiSe.[7]

SnSe, a layered material is garnering rapid recognition due to the complimentary record zT in both conductions (2.6 for p-type and 2.8 for n-type conduction).[71,72,82,83] This huge increment in the zT is a consequence of layer driven anisotropic mobility, which results in a high electrical conductivity and the subsequent power factor towards b-axis as well as highly anharmonic bonds which results in ultralow thermal conductivity. This anharmonicity stems from a long-range resonant network constituting of Se p-states coupled to stereochemically active Sn $5s^2$ lone pairs, and to the zigzag accordion-like shape of the slabs along the b–c plane.[25] These bonds are associated with a soft-mode lattice instability that results in a particular temperature dependence of the force constants and low κ_{lat} (< 1 $Wm^{-1}K^{-1}$) at room temperature.

Similarly, other layered compounds such as bismuth chalcogenides[84] and intergrowth compounds like $SnBi_2Te_4$, $SnBi_4Te_7$, $PbBi_2Te_4$ etc.[85-87] show ultralow κ_{lat} due to enhanced phonon scattering at the layer interfaces and weak van der Waals attraction between the layers.

Based on the above mentioned strategies to optimize thermoelectric performance of a material, we have summarized several state of art thermoelectric materials in Fig. 6.

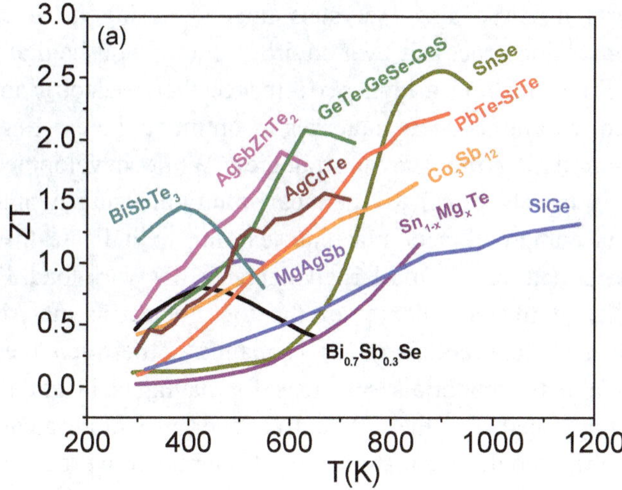

Fig. 6. (a) zT of state of art thermoelectric materials such as GeTe,[6] BiSe,[7] PbTe,[14] Bi$_2$Te$_3$,[15] SnTe,[24] AgCuTe,[62] MgAgSb,[73] AgSbTe$_2$,[77] CoSb$_3$,[79] SnSe,[82] SiGe.[88]

4. Conclusions and Outlook

Due to the low efficiency of market based thermoelectrics, its outreach is limited only to niche applications. But in the past decade the thermoelectric community has seen an unprecedented advancement in this field. The advent of more reliable and robust measurement techniques and synthesis of new materials have led to better efficiency and performances and can be integrated in daily energy conversion processes. Furthermore, the improved rationality of the synthetic solid-state chemists to harness new compounds with desirable properties also paved the way in the development. New and exciting strategies are being processed and developed to reduce the interrelation between the parameters governing zT. Strategies like band engineering, slight symmetry reduction and resonant bonding are well understood and used now for increasing the power factor whereas, all-scale hierarchical approaches decrease the thermal conductivity of the compounds. Taking advantage of the anisotropy, the 2D layered compound have seen an unprecedented zT close to magical number 3 in both *p*-type and *n*-type counterparts in SnSe single crystals thus favouring its integration to energy conversion processes over environmentally non-benign PbTe.

Thus, for furnishing a high performance thermoelectric material we either need to compose new materials or optimize the existing materials by the assimilating the known strategies. While development of new materials is largely down to chemical intuition, using a pre-screening process via computation or machine learning to indicate towards high performance materials would largely decrease the work load. Finally, the bottleneck and the complexity of the thermoelectric materials would suggest that a close collaboration is mandated between the chemists, physicists and the materials scientists for having a bright future. The ever-growing need for the alternative energy would mean that the reliance on thermoelectric materials to efficiently convert energy is more than ever.

Acknowledgments

We would like to acknowledge the support of JNCASR, Sheikh Saqr Laboratory and DST. MS and MD thank UGC, for fellowship.

References

1. G. Tan, L. D. Zhao and M. G. Kanatzidis, Rationally designing high-performance bulk thermoelectric materials. *Chem. Rev.* **116**(19), 12123–12149, (2016).

2. J. R. Sootsman, D. Y. Chung and M. G. Kanatzidis, New and old concepts in thermoelectric Materials. *Angew. Chemie Int. Ed.* **48**(46), 8616–8639, (2009).

3. Z. H. Ge, L. D. Zhao, D. Wu, X. Liu, B. P. Zhang, J. F. Li and J. He, Low-cost, abundant binary sulfides as promising thermoelectric materials. *Mater. Today* **19**(4), 227–239, (2016).

4. L.-D. Zhao, V. P. Dravid and M. G. Kanatzidis, The panoscopic approach to high performance thermoelectrics. *Energy Environ. Sci.* **7**(1), 251–268, (2014).

5. S. Perumal, S. Roychowdhury, D. S. Negi, R. Datta and K. Biswas, High thermoelectric performance and enhanced mechanical stability of p-type $Ge_{1-x}Sb_xTe$. *Chem. Mater.* **27**(20), 7171–7178, (2015).

6. M. Samanta and K. Biswas, Low thermal conductivity and high thermoelectric performance in $(GeTe)_{1-2x}(GeSe)_x(GeS)_x$: Competition between solid solution and phase separation. *J. Am. Chem. Soc.* **139**(27), 9382–9391, (2017).

7. M. Samanta, K. Pal, P. Pal, U. V. Waghmare and K. Biswas, Localized vibrations of Bi bilayer leading to ultralow lattice thermal conductivity and high thermoelectric performance in weak topological insulator n-type BiSe. *J. Am. Chem. Soc.* **140**(17), 5866–5872, (2018).

8. A. Soni, Y. Shen, M. Yin, Y. Zhao, L. Yu, X. Hu, Z. Dong, K. A. Khor, M. S. Dresselhaus and Q. Xiong, Interface driven energy filtering of thermoelectric power in spark plasma sintered $Bi_2Te_{2.7}Se_{0.3}$ nanoplatelet composites. *Nano Lett.* **12**(8), 4305–4310, (2012).

9. J. H. Bahk, Z. Bian and A. Shakouri, Electron energy filtering by a nonplanar potential to enhance the thermoelectric power factor in bulk materials. *Phys. Rev. B - Condens. Matter Mater. Phys.* **87**(7), 075204, (2013).

10. A. Banik and K. Biswas, Lead-free thermoelectrics: promising thermoelectric performance in p-type $SnTe_{1-x}Se_x$ system. *J. Mater. Chem. A* **2**, 9620, (2014).

11. S. Roychowdhury, U. S. Shenoy, U. V Waghmare and K. Biswas, Tailoring of electronic structure and thermoelectric properties of a topological crystalline insulator by chemical doping. *Angew. Chemie - Int. Ed.* **54**(50), 15241–15245, (2015).

12. K. F. Hsu, S. Loo, F. Guo, W. Chen, J. S. Dyck, C. Uher, T. Hogan, E. K. Polychroniadis and M. G. Kanatzidis, Cubic $AgPb_mSbTe_{2+m}$: bulk thermoelectric materials with high figure of merit. *Science.* **303**(5659), 818–821, (2004).

13. K. Biswas, J. He, Q. Zhang, G. Wang, C. Uher, V. P. Dravid and M. G. Kanatzidis, Strained endotaxial nanostructures with high thermoelectric figure of merit. *Nat. Chem.* **3**(2), 160–166, (2011).

14. K. Biswas, J. He, I. D. Blum, C.-I. Wu, T. P. Hogan, D. N. Seidman, V. P. Dravid and M. G. Kanatzidis, High-performance bulk thermoelectrics with all-scale hierarchical architectures. *Nature* **489**(7416), 414–418, (2012).

15. B. Poudel, Q. Hao, Y. Ma, Y. Lan, A. Minnich, B. Yu, X. Yan, D. Wang, A. Muto, D. Vashaee, X. Chen, J. Liu, M. S. Dresselhaus, G. Chen and Z. Ren, High-thermoelectric performance of nanostructured bismuth antimony telluride bulk alloys. *Science.* **320**(5876), 634–638, (2008).

16. S. Il Kim, K. H. Lee, H. A. Mun, H. S. Kim, S. W. Hwang, J. W. Roh, D. J. Yang, W. H. Shin, X. S. Li, Y. H. Lee, G. J. Snyder and S. W. Kim, Dense dislocation arrays embedded in grain boundaries for high-performance bulk thermoelectrics. *Science.* **348**(6230), 109–114, (2015).

17. S. Roychowdhury, M. Samanta, S. Perumal and K. Biswas, Germanium chalcogenide thermoelectrics: electronic structure modulation and low lattice thermal conductivity. *Chem. Mater.* **30**(17), 5799–5813, (2018).

18. S. Perumal, S. Roychowdhury and K. Biswas, High performance thermoelectric materials and devices based on GeTe. *J. Mater. Chem. C* **4**(32), 7520–7536, (2016).

19. A. Banik, S. Roychowdhury and K. Biswas, The journey of tin chalcogenides towards high-performance thermoelectrics and topological materials. *Chem. Commun.* **54**(50), 6573–6590, (2018).

20. R. Moshwan, L. Yang, J. Zou and Z.-G. Chen, Eco-friendly SnTe thermoelectric materials: progress and future challenges. *Adv. Funct. Mater.* **27**(43), 1703278, (2017).

21. M. K. Jana and K. Biswas, Crystalline solids with intrinsically low lattice thermal conductivity for thermoelectric energy conversion. *ACS Energy Lett.* **3**(6), 1315–1324, (2018).

22. G. J. Snyder and E. S. Toberer, Complex thermoelectric materials. *Nat. Mater.* **7**(2), 105–114, (2008).

23. X. Zhou, Y. Yan, X. Lu, H. Zhu, X. Han, G. Chen and Z. Ren, Routes for high-performance thermoelectric materials. *Mater. Today* , 1–15, (2018).

24. A. Banik, U. S. Shenoy, S. Anand, U. V. Waghmare and K. Biswas, Mg alloying in SnTe facilitates valence band convergence and optimizes thermoelectric properties. *Chem. Mater.* **27**(2), 581–587, (2015).

25. L. D. Zhao, H. J. Wu, S. Q. Hao, C. I. Wu, X. Y. Zhou, K. Biswas, J. Q. He, T. P. Hogan, C. Uher, C. Wolverton, V. P. Dravid and M. G. Kanatzidis, All-scale hierarchical thermoelectrics: MgTe in PbTe facilitates valence band convergence and suppresses bipolar thermal transport for high performance. *Energy Environ. Sci.* **6**(11), 3346, (2013).

26. K. Ahn, M.-K. Han, J. He, J. Androulakis, S. Ballikaya, C. Uher, V. P. Dravid and M. G. Kanatzidis, Exploring resonance levels and nanostructuring in the PbTe–CdTe system and enhancement of the thermoelectric figure of merit. *J. Am. Chem. Soc.* **132**(14), 5227–5235, (2010).

27. Y. Pei, H. Wang, Z. M. Gibbs, A. D. LaLonde and J. G. Snyder, Thermopower enhancement in $Pb_{1-x}Mn_xTe$ alloys and its effect on thermoelectric efficiency. *NPG Asia Mater.* **4**(9), 1–6, (2012).

28. H. Wang, Z. M. Gibbs, Y. Takagiwa and G. J. Snyder, Tuning bands of PbSe for better thermoelectric efficiency. *Energy Environ. Sci.* **7**(2), 804–811, (2014).

29. G. Tan, L.-D. Zhao, F. Shi, J. W. Doak, S.-H. Lo, H. Sun, C. Wolverton, V. P. Dravid, C. Uher and M. G. Kanatzidis, High thermoelectric performance of p-type SnTe via a synergistic band engineering and nanostructuring approach. *J. Am. Chem. Soc.* **136**(19), 7006–7017, (2014).

30. G. Tan, F. Shi, J. W. Doak, H. Sun, L.-D. Zhao, P. Wang, C. Uher, C. Wolverton, V. P. Dravid and M. G. Kanatzidis, Extraordinary role of Hg in enhancing the thermoelectric performance of p-type SnTe. *Energy Environ. Sci.* **8**(1), 267–277, (2015).

31. D. Wu, L. D. Zhao, S. Hao, Q. Jiang, F. Zheng, J. W. Doak, H. Wu, H. Chi, Y. Gelbstein, C. Uher, C. Wolverton, M. Kanatzidis and J. He, Origin of the high performance in GeTe-based thermoelectric materials upon Bi_2Te_3 doping. *J. Am. Chem. Soc.* **136**(32), 11412–11419, (2014).

32. S. Perumal, P. Bellare, U. S. Shenoy, U. V. Waghmare and K. Biswas, Low Thermal Conductivity and High Thermoelectric performance in Sb and Bi codoped GeTe: complementary effect of band convergence and nanostructuring. *Chem.*

Mater. **29**(24), 10426–10435, (2017).

33. Z. Zheng, X. Su, R. Deng, C. Stoumpos, H. Xie, W. Liu, Y. Yan, S. Hao, C. Uher, C. Wolverton, M. G. Kanatzidis and X. Tang, Rhombohedral to cubic conversion of GeTe via MnTe alloying leads to ultralow thermal conductivity, electronic band convergence, and high thermoelectric performance. *J. Am. Chem. Soc.* **140**(7), 2673–2686, (2018).

34. W. Liu, X. Tan, K. Yin, H. Liu, X. Tang, J. Shi, Q. Zhang and C. Uher, Convergence of conduction bands as a means of enhancing thermoelectric performance of n-type $Mg_2Si_{1-x}Sn$ x solid solutions. *Phys. Rev. Lett.* **108**(16), 1–5, (2012).

35. S. Roychowdhury and K. Biswas, Slight symmetry reduction in thermoelectrics. *Chem* **4**(5), 939–942, (2018).

36. J. Li, X. Zhang, Z. Chen, S. Lin, W. Li, J. Shen, I. T. Witting, A. Faghaninia, Y. Chen, A. Jain, L. Chen, G. J. Snyder and Y. Pei, Low-symmetry rhombohedral GeTe thermoelectrics. *Joule* **2**(5), 976–987, (2018).

37. J. P. Heremans, V. Jovovic, E. S. Toberer, A. Saramat, K. Kurosaki, A. Charoenphakdee, S. Yamanaka and G. J. Snyder, Enhancement of thermoelectric of the electronic density of states. *Science.* **321**, 1457–1461, (2008).

38. A. Banik, U. S. Shenoy, S. Saha, U. V. Waghmare and K. Biswas, High power factor and enhanced thermoelectric performance of $SnTe$-$AgInTe_2$: synergistic effect of resonance level and valence band convergence. *J. Am. Chem. Soc.* **138**(39), 13068–13075, (2016).

39. Q. Zhang, B. Liao, Y. Lan, K. Lukas, W. Liu, K. Esfarjani, C. Opeil, D. Broido, G. Chen and Z. Ren, High thermoelectric performance by resonant dopant indium in nanostructured SnTe. *Proc. Natl. Acad. Sci.* **110**(33), 13261–13266, (2013).

40. L. Wu, X. Li, S. Wang, T. Zhang, J. Yang, W. Zhang, L. Chen and J. Yang, Resonant level-induced high thermoelectric response in indium-doped GeTe. *NPG Asia Mater.* **9**(1), e343–7, (2017).

41. M. Hong, Z. G. Chen, L. Yang, Y. C. Zou, M. S. Dargusch, H. Wang and J. Zou, Realizing zT of 2.3 in $Ge_{1-x-y}Sb_xIn_yTe$ via reducing the phase-transition temperature and introducing resonant energy doping. *Adv. Mater.* **30**(11), 1–8, (2018).

42. Yu. Xiao and L-D. Zhao, Charge and phonon transport in PbTe-based thermoelectric materials. *npj. Quant. Mater.* **3**(1), 55, (2018).

43. R. J. Korkosz, T. C. Chasapis, S. H. Lo, J. W. Doak, Y. J. Kim, C. I. Wu, E. Hatzikraniotis, T. P. Hogan, D. N. Seidman, C. Wolverton, V. P. Dravid and M. G. Kanatzidis, High ZT in p-type $(PbTe)_{1-2x}(PbSe)_x(PbS)x$ thermoelectric materials. *J. Am. Chem. Soc.* **136**(8), 3225–3227, (2014).

44. M. Samanta, S. Roychowdhury, J. Ghatak, S. Perumal and K. Biswas, Ultrahigh average thermoelectric figure of merit, low lattice thermal conductivity and enhanced microhardness in nanostructured $(GeTe)_x(AgSbSe_2)_{100-x}$. *Chem. – A Eur. J.* **23**(31), 7438–7443, (2017).

45. S. N. Girard, J. He, X. Zhou, D. Shoemaker, C. M. Jaworski, C. Uher, V. P. Dravid,

J. P. Heremans and M. G. Kanatzidis, High performance Na-doped PbTe–PbS thermoelectric materials: electronic density of states modification and shape-controlled nanostructures. *J. Am. Chem. Soc.* **133**(41), 16588–16597, (2011).

46. G. Tan, C. C. Stoumpos, S. Wang, T. P. Bailey, L.-D. Zhao, C. Uher and M. G. Kanatzidis, Subtle roles of Sb and S in regulating the thermoelectric properties of n-Type PbTe to high performance. *Adv. Energy Mater.* **7**(18), 1700099, (2017).

47. T. Zhu, Y. Liu, C. Fu, J. P. Heremans, J. G. Snyder and X. Zhao, Compromise and synergy in high-efficiency thermoelectric materials *Adv. Mater.* **29**(14), 1605884, (2017).

48. J. P. Heremans, The anharmonicity blacksmith. *Nat. Phys.* **11**(12), 990–991, (2015).

49. J. Callaway, Model for lattice thermal conductivity at low temperatures. *Phys. Rev.* **113**(4), 1046–1051, (1959).

50. J. He, S. N. Girard, M. G. Kanatzidis and V. P. Dravid, Microstructure-lattice thermal conductivity correlation in nanostructured $PbTe_{0.7}S_{0.3}$ thermoelectric materials. *Adv. Funct. Mater.* **20**(5), 764–772, (2010).

51. P. G. Klemens, Thermal resistance due to point defects at high temperatures. *Phys. Rev.* **119**(2), 507–509, (1960).

52. B. Abeles, Lattice thermal conductivity of disordered semiconductor alloys at high temperatures. *Phys. Rev.* **131**(5), 1906–1911, (1963).

53. Y. Pei, X. Shi, A. LaLonde, H. Wang, L. Chen and G. J. Snyder, Convergence of electronic bands for high performance bulk thermoelectrics. *Nature* **473**(7345), 66–69, (2011).

54. F. Yang and C. Dames, Mean free path spectra as a tool to understand thermal conductivity in bulk and nanostructures. *Phys. Rev. B* **87**(3), 035437, (2013).

55. J. Li, Q. Tan, J.-F. Li, D.-W. Liu, F. Li, Z.-Y. Li, M. Zou and K. Wang, BiSbTe-based nanocomposites with high ZT: the effect of SiC nanodispersion on thermoelectric properties. *Adv. Funct. Mater.* **23**(35), 4317–4323, (2013).

56. M. Samanta, S. Roychowdhury, J. Ghatak, S. Perumal and K. Biswas, Ultrahigh average thermoelectric figure of merit, low lattice thermal conductivity and enhanced microhardness in nanostructured $(GeTe)_x(AgSbSe_2)_{100-x}$. *Chem. - A Eur. J.* **23**(31), 7438–7443, (2017).

57. S. Lee, K. Esfarjani, T. Luo, J. Zhou, Z. Tian and G. Chen, Resonant bonding leads to low lattice thermal conductivity. *Nat. Commun.* **5**, 3525, (2014).

58. J. Davidow and Y. Gelbstein, A comparison between the mechanical and thermoelectric properties of three highly efficient p-type GeTe-Rich compositions: TAGS-80, TAGS-85 and 3% Bi_2Te_3-Doped $Ge_{0.87}Pb_{0.13}Te$. *J. Electron. Mater.* **42**(7), 1542–1549, (2013).

59. M. K. Jana, K. Pal, U. V Waghmare and K. Biswas, The origin of ultralow thermal conductivity in InTe: lone-pair-induced anharmonic rattling. *Angew. Chemie - Int. Ed.* **55**(27), 7792–7796, (2016).

60. M. K. Jana, K. Pal, A. Warankar, P. Mandal, U. V Waghmare and K. Biswas,

Intrinsic rattler-induced low thermal conductivity in Zintl type $TlInTe_2$. *J. Am. Chem. Soc.* **139**, 4350–4353, (2017).

61. S. N. Guin, J. Pan, A. Bhowmik, D. Sanyal, U. V. Waghmare and K. Biswas, Temperature dependent reversible p - N - p type conduction switching with colossal change in thermopower of semiconducting AgCuS. *J. Am. Chem. Soc.* **136**(36), 12712–12720, (2014).

62. S. Roychowdhury, M. K. Jana, J. Pan, S. N. Guin, D. Sanyal, U. V. Waghmare and K. Biswas, Soft phonon modes leading to ultralow thermal conductivity and high thermoelectric performance in AgCuTe. *Angew. Chemie - Int. Ed.* **57**(15), 4043–4047, (2018).

63. S. N. Guin, A. Chatterjee, D. S. Negi, R. Datta and K. Biswas, High thermoelectric performance in tellurium free p-type $AgSbSe_2$. *Energy Environ. Sci.* **6**, 2603–2608, (2013).

64. F. Gascoin, S. Ottensmann, D. Stark, S. M. Haïle and G. J. Snyder, Zintl phases as thermoelectric materials: Tuned transport properties of the compounds $Ca_xYb_{1-x}Zn_2Sb_2$. *Adv. Funct. Mater.* **15**(11), 1860–1864, (2005).

65. S. Mukhopadhyay, D. S. Parker, B. C. Sales, A. A. Puretzky, M. A. McGuire and L. Lindsay, Two-channel model for ultralow thermal conductivity of crystalline Tl_3VSe_4. *Science.* **360**(6396), 1455–1458, (2018).

66. H. Liu, X. Shi, F. Xu, L. Zhang, W. Zhang, L. Chen, Q. Li, C. Uher, T. Day and G. J. Snyder, Copper ion liquid-like thermoelectrics. *Nat. Mater.* **11**(5), 422–425, (2012).

67. S. Ishiwata, Y. Shiomi, J. S. Lee, M. S. Bahramy, T. Suzuki, M. Uchida, R. Arita, Y. Taguchi and Y. Tokura, Extremely high electron mobility in a phonon-glass semimetal. *Nat. Mater.* **12**(6), 512–517, (2013).

68. B. Li, H. Wang, Y. Kawakita, Q. Zhang, M. Feygenson, H. L. Yu, D. Wu, K. Ohara, T. Kikuchi, K. Shibata, T. Yamada, X. K. Ning, Y. Chen, J. Q. He, D. Vaknin, R. Q. Wu, K. Nakajima and M. G. Kanatzidis, Liquid-like thermal conduction in intercalated layered crystalline solids. *Nat. Mater.* **17**(3), 226–230, (2018).

69. D. T. Morelli, V. Jovovic and J. P. Heremans, Intrinsically minimal thermal conductivity in cubic $I\text{-}V\text{-}VI_2$ semiconductors. *Phys. Rev. Lett.* **101**(3), 16–19, (2008).

70. M. D. Nielsen, V. Ozolins and J. P. Heremans, Lone pair electrons minimize lattice thermal conductivity. *Energy Environ. Sci.* **6**(2), 570–578, (2013).

71. C. Chang, M. Wu, D. He, Y. Pei, C.-F. Wu, X. Wu, H. Yu, F. Zhu, K. Wang, Y. Chen, L. Huang, J.-F. Li, J. He and L.-D. Zhao, 3D charge and 2D phonon transports leading to high out-of-plane ZT in n-type SnSe crystals. *Science* **360**(6390), 778–783, (2018).

72. L.-D. Zhao, S.-H. Lo, Y. Zhang, H. Sun, G. Tan, C. Uher, C. Wolverton, V. P. Dravid and M. G. Kanatzidis, Ultralow thermal conductivity and high thermoelectric figure of merit in SnSe crystals. *Nature* **508**(7496), 373–377, (2014).

73. P. Ying, X. Liu, C. Fu, X. Yue, H. Xie, X. Zhao, W. Zhang and T. Zhu, High performance α-MgAgSb thermoelectric materials for low temperature power generation. *Chem. Mater.* **27**(3), 909–913, (2015).

74. S. N. Guin, V. Srihari and K. Biswas, Promising thermoelectric performance in n-type $AgBiSe_2$:effect of aliovalent anion doping. *J. Mater. Chem. A*, **3**(2), 648–655, (2015).

75. S. N. Guin and K. Biswas, Cation Disorder and Bond Anharmonicity Optimize the thermoelectric properties in kinetically stabilized rocksalt $AgBiS_2$ Nanocrystals. *Chem. Mater.* **25**(15), 3225–3231, (2013).

76. J. Ma, O. Delaire, A. F. May, C. E. Carlton, M. A. McGuire, L. H. VanBebber, D. L. Abernathy, G. Ehlers, T. Hong, A. Huq, W. Tian, V. M. Keppens, Y. Shao-Horn and B. C. Sales, Glass-like phonon scattering from a spontaneous nanostructure in $AgSbTe_2$. *Nat. Nanotechnol.* **8**, 445, (2013).

77. S. Roychowdhury, R. Panigrahi, S. Perumal and K. Biswas, Ultrahigh thermoelectric figure of merit and enhanced mechanical stability of p -type $AgSb_{1-x}Zn_xTe_2$. *ACS Energy Lett.*, 349–356, (2017).

78. B. C. Sales, D. Mandrus, B. C. Chakoumakos, V. Keppens and J. R. Thompson, Filled skutterudite antimonides: Electron crystals and phonon glasses. *Phys. Rev. B* **56**(23), 15081–15089, (1997).

79. X. Shi, J. Yang, J. R. Salvador, M. Chi, J. Y. Cho, H. Wang, S. Bai, J. Yang, W. Zhang and L. Chen, Multiple-filled skutterudites: high thermoelectric figure of merit through separately optimizing electrical and thermal transports. *J. Am. Chem. Soc.* **133**, 7837–7846, (2011).

80. T. Takabatake and K. Suekuni, Phonon-glass electron-crystal thermoelectric clathrates: experiments and theory. *Rev. Mod. Phys.* **86**(June), 669–716, (2014).

81. Y. He, T. Day, T. Zhang, H. Liu, X. Shi, L. Chen and G. J. Snyder, High thermoelectric performance in non-toxic earth-abundant copper sulfide. *Adv. Mater.* **26**(23), 3974–3978, (2014).

82. L.-D. Zhao, G. Tan, S. Hao, J. He, Y. Pei, H. Chi, H. Wang, S. Gong, H. Xu, V. P. Dravid, C. Uher, G. J. Snyder, C. Wolverton and M. G. Kanatzidis, Ultrahigh power factor and thermoelectric performance in hole-doped single-crystal SnSe. *Science* **351**(6269), 141–144, (2016).

83. S. Chandra, A. Banik and K. Biswas, n -type ultrathin few-layer nanosheets of Bi-doped SnSe: synthesis and thermoelectric properties. *ACS Energy Lett.* **3**(5), 1153–1158, (2018).

84. J. P. Heremans, R. J. Cava and N. Samarth, Tetradymites as thermoelectrics and topological insulators. *Nat. Rev. Mater.* **2** (10), 17049, (2017).

85. M. G. Kanatzidis, Structural evolution and phase homologies for "design" and prediction of solid-state compounds. *Acc. Chem. Res.* **38**(4), 359–368, (2005).

86. A. Banik and K. Biswas, Synthetic nanosheets of natural van der Waals heterostructures. *Angew. Chemie Int. Ed.* **56**(46), 14561–14566, (2017).

87. A. Chatterjee and K. Biswas, Solution-based synthesis of layered intergrowth compounds of the homologous $Pb_mBi_{2n}Te_{3n+m}$ series as nanosheets. *Angew. Chemie Int. Ed.* **54**(19), 5623–5627, (2015).

88. X. W. Wang, H. Lee, Y. C. Lan, G. H. Zhu, G. Joshi, D. Z. Wang, J. Yang, A. J. Muto, M. Y. Tang, J. Klatsky, S. Song, M. S. Dresselhaus, G. Chen and Z. F. Ren, Enhanced thermoelectric figure of merit in nanostructured n-type silicon germanium bulk alloy. *Appl. Phys.* **8**(12), 4670–4674, (2008).

Chapter 16

Generation of Hydrogen by Water Splitting

Anand Roy, M. Chhetri and C. N. R. Rao[*]

New Chemistry Unit, International Centre for Materials Science and School of Advanced Materials, Jawaharlal Nehru Centre for Advanced Scientific Research, Jakkur, Bangalore 560064, Karnataka, India
[*]*cnrrao@jncasr.ac.in*

Various approaches to accomplish sun-light driven splitting of water to produce high energy density hydrogen fuel have been described in the chapter. The pros and cons of the use of different oxides and non-oxide-based photocatalysts have been discussed. Improving hydrogen generation by means of anion doping, heterostructures, co-catalyst deposition and surface modification have been described. Thermochemical water splitting using two or multi-step approaches and the role of the material involved have been discussed. For enhanced hydrogen production, coupling of electrochemical and photochemical by photoelectrochemical water splitting can be employed. The principles of the various processes and the rationale behind catalyst selection have been examined here in the chapter.

1. Introduction

The need for clean and renewable sources of energy is one of the top concerns of humankind considering the limited stock and undesirable effect of coal-based unrenewable energy resources.[1] Among the various renewable energy sources, hydrogen is promising because of its high energy density and zero environmental hazards.[2] Though H_2 is an ideal choice to be used as a clean source of fuel, one of the challenges lies in its production. Almost 85 to 90% of H_2 is being produced by natural gas reforming which requires high temperature and pressure and results in the evolution of hazardous by-products such as CO and CO_2.[3] It is most

376

desirable to move towards H_2 production using renewable sources of energy or using the abundant natural energy of sunlight. In doing so, natural photosynthesis provides us reason and motivation to transform solar energy into chemical energy. Plants synthesize carbohydrates using natural photosynthesis which is a redox process wherein CO_2 gets reduced to carbohydrates and H_2O gets oxidized to O_2 in a series of events, started by the absorption of sun-light by the chlorophyll pigments. Such a process is accomplished using single or two photosystems (Z-scheme). Several efforts have been made to mimic natural photosynthesis in order to transform solar energy into chemical energy. In this regard, four possible methods can be envisioned which are powered by abundant energy input from the sun: photochemical, thermochemical, photoelectrochemical and electrochemical (PV electrolyzer). The source of hydrogen is water which is in plenty (75% on earth). It should be noted however, that splitting of H_2O into H_2 and O_2 is a thermodynamically uphill reaction (ΔG =237 kJ/mol) and requires careful experimental strategies.

2. Photocatalytic Water Splitting

Photocatalysts are semiconductor materials which absorb solar radiation corresponding to their band gap and generate electrons (e^-) and holes (h^+) in the conduction and valence bands respectively. A suitable semiconductor (more negative conduction band minimum (CBM) and more positive valence band maxima (VBM) than the water redox potential) with electrons and holes results in the production of H_2 and O_2 respectively (Figure 1(a)).[3] Generated charge carriers migrate to the surface of the semiconductor and process the reduction/oxidation of H_2O. The minimum theoretical band gap of a semiconductor photocatalyst should be 1.23 eV. It should be noted that reduction of H_2O is a two-electron process whereas oxidation of H_2O is a four-electron process. Photocatalytic H_2O splitting involves a sequence of event initiated by the absorption of light (Figure 1(b)).[4] Wide band gap semiconductors such as TiO_2, $SrTiO_3$, GaN, ZnO can harvest only ultraviolet (UV) region of the solar spectrum whereas CdS, $CdIn_2S_4$, $ZnIn_2S_4$, Ta_3N_5, $BiVO_4$, WO_3,

C_3N_4 can harvest a large fraction of visible radiation because of their narrow band gap. Since solar radiation available to the earth surface constitutes 45–50% of the visible-light and only 4–5% of ultraviolet-light it is advisable to use semiconductors with visible light harvesting ability.

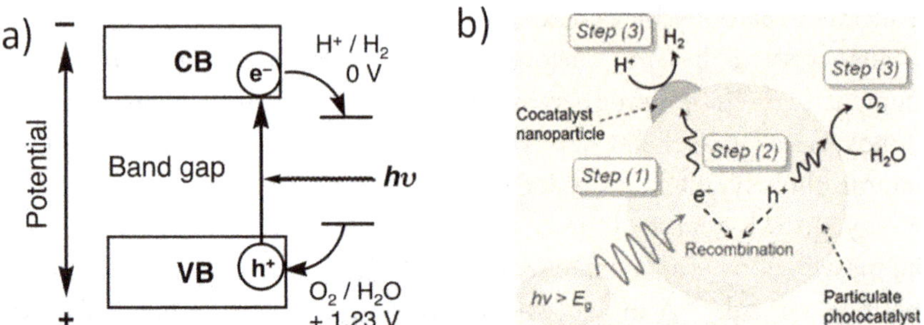

Fig. 1. (a) Principle of photoreduction and oxidation of H_2O using semiconductor, adapted with permission from Ref. 3. Copyright The Royal Society of Chemistry 2009 (b) Process involved in the water splitting. Reproduced with permission from Ref. 4. Copyright 2011 Elsevier B.V.

Various parameters influence the photocatalytic activity of a material: crystallinity, surface area, porosity, particle size, morphology, surface defects, band gap and band edge position.[3] A high crystalline quality of material reduces the number of defect sites which is beneficial for catalysis. Since, photocatalysis is a surface phenomenon design of porous material with higher surface area it is known to enhance the catalysis.

2.1. *Photoreduction of H_2O to produce H_2*

The photoexcited electrons in the CB of semiconductors facilitate the reduction of H_2O. The holes generated in the VB need to be quenched to reduce the electron-hole recombination. There are several strategies to quench the holes; among them the use of sacrificial agents such as Na_2S-Na_2SO_3, ethanol, isopropanol, triethanolamine and lactic acid have been popular.[3]

Oxide semiconductors such as TiO_2, $SrTiO_3$, ZnO, Ga_2O_3 and Ta_2O_5 have been explored for the photoreduction of H_2O, because of their stability in aqueous media.[3] However, most of the metal oxides possess wide band gap which limit their application. Tuning the band gap by means of cation or anion substitution is an important strategy to alter the absorption properties. Doping Fe and other cations in TiO_2 has shown a positive impact on the photocatalytic activity.[3,5] Apart from cation substitution, altering the valence band by anionic substitution is also a known means of photoreduction.

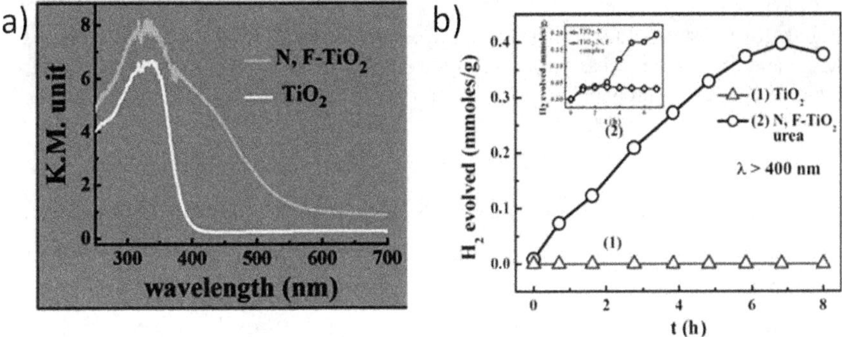

Fig. 2. (a) UV-Visible absorption spectra of undoped and N, F doped TiO_2, (b) Time dependent H_2 generation from undoped and N, F co-doped TiO_2 under visible light irradiation. Reproduced with permission from Ref. 6. Copyright 2013 American Chemical Society.

Rao and coworkers have reported the shift in the absorption edge of TiO_2 from UV to a visible region by the co-doping of N, F.[6] These co-doped samples exhibit superior hydrogen evolution (Figure 2).[6] A similar effect of N, F co-doping in ZnO has also been reported.[7]

Apart from metal oxides, metal sulfides such as CdS and ZnS are well explored materials for photoreduction of H_2O.[3] Visible-light harvesting ability of CdS (band gap of 2.4 eV) and availability of highly reductive electrons make it a very important catalyst for HER. Unlike CdS, wide band gap (3.6 eV) of ZnS, allows only UV-light absorption. Substitution of Zn^{2+} by Cd^{2+} in ZnS, results in the solid solution of $Zn_{1-x}Cd_xS$ wherein the band gap can be tuned from UV-region to the visible region.[8] Although metal sulfides have been intensively studied for HER, a serious drawback of sulfide based materials is photocorossion.[3]

 A. Roy, M. Chhetri & C. N. R. Rao

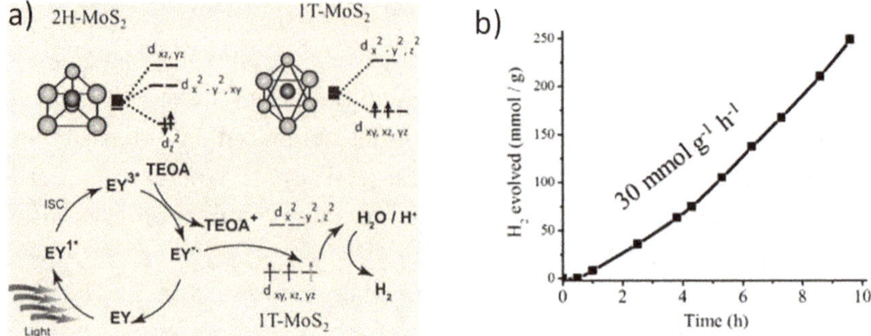

Fig. 3. (a) Crystal-field-splitting induced electronic configuration of 2H-MoS$_2$ and 1T-MoS$_2$ and proposed mechanism for catalytic activity of 1T-MoS$_2$. (b) Time course of H$_2$ evolved by freshly prepared 1T-MoS$_2$. Reproduced with permission Ref. 18. Copyright 2013 Wiley-VCH Verlag GmbH & Co. kGaA, Weinheim.

Photocorossion is a process in which the photocatalyst is degraded by its own photogenerated electrons and holes. There have been various strategies to overcome the problem of photocorossion like the use of sacrificial agents, co-catalysts, and heterostructures.[3,9] A presence of a co-catalyst on the surface of the photocatalyst reduces the charge carriers recombination and lowers the activation energy barrier for H$_2$O reduction.[10] Pt is known to be a very efficient co-catalyst for hydrogen evolution, however, scarcity and high cost limit its use.[10] Various transition metal oxides and sulfide based materials such as NiO, NiS, MoS$_2$, CuS, CuO, etc. have been reported as efficient co-catalysts.[10] In the recent years, transition metal phosphides such as Ni$_2$P, CuP, and Co$_2$P have also come to be recognised as excellent HER co-catalysts.[11] The use of semiconductor based heterostructures are also beneficial for the separation of electrons and holes, resulting in the enhancement of catalytic activity as well as stability. Alivisatos and coworkers have achieved a highly efficient H$_2$ production by the CdSe/CdS/Pt-based nanoheterostructure.[12] A remarkable H$_2$ yield has been reported from the ZnO/Pt/CdS heterostructure (Figure 3).[13] Other than CdS-based semiconductors, there have been reports of ternary semiconductors such as ZnIn$_2$S$_4$, CuGaS$_2$ and CuInS$_2$ for visible-light induced HER.[3] New ternary semiconductors with formula Cd$_4$P$_2$X$_3$ (X= Cl, Br and I) which are analogues of CdS exhibit good HER properties even in the absence of a co-catalyst or a sacrificial agent.[14] Apart from a metal-based photocatalyst, a metal-free graphitic carbon nitride (C_3N_4) has attained enormous attention. Wang et al. have reported steady hydrogen evolution for 75 h by using C_3N_4.[15] Doping non-metals (S, P, B, etc) in C_3N_4 has been considerd to be beneficial.[16] Covalent cross-linking between C_3N_4 and MoS_2, has also been reported to exhibit significant enhancement in the HER activity.[17]

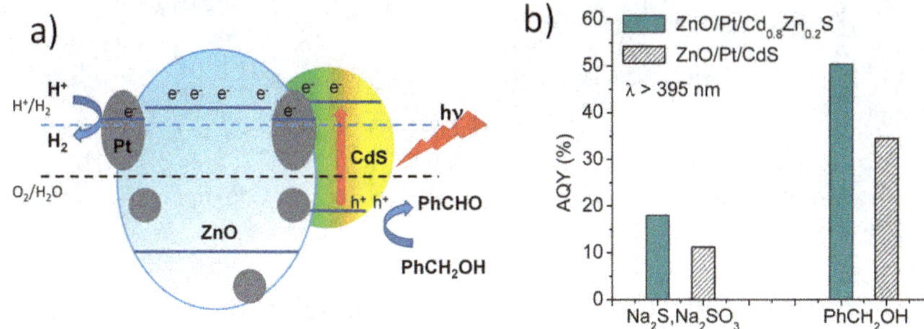

Fig. 4. (a) Mechanism of hydrogen evolution by ZnO/Pt/CdS and (b) apparent quantum yield (% AQY) of ZnO/Pt/$Cd_{1-x}Zn_xS$ in the presence of different sacrificial agents. Adapted with permission from Ref. 13. Copyright The Royal Society of Chemistry 2013.

A different approach is to use dye, as a photosensitizer in the presence of catalytic materials such as TiO_2, MoS_2, $MoSe_2$, graphene etc.[18] Photoexcitation of dye (Eosin Y) generates a singlet excited state (EY^{1*}), which transforms into a triplet state (EY^{3*}) followed by the formation of EY^- after accepting electrons from sacrificial electron donors. This electron migrates to the surface of the catalytic material wherein it reduces the adsorbed H^+ into H_2 (Figure 4).[18] MoS_2 (p-type) composite with N-doped graphene (n-type) significantly enhances HER due to the efficient charge separation.[18] $1T$-MoS_2 is known to be catalytically better than the 2H form. There have been recent reports wherein 2H and 1T form of $MoSe_2$ have been used with Eosin-Y. It is noteworthy to mention that $1T$-$MoSe_2$ showed superior activity than 2H $MoSe_2$ and MoS_2 analogues.[19]

 A. Roy, M. Chhetri & C. N. R. Rao

2.2. *Photooxidation of H_2O to produce O_2*

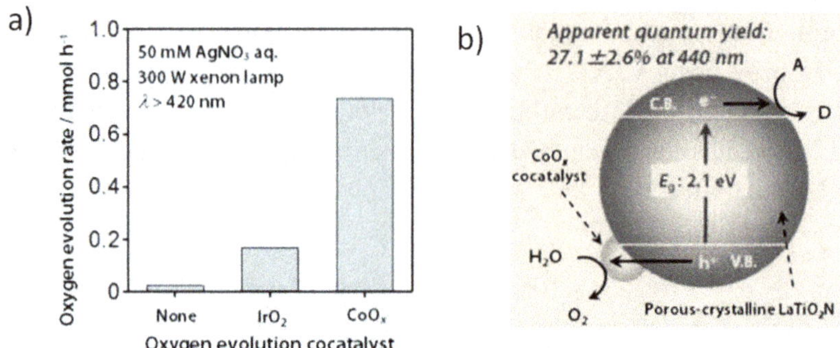

Fig. 5. (a) Oxygen evolution activity of LaTiO2N loaded with IrO2 or CoOx co-catalyst, (b) graphical representation of oxygen evolution from CoOx loaded LaTiO2N under visible light irradiation. Reproduced with permission Ref. 23. Copyright 2012 American Chemical Society.

Photooxidation of H_2O into O_2 is a four-electron process, hence kinetically challenging.[20] In the case of natural photosynthesis, Mn_4CaO_4 complex facilitates the oxidation of water.[20] Semiconductor materials such as $BiVO_4$, WO_3, Fe_2O_3, TaON, $LaTiO_2N$, C_3N_4 etc. possess appropriate VBM position and light harvesting ability to catalyze water oxidation[3] resulting in the generation of charge carriers (electrons and holes). Holes take part in this reaction and therefore electron acceptor sacrificial agents such as $AgNO_3$, $Na_2S_2O_8$, etc. are being used to quench the electrons. Transition metal oxide-based material such as CoO_x, MnO_x, NiO_x, IrO_2, etc. exhibit good co-catalytic activity for oxygen evolution.[14]

In recent times, cobalt phosphate (CoPi) based co-catalyst has received great attention.[3] Can and coworkers have reported the superior performance of the CoPi co-catalyst over a series of H_2O oxidation co-catalysts by depositing them on $BiVO_4$.[21] Kazunari and co-workers have reported stable O_2 production from $TaON/CoO_x$ under visible light irradiation in the presence of $AgNO_3$ as an electron scavenger.[22] Efficient oxidation of water using cobalt-modified $LaTiO_2N$ (band gap 2.1 eV) is reported by the same author which shows a record high AQY of ~ 28 % (Figure 5).[23] Metal-free C_3N_4 also exhibits the potential to perform water oxidation.[24] Rao and coworkers have reported H_2O oxidation using a series of Mn and Co-based compounds wherein the effect of e_g electrons on catalysis has been described.[25]

2.3. *Overall decomposition of water to produce stoichiometric H_2 and O_2*

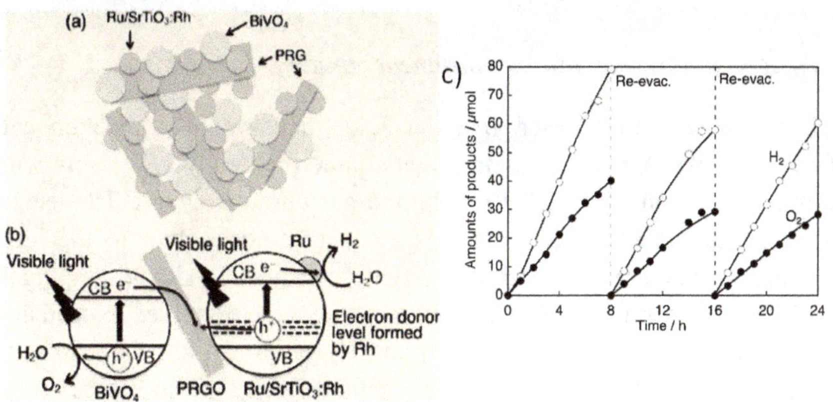

Fig. 6. (a) Schematic of a suspension of Ru/SrTiO₃ and PRGO/BiVO₄ in water (pH 3.5). (b) Mechanism of Z-scheme water splitting under visible light using Ru/SrTiO₃:Rh and PRGO/BiVO₄ (c) Overall water splitting under visible-light irradiation using (Ru/SrTiO₃: Rh)-(PRGO/BiVO₄) system. Reproduced with permission from Ref. 29. Copyright 2011 American Chemical Society.

There have been limited reports on overall water splitting. Since photogenerated charge carriers are involved in the reduction and oxidation of H_2O, sacrificial agents are not needed. There are only a few semiconductor materials such as TiO_2, Ge_3N_4, C_3N_4, TiON, $(Ga_{1-x}Zn_x)$-$(N_{1-x}O_x)$ and InGaN, which are known to perform overall H_2O splitting. Among these materials TiO_2 and Ge_3N_4 work under UV-light.[3]One of the initial reports on overall H_2O splitting by Kazunari and co-workers involves the use of solid solution of gallium nitride-zinc oxide $((Ga_{1-x}Zn_x)$-$(N_{1-x}O_x))$.[26] The design of CoP and Pt dual co-catalyst on C_3N_4 exhibited overall decomposition of H_2O under visible light.[27] Not long ago, Kibria *et al.* have demonstrated an overall H_2O splitting under 400-475 nm irradiation, using Mg-doped InGaN metal wires with a quantum yield of 12.3%.[28] Apart from the use of single semiconductors, there have been strategies to use two semiconductor-based Z-scheme systems, wherein oxidation and reduction take place at different semiconductor surfaces.[3] Rose and co-workers have used $Ru/SrTiO_3$:Rh and $BiVO_4$ as the H_2 and O_2 evolution photocatalysts respectively wherein photoreduced graphene oxide (pRGO) has been used as an electron mediator (Figure 6).[29]

2.4. *Reaction set-up and measurements units*

The most commonly used reaction setup consists of a reaction cell (quartz or glass), vacuum pump, light source (generally a Xe-lamp with power in the range of 300–400 W), and gas chromatograph. The use of solar simulator as a light source is also in practice. Prior to the light irradiation, reaction cell with photocatalyst dispersion should be purged with the inert gas (N_2 or Ar) in order to remove the dissolved oxygen and the should be air tight. Evolved O_2 and H_2 gases can also be measured using gas sensors or volumetric methods. The performance of a photocatalyst can be reported using different parameters.

$$\text{Turn over number} = \frac{Number\ of\ reacted\ electrons}{Number\ of\ atoms\ at\ the\ surface\ of\ catalyst}$$

$$\text{Apparent quantum yield (\%)} = 2 \times \frac{Number\ of\ reacted\ electrons}{Number\ of\ incident\ photons} * 100$$

3. Thermochemical Water Splitting

Another strategy to convert solar energy into chemical energy is the thermochemical water splitting which involves the use of solar concentrators. The ease of reduction at an elevated temperature followed by the reversible oxidation are the required properties of a material for its application in the thermochemical splitting of H_2O and CO_2. Direct (one step) thermochemical water splitting is not favored because of a large energy requirement (T>2500 K), however, it can be made feasible by the use of two or multi-step strategy.

3.1. *Two-step thermochemical approach*

The two-step approach involves an endothermic step (T_{RED}). Metal oxide (MO_{OXD}) reduces to metal or lower valent metal oxide (MO_{RED}), with the release of O_2 (g), followed by the re-oxidation (T_{OXD}) of the metal oxides upon reaction with H_2O (g), leading to release of the stoichiometric amount of H_2 (g).[30]

$$\text{Endothermal step: } MO_{OXD} \rightarrow MO_{RED} + \tfrac{1}{2} O_2$$
$$\text{Exothermal step: } MO_{RED} + H_2O \rightarrow MO_{OXD} + H_2$$

$T_{RED} > T_{OXD}$ is a driving force to make this approach feasible. It should be noted that splitting of CO_2 is analogous to the splitting of H_2O, which releases a stoichiometric amount of CO. Two-step cycles can involve stoichiometric (Fe_3O_4/FeO cycle) or non-stoichiometric ($CeO_2/CeO_{2-\delta}$) paths.[30]

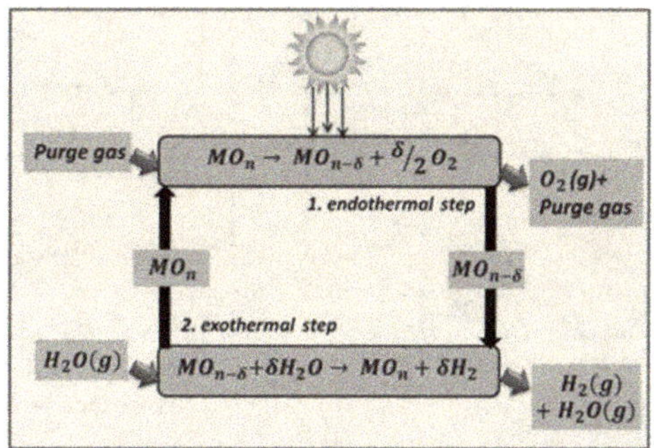

Fig. 7. Two-step solar thermochemical splitting of H_2O using nonstoichiometric metal oxide redox pairs. Reproduced with permission from Ref. 30. Copyright PNAS 2017.

Though stoichiometric path produces more amounts of H_2 but sintering and melting of materials due to the high operating temperature (Fe_2O_3/FeO; 2500 K) limits its application. Substitution of divalent cations such as Zn, Ni, Co and Mn in Fe_2O_3 has been reported to lower the operational temperature.[30] Incorporation of NiO in Fe_2O_3 is known to greatly enhance the performance of this cycle.[30] In the case of nonstoichiometric process (Figure 7), step 1 is involved in the generation of nonstoichiometric oxide by thermal reduction which is followed by step 2 wherein oxygen vacancies are compensated by the reaction of H_2O or CO_2.[30] The oxygen exchange is driven by nonstoichiometry which finally influences the H_2 yield and solar to fuel conversion efficiency ($\eta_{solar/fuel}$).

3.1.1. *Based on CeO₂ and perovskites*

CeO_2 is the most explored material for this operation which involves a reduction temperature of 2273 K. Such high operational temperature for CeO_2/Ce_2O_3 results in the sublimation of CeO_2. Doping trivalent (La^{3+}, Cr^{3+}) and divalent (Ca^{2+}, Mg^{2+}) metal ions enhances the performance of CeO_2.[30]Among various tetravalent dopants, doping of Zr^{4+} in CeO_2, is known to affect reduction capability to a larger extent because of its small size.[30]

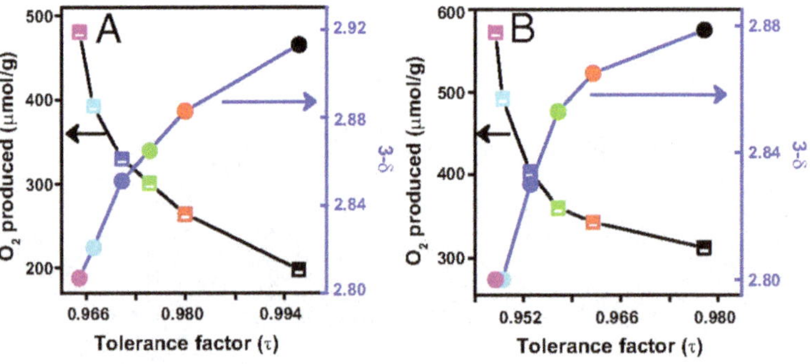

Fig. 8. Variation of O_2 production and O_2 nonstoichiometry generated (3-δ) at 1,400 °C reduction temperature as a function of tolerance factor of (a) $Ln_{0.5}Sr_{0.5}MnO_3$ and (b) $Ln_{0.5}Ca_{0.5}MnO_3$ (Ln = La, Nd, Sm, Gd, Dy, Y). Black arrows show production of O_2 with τ of perovskites; blue arrows show O_2 nonstoichiometry. Reproduced with permission from Ref. 30. Copyright PNAS 2017.

A large number of perovskite oxide ($ABO_{3-\delta}$) materials show oxygen non-stoichiometry. $La_{1-x}Sr_xMO_3$ (M= Mn, Fe: x= 0-1) has been explored for the thermochemical syn-gas production.[31] $La_{1-x}Sr_xMnO_3$ (LSM) family of perovskites have been used for the thermochemical reduction of H_2O and CO_2.[31] In LSM perovskites, substitution of trivalent Ln^{3+} by a divalent A^{2+} creates Mn^{3+}/Mn^{4+} redox pair which assists the water splitting. $La_{0.65}Sr_{0.35}MnO_3$ reduces to the larger extent than CeO_2 does under the same pO_2 and temperature.[30] $La_{1-x}Ca_xMnO_3$ (LCM) (x = 0.35, 0.5 or 0.65) perovskites have been reported to show better activity than LSM-based systems.[30] The tolerance factor of LaCM50 (τ = 0.978) is lesser than that of LaSM50 (τ = 0.996). The superior activity of LaCM50 over LaSM50 is related to the smaller tolerance factor (τ) due to smaller size of Ca^{2+}. Lower tolerance factor causes greater structural distortion, resulting in the increased O_2 evolution. Similarly the effect of τ has been investigated in the case of $Ln_{0.5}Ca_{0.5}MnO_3$ and $Ln_{0.5}Sr_{0.5}MnO_3$ (Ln = La, Nd, Sm, Gd, Dy and Y) (Figure 8).[30]

3.2. *Low temperature multiphase cycle*

Of late, transition metal oxide-based multistep cycles have received significant attention because of the need of low-temperature process. Mn_3O_4/MnO thermochemical cycle proposed by Davis and coworkers involves a series of reactions as shown in Figure 9.[30] It is noteworthy that the production of H_2 occurs at 850 °C. Use of Na_2CO_3 greatly decreases the thermodynamic barrier of Mn_3O_4 to Mn_2O_3 conversion and results in the formation of MnO and α-$NaMnO_2$ at 850 °C. Introduction of H_2O (g) at 850 °C oxidizes MnO to α-$NaMnO_2$ with the evolution of H_2 (g). It has been shown that the rate of H_2 evolution is enhanced to a larger extent by the use of high surface area nanoparticles.[30]

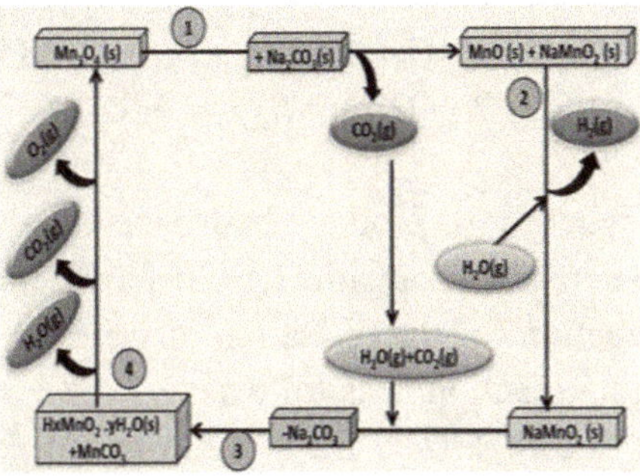

Fig. 9. Schematic of Mn(II)/Mn(III) based low temperature multistep thermochemical cycle. Adapted with permission from Ref. 30. Copyright PNAS 2017.

4. Electrochemical Water Splitting

Water splitting through electrolysis can also be achieved via electrochemical reduction of hydronium and hydroxide ions in H_2O.[32-33] For the evolution of H_2 the source of hydrogen depends on the acidic or alkaline medium used.

$$2H^+(aq) + 2e^- \rightarrow H_2(gas) \qquad (acidic)$$
$$2H_2O + 2e^- \rightarrow 2OH^-(aq) + H_2(gas) \ (alkaline)$$

Unlike photochemical and thermochemical methods, an additional input energy in the form of electricity (applied voltage or current) is needed in electrochemical water splitting. However if the input energy comes from the solar powered photovoltaic cells then electrolysis can be rendered very cheap and hydrogen production at a large scale is possible.

4.1. *Electrochemistry of water splitting reaction*

The Gibbs free energy associated with water splitting is $\Delta G^0 = +237$ kJ/mol. This corresponds to the redox potential of $\Delta E^0 = 1.23$ V [$\Delta G^0 = -nF\Delta E^0$]. The half reactions at cathode and anode can be represented as,

In acidic medium,
$$2H^+ + 2e^- \rightarrow H_2 \qquad E^0_{red} = 0.000\,V \qquad vs\ NHE$$
$$2H_2O + 4h^+ \rightarrow O_2 + 4H^+ \qquad E^0_{ox} = 1.229\,V \qquad vs\ NHE$$

In alkaline medium,
$$4H_2O + 4e^- \rightarrow 2H_2 + 4OH^- \qquad E^0_{red} = -0.828\,V \qquad vs\ NHE$$
$$4OH^- + 4h^+ \rightarrow 2H_2O + O_2 \qquad E^0_{ox} = -0.401\,V \qquad vs\ NHE$$

The cathodic half-reaction to produce hydrogen involves a two electron transfer process whereas the anodic half reaction proceeds via a four electron transfer reaction for the oxidation of water to form oxygen. The overall water splitting reaction is as follows:
$$2H_2O \rightarrow 2H_2 + O_2 \quad \Delta G^0 = 237\,kJmol^{-1}$$
In an experimental set-up when the two electrodes are submerged in an aqueous electrolyte solution, upon application of required voltage (equal

to or more than 1.23 V vs RHE), hydrogen and oxygen are produced at the respective electrodes. Considering the Nernst equations for the two half cells, theoretically in order to split water we would only need a voltage difference of 1.23 V between the anode and the cathode, but in practice we find it is necessary to apply a larger voltage due to the internal resistances accompanying the cell as well as the need to surpass kinetic barriers at each electrode. This difference in the required voltage more than the thermodynamically determined voltage is called overpotential (η).

$$\eta_{required} = 1.23 + \eta_{cathode} + \eta_{anode} + iR_{internal\ resistance}$$

where $\eta_{cathode}$ and η_{anode} are the kinetic activation barriers imposed at the surface of electrode to the electron transfer reactions at cathode and anode.

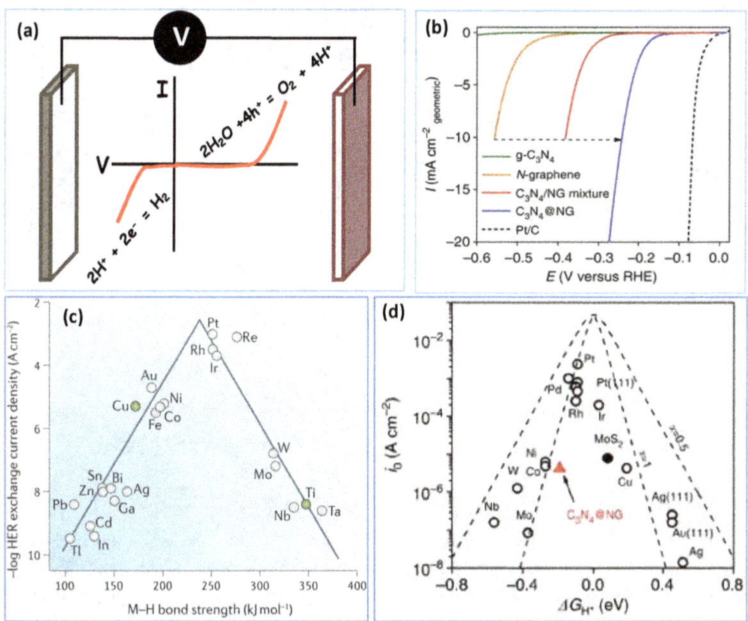

Fig. 10. (a) Representative schematic of principle of electrochemical water splitting. (b) HER polarization curves for various carbon based electrocatlaysts in comparison to Pt/C. Copyright[32] 2014 Springer Nature Limited. (c) Volcano plot for HER activity, comparison of transition metal electrocatalysts based on Sebatier's principle. Copyright[33] 2017, Springer Nature Ltd. (d) Plot of exchange current density as a function of ΔG_{H^*}.

This extra voltage can be minimized by the rational optimization of many parameters. The overall target is to reduce the overpotential as low as zero. For the evaluation of catalytic activity of an electrocatalyst, electrochemical techniques like linear sweep voltametry and cyclic voltametry (polarization method), tafel analysis (tafel slope), electrochemical impedance spectroscopy (EIS) and stability tests like amperometric *I-t* test along with accelerated cycling tests are employed.

4.1.1. *Catalyst selection*

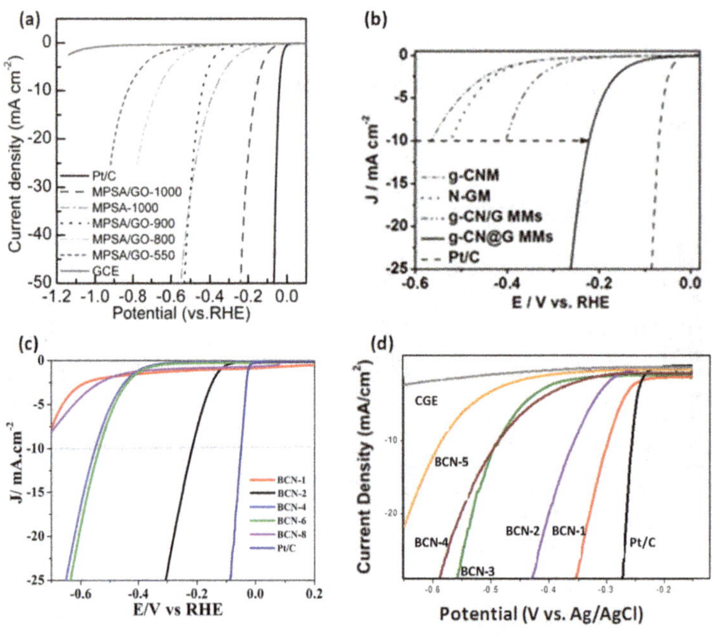

Fig. 11. (a) LSV of MSPA/GO samples along with Pt/C. Copyright[55] 2016 Wiley-VCH Verlag GmbH & Co. KGaA, Weinheim. (b) Comparison of HER activity of BCN electrodes with Pt/C. Copyright[36] 2016 Wiley-VCH Verlag GmbH & Co. KGaA, Weinheim. (c) LSV of BCN nanotube alloptropes in comparison with Pt/C for HER test. Copyright[37] 2016 The Royal Society of Chemistry. (d) LSV of different carbon content $B_xC_yN_z$ in comparison with Pt/C. Copyright[57] 2016 The Royal Society of Chemistry.

In order to compare the catalytic activity of different electrode materials, the current density at a constant overpotential or the overpotential at a constant current density is measured. A good electrocatalyst should result in a high current density at a low overpotential. To understand the activity, Sebatier's principle is invoked according to which if one plots some adsorption property (ΔG^0_{ad}) [for instance, in case of hydrogen evolution reaction, hydrogen binding energy (HBE) is used] of the prospective electrocatalysts as a function of obtainable exchange current density or reaction rate, one can get a volcano shaped plot. [34] This helps in identifying the catalysts with the best possible activity. For example the one occupying the top of the volcano plot is the best catalyst for that particular redox reaction. Pt or Pt group metal (PGM) occupies the top of the volcano plot, thus making it the most desirable catalyst for electrochemical hydrogen evolution reaction (HER) (Figure 10).[35] There has been much progress in fabrication of non-metal catalysts including graphene, doped graphenes and graphitic carbon nitride (g-C_3N_4).[36] But the HBE as a descriptor is unfavourable for HER process for graphene and g-C_3N_4 and hence it results in poor HER activity. Thus nanocomposites of g-C_3N_4@graphene and g-C_3N_4@nitrogenated graphene have been tested for HER.[36,33] Their activity was shown to be comparable to many active metal based electrocatalysts (Figure 11).

Lately, there have been reports of various metal alloys and carbon based electrocatalysts for HER whose activity is very close to that of the commercial Pt/C.[37-39] Dual doped carbon matrices show enhancement in HER activity. For example, nitrogen and phosphorous co-doped carbon gives an onset potential of 60 mV and $\eta@30mA/cm^2$ as 210 mV. [37] Similarly, Nitrogen and sulfur co-doping produces an onset potential as low as 27 mV and a small Tafel slope of 68 mV/dec.[38] Onset overpotential as low as 8 mV have been demonstrated using the co-doping strategy. Another example of low cost materials for HER is the carbon rich borocarbonitrides. For instance, BC_7N_2 shows an overpotential of 298 mV and 330 mV to obtain current densities of 10 and 20mA/cm^2 respectively (Figure 11).[39] An appropriate combination of an active 2D electrocatalyst such as BCN and MoS_2 have been covalently cross linked to render even higher HER activity.[40]

5. Photoelectrochemical Water Splitting

One of the feasible ways to reduce the cost of hydrogen production is to couple electrochemical and photochemical techniques in water splitting. This can be performed by photoelectrochemical (PEC) cells.[41] In a PEC cell, two electrodes (appropriate semiconductor materials) when in contact with aqueous electrolyte, upon light illumination either generates a potential (under the open circuit conditions) or produces a current flow in the cell (under short circuit conditions).[41] Thus water can be successfully split to yield hydrogen and oxygen. For this to happen one of the most vital phenomena is the formation of a hetero-interface of electrode/electrolyte which is reminiscent of a Schottky barrier. The semiconductor/electrolyte interface develops a space charge region in order to attain equilibrium between the difference in the Fermi level of the semiconductor and the redox potential of electrolyte. This results into band bending at the interface as shown in the schematic Figure 12.[41]

In PEC system, similar to photochemical approach the photoelectrode (semiconductor) absorbs sunlight and generates excited electrons (e_{CB}^-) and holes (h_{VB}^+) in the conduction band and the valence band respectively. These two participate in water splitting and thus minimizes the input energy (usually in terms of applied bias) required for hydrogen production. The PEC reaction is shown below,

$$Photocathode: \; 4H^+ + 4e_{CB}^- = 2H_2$$
$$Photoanode: \quad 2H_2O + 4h_{VB}^+ = O_2 + 4H^+$$
$$Overall: \; 2H_2O + 4hv\,(e_{CB}^- + h_{VB}^+) = 2H_2 + O_2$$

The photogenerated charges can be separated by two mechanisms- drift and diffusion. While diffusion is related to the intrinsic nature of the semiconductor (the concentration gradient of the local charge particles), drift is provided by the applied electrical bias. For example, in n-type semiconductor, the photogenerated electrons are driven towards the semiconductor/conducting substrate interface and holes are driven towards the semiconductor-electrolyte interface (depletion layer). The ratio of concentration of charges reaching the respective interfaces and the charges recombined is known as separation efficiency (η_{sep}). The efficiency of a semiconductor as the photoelectrode is defined by various parameters. The ratio of the photogenerated charges undergoing water splitting to all the photogenerated charges reaching the semiconductor-electrolyte interface is termed as the catalytic efficiency (η_k). Based on the half-reactions at the surface, an n-type semiconductor in PEC water splitting is called a photoanode (oxidation) and a p-type semiconductor is called a photocathode (reduction) (Figure 12).

To evaluate the efficiency of the photoelectrode for the PEC water splitting, there are two efficiency definitions depending on the mode of the PEC study. One is the benchmark efficiency definition: solar to hydrogen conversion efficiency (STH).[40] This is the most important of all efficiency measurements for a PEC device operating in two electrode configurations where no bias is applied.

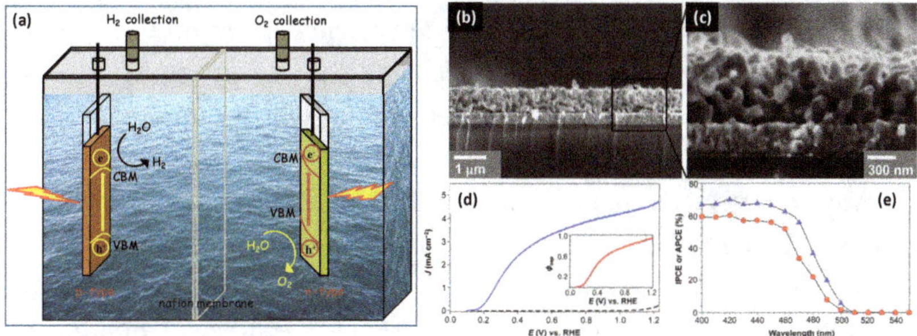

Fig. 12. (a) Schematic representation of photoelectrochemical device in parallel configuration with p-type and n-type semiconductors. (b) Side view FESEM image of FTO/BiVO$_4$ electrode for PEC water splitting. (c) The magnified image of nanoporous BiVO$_4$ film. (d) Photocurrent density (J) vs voltage curve for FTO/BiVO$_4$ electrode under 1.5 G, 100 mW/cm^2 light illumination. (e) IPCE and APCE measurements of FTO/BiVO$_4$ electrode represented by red dot and blue triangle respectively measured at 0.6 V vs RHE. Copyright[41] 2014 American Association for the Advancement of Science.

$$STH = \left[\frac{(mmol\ H_2\ s^{-1}) \times (237\ kJ\,mol^{-1})}{P_{total}(mW\,cm^{-2}) \times Area(cm^2)} \right]$$

where, P_{total} is the total integrated power input density. STH is simply the chemical energy produced as the function of solar energy absorbed. The measurement of STH involves exposing the PEC device to the broadband solar Air Mass 1.5 Global (AM 1.5 G) under zero bias condition. The other efficiency definition includes indicative efficiencies (where an applied bias is provided to carry out water splitting). This helps in understanding the photoelectrode material characterizations before it is used in the PEC devices. The most frequently used diagnostic efficiency is the applied bias photon to current efficiency (ABPE) and incident photon to current efficiency (IPCE).

$$ABPE = \left[\frac{j(mAcm^{-2}) \times (1.23 - V_{appl.})(V)}{P_{total}(mW\,cm^{-2})}\right] IPCE$$

$$= \eta_{sep.} \times \eta_{trans.} \times \eta_{interface} = \left[\frac{j(mAcm^{-2}) \times 1239.8(V \cdot nm)}{P_{mono}(mW\,cm^{-2}) \times \lambda(nm)}\right]$$

where j is the photocurrent density in an applied bias $V_{appl.}$. The η terms stands for efficiency for photogenerated charge separation (sep.), charge transport (trans.) and the kinetic efficiency for charge transfer reaction (interface) respectively. λ is the wavelength at which the photoelectrode is monochromatically illuminated with a power (P_{mono}).

Any prospective material to be used in PEC water splitting as a photoelectrode must be a good absorber of sunlight. The theoretical values of STH or any efficiency is governed by the bandgap of material. For example small bandgap materials (Eg<1.9eV) can have a higher STH whereas, wide band gap materials have decreased maximal theoretical STH. Additionally the effect of kinetic overpotential contributes to the overall efficiency as wider band gap materials are more likely to convert absorbed photons to current.

6. Conclusion

From the preceding sections it should be clear that today there are many ways of producing hydrogen using solar energy. From the recent developments in the various methods of H_2 generation based on research on catalyst systems, it appears as though H_2 generation is not a difficult task to achieve. While solar photochemical approach is an obvious means of doing so, thermochemical approach also appears to be reasonably attractive. A noteworthy advancement in using solar energy to split water is the use of sea water instead of normal water as the hydrogen source, particularly considering the acute shortage of clean water in the present times. Due to the involvement of various salts, splitting sea water is challenging. However, it may be possible that with a suitable choice of catalysts, one can split sea water without any prior treatment.

References

1. W. Wang, S. Wang, X. Ma and J. Gong, Recent advances in catalytic hydrogenation of carbon dioxide. *Chem. Soc. Rev.* **40** (7), 3703-3727 (2011).
2. T. Jafari, E. Moharreri, A. Amin, R.Miao, W.Song and S. Suib, Photocatalytic water splitting—the untamed dream: A review of recent advances. *Molecules,* **21** (7), 900 (**2016**).
3. A. Kudo and Y. Miseki, Heterogeneous photocatalyst materials for water splitting. *Chem. Soc. Rev.* **38** (1), 253-278 (2009).
4. K. Maeda, Photocatalytic water splitting using semiconductor particles: History and recent developments. *J. Photochem. Photobiol. C.* **12** (4), 237-268(2011).
5. S. Piskunov, O. Lisovski, J. Begens, D. Bocharov, F. Zhukovskii, M. Wessel, E. Spohr, C-, N-, S-, and Fe-Doped TiO_2 and $SrTiO_3$ nanotubes for visible-light-driven photocatalytic water splitting: prediction from first principles. *J. Phys. Chem. C.* **119** (32), 18686-18696(2015).
6. N. Kumar, U. Maitra, V. I. Hegde, U. V. Waghmare, A. Sundaresan and C. N. R. Rao, Synthesis, characterization, photocatalysis, and varied properties of TiO_2 cosubstituted with nitrogen and fluorine. *Inorg. Chem.* **52** (18), 10512-10519 (2013).
7. S. R. Lingampalli and C. N. R. Rao, Remarkable improvement in visible-light induced hydrogen generation by $ZnO/Pt/Cd_{1-y}Zn_yS$ heterostructures through substitution of N and F in ZnO. *J. Mat. Chem. A.* **2**(21), 7702-7705 (**2014**).
8. Q. Li, H. Meng, P. Zhou, Y. Zheng, J. Wang, J. Yu and J. Gong, $Zn_{1-x}Cd_xS$ solid solutions with controlled bandgap and enhanced visible-light photocatalytic H_2-production activity. *ACS Catal.* **3** (5), 882-889 (2013).
9. P. Fatahi, A. Roy, M. Bahrami and S. J. Hoseini, Visible-light-driven efficient hydrogen production from CdS nanoRods anchored with co-catalysts based on transition metal alloy nanosheets of NiPd, NiZn, and NiPdZn. *ACS Appl. Energy Mater.* **1** (10), 5318-5327, (2018).
10. J. Ran, J. Zhang, J.Yu, M. Jaroniec and S. Z.Qiao, Earth-abundant cocatalysts for semiconductor-based photocatalytic water splitting. *Chem. Soc. Rev.* **43** (22), 7787-7812 (2014).

11. Z. Sun, Q. Yue, J. Li, J. Xu, H. Zheng and P. Du, Copper phosphide modified cadmium sulfide nanorods as a novel p–n heterojunction for highly efficient visible-light-driven hydrogen production in water. *J. Mater. Chem. A.* **3** (19), 10243-10247 (2015).

12. D. V. Talapin, J. H. Nelson, E. V. Shevchenko, S. Aloni, B. Sadtler and A. P. Alivisatos, Seeded growth of highly luminescent CdSe/CdS nanoheterostructures with rod and tetrapod morphologies. *Nano Lett.* **7** (10), 2951-2959 (2007).

13. S. R. Lingampalli, U. K. Gautam and C. N. R. Rao, Highly efficient photocatalytic hydrogen generation by solution-processed ZnO/Pt/CdS, ZnO/Pt/Cd$_{1-x}$Zn$_x$S and ZnO/Pt/CdS$_{1-x}$Se$_x$ hybrid nanostructures. *Energy Environ. Sci.* **6** (12), 3589-3594 (2013).

14. A. Roy, M. Chhetri, S. Prasad, U.V. Waghmare and C. N. R. Rao, Unique features of the photocatalytic reduction of H_2O and CO_2 by new catalysts based on the analogues of CdS, $Cd_4P_2X_3$ (X = Cl, Br, I). *ACS Appl. Mater. Interfaces.* **10** (3), 2526-2536 (2018).

15. X. Wang, K. Maeda, A. Thomas, K. Takanabe, G. Xin, J. M. Carlsson, K. Domen and M. Antonietti, A metal-free polymeric photocatalyst for hydrogen production from water under visible light. *Nat. Mater.* **8** (1), 76-80 (2008).

16. G. Zhang, M. Zhang, X.Ye, X. Qiu, S. Lin and X. Wang, Iodine modified carbon nitride semiconductors as visible light photocatalysts for hydrogen evolution. *Adv. Mater.* **26** (5), 805-809 (2014).

17. K. Pramoda, U. Gupta, M. Chhetri, A. Bandyopadhyay, S. K. Pati and C. N. R. Rao, Nanocomposites of C_3N_4 with layers of MoS_2 and nitrogenated RGO, obtained by covalent cross-linking: synthesis, characterization, and HER activity. *ACS Appl. Mater. Interfaces.* **9**(12), 10664-10672 (2017).

18. U. Maitra, U. Gupta, M. De, R. Datta, A. Govindaraj and C. N. R. Rao, Highly effective visible-light-induced H_2 generation by single-layer 1T-MoS_2 and a nanocomposite of few-layer 2H-MoS_2 with heavily nitrogenated graphene. *Angew Chem. Int. Ed.* **52** (49), 13057-13061 (2013).

19. U. Gupta, B. S.Naidu, U.Maitra, A.Singh, S. N. Shirodkar, U. V. Waghmare and C. N. R. Rao, Characterization of few-layer 1T-$MoSe_2$ and its superior performance in the visible-light induced hydrogen evolution reaction. *APL Mater.* **2** (9), 092802 (2014).

20. D. M. Robinson, Y. B. Go, M. Greenblatt and G. C. Dismukes, Water oxidation by λ-MnO_2: catalysis by the cubical Mn_4O_4 subcluster obtained by delithiation of spinel $LiMn_2O_4$. *J. Am. Chem. Soc.* **132** (33), 11467-11469 (2010).

21. D. Wang, R. Li, J. Zhu, J. Shi, J. Han, X. Zong and C. Li, Photocatalytic water oxidation on $BiVO_4$ with the electrocatalyst as an oxidation cocatalyst: essential relations between electrocatalyst and photocatalyst. *J. Phys.Chem. C* **116** (8), 5082-5089 (2012).

22. G. Hitoki, T. Takata, J. N. Kondo, M. Hara, H. Kobayashi and K. Domen, An oxynitride, TaON, as an efficient water oxidation photocatalyst under visible light irradiation ($\lambda \leq 500$ nm). *Chem. Commun.* **0** (16), 1698-1699 (2002).

23. F. Zhang, A. Yamakata, K. Maeda, Y. Moriya, T. Takata, J. Kubota, K. Teshima, S. Oishi and K. Domen, Cobalt-modified porous single-crystalline LaTiO$_2$N for highly efficient water oxidation under visible light. *J. Am. Chem. Soc.* **134** (20), 8348-8351 (2012).

24. R.-L. Lee, P. D. Tran, S. S. Pramana, S. Y. Chiam, Y. Ren, S. Meng, L. H. Wong and J. Barber, Assembling graphitic-carbon-nitride with cobalt-oxide-phosphate to construct an efficient hybrid photocatalyst for water splitting application. *Catal. Sci. Technol.* **3** (7), 1694-1698(2013).

25. U. Maitra, B. S. Naidu, A. Govindaraj and C. N. R. Rao, Importance of trivalency and the eg1 configuration in the photocatalytic oxidation of water by Mn and Co *Proc. Natl. Acad. Sci.U. S.A.* **110** (29), 11704 (2013).

26. K. Maeda, T. Takata, M. Hara, N. Saito, Y. Inoue, H. Kobayashi and K. Domen, GaN:ZnO solid solution as a photocatalyst for visible-light-driven overall water splitting. *J. Am. Chem. Soc.* **127** (23), 8286-8287 (2005).

27. Z. Pan, Y. Zheng, F. Guo, P. Niu and X. Wang, Decorating CoP and Pt nanoparticles on graphitic carbon nitride nanosheets to promote overall water splitting by conjugated polymers. *ChemSusChem.* **10** (1), 87-90 (2017).

28. M. G. Kibria, F. A. Chowdhury, S. Zhao, B. AlOtaibi, M. L. Trudeau, H. Guo and Z. Mi,Visible light-driven efficient overall water splitting using p-type metal-nitride nanowire arrays. *Nat. Commun.* **6**, 6797 (2015).

29. A. Iwase, Y. H. Ng, Y. Ishiguro, A. Kudo and R. Amal, Reduced graphene oxide as a solid-state electron mediator in Z-scheme photocatalytic water splitting under visible light. *J. Am. Chem. Soc.* **133** (29), 11054-11057 (2011).

30. C. N. R. Rao and S. Dey, Solar thermochemical splitting of water to generate hydrogen. *Proc. Natl. Acad. Sci.* **114** (51), 13385 (2017).

31. J. R Scheffe and D. Weibel, A. Steinfeld, Lanthanum–Strontium–Manganese perovskites as redox materials for solar thermochemical splitting of H$_2$O and CO$_2$. *Energy Fuels.* 27, 4250-4257 (2013).

32. E. J. Popczun, C. G. Read, C. W. Roske, N. S. Lewis, R. E. Schaak, Highly active electrocatalysis of the hydrogen evolution reaction by cobalt phosphide nanoparticles. *Angew. Chem. Int. Ed.* **53** (21), 5427-5430 (2014).

33. Y. Zheng, Y. Jiao, Y. Zhu, L. H. Li, Y. Han, Y. Chen, A. Du, M. Jaroniec, S. Z. Qiao, Hydrogen evolution by a metal-free electrocatalyst. *Nat. Commun.* **5**, 3783 (2014).

34. J. K. Nørskov, T. Bligaard, J. Rossmeisl, C. H. Christensen, Towards the computational design of solid catalysts. *Nat. Chem.* **1**, 37 (2009).

35. B. Konkena, J. Masa, W. Xia, M. Muhler, W. Schuhmann, MoSSe@reduced graphene oxide nanocomposite heterostructures as efficient and stable

electrocatalysts for the hydrogen evolution reaction. *Nano Energy,* **29**, 46-53 (2016).

36. C. Guio, L. Stren, X. Hu, Nanostructured hydrogenating catalysts for electrochemical hydrogen evolution. *Chem. Soc. Rev.,* **43**, 6555-6569 (2014).

37. J. Zhang, J, L. Qu, G. Shi, J. Liu, J. Chen, L. Dai, N, P-Codoped carbon networks as efficient metal-free bifunctional catalysts for oxygen reduction and hydrogen evolution reactions. *Angew. Chem. Int. Ed.,* **55** (6), 2230-2234 (2016).

38. H. Tabassum, R. Zou, A. Mahmood, Z. Liang, S. Guo, A catalyst-free synthesis of B, N co-doped graphene nanostructures with tunable dimensions as highly efficient metal free dual electrocatalysts. *J. Mater. Chem. A,* **4** (42), 16469-16475 (2016).

39. M. Chhetri, S. Maitra, H. Chakraborty, U. V. Waghmare, C. N. R. Rao, Superior performance of borocarbonitrides, $B_xC_yN_z$, as stable, low-cost metal-free electrocatalysts for the hydrogen evolution reaction. *Energy Environ. Sci.,* **9** (1), 95-101 (2016).

40. K. Parmoda, M. Ayyub, N. Singh, M. Chhetri, U. Gupta, A. Soni, Covalently bounded MoS_2-borocarbonitride nanocomposites generated by using surface functionalities on the nanosheets and their remarkable HER activity. *J. Phys. Chem. C,* **122** (25), 13376-13384 (2018).

41. Z. Chen, T. G. Deutsch, H. N. Dinh, K. Domen, K. Emery, A. J. Forman, N. Gaillard, R. Garland, C. Heske, T. F. Jaramillo, A. Kleiman-Shwarsctein, E. Miller, K. Takanabe, J. Turner, *Photoelectrochemical Water Splitting: Standards, Experimental Methods, and Protocols*, Springer New York, New York, NY (2013).

Chapter 17

Thermochemical CO_2 Reduction

Soumyabrata Roy and Sebastian C. Peter[*]

New Chemistry Unit and School of Advanced Materials, Jawaharlal Nehru Centre for Advanced Scientific Research, Bangalore 560064, Karnataka, India
[*]*sebastiancp@jncasr.ac.in*

The intertwined problems of increasing energy demands and the resultant environmental crisis, arising from increasing atmospheric Carbon Dioxide (CO_2) levels (> 400 ppm), pose deep threats to the existence of our society. Fixing a portion of the CO_2 back in the form of chemicals and fuels using renewable energy sources is envisioned to be one of the most practical solutions to this crisis. However, owing to the high thermodynamic stability of CO_2, its reduction is an energetically uphill process which needs suitable reductants, energy intensive processes and efficient catalysts. Among various pathways, thermocatalytic hydrogenation stands out singularly because of its scalability, adaptability to the existing industrial infrastructure and wide product scope. The critical challenges in thermocatalytic CO_2 reduction can be solved only through a detailed understanding of the individual transformation steps, logical catalytic designs and energy optimization via reactor innovations. Thus, important findings and advances in these three aspects of CO_2 conversion to C1 products have been presented in this chapter with a special emphasis on catalyst design and mechanistic understanding.

1. Introduction

1.1. *The carbon cycle, GHG (Green House Gases) emissions and environmental crisis*

Carbon, the ubiquitous building element of all living things, is the fourth most abundant element on earth. Carbon flows spontaneously between various sources, reservoirs and sinks through a variety of chemical, physical, geological and biological processes in a natural biogeochemical loop, known as the carbon cycle. Lithosphere (rocks) stores most of this carbon on the earth's surface followed by ocean bodies (inorganic carbon and marine ecology), plants, fossil fuels, soil and atmosphere. A large amount of carbon exists in the form of its highest oxidized form, CO_2 (3 billion tons), assimilated into ocean bodies and living biomass, which constitute the major sinks. In an equilibrium state, the natural carbon cycle broadly balances the global carbon flux to maintain an optimal atmospheric concentration of CO_2 that along with other greenhouse gases traps the required amount of back-radiated solar energy to maintain a bio-habitable temperature on the earth's surface. The slow natural carbon cycle, which acts as a thermostat to the earth takes about 100–200 million years to transfer carbon from one reservoir to the other. This cycle is hampered by a fast carbon cycle originating from anthropogenic activities (adding 10^{15} grams carbon annually), which builds up carbon at a faster rate over certain reservoirs, leading to warmer temperatures on the earth. Human activities accumulate to the addition of about 30 billion tons of CO_2 annually which is ~200 times the amount of CO_2 added due to annual volcanic eruptions.

GHGs comprising of methane (CH_4), water vapor (H_2O), CO_2, ozone (O_3), nitrous oxide (N_2O) and chlorofluorocarbons (CFCs), trap the back-radiated solar energy from the earth's surface and make the planet warmer. By analyzing the ice cores on the earth's crust, the fluctuations in the GHG concentrations between the glacial and interglacial phases can be directly correlated with atmospheric temperatures.[1] This analysis has clearly shown that since the last 80000 years the average global atmospheric CO_2 concentration ranged between 180–280 ppm until the beginning of the industrial revolution in 1750. Since then, the atmospheric CO_2 level has increased steadily and crossed the deadly 400 ppm mark. There have been various studies and data analyses which conclusively proved the relation between anthropogenic activity and the increase in CO_2 levels in the atmosphere.[2] In a similar way, the link between atmospheric CO_2 concentration and the average global temperature have been proved through many extensive studies, one of which showed the sudden congruent increase in both CO_2 levels and the average global temperature since the advent of industrialization.[3] Studies have indicated that an increase in the global mean temperature by more than 2 °C (correspond to an atmospheric CO$_2$ level of ~450 ppm) will result in catastrophic changes in the environment. At the current global emission rates, that value of critical atmospheric CO$_2$ concentration is predicted to be reached within 25–30 years. Thus, the urgent need to mitigate CO$_2$ emission has led various global initiatives like the Intergovernmental Panel on Climate Change (IPCC) and the United Nations Climate Change Conference (COP21, Paris, 2015) to formulate strategies to curb CO$_2$ emissions by at least one-half of the current value by 2050.[4]

1.2. *The solution to the problem*

The straightforward technical solution to the CO$_2$ crisis is to capture carbon in the form of CO$_2$ from any source and eventually from the atmosphere and transform it to renewed fuels and chemical utilizing renewable sources of energy.[5, 6] CO$_2$ emission can be envisaged to be curbed through three possible strategies: (a) shifting the global energy infrastructure to greener fuels like H$_2$ and renewable energy grid, (b) CO$_2$ storage, and (c) utilization/conversion of CO$_2$.[7] Strategy (a) requires huge changes in the transportation infrastructure and such shifts in major countries with large fossil fuel deposits is very unlikely to happen. CO$_2$ storage (strategy (b)) on the other hand, is largely cost and energy intensive, has a large carbon footprint, and permanency of the stored CO$_2$ is not as high as desired.[8] Carbon utilization is 20–40 times more efficient over sequestration in long time spans[9] and thus has emerged as the most practically viable solution to mitigate climate change and curb atmospheric CO$_2$ levels. There are various pathways of CO$_2$ conversion depending on the energy sources being used, which can be termed as electrochemical, photochemical and thermochemical. Out of these, the thermochemical pathway has serious advantages over the other processes from an industrial perspective, as it offers a scalable, economical and sustainable solution[10] to the intertwined energy and environmental crisis.[7]

1.3. *CO$_2$ chemistry*

1.3.1. *Energetics of CO$_2$ utilization*

CO$_2$ is a thermodynamically stable molecule with $\Delta G^0_f = -396$ kJ mol^{-1}. It is a linear, stable and chemically inert molecule with ultralow electron affinity and a large HOMO-LUMO energy gap (13.7 eV). Its transformation is generally dominated by nucleophilic attacks at the carbon centre, which is an uphill process demanding a substantial input of energy (750 kJ mol^{-1} for the dissociation of the C=O bond).[11] The quantitative energy required for its conversion depends on the number of electrons required to form a particular desired product. Depending on the changes in carbon oxidation states, CO$_2$ conversion can be classified into two categories. (a) where +4 oxidation state usually remains intact (low energy process) and (b) where the carbon oxidation state is reduced (high energy process). In category (a) CO$_2$ usually reacts with electron rich molecules like H$_2$O, RR'R''C-, RR'NH, OH-, alkynes and olefins under mild temperature conditions (< 275 K) producing carboxylates – RCOOH(R), carbonates –(RO)$_2$, urea –(RHN)$_2$CO, linear esters, carbamates –RR'NCOOR'', lactones, polymeric materials such as polycarbonates and polyurethanes, hydrogen carbonates, inorganic carbonates and other similar compounds. In high energy processes (b), carbon undergoes reduction to lower oxidation states (at least two units below +4), and the examples include: CO, CH$_3$OH, HCOOH, HCHO, C$_2$H$_6$O, CH$_4$, hydrocarbons, oxygenates, higher alcohols, etc. In this chapter, we will be primarily presenting an overview on the catalytic aspects and reaction mechanisms, of the high energy reduction processes producing various important fuels and valuable chemicals by thermochemical CO$_2$ hydrogenation.

1.3.2. *Thermodynamics and kinetics of the CO$_2$ reduction*

The high thermodynamic stability of CO$_2$ is the biggest bottleneck for its extensive chemical utilization. The Gibbs-Helmholtz relationship, $\Delta G = \Delta H - T\Delta S$, provides a perfect basis for understanding the thermodynamic limitations of the CO_2 conversion. The strong C=O bonds require substantial energy to be broken, and the entropy factor ($-T\Delta S$) cannot drive the thermodynamic feasibility overcoming the large positive enthalpy requirements. The direct CO_2 dissociation to form CO and O_2 requires much higher enthalpies (ΔH = 293.0 kJ mol^{-1}) than hydrogenation (ΔH = 41.2 kJ mol^{-1}), as in the latter case, less stable reactants react to produce more stable products. Thus, the more thermodynamically favorable reverse water gas shift reaction (RWGS, H_2 (g) + CO_2 (g) → CO (g) + H_2O (g)) at elevated temperatures (~1500 K) can be operated at entropically dominated exergonic conditions ($\Delta G°$ = -1.7 kJ mol^{-1}). In a similar way carbonates, both organic and inorganic are more stable than CO_2 and hence thermodynamically favourable. CO_2 being the highest oxidized molecule of carbon, is generally inert to combustion with O_2. On the contrary, reactions of CO_2 with high energy reactants like hydroxides, olefins and amines are thermodynamically favourable and occur spontaneously at normal temperature and pressure conditions. Though thermodynamics of a reaction are critical to the energetics of a chemical transformation, the rates at which the reaction occurs is primarily governed by kinetics. For example many exergonic reactions may be severely limited by high kinetic barriers and thus may occur at rates so slow, that it hampers their commercial applications. Formation of inorganic carbonates occurring naturally from silicates, but at extremely slow rates is a typical example of such kinetically hindered processes. Thus it is important to emphasize that any CO_2 transformation occurs through interplay of thermodynamic and kinetic considerations.

2. Brief Overview on Thermochemical CO_2 Hydrogenation Process

The thermodynamic constraints of CO_2 reduction necessitates the use of efficient catalysts and optimal reaction conditions for minimizing the input energy requirements for the process which have been elaborated by Song et al.[12] Considering the insufficient market size of the CO_2-derived chemical industry, tapping in to the fuel generation sector is very important to have an impactful effect on the growing atmospheric CO_2

levels.[13] From a quantitative perspective, chemicals produced from CO_2 can account for a mere 4% of CO_2 emissions, while fuels can account for about 30% of the total CO_2 emissions and 100% of that from power plants.[14] Thus, a combination of chemical, petrochemical and fuel industries exploiting CO_2 as a raw material for heterogenous catalysis, can potentially form an artificial carbon cycle which can recycle waste CO_2 in a sustainable process.[15] Thermochemical hydrogenation of CO_2 is one of the most sustainable and industrially relevant reduction processes for fixing CO_2 into useful chemicals and fuel products, which can have a remarkable effect on the energy and environmental sectors.[7] The first industrially developed CO_2 hydrogenation reaction was the "Sabatier reaction", producing CH_4, which was discovered as early as the 1910s[16] and acted as a crucial basis for understanding the phenomenon of catalysis. The next important industrial process developed in this area was the Fischer–Tropsch (FT) process, which could produce hydrocarbons from syngas (mixture of CO and H_2). The development of the FT process diminished the industrial importance of the Sabatier reaction.[16] The product of the Sabatier reaction, CH_4, is one of the easiest hydrocarbons to produce through CO_2 hydrogenation. However, its low chemical reactivity impedes its conversion to further derivatives and chemicals.[16] Methanol in comparison is much more reactive to yield several downstream processes and thus is regarded as one of the most important CO_2 hydrogenation products.[16] Important examples of these processes include HCHO (formaldehyde) and HCOOH (formic acid) formation and their subsequent conversion to formamide, sugars, amino acids, production of higher hydrocarbons or fuels through the MTO (methanol to olefins) and MTG (methanol to gasoline) processes.[16] One of the most important gaseous products of CO_2 reduction is CO, which is heavily used in the FT process for producing hydrocarbons and fuels and also as a major ingredient in the syngas. Furthermore, it has extensive applications in the metal fabrication, steel and pharmaceutical industry. Insufficient conversion and product selectivity, which originate from competitive and unfavourable thermodynamic and kinetic factors, present one of the most critical challenges in catalytic CO_2 hydrogenation. Thus, the development of more efficient, product selective and durable catalysts along with innovations in energy efficient integrated reactor systems to increase selective catalytic rates are the only ways to solve the major bottlenecks of this technology. The development in catalytic designs for better performance can only be achieved through systematic and iterative structure-activity relationship studies by a detailed understanding of reaction mechanisms.[9] However, in situ mechanistic studies are challenging to be carried out in this thermochemical process due to the lack of suitable operando techniques under the extreme reaction conditions in spectroscopically opaque reactor designs.

A plethora of catalyst engineering approaches for CO_2 activation and its subsequent conversion to value-added products have been discussed in the light of mechanistic understandings in a number of recent reviews and perspective articles.[5, 8, 17] An important account of catalytic insights into CO_2 conversion to fuels via the reverse water gas shift reaction was presented by Daza et al.[18] The review by Porosoff et al. provided a comprehensive overview of catalyst designs for CO_2 conversion to a wide range of products.[9] Dorner et al.[19] presented an extensive overview of hydrocarbon production from CO_2 by modified FT catalysts. Rodriguez et al. elucidated and emphasized the role and the promotional effect of interfaces (metal/oxide or carbide) on the catalytic activity of CO_2 hydrogenation to methanol (MeOH).[20] Li et al. in a crucial theoretical review[20] elucidated the important reaction mechanisms for C1 fuel formation pathways of CO_2 conversion. The complicated reaction mechanisms and competing thermodynamic and kinetic aspects of CO_2 conversion pathways often lead to the formation of multiple products which decreases the overall efficiency of product formation. Selectivity is a very important consideration in industrial applications as the separation cost of chemically similar products affects the commercial dividend of a catalytic process. Selectivity is a critical function of the catalyst composition & structure and reaction conditions, which typically governs the reaction pathways in the overall energy profile landscape to CO_2 conversion to multiple products. Catalytic interfaces, substrate-reactant binding energies and electronic nature[21] of multi-functional composite catalysts are often found to play a crucial role in the various classes of CO_2 hydrogenation catalysts by tuning reaction mechanisms which have been comprehensively investigated in many operando studies corroborated by theoretical insights.[17] Such critical structural aspects of the catalyst design tunes the electronic properties of the active site for optimal adsorption/stabilization of key intermediates driving the catalytic transformation through a particular reaction pathway to form desired products with a high catalytic selectivity.[21]

3. Products, Catalysts and Reaction Mechanisms

3.1. *Carbon monoxide*

CO is a product of immense industrial importance having extensive applications in the FT process, pharma and biotechnology and metal & steel fabrication industry. Industrially it is produced through a number of processes involving formation of producer gas, water gas, syngas (from natural gas or other fuels), high temperature electrolysis of CO_2, reduction of metal-oxide ores, direct partial oxidation of carbon etc. The direct hydrogenation of CO_2 to form CO occurs via the reverse water gas shift reaction (RWGS, Eq. 1) which is a highly endothermic reaction (ΔH = 41.2 kJ/mol) requiring high temperatures between 500–800 °C and comparatively low pressures typically between 0-5 Bar.

$$CO_2 + H_2 \rightarrow CO + H_2O - - - - - - \Delta H_{298K} = 41.2 kJ/mol \qquad (1)$$

The RWGS reaction can proceed through (a) the formate dissociation pathway or (b) the redox mechanism depending on the type of catalysts. Depending on the nature and chemical composition, majorly investigated RWGS catalysts can be divided into the following categories (i) Cu based, (ii) noble metal nanoparticles supported on oxides (M−NPs@M−O), (iii) transition metal carbide systems and (iv) reducible oxide supports. Cu based catalysts primarily exhibit the redox mechanism where the facile redox shuttle between Cu^0 to Cu^{+1}, facilitate the reduction of CO_2 to CO, producing H_2O on H_2 oxidation.[22] Contrarily on M−NPs@M−O primarily the formate mechanism is followed, where after CO$_2$ hydrogenation to formate, subsequent dissociation via C=O bond cleaving leads to CO formation. The formate mechanism usually requires a bifunctional catalyst (eg: M−NPs@M−O) where H$_2$ can dissociatively bind on one site (metal site) and spill over to another (M-O site) where CO$_2$ is adsorbed.[9] Some of the exemplary catalysts along with their reaction mechanisms are provided in Table 1.

Some of the initial and very important investigations elucidating the mechanisms of RWGS pathway and the role of the catalyst nature were studied using a combination of in-situ and ex-situ FTIR, ATR and DRIFTS on Pd/Al$_2$O$_3$. Through a series of control studies on bare Al$_2$O$_3$, Pd/Al$_2$O$_3$ using different composition of reactant gases (CO$_2$ & H$_2$), Arunajatesan et al. proved the occurrence of formate mechanism in Pd/Al$_2$O$_3$ by analysing the adsorbates and products like carbonates, CO and formates.[23] The study also concluded that short-residence continuous flow reactors are more suitable for RWGS than batch reactors.[23] The importance of concentration and dispersion of metal loading in M−NPs@M−O catalysts for efficient and selective RWGS was also investigated in another study involving Pd/Al$_2$O$_3$.

This study demonstrated that while an increasing Pd loading enhanced CO$_2$ conversion efficiency it also led to a decreased CO selectivity as methane formation was highly enhanced.[25] This was attributed to the increase of active Pd terrace sites having stronger CO binding affinity, which led to its subsequent hydrogenation to methane (Figure 1). Also, the lower Pd-loaded catalyst showed better durability due to lower CO poisoning of the active metal sites.[25] Ferri et al. through detailed ATR studies on CO adsorbed Pt/Al$_2$O$_3$, compared the CO frequency and the ratio of intensities of CO$_L$:CO$_B$ (linear:bidentate) to conclude that the specific binding affinity of CO for the un-coordinated interfacial sites between Pt and Al$_2$O$_3$ (Figure 2) paved the way for probing the active sites for this class of catalysts.[33] Due to the equilibrium reaction most of the WGS catalysts are also active for the RWGS reaction under suitable conditions.[34] Cu-Zn-Al$_2$O$_3$ (CZA) systems are a classic family of catalysts, which have been extensively explored for RWGS.

Table 1. Examples of the important RWGS catalysts, which were used to elucidate the most dominant catalytic mechanisms of CO formation on various catalytic surfaces. Catalysts compositions, results of conversion and the detected/predicted mechanisms are listed.

Catalyst & Composition	Reaction Condition	Conversion & Selectivity	Mechanism assigned
Fe doped-Cu/SiO$_2$ (Cu/Fe = 10:0.3)[22]	Fixed bed reactor; 40 ml/min (1:1 of CO$_2$:H$_2$); 600°C & atm pressure	15% (conversion)	–
CuO/ZnO/Al$_2$O$_3$ (35% of Cu)[24]	Fixed bed reactor; 493K-523K & atm pressure	-	Redox Mechanism
Pd/Al$_2$O$_3$ (1% Pd loading)[25]	1:1 of CO$_2$:H$_2$; 342K & 138 bar	-	Formate Decomposition
Cu/K/SiO$_2$ (9% Cu +1.9% K)[26]	Fixed bed reactor; 1:1 of CO$_2$:H$_2$ 600°C & atm pressure	12.8% conversion	Formate decomposition
Rh/SiO$_2$[27]	Fixed bed reactor; 100 ml/min at H$_2$/CO$_2$ = 3; 403K 5 MPa	0.52% conversion & 88% selectivity	-
NiO/SBA-15 (10% NiO)[28]	Fixed bed reactor; 1:1 of CO$_2$:H$_2$; 600°C & atm. Pressure	20% and 100% selectivity	-
CuO(5wt%)-NiO(1wt%)/SBA & CuO(10wt%)-CeO$_2$(1wt%)/SBA-15[28]	Fixed bed reactor; 1:1 of CO$_2$:H$_2$; 400-900°C & atm. Pressure	>50% conversion at 900°C and 100% selectivity	-
2%Pt/CeO$_2$[29]	21000 h^{-1} GHSV (1% CO$_2$ + 4% H$_2$ in Ar) 498 K atm pressure	-	Hydrogenation of surface carbonates and H-spill over.
1%wtPt/TiO$_2$[30]	Fixed bed reactor; 100 ml/min; H$_2$/CO$_2$ = 3; 12,000h^{-1}; 873K & atm pressure	55% conversion	Through easy carbonate formation on TiO$_2$ which gets reduced by Ti^{3+}
Cu-Ni/g-Al$_2$O$_3$[31]	Fixed bed reactor; 2000h^{-1} GHSV with CO$_2$/H$_2$ ratio = 1; 873 K and atm. pressure	28.7% conversion and 79.7% selectivity	Redox mechanism by Cu
PtCo/CeO$_2$ (1.7%Pt; 1.5% Co)[32]	Batch reactor; 1:3 partial pressure ratio of CO$_2$:H$_2$; 573K & 30Torr	10% conversion and 259.4 of CO/CH$_4$ selectivity	Formate decomposition

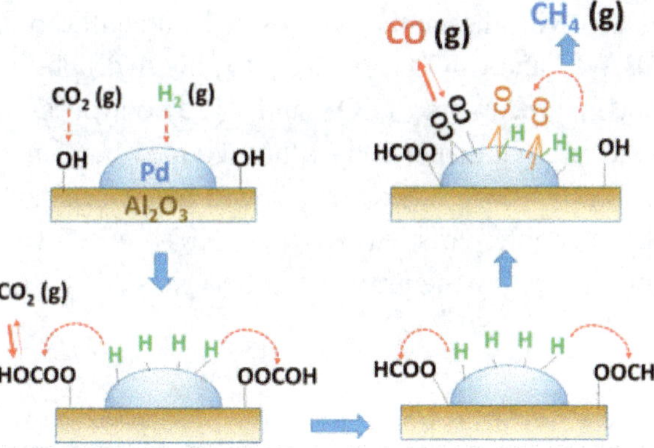

Fig. 1. Schematic representation of CH₄ and CO production on Pd/Al₂O₃ with different loading. Reproduced with permission from Ref. 25 Copyright 2015 of The American Chemical Society.

Optimal Cu:Zn ratio was found be greater than 3, while the Al_2O_3 support led to better dispersion of metal oxide islands and higher surface of active Cu^0 sites, which collectively enhanced the catalytic activity. $Cu-SiO_2$ is another Cu-based system, which yielded a better RWGS activity. An interesting study with $Cu-SiO_2$ showed that the incorporation of potassium (K) promoters in this system can drive the reaction through the formate pathway (over the redox mechanism) due to the formation active Cu-K interfacial sites which favored formate production.[34] In the redox pathway, the reduction rate of Cu_2O is much slower than the CO formation rate, which led to a resultant Cu oxidation over longer timescales leading to catalyst deactivation.

In these catalysts, Cu is also prone to sintering due to high temperature requirements of the endothermic RWGS reaction,[22] which was shown to be prevented by the Fe promotion in an impregnated Cu/SiO_2 catalyst.[35] In a comprehensive catalytic study on dispersed NiO on SBA-15 it was found out that a higher loading led to a better activity but a lower CO selectivity due to the formation of paired NiO sites promoting CH₄ formation.[36] Thus in an attempt to increase Ni loading without NiO site pairing, different Ni precursors were screened to find that $NiSO_4$, due to the formation of bigger sulphate aggregates, did not re-disperse during calcination giving well separated Ni sites even at a

25% loading. In hydrothermally synthesized bimetallic oxides on SBA (CuO-NiO/SBA-15 and CuO-CeO$_2$/SBA-15) the hydrophilicity of the M-nitrates and other precursors (TEOS and P123) played a crucial role in determining the relative dispersion of the two metals.[28] The mixed metal oxide (MMO) systems were generally found to give better activity than their single M-O counterparts. However, in NiO-CeO$_2$/SBA-15 case, CO selectivity got affected by the proximity of Ni sites.[28]

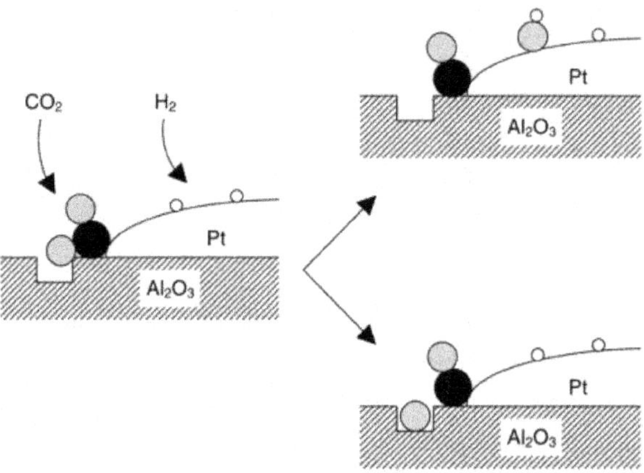

Fig. 2. Pictorial presentation of RWGS mechanism on Pt/Al$_2$O$_3$. Reproduced with permission from Ref. 33. Copyright 2002 Royal Society of Chemistry.

The use of electronically active reducible supports in these MMO systems easily outperformed the inert non-reducible supports like Al$_2$O$_3$ and SiO$_2$.[29] Unlike in Pt/Al$_2$O$_3$, Pt/TiO$_2$ was found to follow the redox mechanism, by creating vacant O sites (Pt-O$_v$-Ti^{+3}) through strong M-support interactions, which was proven through FTIR and XPS studies.[30] In the case of CeO$_2$ supported catalysts, a reduced CeO$_2$ surface was found to facilitate direct CO$_2$ dissociation to generate CO by oxygen exchange from CO$_2$,[37] with a subsequent regeneration of O vacancy on ceria by spilled over hydrogen. By controlling the strength of different bimetallic bonds by an interactive CeO$_2$ support,[32] Porosoff et. al. succeeded in booming the RWGS activity where PtCo/CeO$_2$ was found to excel due to optimal CO adsorption energy dictated by d-band centres.[32]

3.2. *CO$_2$ to CH$_3$OH*

Methanol having a current global market of ~70-80 million tons annually, is primarily used for the syntheses of important downstream products like formaldehyde, dimethyl ether, methyl tertbutyl ether (MTBE) and methyl acetic acid etc. It has extensive applications in the MTG and MTO processes and has also been projected as a future transportation fuel or a fuel substituent.[13] Industrially MeOH is produced from syngas, a stoichiometric mixture of CO, H$_2$, and CO$_2$. Thermochemical CO$_2$ to MeOH conversion can be a combined outcome of three reactions, the indirect pathway through Eq. 1 and 3 or through Eq. 2 which is the preferred direct pathway.[38] Reaction conditions determine the relative percentages of the direct and indirect pathways, which is difficult to deconvolute experimentally. From the enthalpy and entropy considerations it can be realized that direct CO$_2$ to MeOH (CTM) conversion is favored under low temperature and high pressure conditions (Table 2). RWGS (Eq. 1) is endothermic while both the subsequent CO hydrogenation (Eq. 3) and CTM (Eq. 2) are highly exothermic and entropically unfavored. This makes the overall process limited by the thermodynamic equilibrium which can be shifted by an in-situ product (water or MeOH) removal.[38]

$$CO_2 + 3H_2 \rightarrow CH_3OH + H_2O \text{ -------------} \Delta H_{298K} = \text{-49.5 kJ/mol} \quad (2)$$

$$CO + 2H_2 \rightarrow CH_3OH \text{ -----------------} \Delta H_{298K} = \text{-90.6 kJ/mol} \quad (3)$$

Typical CTM catalysts can also facilitate RWGS and subsequent CO hydrogenation, where Cu-based systems constitute as much as 75% of the entire catalyst spectrum.[39] CZA is the most common motif, which is extensively used as the core of the industrial catalyst. The CZA systems are usually prepared by a co-precipitation technique following the consecutive steps of precipitation-aging-washing-drying-calcination and reduction.[40] The relative production of CO and methanol over CZA systems can be controlled by the stoichiometric CO$_2$:H$_2$ ratio and the temperature and pressure conditions, exploiting the opposing enthalpy and entropy requirements of the two processes. The exact mechanistic role of each component of the CZA present at the surface is not completely understood.[41] It has been broadly agreed upon that CO_2 is adsorbed on the oxide surface followed by reduction with H_2 spilled over from Cu sites.[42] The interface between Cu and M-O sites is believed to play a crucial role though the nature of the exact catalytic site and the role of ZnO is still elusive.[43]

Two primary CTM pathways are: 1) formate pathway[44] and 2) CO pathway through RWGS and subsequent hydrogenation. Some exemplary catalysts and their pathways are presented in Table 2. In the formate pathway CO_2 and H_2 reacts to initially produce $^*HCOO$, which further hydrogenates to yield $^*H_2COO$, then $^*H_2CO$ via HCOOH. Sequential hydrogenation of $^*H_2CO$ further reduces to $^*H_3CO$ and H_3COH.[45] On both supported (SiO_2) and unsupported Cu catalysts formate was found to act as a spectator to methanol synthesis.[45] In an excellent mechanistic study by Yang et. al.[45] involving mono and bi-dentate formates on the Cu surfaces monitored through simultaneous Mass and IR spectroscopies, the methanol formation from CO_2 was finally attributed to the hydrogenation of mono-dentate formate to form methoxy, in the presence of dry hydrogen and water-derived co-adsorbates.[45]

On supports like ZnO, ZrO_2 or TiO_2-ZrO_2 combinations, the MeOH activity increased proportionally with increasing Cu dispersion and higher CO_2 and H_2 adsorption due to an increased basicity.[46] The effective surface area of Cu was found to increase in the following order: ZnO<ZnO-TiO_2<ZnO-ZrO_2<ZnO-ZrO_2-TiO_2. These catalysts typically followed the well described "dual-site mechanisms" involving the reduction of carbonates or bicarbonates on the oxide supports by spilled over H from Cu, which were reasonably proved through in-situ-DRIFT and DFT studies.[47]

Formate was found to be the lowest energy pathway, which negated the occurrence of other competing pathways on these catalysts.[47] The passivation of the M/M-O interfacial sites by dissociated O* during reaction[48] could be avoided in In_2O_3 based systems where the vacant

In₂O₃ sites (O-vacancy) could trap CO_2 and dissociate C=O to produce methanol with 100% selectivity through the formate pathway.[49] The iterative shuttle between the creation of O-vacancies on In₂O₃ by H, producing water, and its subsequent repairment during methanol formation (Figure 3)[49] led to such a high catalytic efficiency which was highlighted by Ye et al. Later, in 2016 this was experimentally exploited in an In₂O₃ loaded ZrO₂ system which could exhibit a consistent 1000h on stream activity with 100% MeOH selectivity.[49]

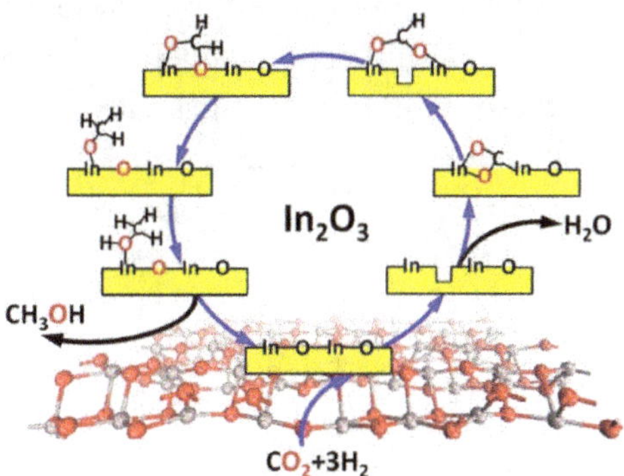

Fig. 3. Schematic representation of shuttling between O vacancy creation – CO₂ trapping – and regeneration of O vacancy. Reproduced with permission from Ref. 49. Copyright 2013 American Chemical Society.

Reducible supports like CeO_2, TiO_2 etc are found to exhibit quite different reaction mechanisms which proceed via the RWGS pathway. An elucidating report by Rodriguez et. al.[50] explained the role of oxygen vacancy clusters, increased roughness and the easy reducibility of Cu(I) to Cu(0) on these supports to drive the pathway through CO formation. DFT calculations showed favorable pathway of subsequent CO hydrogenation to produce MeOH through HCO, H_2CO (formaldehyde) and H_3CO (methoxy) intermediates. The high stability of formate on these surfaces ruled out the possibilities of a formate pathway.[50] In spite of yielding much better activities, the commercial usage of these oxides is limited due to their significantly higher costs.

The O adsorption energy relative to the Cu-211 surface was found to be a powerful theoretical descriptor of CTM catalytic activity on different metal surfaces and led to the construction of a novel activity volcano curve (Figure 4) defining a catalyst's MeOH conversion efficiency.[51] This experimental validity of this theoretical model was ratified by comparative activity studies in Ni-Ga systems on SiO_2. The catalyst followed the RWGS pathway and closely matched with the theoretical model.[51] Unlike a typical CTM reaction, this process was carried out under atmospheric pressures, probably leading to the CO pathway. Similarly Pd_2Ga impregnated on mesoporous SiO_2 could efficiently catalyze the CTM reaction under atmospheric pressures.[52] However none of these works mechanistically explained the high conversion rates at such nominal pressures, which was unprecedented in the CTM technology.

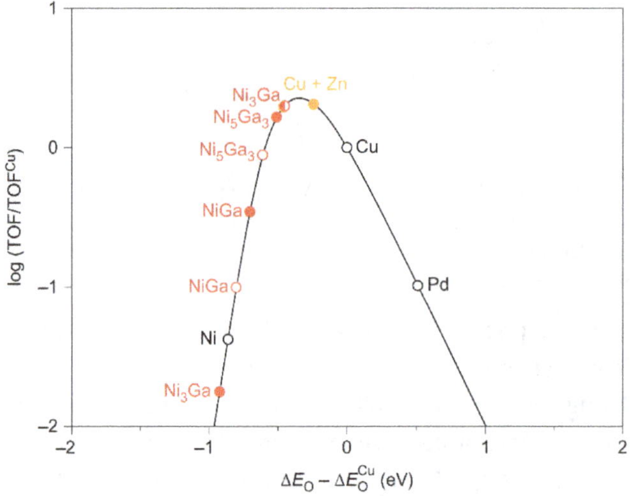

Fig. 4. Theoretical activity volcano for CO_2 hydrogenation to MeOH. Turnover frequency (TOF) is plotted as a function of ΔE_O, relative to Cu(211). ΔE_O for the stepped 211 surfaces of Cu, Ni, and Pd is depicted as open black circles, and Cu + Zn is depicted in orange. ΔE_O for Ni-Ga intermetallic compounds is depicted in red. Closed circles indicate Ni rich sites, open circles Ga rich sites and half-open circles mixed sites. Reaction conditions are 500 K, 1 bar, and a $CO_2:H_2$ ratio of 1:3. Reproduced with permission from Ref. 51. Copyright 2014 Springer Nature.

Table 2. Important examples of mechanistic studies by exemplary catalysts for CO_2 to MeOH conversion. Catalysts compositions, results of conversion and the detected/predicted mechanisms (wherever available) of these selected catalysts for CO_2 to methanol conversion have been provided in this table.

Catalyst & Composition	Reaction Condition	Conversion & Selectivity	Mechanism Assigned
$CuO/ZnO/ZrO_2/Ga_2O_3$ (5:3:1:1)[53]	Fixed Bed reactor; 1200 GHSV with $H_2/(CO_2+CO)$ = 2.4; 513 K & 27.2 atm.	60% selectivity	-
$CuO/ZnO/Al_2O_3$[54]	High throughput 49-channel parallel reactor; $H_2/CO/CO_2$ = 50-60%/30-40%/5-10%; 245 °C & 4.5MPa	30% conversion	-
$CuO/ZnO/TiO_2$-ZrO_2 = 40/40/10/10[46]	SS tubular reactor; $CO_2:H_2:N_2$ = 22:66:12, molar ratio, GHSV = 2400 $mLh^{-1}g_{cat}^{-1}$; 543K & 3 MPa	17.4% conversion & 43.8% selectivity	Dual site-mechanism
(0.8%)Cu/ZrO_2[47]	Fixed bed reactor; $H_2:CO_2$=3:1; 230 °C & 25 bar	6% conversion & 75% selectivity	Formate pathway
(2.3%) Cu/SiO_2[47]	Fixed bed reactor; $H_2:CO_2$=3:1; 230 °C & 25 bar	49% selectivity	Formate pathway
Fe(3wt%)-Cu(10%) /MCM-41[55]	SS tubular reactor; $H_2:CO_2$=3:1; 200 °C & 10 Bar;	2% conversion & 99.7% selectivity	-
In_2O_3/ZrO_2 (9% In)[56]	Fixed bed reactor; $16000h^{-1}$ GHSV $H_2:CO_2$=4:1; 573K & 5 MPa	5.2% conversion & 100%	Cyclic creation & annihilation of O-vacancy on In_2O_3
Ni_5Ga_3/SiO_2[51]	Fixed Bed reactor; 100Nml/min; $6000h^{-1}$, $H_2:CO_2$ = 3:1; atmospheric pressure	0.25mol MeOH /mol active metal hour yield & 100% selectivity (MeOH+DME)	-
$GaPd_2/SiO_2$[52]	Fixed Bed reactor; 100Nml/min; $H_2:CO_2$ = 3:1; 165 °C atm pressure	0.12% methanol yield	-
Pd-Cu/SBA-15[57]	Fixed bed reactor; GHSV = 3600mL (STP) g/cat/h: H_2/CO_2 = 24%/72% vol; 250 °C & 4.1MPa	6.5% conversion & 23% selectivity	-
Unsupported PdIn intermetallic[58]	$H_2:CO_2$ = 3:1; 210 °C & 50 Bar	3% conversion & >80% selectivity	-

3.3. *CO₂ to DME*

With a global market size of 5–7 million tons annually, the applications of Dimethyl ether (DME) currently spans the areas of propane substitution for residential usage, solvents, aerosol sprays, chemical feed stocks etc. It is also projected as an excellent fuel substituent for diesel and LPG because of its fuel efficiency, green nature, processability and safety aspects. The only bottleneck before the extensive usage of DME as a fuel is its physical state at NTP (liquid gas, BP is -24.8 °C) which demands major changes in the energy and transportation infrastructure.[13] Commercially, DME is produced either through solid acid-catalyzed dehydration of MeOH or obtained as an industrial intermediate in the MTO & MTG processes. The direct conversion of CO_2 to DME can be achieved through two subsequent reactions (Eqs. 2 & 4) by the use of bifunctional catalysts.

$$2CH_3OH \ \rightarrow \ CH_3OCH_3 + H_2O \text{------------} \Delta H = -23.4 \text{ kJ/mol (4)}$$

In typical bifunctional catalysts for DME, CO_2 is first converted to MeOH on one site following Eq. 2 and then two MeOH molecules are subsequently condensed (Eq. 4) on another site, optimally spaced from the former. Direct DME synthesis strongly shifts the thermodynamic equilibrium for CO_2 conversion by in-situ removal of MeOH, which results in high conversion rates.[59] The economic dividend of lower DME production cost through this process, makes it industrially important.

γ-Al_2O_3 is widely considered as one of the best solid acid component in the bifunctional DME catalyst. Unfortunately, coke formation due to the strong acidic sites in Al_2O_3 matrix drives the reaction to form secondary hydrocarbon products, which in-turn blocks the active catalytic sites.[60] The modulation of site acidity in γ-Al_2O_3 could be achieved by modifying Al with P (Phosphorous), where the Al/P ratio of 1.5 gave the optimum acidity resulting in a decreased formation of coke and HCs.[60] Thus it could be conclusively inferred that DME formation is most efficiently catalyzed at moderate to weak acidic sites.[60] The next issue with γ-alumina was the competitive affinity of water and MeOH for the Lewis acidic sites which decreases the DME yields highly.[61] It was later found out that the negative H_2O poisoning effect was specific to the presence of Lewis acid sites and could be substantially avoided by using zeolites which have a high Bronsted to Lewis acid site ratio and good water repellence.[62] The coke formation on acidic sites could be further reduced by co-feeding water in the reactant gas, which prevents the adsorption of coke on the acidic sites.[63] Water co-feeding improved Na-H-ZSM's catalytic performance highly, ratifying that water cannot deactivate acidic sites of zeolites[64] because of their hydrophobicity and high Bronsted acidity.[61] H-ZSM-5 has proved to be one of the best solid-acid DME catalysts owing to its high Bronsted site content and good water repellence.[65]

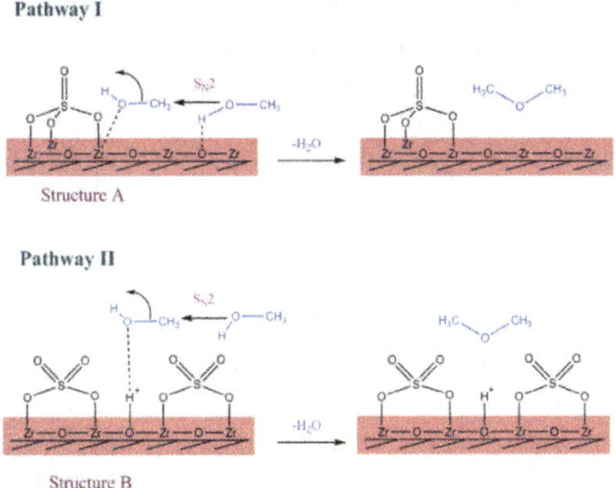

Fig. 5. Mechanistic scheme for the formation of DME by methanol dehydration on low (pathway I) and highly (pathway II) sulphated zirconia. Reproduced with permission from Ref. 66. Copyright 2015 of Royal Society of Chemistry.

In the case of MMO systems, bifunctionalized with zeolites for DME synthesis, most structural modification has been attempted on the MeOH component of the catalyst. In an attempt to increase the MeOH yield and subsequent DME production, the dispersion of Cu sites was increased by loading ZrO_2 in the MeOH component of $Cu/ZnO/Al_2O_3/HZSM$-5, where a 5% ZrO_2 loading was found to yield the best activity for DME production.[67] In another study, La-oxide loading, again, on the MeOH - active part was found to increase both Cu site dispersion and the strength and concentration of acidic sites in HZSM-5 to produce a high DME activity.[68] Parallel to H-ZSM-5, few recent studies have identified sulphated zirconia as the dehydrating solid acid where the sulphur content significantly modulated the extent and mechanism of MeOH condensation (Figure 5) with 20% S-content showing best activity.[66]

3.4. CO_2 to CH_4

Over the years, the CO_2 methanation process has been of tremendous interest to both academia and industry. With its extensive applications as LNG (Liquefied natural gas), fuels and chemical feedstock,[69] the industry has shown a high demand for CO free CH_4.[69] Though thermodynamically favored (Eq. 5), CO_2 methanation is severely limited in kinetics due to high electron transfer requirements.[70] Supported Ni catalysts are the most explored class of materials for CO_2 methanation.[69, 71-73]

$$CO_2 + 4H_2 \rightarrow CH_4 + 2H_2O \quad \Delta H = -252.9 \text{ kJ/mol} \,\&\, \Delta G = -130.8 \text{ kJ/mol-}$$

$$(5)$$

Methane formation from CO_2 are usually carried out in fixed bed reactors with a 4:1 H_2:CO_2 feed, typically between 320-400°C at atmospheric pressure. Ni catalysts, which are ubiquitously used in CO_2 methanation are prone to deactivation due to CO poisoning avoiding which, sometimes, necessitate the use of Ru, Rh, Pd etc. Majority of the catalysts, particularly those containing Ni, catalyzes the CH_4 formation through the indirect RWGS pathway. This involves the subsequent hydrogenation of CO, whose formation is unavoidable over these metals. The dissociation of CO by formation of the first C-H bond is generally believed to be the r.d.s., though this hypothesis haven't been unanimously experimentally confirmed.[74] This is because of the strong dependence of the catalytic activity and selectivity on aspects like the choice of active metals, supports, promoters and synthetic conditions.[13] The unsupported metal catalysts show CO_2 methanation activity as per the following trend of Ru > Ir > Rh > Ni > Co > Os> Pt > Fe > Mo > Pd > Ag[75], while the trend slightly alters (Ru > Fe > Ni > Co > Rh > Pd > Pt > Ir)[76] when a particular active metal surface area is considered on supports. The order for selectivity varies as Ni > Co > Fe > Ru.[77] SiO_2,[78] TiO_2,[71] Al_2O_3,[73] ZrO_2,[72] and CeO_2,[72] are among the commonly used oxide supports.[79] The most used and suitable supports for methanation are γ-Al$_2$O$_3$, SiO$_2$, and TiO$_2$ owing to their remarkable CO$_2$ adsorption properties and high thermal stability at elevated temperatures.[80] Mesoporous supports like mesostructured silica nanoparticles (MSNs), HY-Zeolites, MCM-41, SBA-15, are another class of interesting materials. Among these MSNs have the most promising activity and stability, owing to the large concentration of accessible basic sites, high porous surface area and stable architectures.[81] Na, K, V, or La[80] are some typical promoters of methanation activity.[82] Ceria doping increased reducibility and stability,[83] while MgO doping was found to enhance the thermal stability and resist coke formation[84] while both promoting the overall Ni activity. Table 3 enlists the details of some of the important CO$_2$ methanation catalysts and their reaction mechanisms. The primary challenge for further catalytic designs for this process mainly involves stability enhancement as the catalysts occasionally get affected by coke formation, sulfur poisoning (from reactant), thermal stress, and attrition.

The mechanism for CO$_2$ methanation process can be classified as the (1) mechanism involving CO as intermediate and the (2) mechanism which does not involve CO. Majority of the recent spectroscopic studies, involving IR-operando studies and in-situ DRIFTS have reported the absence of CO in the reaction pathway, particularly in the case of CeO$_2$-ZrO$_2$. The mechanism of methanation on Ni/CeO$_2$-ZrO$_2$ was determined by IR as a formate pathway without any CO involvement, the proposed steps of which has been shown in Figure 6.[85] Studies involving the

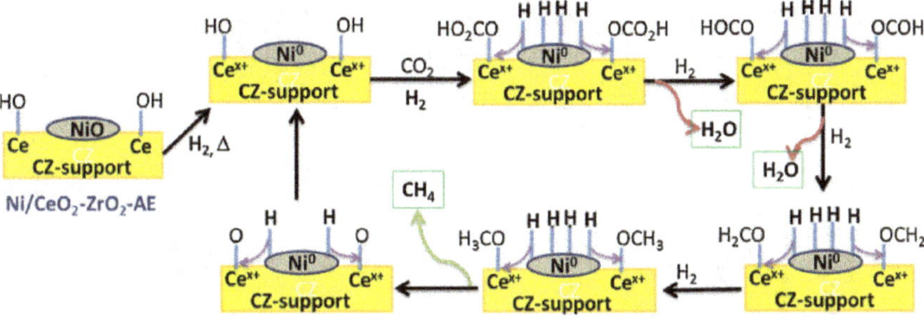

Fig. 6. Reaction mechanism proposed over Ni/CZ-AE catalysis for CO$_2$ methanation reaction. Reproduced with permission from Ref. 86. Copyright 2017 Elsevier.

Table 3. Important catalysts and exemplary mechanistic studies of CO_2 methanation. Catalysts compositions, results of conversion and the detected/predicted mechanisms (wherever available) of these selected catalysts for CO_2 to methane conversion have been provided in this table.

Catalyst & Composition	Reaction Condition	Conversion & Selectivity	Mechanism assigned
(5%)Ni/MSN[69]	Fixed Bed reactor; 50,000ml/gh with H_2/CO_2 = 4/1; 300 °C & atm pressure	99.9% conversion & 95% selectivity	Adsorption of CO_2 by MSN as carbonates & reduction of it by Ni by spillover
(20%)Ni/g-alumina[73]	Fixed bed reactor; 6000ml/g/h H_2/CO_2 = 4/1; 350 °C and atmospheric pressure	>95% conversion & 100% selectivity	-
Ni/CeO$_2$[87]	Fixed bed reactor (4mm dia); 10000h^{-1} space velocity H_2/CO_2 = 4/1; 300 °C atmospheric pressure	90% conversion & 100% selectivity	-
Ni/CeO$_2$[88]	Fixed bed reactor; (6mm dia) CO_2:H_2:Ar = 10%:46%:44% by volume; 340 °C & atmospheric pressure	91.1% conversion & 100% selectivity	Via CO route
(5wt%)Ni/CeO$_2$-ZrO$_2$[85]	Fixed bed reactor: H_2/CO_2=4/1; 350 °C & atmospheric pressure	80% conversion & 100% selectivity	Via carbonates and formates
(10wt%)Ni/Ce$_{0.72}$Zr$_{0.28}$[89]	Fixed bed reactor; H_2/CO_2 = 4/1; H_2:CO_2:N_2 = 36:9:10; 350 °C, atmospheric pressure	Almost 76% conversion & >98% selectivity	-
(3%wt)Ru/g-alumina[90]	Fixed bed reactor; GHSV=55000h^{-1} H_2/CO_2 = 5/1; 673 K & 1atm	85% conversion and 85% selectivity	Via carbonates and formates
Rh/g-alumina[91]	Fixed bed reactor; H_2:CO_2 = 1:1; 25 °C and atmospheric pressure	17% conversion and 100% selectivity	Via CO route

influence of Ni loading (between 0.5–10%) on activity & selectivity showed that a low Ni loading yielded a higher activity towards CO_2 hydrogenation but with high CO selectivity, while high loading favored CH_4.[78] The reason for this was attributed to the lower H_2 concentration around CO, in case of low Ni loading, which was insufficient to reduce m-HCOO all the way to CH_4. Both parallel and consecutive pathways were followed in case of higher Ni loading. But irrespective of the loading, FTIR studies proved, that all the pathways in supported systems proceed through m-HCOO, whose reduction was identified as the crucial step.[78]

4. Summary and Outlook

Innovation in catalytic designs, reactor technology, carbon-free hydrogen production, carbon capture technology and renewable electricity development are the key components to achieve commercial maturity level for CO_2 hydrogenation technology. Among them, catalysis design has a peculiar role as it can dictate the production of a desired product with higher efficiency. The design of a cheap, scalable and a durable catalyst necessitates a deep understanding of reaction mechanisms, the formation of key intermediates and their binding energies, by understanding the role of identified active catalytic components, and the poisoning effects of reaction environments on the catalysts. The choice of variable parameters such as temperature and pressure conditions becomes very critical in various CO_2 hydrogenation processes like CO_2-FT (exothermic), CTM (exothermic) and RWGS (endothermic). Higher temperature favors methanation and lower pressure favors RWGS. It can be concluded that conversion and selectivity for the CO_2 reduction is associated with complex mechanisms involving many descriptors which is subjectively governed by the thermodynamic aspects, reaction conditions and nature and design of the catalyst in particular.

One other important critical aspect of CO_2 hydrogenation technology is the techno-economic viability. The two bottlenecks of the thermochemical CO_2 reduction are the requirement of a large amount of H_2 and expensive electricity. The cost towards CO_2 capture also cannot be neglected although there are substantial improvements made in this direction. The introduction of renewable energy such as solar, geothermal and wind are absolute necessary to make it as an economically viable process. Efficient catalyst design, innovative reactor designs, low cost H_2 production are the crucial factors governing the economic viability of this process. An annual global increase in the chemicals and fuel's demand and their utilization in different sectors will prove beneficial for the immediate market penetration of this technology. Finally, initiative from the public sectors for curbing CO_2 emissions will foster the market demand growth of such non-conventional technologies based on the CO_2 utilization.

References

1. J. R. Petit, J. Jouzel, D. Raynaud, N. I. Barkov, J. M. Barnola, I. Basile, M. Bender, J. Chappellaz, M. Davis, G. Delaygue, M. Delmotte, V. M. Kotlyakov, M. Legrand, V. Y. Lipenkov, C. Lorius, L. PÉpin, C. Ritz, E. Saltzman and M. Stievenard, Climate and atmospheric history of the past 420,000 years from the Vostok ice core, Antarctica, *Nature* **399**(429) (1999).
2. https://zfacts.com/p/194.html, in.
3. A report by Government of South Australia's Primary Industries and Regions SA - http://www.pir.sa.gov.au/pirsa/nrm/climate change/a guide to climate change and adaption in agriculture in south australia/chapter 1, in.
4. E. Catizzone, G. Bonura, M. Migliori, F. Frusteri and G. Giordano, CO_2 recycling to dimethyl ether: state-of-the-art and perspectives, *Molecules* **23**(31), 1-28 (2018).
5. A. Goeppert, M. Czaun, J. P. Jones, G. K. S. Prakash and G. A. Olah, Recycling of carbon dioxide to methanol and derived products — closing the loop, *Chem. Soc. Rev.* **43**(23), 7995-8048 (2014).
6. S. C. Peter, Reduction of CO2 to Chemicals and Fuels: A solution to global warming and energy crisis, *ACS Energy Lett.* **3**(7), 1557-1561 (2018).
7. S. Saeidi, N. A. S. Amin and M. R. Rahimpour, Hydrogenation of CO_2 to value-added products—A review and potential future developments, *J. CO2. Util.* **5**(66-81) (2014).
8. M. Aresta, A. Dibenedetto and A. Angelini, Catalysis for the valorization of exhaust carbon: from CO_2 to chemicals, materials, and fuels. technological use of CO_2, *Chem. Rev.* **114**(3), 1709-1742 (2014).
9. M. D. Porosoff, B. H. Yan and J. G. G. Chen, Catalytic reduction of CO_2 by H_2 for synthesis of CO, methanol and hydrocarbons: challenges and opportunities, *Energ. Environ. Sci.* **9**(1), 62-73 (2016).

10. S. Roy, A. Cherevotan and S. C. Peter, Thermochemical CO_2 hydrogenation to single carbon products: scientific and technological challenges, *ACS Energy Lett.* **3**(8), 1938-1966 (2018).

11. Z. Y. Sun, T. Ma, H. C. Tao, Q. Fan and B. X. Han, Fundamentals and challenges of electrochemical CO_2 reduction using two-dimensional materials, *Chem.* **3**(4), 560-587 (2017).

12. S. G. Jadhav, P. D. Vaidya, B. M. Bhanage and J. B. Joshi, Catalytic carbon dioxide hydrogenation to methanol: A review of recent studies, *Chem. Eng. Res. Des.* **92**(11), 2557-2567 (2014).

13. J. Artz, T. E. Muller, K. Thenert, J. Kleinekorte, R. Meys, A. Sternberg, A. Bardow and W. Leitner, Sustainable conversion of carbon dioxide: an integrated review of catalysis and life cycle assessment, *Chem. Rev.* **118**(2), 434-504 (2018).

14. P. Kaiser, R. B. Unde, C. Kern and A. Jess, Production of liquid hydrocarbons with CO_2 as carbon source based on reverse water-gas shift and fischer-tropsch synthesis, *Chem. Eng. Technol.* **85**(4), 489-499 (2013).

15. J. Jia, C. X. Qian, Y. C. Dong, Y. F. Li, H. Wang, M. Ghoussoub, K. T. Butler, A. Walsh and G. A. Ozin, Heterogeneous catalytic hydrogenation of CO_2 by metal oxides: defect engineering - perfecting imperfection, *Chem. Soc. Rev.* **46**(15), 4631-4644 (2017).

16. G. A. Olah, T. Mathew and G. K. S. Prakash, Chemical formation of methanol and hydrocarbon ("Organic") derivatives from CO_2 and H_2-carbon sources for subsequent biological cell evolution and life's origin, *J. Am. Chem. Soc.* **139**(2), 566-570 (2017).

17. S. Kattel, P. Liu and J. G. G. Chen, Tuning selectivity of CO_2 hydrogenation reactions at the metal/oxide interface, *J. Am. Chem. Soc.* **139**(29), 9739-9754 (2017).

18. Y. A. Daza and J. N. Kuhn, CO_2 conversion by reverse water gas shift catalysis: comparison of catalysts, mechanisms and their consequences for CO_2 conversion to liquid fuels, *Rsc Adv.* **6**(55), 49675-49691 (2016).

19. R. W. Dorner, D. R. Hardy, F. W. Williams and H. D. Willauer, Heterogeneous catalytic CO_2 conversion to value-added hydrocarbons, *Energ. Environ. Sci.* **3**(7), 884-890 (2010).

20. X. F. Yang, S. Kattel, S. D. Senanayake, J. A. Boscoboinik, X. W. Nie, J. Graciani, J. A. Rodriguez, P. Liu, D. J. Stacchiola and J. G. G. Chen, Low pressure CO_2 hydrogenation to methanol over gold nanoparticles activated on a CeOx/TiO₂ interface, *J. Am. Chem. Soc.* **137**(32), 10104-10107 (2015).

21. S. Kattel, B. H. Yan, Y. X. Yang, J. G. G. Chen and P. Liu, Optimizing binding energies of key intermediates for CO_2 hydrogenation to methanol over oxide-supported copper, *J. Am. Chem. Soc.* **138**(38), 12440-12450 (2016).

22. C. S. Chen, W. H. Cheng and S. S. Lin, Study of iron-promoted Cu/SiO₂ catalyst on high temperature reverse water gas shift reaction, *Appl. Catal. A-Gen.* **257**(1), 97-106 (2004).

23. V. Arunajatesan, B. Subramaniam, K. W. Hutchenson and F. E. Herkes, In situ FTIR investigations of reverse water gas shift reaction activity at supercritical conditions, *Chem. Eng. Sci.* **62**(18-20), 5062-5069 (2007).

24. M. J. L. Gines, A. J. Marchi and C. R. Apesteguia, Kinetic study of the reverse water-gas shift reaction over $CuO/ZnO/Al_2O_3$ catalysts, *Appl. Catal. A-Gen.* **154** (1-2), 155-171 (1997).

25. X. Wang, H. Shi, J. H. Kwak and J. Szanyi, Mechanism of CO_2 hydrogenation on Pd/Al_2O_3 catalysts: kinetics and transient DRIFTS-MS studies, *Acs Catal.* **5**(11), 6337-6349 (2015).

26. C. S. Chen, W. H. Cheng and S. S. Lin, Study of reverse water gas shift reaction by TPD, TPR and CO_2 hydrogenation over potassium-promoted Cu/SiO_2 catalyst, *Appl. Catal. A-Gen.* **238**(1), 55-67 (2003).

27. Kusama, K. K. Bando, K. Okabe and H. Arakawa, CO_2 hydrogenation reactivity and structure of Rh/SiO_2 catalysts prepared from acetate, chloride and nitrate precursors, *Appl. Catal. A-Gen.* **205**(1-2), 285-294 (2001).

28. B. W. Lu, Y. W. Ju, T. Abe and K. Kawamoto, Dispersion and distribution of bimetallic oxides in SBA-15, and their enhanced activity for reverse water gas shift reaction, *Inorg. Chem. Front.* **2**(8), 741-748 (2015).

29. A. Goguet, F. C. Meunier, D. Tibiletti, J. P. Breen and R. Burch, Spectrokinetic investigation of reverse water-gas-shift reaction intermediates over a Pt/CeO_2 catalyst, *J. Phys. Chem. B* **108**(52), 20240-20246 (2004).

30. S. S. Kim, H. H. Lee and S. C. Hong, A study on the effect of support's reducibility on the reverse water-gas shift reaction over Pt catalysts, *Appl. Catal. A-Gen.* **423** (100-107 (2012).

31. Y. Liu and D. Z. Liu, Study of bimetallic $Cu-Ni/gamma-Al_2O_3$ catalysts for carbon dioxide hydrogenation, *Int. J. Hydrogen. Energ.* **24**(4), 351-354 (1999).

32. M. D. Porosoff and J. G. G. Chen, Trends in the catalytic reduction of CO_2 by hydrogen over supported monometallic and bimetallic catalysts, *J. Catal.* **301** (30-37) (2013).

33. D. Ferri, T. Burgi and A. Baiker, Probing boundary sites on a Pt/Al_2O_3 model catalyst by CO_2 hydrogenation and in situ ATR-IR spectroscopy of catalytic solid-liquid interfaces, *Phys. Chem. Chem. Phys.* **4**(12), 2667-2672 (2002).

34. F. S. Stone and D. Waller, $Cu-ZnO$ and $Cu-ZnO/Al_2O_3$ catalysts for the reverse water-gas shift reaction. The effect of the Cu/Zn ratio on precursor characteristics and on the activity of the derived catalysts, *Top. Catal.* **22**(3-4), 305-318 (2003).

35. C. S. Chen, W. H. Cheng and S. S. Lin, Enhanced activity and stability of a Cu/SiO_2 catalyst for the reverse water gas shift reaction by an iron promoter, *Chem. Commun.* **18**, 1770-1771 (2001).

36. B. W. Lu and K. Kawamoto, Direct synthesis of highly loaded and well-dispersed NiO/SBA-15 for producer gas conversion, *Rsc Adv.* **2**(17), 6800-6805 (2012).

37. T. Staudt, Y. Lykhach, N. Tsud, T. Skala, K. C. Prince, V. Matolin and J. Libuda, Ceria reoxidation by CO_2: A model study, *J. Catal.* **275**(1), 181-185 (2010).

38. J. L. Dubois, K. Sayama and H. Arakawa, Conversion of CO_2 to dimethylether and methanol over hybrid catalysts, *Chem. Lett.* **7**, 1115-1118 (1992).

39. A. Alvarez, A. Bansode, A. Urakawa, A. V. Bavykina, T. A. Wezendonk, M. Makkee, J. Gascon and F. Kapteijn, Challenges in the greener production of formates/formic acid, methanol, and DME by heterogeneously catalyzed CO_2 hydrogenation processes, *Chem. Rev.* **117**(14), 9804-9838 (2017).

40. M. Behrens, Coprecipitation: An excellent tool for the synthesis of supported metal catalysts - From the understanding of the well known recipes to new materials, *Catal. Today* **246**(46-54 (2015).

41. W. P. A. Jansen, J. Beckers, J. C. Van der Heuvel, A. W. D. Van der Gon, A. Bliek and H. H. Brongersma, Dynamic behavior of the surface structure of $Cu/ZnO/SiO_2$ catalysis, *J. Catal.* **210**(1), 229-236 (2002).

42. I. A. Fisher and A. T. Bell, In-situ infrared study of methanol synthesis from H_2/CO_2 over Cu/SiO_2 and $Cu/ZrO_2/SiO_2$, *J. Catal.* **172**(1), 222-237 (1997).

43. S. Kuld, C. Conradsen, P. G. Moses, I. Chorkendorff and J. Sehested, Quantification of zinc atoms in a surface alloy on copper in an industrial-type methanol synthesis catalyst, *Angew. Chem. Int. Edit.* **53**(23), 5941-5945 (2014).

44. J. Tabatabaei, B. H. Sakakini and K. C. Waugh, On the mechanism of methanol synthesis and the water-gas shift reaction on ZnO, *Catal. Lett.* **110**(1-2), 77-84 (2006).

45. Y. Yang, C. A. Mims, R. S. Disselkamp, J. H. Kwak, C. H. F. Peden and C. T. Campbell, (Non)formation of methanol by direct hydrogenation of formate on copper catalysts, *J. Phys. Chem. C.* **114**(40), 17205-17211 (2010).

46. J. Xiao, D. S. Mao, X. M. Guo and J. Yu, Effect of TiO_2, ZrO_2, and TiO_2-ZrO_2 on the performance of CuO-ZnO catalyst for CO_2 hydrogenation to methanol, *Appl. Surf. Sci.* **338**(146-153 (2015).

47. K. Larmier, W. C. Liao, S. Tada, E. Lam, R. Verel, A. Bansode, A. Urakawa, A. Comas-Vives and C. Coperet, CO_2-to-methanol hydrogenation on zirconia-supported copper nanoparticles: reaction intermediates and the role of the metal-support interface, *Angew. Chem. Int. Edit.* **56**(9), 2318-2323 (2017).

48. Q. L. Tang, Q. J. Hong and Z. P. Liu, CO_2 fixation into methanol at Cu/ZrO_2 interface from first principles kinetic Monte Carlo, *J. Catal.* **263**(1), 114-122 (2009).

49. J. Y. Ye, C. J. Liu, D. H. Mei and Q. F. Ge, Active Oxygen vacancy site for methanol synthesis from CO_2 hydrogenation on In_2O_3(110): a DFT study, *Acs. Catal.* **3**(6), 1296-1306 (2013).

50. J. Graciani, K. Mudiyanselage, F. Xu, A. E. Baber, J. Evans, S. D. Senanayake, D. J. Stacchiola, P. Liu, J. Hrbek, J. F. Sanz and J. A. Rodriguez, Highly active copper-ceria and copper-ceria-titania catalysts for methanol synthesis from CO_2, *Science* **345**(6196), 546-550 (2014).

51. F. Studt, I. Sharafutdinov, F. Abild-Pedersen, C. F. Elkjaer, J. S. Hummelshoj, S. Dahl, I. Chorkendorff and J. K. Norskov, Discovery of a Ni-Ga catalyst for carbon dioxide reduction to methanol, *Nat. Chem.* **6**(4), 320-324 (2014).

52. E. M. Fiordaliso, I. Sharafutdinov, H. W. P. Carvalho, J. D. Grunwaldt, T. W. Hansen, I. Chorkendorff, J. B. Wagner and C. D. Damsgaard, Intermetallic GaPd$_2$ nanoparticles on SiO$_2$ for low-pressure CO$_2$ hydrogenation to methanol: catalytic performance and in situ characterization, *ACS Catalysis* **5**(10), 5827-5836 (2015).

53. O. S. Joo, K. D. Jung, I. Moon, A. Y. Rozovskii, G. I. Lin, S. H. Han and S. J. Uhm, Carbon dioxide hydrogenation to form methanol via a reverse-water-gas-shift reaction (the CAMERE process), *Ind. Eng. Chem. Res.* **38**(5), 1808-1812 (1999).

54. C. Baltes, S. Vukojevic and F. Schuth, Correlations between synthesis, precursor, and catalyst structure and activity of a large set of CuO/ZnO/Al$_2$O$_2$ catalysts for methanol synthesis, *J. Catal.* **258**(2), 334-344 (2008).

55. S. Kiatphuengporn, M. Chareonpanich and J. Limtrakul, Effect of unimodal and bimodal MCM-41 mesoporous silica supports on activity of Fe-Cu catalysts for CO$_2$ hydrogenation, *Chem. Eng. J.* **240**(527-533) (2014).

56. O. Martin, A. J. Martin, C. Mondelli, S. Mitchell, T. F. Segawa, R. Hauert, C. Drouilly, D. Curulla-Ferre and J. Perez-Ramirez, Indium Oxide as a superior catalyst for methanol synthesis by CO$_2$ hydrogenation, *Angew. Chem. Int. Edit.* **55**(21), 6261-6265 (2016).

57. X. Jiang, N. Koizumi, X. W. Guo and C. S. Song, Bimetallic Pd-Cu catalysts for selective CO$_2$ hydrogenation to methanol, *Appl. Catal. B-Environ.* **170**(173-185) (2015).

58. A. Garcia-Trenco, A. Regoutz, E. R. White, D. J. Payne, M. S. P. Shaffer and C. K. Williams, PdIn intermetallic nanoparticles for the hydrogenation of CO$_2$ to methanol, *Appl. Catal. B-Environ.* **220**(9-18) (2018).

59. K. P. Sun, W. W. Lu, M. Wang and X. L. Xu, Low-temperature synthesis of DME from CO$_2$/H$_2$ over Pd-modified CuO-ZnO-Al$_2$O$_3$,-ZrO$_2$/HZSM-5 catalysts, *Catal. Commun.* **5**(7), 367-370 (2004).

60. F. Yaripour, F. Baghaei, I. Schmidt and J. Perregaard, Catalytic dehydration of methanol to dimethyl ether (DME) over solid-acid catalysts, *Catal. Commun.* **6**(2), 147-152 (2005).

61. V. Vishwanathan, K. W. Jun, J. W. Kim and H. S. Roh, Vapour phase dehydration of crude methanol to dimethyl ether over Na-modified H-ZSM-5 catalysts, *Appl. Catal. A-Gen.* **276**(1-2), 251-255 (2004).

62. M. T. Xu, J. H. Lunsford, D. W. Goodman and A. Bhattacharyya, Synthesis of dimethyl ether (DME) from methanol over solid-acid catalysts, *Appl. Catal. A-Gen.* **149**(2), 289-301 (1997).

63. A. G. Gayubo, A. T. Aguayo, M. Castilla, A. L. Moran and J. Bilbao, Role of water in the kinetic modeling of methanol transformation into hydrocarbons on HZSM-5 zeolite, *Chem. Eng. Commun.* **191**(7), 944-967 (2004).

64 T. Takeguchi, K. Yanagisawa, T. Inui and M. Inoue, Effect of the property of solid acid upon syngas-to-dimethyl ether conversion on the hybrid catalysts composed of Cu-Zn-Ga and solid acids, *Appl. Catal. A-Gen.* **192**(2), 201-209 (2000).

65. G. Bonura, M. Cordaro, L. Spadaro, C. Cannilla, F. Arena and F. Frusteri, Hybrid $Cu-ZnO-ZrO_2/H-ZSM_5$ system for the direct synthesis of DME by CO_2 hydrogenation, *Appl. Catal. B-Environ.* **140**(16-24) (2013).

66. T. Witoon, T. Permsirivanich, N. Kanjanasoontorn, C. Akkaraphataworn, A. Seubsai, K. Faungnawakij, C. Warakulwit, M. Chareonpanich and J. Limtrakul, Direct synthesis of dimethyl ether from CO_2 hydrogenation over $Cu-ZnO-ZrO_2/SO_4^{2-}-ZrO_2$ hybrid catalysts: effects of sulfur-to-zirconia ratios, *Catal. Sci. Technol.* **5**(4), 2347-2357 (2015).

67. Y. Q. Zhao, J. X. Chen and J. Y. Zhang, Effects of ZrO_2 on the performance of $CuO-ZnO-Al_2O_3/HZSM-5$ catalyst for dimethyl ether synthesis from CO_2 hydrogenation, *J. Nat. Gas. Chem.* **16**(4), 389-392 (2007).

68. W. H. Gao Wengui, Wang Yuhao, Guo Wei, Jia Miaoyao Dimethyl ether synthesis from CO2 hydrogenation on La-modified $CuO-ZnO-Al_2O_3/HZSM-5$ bifunctional catalysts, *J. Rare. Earth.* **31**(470) (2013).

69. M. A. A. Aziz, A. A. Jalil, S. Triwahyono, R. R. Mukti, Y. H. Taufiq-Yap and M. R. Sazegar, Highly active Ni-promoted mesostructured silica nanoparticles for CO_2 methanation, *Appl. Catal. B-Environ.* **147**(359-368) (2014).

70. J. N. Park and E. W. McFarland, A highly dispersed $Pd-Mg/SiO_2$ catalyst active for methanation of CO_2, *J. Catal.* **266**(1), 92-97 (2009).

71. S. Akamaru, T. Shimazaki, M. Kubo and T. Abe, Density functional theory analysis of methanation reaction of CO_2 on Ru nanoparticle supported on TiO_2 (101), *Appl. Catal. A-Gen.* **470**(405-411) (2014).

72. P. A. U. Aldana, F. Ocampo, K. Kobl, B. Louis, F. Thibault-Starzyk, M. Daturi, P. Bazin, S. Thomas and A. C. Roger, Catalytic CO_2 valorization into CH_4 on Ni-based ceria-zirconia. Reaction mechanism by operando IR spectroscopy, *Catal. Today* **215**(201-207) (2013).

73. S. Rahmani, M. Rezaei and F. Meshkani, Preparation of highly active nickel catalysts supported on mesoporous nanocrystalline gamma-Al_2O_3 for CO_2 methanation, *J. Ind. Eng. Chem.* **20**(4), 1346-1352 (2014).

74. G. D. B. Weatherbee, C. H., Hydrogenation of CO_2 on Group VIII Metals: II. Kinetics and Mechanism of CO_2 Hydrogenation on nickel., *J. Catal.* **77**(460−472) (1982).

75. F. T. Fischer, H.; Dilthey, P., Über die reduktion von kohlenoxyd zu methan an verschiedenen metallen, *Brennst.-Chem.* **6** (265−271) (1925).

76. M. A. Vannice, The Catalytic synthesis of hydrocarbons from carbon monoxide and hydrogen, *Catal. Rev.: Sci. Eng.* **14**(153−191) (1976).

77. G. A. S. Mills, F. W., Catalytic methanation, *Catal. Rev.: Sci. Eng.* **8**(159−210) (1974).

78. H. C. Wu, Y. C. Chang, J. H. Wu, J. H. Lin, I. K. Lin and C. S. Chen, Methanation of CO_2 and reverse water gas shift reactions on Ni/SiO_2 catalysts: the influence of particle size on selectivity and reaction pathway, *Catal. Sci. Technol.* **5**(8), 4154-4163 (2015).

79. D. C. D. da Silva, S. Letichevsky, L. E. P. Borges and L. G. Appel, The Ni/ZrO$_2$ catalyst and the methanation of CO and CO$_2$, *Int. J. Hydrogen. Energ.* **37**(11), 8923-8928 (2012).

80. S. Tada and R. Kikuchi, Mechanistic study and catalyst development for selective carbon monoxide methanation, *Catalysis Sci. & Tech.* **5**(6), 3061-3070 (2015).

81. M. A. A. Aziz, A. A. Jalil, S. Triwahyono and S. M. Sidika, Methanation of carbon dioxide on metal-promoted mesostructured silica nanoparticles, *Applied Catalysis a-General* **486**(115-122) (2014).

82. C. T. G. Campbell, D. W., A Surface science investigation of the role of potassium promoters in nickel catalysts for CO hydrogenation., *Surf. Sci.* **123**(413−426) (1982).

83. H. Z. Liu, X. J. Zou, X. G. Wang, X. G. Lu and W. Z. Ding, Effect of CeO$_2$ addition on Ni/Al$_2$O$_3$ catalysts for methanation of carbon dioxide with hydrogen, *J. Natural Gas Chem.* **21**(6), 703-707 (2012).

84. M. T. Fan, K. P. Miao, J. D. Lin, H. B. Zhang and D. W. Liao, Mg-Al oxide supported Ni catalysts with enhanced stability for efficient synthetic natural gas from syngas, *Applied Surf. Sci.* **307**(682-688) (2014).

85. P. A. U. Aldana, F. Ocampo, K. Kobl, B. Louis, F. Thibault-Starzyk, M. Daturi, P. Bazin, S. Thomas and A. C. Roger, Catalytic CO$_2$ valorization into CH$_4$ on Ni-based ceria-zirconia. Reaction mechanism by operando IR spectroscopy, *Catalysis Today* **215**(201-207) (2013).

86. J. Ashok, M. L. Ang and S. Kawi, Enhanced activity of CO$_2$ methanation over Ni/CeO$_2$-ZrO$_2$ catalysts: influence of preparation methods, *Catal. Today* **281**(304-311) (2017).

87. S. Tada, T. Shimizu, H. Kameyama, T. Haneda and R. Kikuchi, Ni/CeO$_2$ catalysts with high CO$_2$ methanation activity and high CH$_4$ selectivity at low temperatures, *Int. J. Hydrogen. Energ.* **37**(7), 5527-5531 (2012).

88. G. L. Zhou, H. R. Liu, K. K. Cui, A. P. Jia, G. S. Hu, Z. J. Jiao, Y. Q. Liu and X. M. Zhang, Role of surface Ni and Ce species of Ni/CeO$_2$ catalyst in CO$_2$ methanation, *Appl. Surf. Sci.* **383**(248-252) (2016).

89. F. Ocampo, B. Louis and A. C. Roger, Methanation of carbon dioxide over nickel-based Ce$_{0.72}$Zr$_{0.28}$O$_2$ mixed oxide catalysts prepared by sol-gel method, *Appl. Catal., A-Gen.* **369**(1-2), 90-96 (2009).

90. G. Garbarino, D. Bellotti, E. Finocchio, L. Magistri and G. Busca, Methanation of carbon dioxide on Ru/Al$_2$O$_3$: Catalytic activity and infrared study, *Catal. Today* **277**(21-28) (2016).

91. M. Jacquemin, A. Beuls and P. Ruiz, Catalytic production of methane from CO$_2$ and H$_2$ at low temperature Insight on the reaction mechanism, *Catal. Today* **157** (1-4), 462-466 (2010).

Chapter 18

Computational Modelling of Charge Transport through Molecular Devices

Sandhya Rai[a], Sundaram Balasubramanian[b*] and Swapan K. Pati[a†]

[a]Theoretical Science Unit,
[b]Chemistry and Physics of Materials Unit and School of Advanced Materials,
Jawaharlal Nehru Centre for Advanced Scientific Research,
Jakkur, Bengaluru-560064, Karnataka, India
[*]bala@jncasr.ac.in; [†]pati@jncasr.ac.in

We have outlined a few first principles based on computational approaches, which are used to describe various molecular transport properties. A detailed description has been given on how from crystal structures of molecules, one can calculate charge carrier mobility values, which are useful for field effect transistors. In this respect, we have outlined the strategies one uses to calculate the inner and outer sphere reorganization energies and charge transfer integrals for both the electron and hole. We also have outlined a computational algorithm on how charges tunnel through molecular devices when they are connected between two metal electrodes on either side. Landauer-Buttiker approach coupled with self-consistent calculations has been detailed for calculating the current-voltage characteristics of several prototype molecular devices. The strategies on how to increase the charge carrier mobility in these systems have been described. We also have outlined the microscopic reasons behind the Coulomb-blockade, rectification and negative differential conductance behaviour that is often observed in molecular electronics. The use of room temperature ionic liquids as dielectric medium in single molecule transistors and similar devices has enhanced their characteristics. A survey of such devices and the reasons for the improved performance in terms of intermolecular ordering is also presented.

1. Introduction

Utilizing molecules and molecular assemblies in technological applications currently serves as an exciting and a dynamic area of research. The ease of synthetic modification, fabrication, processing and fine-tuning are some of the fascinating properties exhibited by these molecular systems that provide them an upper hand over their traditional inorganic counterparts. Their conformational flexibility can also give rise to interesting transport properties and a simple manipulation of their composition and geometry can lead to a wide variety of binding, optical and structural properties, which can be efficiently tuned to our needs. The applications span from nanotechnology to biotechnology to medicine owing to their peculiarly interesting physical and chemical properties. These materials are designed in such a way that allows for the overlap of unhybridized p_z orbitals to form an extended conjugation with delocalized π-electrons. It is this interplay between the π-electron and the geometric structure that makes room for a variety of novel concepts, giving rise to many fascinating properties. This chapter discusses in detail about the peculiar electronic properties of such systems that serve as the road to improvement in their properties and thus better performance in their respective technological application.

The molecular structures, which are typically of the order of a few nanometers, depict almost all electronic processes occurring in nature and hence, already possess a natural scale for use as functional nanodevices. Owing to the fact that silicon electronics is fast approaching a roadblock, dictated by both the laws of physics as well as the cost of production, organic molecular semiconductors have gained considerable attention for their application in electronics and opto-electronic devices such as organic field-effect transistors (OFETs), organic light emitting diodes (OLEDs), organic photovoltaic cells (OPVs) and various types of sensors.[1–3] The first suggestion that molecules could indeed be used as alternatives to silicon chips came from Aviram and Ratner,[4] who, discussed theoretically the possibility to construct a molecular rectifier, based on a single organic molecule. They suggested that a single molecule with a donor (D)-spacer-acceptor (A) structure would behave as a p-n junction diode when placed between two electrodes. The single electron transistor (SET) behaviour has already been observed in transport through these molecular materials like

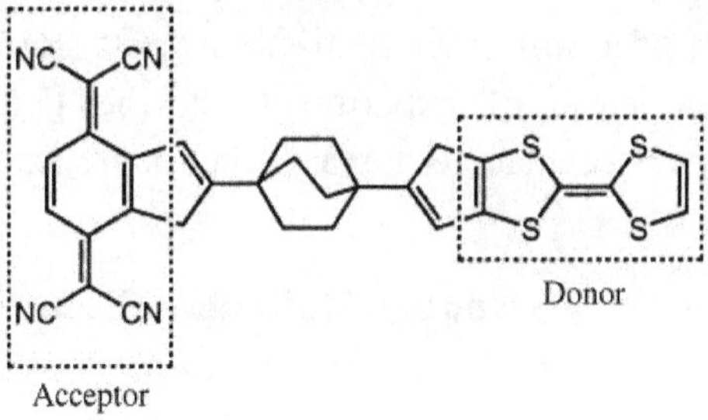

Fig. 1. Molecular analogue of a p-n junction, composed of a donor moiety tetrathiafulvalene (TTF) connected by a methylene bridge to an acceptor moiety tetracyanoquinodimethane (TCNQ) (taken from Ref. 12).

quantum dots, metallic and semiconducting nanoparticles and even in single π-conjugated organic molecules with several distinct charged states which can control its transport properties.[5,6] These systems have great advantage over conventional devices such as the low cost of production, easy fabrication, lightweight and large surface area production. A critical feature of any semiconducting device is the ability to control the electrical conductance, which in turn is characterized by the carrier mobility (μ) and is defined as the ratio between the charge drift velocity (v) and the driving electric field (F):

$$\mu = v/F \qquad (1)$$

At room temperature, the mobility for inorganic semiconductors like the single-crystal silicon can reach as high as 10^2–10^3 cm^2 V^{-1}s^{-1}, where as in that of organic materials it is around 10^{-5} cm^2 V^{-1}s^{-1}, which is too small for practical applications.[7–9] A variety of new materials are being designed to bring up the mobility to 0.1 cm^2 V^{-1}s^{-1} in thin films and 10 cm^2 V^{-1}s^{-1} in crystals.[10,11]. However, creating functional organic materials with large mobilities still remains as a central challenge in the field of organic electronics. In this respect, understanding the charge transport mechanism would give useful insights in order to build up the designing strategies for this class of materials. Though there have been numerous studies over the past decade, the theoretical understanding is still limited owing to the wide variety of structures involved and the complexities of organic materials. Another most challenging and yet most essential step towards the ultimate goal of molecular electronics, is the demonstration of a molecular field-effect transistor (FET). Although there has been a surge of experiments in the field, a number of fundamental issues pose a barrier towards the theoretical understanding.

2. Challenges in Understanding Molecular Electronics and Transport

2.1. *Transport*

Ultra-pure organic single crystals serve as a prototype to understand the charge transport properties.[11] For such ideal systems, the charge transport is limited by only thermal nuclear vibrations, and most theoretical studies are based on the tight-binding Hamiltonian.[12,13]

$$H = \sum_m \epsilon a_m^+ a_m + \sum_{m \neq n} V_{mn} a_m^+ a_n + \sum_\lambda \hbar \omega_\lambda (b_\lambda^+ b_\lambda + 1) \tag{2}$$

$$+ \sum_{m\lambda} g_{\lambda mm} \hbar \omega_\lambda (b_\lambda + b_{-\lambda}^+) a_m^+ a_m$$

$$+ \sum_{m \neq n, \lambda} f_{\lambda mn} \hbar \omega_\lambda (b_\lambda + b_{-\lambda}^+) a_m^+ a_n$$

The operators a_m^+ (a_m) and b_λ^+ (b_λ) are the creation (annihilation) operators for an electron at site m with an onsite energy ϵ_{mm} or a phonon belonging to mode λ with frequency ω_λ. V_{mn} is the electronic transfer integral coupling two adjacent molecules m and n. $g_{\lambda mm}(f_{\lambda mn})$ is a dimensionless coupling constant between phonon λ and the electronic term $\epsilon_{mm}(V_{mn})$. The main difficulty in describing the charge transport mechanism is that the relative magnitude of several factors in the Hamiltonian are not well understood for real systems.

2.2. Molecular electronics

The molecular electronics in application can be seen as a nanoscale molecule between macroscopic electrodes in a complete state of nonequilibrium, with each of the source and drain contacts trying to bring the molecule into equilibrium with its electrochemical potential, thus driving current through the system. One of the major problems is related to the length scales as it deals with a variety of length scales to determine where the classical regime ends and the quantum regime begins. The transport basically is divided into three regimes:[14,15]

2.2.1. Classical transport

Occurs when the length scale of the conductor is greater than both the momentum and phase relaxation lengths and resembles the transport in the macroscopic regime, obeying Ohm's Law.

2.2.2. Ballistic transport

This is the other extreme when the system under consideration is smaller than the momentum and phase relaxation length scales, the transport becomes ballistic, where the rate of transport is independent of the length of the system, and is described in terms of the transmission probabilities given by Landauer's formula:[14,16]

$$G_c = \frac{2e^2}{h} MT \tag{3}$$

Here, the conductance scales linearly with the transmission (T) and the number of eigenmodes (M) in the wire.

2.2.3. Coherent and diffusive transport

Another regime is where the momentum changes occur from impurity scatterings, but with the phase maintenance resulting in a coherent and a diffusive transport. The rate of the electron transfer is exponentially

            S. Rai, S. Balasubramanian & S. K. Pati

dependent on the length of the molecular bridge. This mechanism holds good for short wires with large HOMO-LUMO gaps such as oligoalkanes. It is also known as "superexchange" where the electron transfer proceeds through "virtual" orbitals, which are energetically well–separated from the Fermi levels of the electrodes.[17]

Apart from this, the energy scales of the system consisting of molecules with discrete energy levels connected to a continuous band structure of metallic electrodes pose a lot of challenges in understanding the molecular electronic transport. Some of the related issues are electrode-molecule coupling and contact surface physics, electronic structure of the molecules, device electrostatics and inelastic and thermal effects.[15,18,19]

3. Proposed Approaches and Models

3.1. *Charge transport*

The temperature dependence of mobility is a widely studied mechanism to understand the charge transport in semiconductors. Experiments show that the intrinsic charge mobility decreases with temperature, following a power–law behaviour: $\mu(T)\ T^{-\alpha}$.[20] Several models have been proposed to explain this aspect of transport behaviour. Wide-band theory[21], the polaron model[22-24] and the hopping model[25-33] have been used to study charge transport.

3.2. *Non-equilibrium transport*

The most widely used method for calculating the nonequilibrium transport in the nanoscale systems, considering all the aforementioned issues is the nonequilibrium Green's function formalism (NEGF), which for the case of the coherent transport boils down to the Landauer's formalism for calculating the current.[15]

$$I = \frac{2e}{h} \int_{-\infty}^{\infty} d\,E Tr[\Gamma_L G \Gamma_R G^\dagger](f(E,\mu_L) - f(E,\mu_R)) \qquad (4)$$

G represents the device Green's function, $\Gamma_{L,R}$, the imaginary part of the self-energies correspond to the broadening of the molecular energies, and $f(E)$ represents the Fermi–Dirac distributions at the two electrodes with electrochemical potentials μ_L and μ_R. In this chapter, we discuss how to calculate the charge mobility in the organic molecule semi conductor using the computational chemistry tools. We also discuss various aspects of the molecule connected between the macroscopic electrodes.

4. Theoretical Formulation for the Estimation of the Charge Carrier Transport

Various models such as the hopping model, charge diffusion through random walk and many others exist to describe carrier transport in molecular solids.[31–34] However, we have used the Marcus theory which is described below.

4.1. *Marcus theory of charge transfer*

In order to do a random walk simulation, the charge transfer rates between the molecular dimers is required. Most of the molecular semiconductors contain only one kind of molecule, the charge transfer in an adjacent molecular dimer, M_1 and M_2, is a self-exchange reaction process. The initial and final states can then be represented as $(M_1^+ M_2)$ and $(M_1 M_2^+)$, respectively, with M^+ denoting the charge on molecule M. As per the classical Marcus theory, the charge transfer rate is given as:[35]

$$k = \frac{V}{\hbar}\sqrt{\frac{\pi}{\lambda k_B T}}\,exp\left(-\frac{(\lambda + \Delta G^o)^2}{4\lambda k_B T}\right) \tag{5}$$

Here, V is the transfer integral between the initial and final states, λ is the reorganization energy, which is defined as the energy associated with the geometry relaxation during the charge transfer, and ΔG^o is the relevant change in the Gibbs free energy during the process of charge transfer. For a self-exchange reaction, ΔG^o becomes zero, and hence, it can be seen that the charge transport is modelled as a thermal activation process over a barrier of $\lambda/4$.

4.2. *Estimation of reorganization energy and transfer integral from first principles*

Equation (5) clearly suggests that there are two key factors that modulate the charge transfer rate, i.e., the reorganization energy and the transfer integral.

4.2.1. *Reorganization energy*

The reorganization energy has both internal and external energy contributions. The internal energy contributions arise from changes in the geometry of the molecular dimer when the electron transfer takes place whereas the changes in the surrounding media occurring due to charge transfer contribute to external energy contributions. In the case of

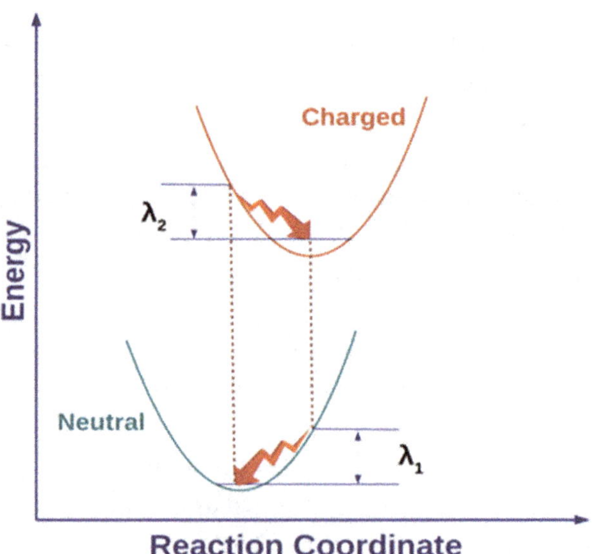

Fig. 2. The potential energy surfaces of the neutral and ionic molecules represent the charge transfer. (reproduced from Ref. 1).

organic dimer molecules, the external contribution is found to be negligible and is thus ignored.[13,31–34] The internal energy contribution is further divided into two parts: (a) The energy difference of the neutral molecule in the optimal charged geometry and in the equilibrium neutral geometry (λ_1) and (b) the energy difference of the charged molecule in these two geometries, (λ_2). The reorganization energies of the hole transport (λ^+) and electron transport (λ^-) are calculated from positively and negatively charged molecules, respectively (Figure 1).

4.2.2. *Transfer integral*

The charge transfer integral reflecting the strength of interaction between the molecular pairs plays a key role in understanding the charge transport properties. These integrals are greatly affected by the nature, size and relative orientations of the interacting monomer units, and thus are capable of establishing the structure–property relationship at a molecular level. The charge transfer integral is defined by the matrix element $H_{mn} = \langle \psi_m H \psi_n \rangle$, where H is the electronic Hamiltonian of the system and ψ_m and ψ_n are the wave functions of the two charge localized states.[33,36,37] Although accurate determination of these coupling matrix elements is a very tedious and challenging issue in this area of research, there have been reports on a few simplified approaches that provide the most reliable estimation of this parameter.

4.2.3. *Dimer–Splitting method*

It is based on the Koopman's Theorem and realization that the absolute value of the transfer integral for the hole (electron) is half of the valence (conduction) bandwidth, that is, the energy difference between the two highest occupied (lowest unoccupied) molecular orbitals [HOMO and HOMO-1 (LUMO and LUMO+1)] in a dimer. This approach provides a reasonable estimation of the charge transfer integrals on symmetric dimers, where the spatial overlap between the molecular orbitals is negligible.

4.2.4. *Fragment orbital approach*

This approach, considering the spatial overlap between the molecular orbitals, provides an accurate estimation of the charge transfer integrals.[38] Within this approach, the dimer molecular levels are expressed as the linear combination of individual monomer molecular levels (fragment orbitals), and the charge transfer integral (H_{mn}) can be obtained as the off–diagonal elements of the Kohn–Sham Hamiltonian matrix, which is expressed as:

$$H_{KS} = SCEC^{-1} \tag{6}$$

where S is the intermolecular overlap matrix, C is the molecular orbital coefficient, and E is the molecular orbital energy. This procedure allows direct calculations of the charge transfer integrals, including signs, without invoking the assumption of negligible spatial overlap. The generalized charge transfer integral in the orthogonal basis can then be calculated using the Lowdin transformation, which is expressed as

$$H_{mn} = H'_{mn} - \frac{1}{2}S(E_m + E_n) \tag{7}$$

5. Various Aspects of Molecules Connected between Macroscopic Electrodes

5.1. *The condition of negative differential conductance*

In certain electrical circuits, it is found that for a certain range of voltage, the current is a decreasing function of voltage. This behaviour is known as negative differential conductance (NDC). This behaviour is exploited in making amplifiers and oscillators and was reported for the first time in oligo (phenylene-ethylenes), functionalized by amine and nitro substituents by Reed and Tour.[39-43] The onset of the NDC behaviour has been studied theoretically and the cause of the phenomenon is attributed to charging, reduction of acceptor moiety, twisting of the ring structure leading to conformational changes, bias driven changes in the molecule–

electrode coupling etc.[44–49] This is heavily affected by the external factors which are needed to be incorporated like the conjugation/dimerization and the presence of donor–acceptor substituents and their roles in taking the molecule through a negative differential conductance which is subjected to an external electrical field.

5.2. *The role of donor–acceptor groups*

It is found that not only the NDC, but the asymmetry of the molecule also causes a modification in the I–V characteristics.[50,51]

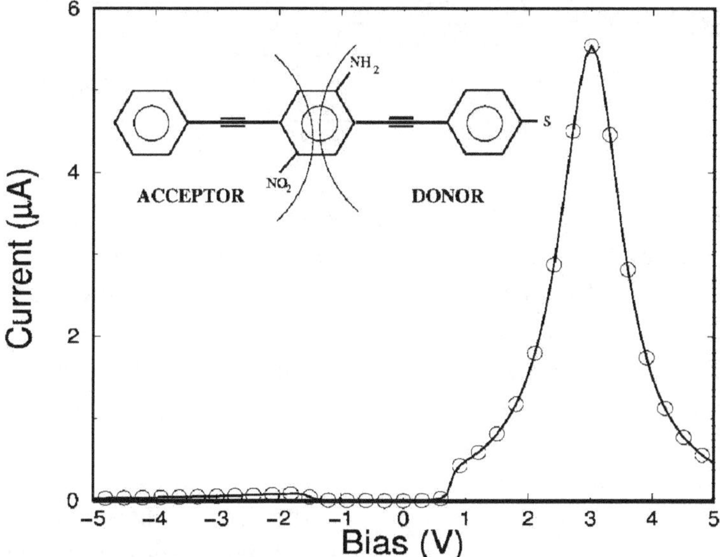

Fig. 3. Current-voltage characteristics for the 2-site system for $\epsilon_2 = -\epsilon_1 = 0.5$ eV and $t = 0.1$ eV. Inset is the Tour molecule (Reprinted (adapted) with permission from Ref. 17. Copyright (2008) American Chemical Society).

It is assumed that a part of the Tour molecule containing the donor group (NH_2) has a negative on-site energy and the part having the acceptor group (NO_2) has a positive on-site energy. The spatial variation of the bias on the structure is considered to drop as a ramp function, varying linearly from one electrode to other. With this potential, the energies for the dimer with Hamiltonian $H = \sum_{i=1,2} \epsilon_i (a_i^\dagger a_i) + \sum_i -t(a_i^\dagger a_{i+1} + hc)$, where ϵ_i is the on-site energy of site i and t is the hopping integral between the donor and acceptor, can be easily derived as

$$E_\pm = \frac{\epsilon_1 + \epsilon_2 - V}{2} \mp \frac{\sqrt{9(\epsilon_1 - \epsilon_2)^2 + 36t^2 + V^2 + 6V(\epsilon_1 - \epsilon_2)}}{6} \tag{14}$$

The coupling to the electrode modifies the bare Greens function of the molecule, which can be written as,

$$G_{1.2}(E,V) = \frac{\sqrt{V - 3(\epsilon_2 - \epsilon_1 - V)^2 + 36t^2}}{6t(E - E_1 + i\sum_L \underset{i}{+} \sum_R)} + \frac{V - 3(\epsilon_2 - \epsilon_1) + \sqrt{2(\epsilon_2 - \epsilon_1) - V^2 + 36t^2}}{6t(E - E_2 + i\sum_L \underset{i}{+} \sum_R))} \tag{15}$$

where $\sum_L$ and $\sum_R$ are the self-energies corresponding to the left and the right electrodes. These quantities are calculated within the Newns–Anderson model, and subsequently, current from the Landauer's formula. At a zero bias, with V=0, the presence of different on-site energies opens up a gap larger than that for a purely hopping model ($\epsilon_2 = \epsilon_1 = 0eV$) near the zero of energy indicating the preference of the electrons to stay at the atomic site with a negative on—site energy.[52] The equilibrium transmission is found to be large for a purely hopping model since it corresponds to an equal distribution of charge. The system becomes insulating with the inclusion of different on–site energies. This happens because the charge transfer and the zero–bias transmission reduces the preferential charge localization. As shown in Figure 3, the current is negligible around the zero of energy and around a bias of 1 V, there is a small jump in the current. With an increase in the forward bias, around a bias of 3 V, the current shows a sharp rise and fall, indicating a strong negative differential conductance (NDC). On the other hand, with an increase in the reverse bias, the system continues to remain insulating with a negligible current.

For speculating the reason of the NDC, the variation of the energy levels (E_k) of the bare molecular dimer with a bias and the numerator of the Green's function, $\langle 1|k \rangle \langle 2|k \rangle$ where k=1, 2 are the eigenstates are studied as a function of bias (Figure 4). With an increase in the forward bias, the energy levels come close to one another up to the critical bias V_c at which the NDC is seen, after which they move farther away. The contribution to the eigenstate dimer coefficient around this V_c, increases quite sharply suggesting a more delocalized state. The average inverse participation ratio is calculated, which is defined as the extent of delocalization for a given eigenstate with energy E_k.

$$IPR_{av} = \frac{1}{D(E)} \frac{1}{N} \sum_k P_k^{-1} \, \delta(E - E_k) \tag{16}$$

where P_k^{-1} is the IPR, defined as $P_k^{-1} = (1/N) \sum_j |\psi(j,k)|^4$ where j is the atomic index and $D(E)$ is the density of states. Figure 8 shows a strong dip in the values of IPR$_{av}$ around the critical bias confirming a complete delocalization in the system, while at other values of the bias, the IPR$_{av}$ is much larger due to the localized nature of the eigenstates. The critical bias V_c, can be quantified by minimizing the gap between the energies with respect to the applied bias, and obtain $V_c = 3(\epsilon_2 - \epsilon_1)$. At this critical bias, the energies take the values $\mp t$, precisely the energies of the system with $\epsilon_2 = \epsilon_1 = 0$.[52] However, with an increase in the reverse bias, the energy levels start diverging away from their zero bias gap making the system more insulating, explaining the small current that is observed in Figure 3 for a negative bias. This is how the observations are explained. Initially for a small bias, as noted before, the system tends to accumulate its charge density at the site with a lower on-site energy. Such localization makes the system insulating. If this site is closer to the electrode with a higher chemical potential, an increase in the bias makes the charges tend to move towards the other site. When the bias equals the critical bias V_c, where the NDC is seen, the charge densities are equally distributed at both sites with no preference of one site over another, describing a situation where both the on-site energies are equal. Further increase of the bias would localize the charges on the other site resembling an insulating dimer with its on-site energies interchanged, precisely the case as in the reverse bias situation.

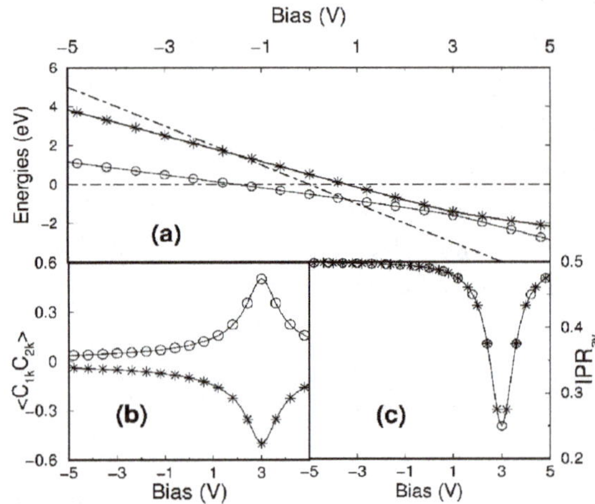

Fig. 4. (a) Variation of the two levels (circles and stars) of the 2-site system with the applied bias, for $\epsilon_2 = -\epsilon_1 = 0.5$ eV and $t = 0.1$ eV. The dotted lines indicate the variation of the Fermi energies of the electrodes with bias. (b) The numerator of the Greens function matrix element for the corresponding energy levels shown in (a). 1 and 2 represent the site index and k specifies the corresponding level. (c) The IPR$_{av}$ for the levels shown in (a). (Reprinted (adapted) with permission from Ref. 17. Copyright (2008) American Chemical Society).

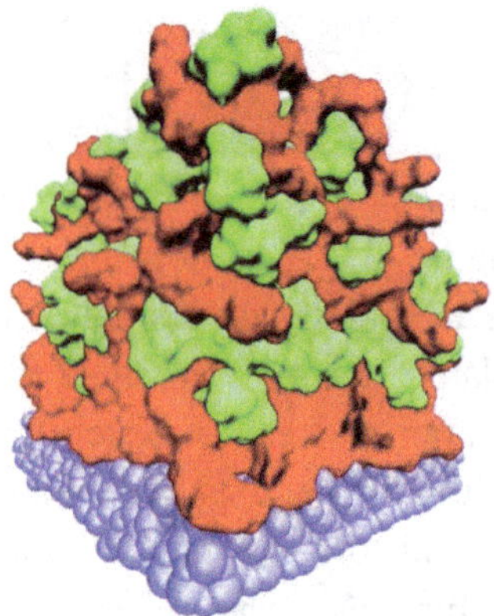

Fig. 5. Double layer structuring of [C$_8$mim][NTf$_2$] ionic liquid on a charged mica surface. Red colour denotes cation locations while green colour denotes anion locations. The capacitance arising from this double layer enhances the electric field experienced by a molecule in a three-terminal molecular device. (Adapted from Ref. 53.)

6.   Three Terminal Single Molecule Devices and the Role of Dielectric

There has been much interest of late, in the design of the three terminal (source, drain and gate) devices using single molecules for charge transport[54]. In such devices, the source and drain electrodes are constituted by graphene or similar two dimensional, conductors to which the molecule can be covalently bonded through functional groups. These three-terminal devices behave as transistors, in the sense that the energy levels of the molecule and thus the current through the device can be controlled by the gate voltage. Typically, an array of such devices can be constructed using lithographic methods[55]. In conventional single molecule transistors, either a back gate (Al/Al_2O_3)[56] or an electrochemical gate[57] are employed. These suffer from the handicaps of requiring high gate voltages to generate reasonably large electric fields at the molecular junction while the latter cannot be used at low temperatures due to the use of water as the solvent. Thus, real applications of these devices have not yet seen the light of the day. A recent advancement towards enhancing the performance of these single molecule transistors has been the use of room temperature ionic liquids (RTILs) as the dielectric medium[58]. RTILs are liquids at or near ambient conditions and are composed purely of ions. Unlike electrolyte solutions which are salts dissolved in a solvent (more often than not, water), RTILs are just salts whose thermodynamically stable phase at room temperature and pressure is the liquid. Invariably, RTILs are composed of inorganic molecular anions and organic cations. The spread or the distribution of the charge over a volume larger than in, say, halides or alkali ions make the crystalline phase of RTILs to have lower cohesive energies than traditional alkali halide salts. RTILs have been used in the synthesis of nanoparticles[59], as chemical reaction media[60], product extraction[61], electrolytes[62], in the exfoliation of graphene[63] and so on. The innumerable number of cation-anion combinations possible make RTILs to be considered as designer solvents – in principle, one can find RTILs with desired properties by a suitable combination of anion and cation. RTILs have been modelled using various computational methods such as, quantum chemical calculations, atomistic molecular dynamics (MD) simulations, coarse grained MD simulations, and ab initio MD simulations.[64] Effective force fields to model RTILs with capabilities to quantitatively predict several physical properties have been developed over the years.[65,66] These have been employed to study the behavior of RTILs near a charged electrode; the formation of an electrical double layer at such a solid interface is observed[67] (Figure 5), the capacitance of which has been used to increase the performance of energy storage devices. In particular, the use of oligomeric ionic liquids, wherein multiple charge centers constituting the cation are linked by ether groups (see Figure 6) have been shown experimentally to offer extremely good I-V characteristics in a FET device in contrast to devices using monomeric ILs.[68] The features of the double layer formed out of the ions near the electrodes can be tuned by a choice of IL, which then can be used to enhance the device performance. In a similar vein as above, ionic liquids have been employed as a dielectric in three-terminal single molecule transistors. Unlike aqueous electrolyte solutions, RTILs exhibit a large liquid range (close to $200^{o}C$) and a wide electrochemical window (typically around 3 V).[69] Thus, RTILs are ideally suited to be used as electrolytes in electrochemical cells and in such three-terminal devices. Xin et al. recently studied the device characteristics of three molecular junctions. The molecules considered biphenyl, triphenyl and hexaphenyl were functionalized at their terminii for covalent bonding to the graphene electrodes (source and drain).[55] The gate was patterned by chromium on platinum. Ionic liquid was used as the electrolyte. While the phenyl molecules themselves have large bandgaps (more than 3.5 eV) the device showed exceptional characteristics. The Fermi level of graphene lies in the HOMO-LUMO gap of each of these aromatic molecules. With no bias from the gate, the hexaphenyl device shows a higher slope in the I-V plot than the other two molecules. The advantage of using ionic liquids as the electrolyte in such a three-terminal device is that the formation of a double layer at the gate electrode enhances the electric field experienced by the single molecule. Thus, for the same current, the device can be operated at lower gate voltages, which is an advantage.

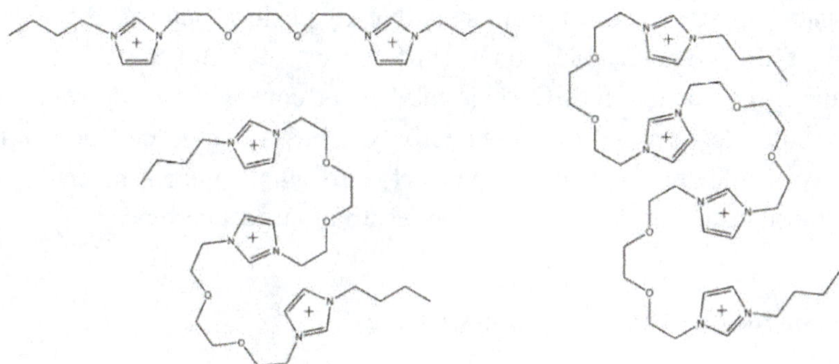

Fig. 6. Molecular structures of oligomeric ionic liquids whose use in molecular electronic devices provides superior I-V characteristics.

Not only do the charges of the ionic liquid accumulate at the gate electrode, they organize themselves around the molecule as well.[55] Atomistic MD simulations of triphenyl soaked in a RTIL show considerable structuring of both the anions and the cations around the molecule (see Figure 7) for the respective pair correlation functions). For a given drain voltage, as a function of gate voltage, the device with either triphenyl or hexaphenyl molecules show an ambipolar conductivity. As

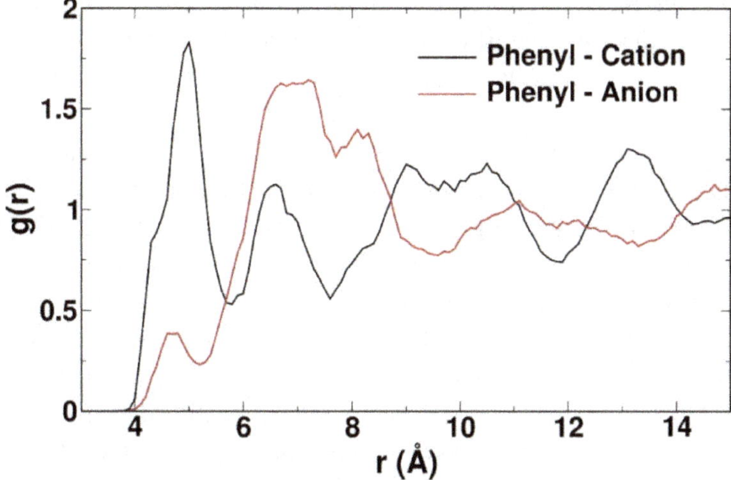

Fig. 7. Pair correlation functions between the central phenyl ring of triphenyl soaked in [DEME][TFSI] ionic liquid at ambient conditions, obtained from atomistic MD simulations.

the gate voltage changed from negative to positive values, the Fermi level of the graphene electrodes shift towards the molecule's LUMO (relatively) from its HOMO. This movement causes these molecules to show both electron and hole conductivity. Thus, the gate voltage can be used to modulate the charge transport with gate voltages much lower than those used in conventional three-terminal molecular devices.[55]

7. Conclusions and Future Perspective

In this chapter, we outlined the first principles based computational scheme and theories involved in describing the carrier mobility and transport properties of small organic molecular semiconductors. Based on these studies, one can come up with several design strategies aimed at a high mobility and an efficient charge transport within the material.

High mobility in materials is favoured by small reorganization energies and a large intermolecular electronic coupling. Computations suggest that rigid structures with an extended conjugation, and substitutions like cynation, which favour delocalization, favour small reorganization energy. Structural modification of the molecule or modification in the environmental conditions favours different phases of the crystal, resulting in larger intermolecular transfer integrals. It is concluded that for an efficient charge transfer network, a uniformly distributed intermolecular charge transfer is necessary. We have also elaborated the importance of the potential profile, i.e., the way an external applied field would fall across the molecule, and how the presence of electron correlations in the molecule can drastically change its shape, thereby changing the device response along with it. The NDC phenomenon has captured the attention of scientists in the recent years, due to its immediate application in switching devices. It has been shown that the presence of a two sublattice structure, caused either a lattice distortion or substituents with large dipolar strength, can result in the NDC, when the external field has a strong spatial dependence. It is reported that the ratio t (hopping integral): $\epsilon_2 - \epsilon_1$ (measure of the dipolar character), is very crucial in determining the nature of the I–V characteristics. Three features, namely (1) the critical bias (2) the sharpness of the NDC peak and (3) the extent of asymmetry in the I–V curves is sensitive to this ratio.

It should be noted that the thermal fluctuations in the molecular orientations may be strong, in organic molecular systems. This could result in large fluctuations in the intermolecular transfer integrals, which needs to be treated accordingly. This kind of a dynamic disorder is responsible for the origin of the band–like behaviour, and has been incorporated in the one–dimensional system via the Su–Schrieffer–Heeger model. However, for higher dimensions it deserves further studies. Although, only one of the technologically useful nonlinear phenomenons, i.e., NDC, has been discussed in detail, the other phenomena such as multistability and hysteresis, which arise out of the structural changes in a molecule during transport together with inelastic processes due to the interactions at different energy and length scales in organic systems, deserve attention.

Room temperature ionic liquids enhance the electrical characteristics of single molecule devices through the formation of electrical double layer at the electrodes as well as by the structuring of ions around the single molecule. These systems are amenable to molecular simulations and ab initio MD simulations which are likely to be pursued by various research groups in the near future.

Acknowledgements

SKP thanks Dr. S. Lakshmi, Dr. Ayan Datta, Dr. Sudipta Dutta and Dr. Sasmita Mohakud, who have contributed to various aspects of this topic as part of their graduate research work in his group. SB and SKP thank DST for support.

References

1. P. K. Chattaraj and S. K. Ghosh, *Concepts and Methods in Modern Theoretical Chemistry: Statistical Mechanics*. CRC Press, New York (2013).
2. Y. Shirota and H. Kageyama, Charge carrier transporting molecular materials and their applications in devices, *Chem. Rev.,* **107**(4), 953–1010 (2007).

3.   A. R. Murphy and J. M. Frechet, "Organic semiconducting oligomers for use in thin film transistors," *Chem. Rev.*, **107**(4), 1066–1096 (2007).

4.   A. Aviram and M. A. Ratner, Molecular rectifiers, *Chem. Phys. Lett.*, **29**(2), 277–283 (1974).

5.   F. Chen, J. He, C. Nuckolls, T. Roberts, J. E. Klare, and S. Lindsay, A molecular switch based on potential-induced changes of oxidation state, *Nano Lett.*, **5**(3), 503–506 (2005).

6.   D. Gust, T. A. Moore, and A. L. Moore, Molecular switches controlled by light, *Chem. Commun.*, 1169–1178 (2006).

7.   R. Logan and A. Peters, Impurity effects upon mobility in silicon, *J. App. Phys.*, **31**(1), 122–124 (1960).

8.   T. Dürkop, S. Getty, E. Cobas, and M. Fuhrer, Extraordinary mobility in semiconducting carbon nanotubes, *Nano Lett.*, **4**(1), 35–39 (2004).

9.   A. Tsumura, H. Koezuka, and T. Ando, Macromolecular electronic device: Field-effect transistor with a polythiophene thin film, *Appl. Phys. Lett.*, **49**(18), 1210–1212 (1986).

10.   C. Reese and Z. Bao, Organic single-crystal field-effect transistors, *Materials Today*, **10**(3), 20–27 (2007).

11.   M. Gershenson, V. Podzorov, and A. Morpurgo, Colloquium: Electronic transport in single-crystal organic transistors, *Rev. Mod. Phys.*, **78**(3), 973 (2006).

12.   M. Pope and C. E. Swenberg, *Electronic Processes in Organic Molecular Crystals*, Oxford University Press, New York (1987); C.W. Tang, S.A. VanSlyke, Organic Electroluminescent Diodes, *Appl. Phys. Lett.*, **51**(913), 89–103 (1982).

13.   V. Coropceanu, J. Cornil, D. A. da Silva Filho, Y. Olivier, R. Silbey, and J.-L. Brédas, Charge transport in organic semiconductors, *Chem. Rev.* **107**(4), 926–952 (2007).

14.   N. Tao, Electrochemical fabrication of metallic quantum wires, *J. Chem. Education*, **82**(5), 720 (2005).

15.   S. Datta, *Electronic Transport in Mesoscopic Systems*, Cambridge University Press, Cambridge (1997).

16.   R. Landauer, R. Landauer, *IBM J. Res. Dev.*, **1**, p. 223, (1957).

17.   S. Lakshmi, S. Dutta, and S. K. Pati, Molecular electronics: Effect of external electric field, *The J. Phys. Chem. C*, **112**(38), 14718–14730 (2008).

18.   F. Zahid, M. Paulsson, and S. Datta, Electrical conduction through molecules, in *Advanced Semiconductor and Organic Nano-techniques*, Elsevier, New York (2003), pp. 1–41.

19.   V. Mujica, M. Kemp, and M. A. Ratner, Electron conduction in molecular wires. I. A scattering formalism, *The J. Chem. Phys.*, **101**(8), 6849–6855 (1994).

20.   O. D. Jurchescu, J. Baas, and T. T. Palstra, Effect of impurities on the mobility of single crystal pentacene, *Appl. Phys. Lett.*, **84**(16), 3061–3063 (2004).

21. V. Podzorov, E. Menard, A. Borissov, V. Kiryukhin, J. Rogers, and M. Gershenson, Intrinsic charge transport on the surface of organic semiconductors, *Phys. Rev. Lett.*, **93**(8), 086602 (2004).

22. K. Hannewald and P. Bobbert, Anisotropy effects in phonon-assisted charge-carrier transport in organic molecular crystals, *Phys. Rev. B*, **69**(7), 075212 (2004).

23. L. Wang, Q. Peng, Q. Li, and Z. Shuai, Roles of inter-and intramolecular vibrations and band-hopping crossover in the charge transport in naphthalene crystal, *The J. Chem. Phys.*, **127**(4), 044506 (2007).

24. L. Wang, Q. Li, and Z. Shuai, Effects of pressure and temperature on the carrier transports in organic crystal: A first-principles study, *The J. Chem. Phys.*, **128**(19), 194706 (2008).

25. H. Bässler, Charge transport in disordered organic photoconductors a Monte Carlo simulation study, *Physica Status Solidi (b)*, **175**(1), 15–56 (1993).

26. G. Nan, X. Yang, L. Wang, Z. Shuai, and Y. Zhao, Nuclear tunneling effects of charge transport in rubrene, tetracene, and pentacene, *Phys. Rev. B*, **79**(11), 115203 (2009).

27. D. Eley, H. Inokuchi, and M. Willis, The semi-conductivity of organic substances. Part 4. Semi-quinone type molecular complexes, *Discussions of the Faraday Soc.*, **28**, 54–63 (1959).

28. Y. Song *et al.*, A cyclic triphenylamine dimer for organic field-effect transistors with high performance, *J. American Chem. Soc.*, **128**(50), 15940–15941 (2006).

29. X. Yang, Q. Li, and Z. Shuai, Theoretical modelling of carrier transports in molecular semiconductors: Molecular design of triphenylamine dimer systems, *Nanotechnology*, **18**(42), 424029 (2007).

30. X. Yang, L. Wang, C. Wang, W. Long, and Z. Shuai, Influences of crystal structures and molecular sizes on the charge mobility of organic semiconductors: Oligothiophenes, *Chem. Mater.*, **20**(9), 3205–3211 (2008).

31. W.-Q. Deng and W. A. Goddard, Predictions of hole mobilities in oligoacene organic semiconductors from quantum mechanical calculations, *The J. Phys. Chem. B*, **108**(25), 8614–8621 (2004).

32. Y. Zhang, X. Cai, Y. Bian, X. Li, and J. Jiang, "Heteroatom substitution of oligothienoacenes: From good p-type semiconductors to good ambipolar semiconductors for organic field-effect transistors," *The J. Phys. Chem. C*, **112**(13), 5148–5159 (2008).

33. G. R. Hutchison, M. A. Ratner, and T. J. Marks, Hopping transport in conductive heterocyclic oligomers: Reorganization energies and substituent effects, *J. American Chem. Soc.*, **127**(7) 2339–2350 (2005).

34. G. Nan, L. Wang, X. Yang, Z. Shuai, and Y. Zhao, Charge transfer rates in organic semiconductors beyond first-order perturbation: From weak to strong coupling regimes, *The J. Chem. Phys.*, **130**(2), 024704 (2009).

35. R. A. Marcus, Electron transfer reactions in chemistry: Theory and experiment (nobel lecture), *Angewandte Chemie International Edition in English*, **32**(8), 1111–1121, (1993).

36. X.-Y. Li, Electron transfer between tryptophan and tyrosine: Theoretical calculation of electron transfer matrix element for intramolecular hole transfer, *J. Computa. Chem.*, **22**(6), 565–579 (2001).

37. E. F. Valeev, V. Coropceanu, D. A. da Silva Filho, S. Salman, and J.-L. Brédas, Effect of electronic polarization on charge-transport parameters in molecular organic semiconductors, *J. American Chem. Soc.*, **128**(30), 9882–9886 (2006).

38. A. Troisi and G. Orlandi, Hole migration in dna: A theoretical analysis of the role of structural fluctuations, *The J. Phys. Chem. B*, **106**(8), 2093–2101 (2002).

39. J. Chen, W. Wang, M. Reed, A. Rawlett, D. Price, and J. Tour, Room-temperature negative differential resistance in nanoscale molecular junctions, *Appl. Phys. Lett.*, **77**(8), 1224–1226 (2000).

40. M. Reed, J. Chen, A. Rawlett, D. Price, and J. Tour, Molecular random access memory cell, *Appl. Phys. Lett.*, **78**(23), 3735–3737 (2001).

41. X. Xiao, L. A. Nagahara, A. M. Rawlett, and N. Tao, Electrochemical gate-controlled conductance of single oligo (phenylene ethynylene)s, *J. American Chem. Soc.*, **127**(25), 9235–9240 (2005).

42. J. He and S. M. Lindsay, On the mechanism of negative differential resistance in ferrocenylundecanethiol self-assembled monolayers, *J. American Chem. Soc.*, **127**(34), 11932–11933 (2005).

43. C. Zhang, M.-H. Du, H.-P. Cheng, X.-G. Zhang, A. Roitberg, and J. Krause, Coherent electron transport through an azobenzene molecule: A light-driven molecular switch, *Phys. Rev. Lett.*, **92**(15), 158301 (2004).

44. J. M. Seminario, R. A. Araujo, and L. Yan, Negative differential resistance in metallic and semiconducting clusters, *The J. Phys. Chem. B*, **108**(22), 6915–6918 (2004).

45. H. Dalgleish and G. Kirczenow, A new approach to the realization and control of negative differential resistance in single-molecule nanoelectronic devices: Designer transition metal- thiol interface states, *Nano Lett.*, **6**(6), pp. 1274–1278 (2006).

46. R. Liu, S.-H. Ke, H. U. Baranger, and W. Yang, Negative differential resistance and hysteresis through an organometallic molecule from molecular-level crossing, *J. American Chem. Soc.*, **128**(19), 6274–6275 (2006).

47. J. M. Seminario, A. G. Zacarias, and J. M. Tour, Theoretical study of a molecular resonant tunneling diode, *J. American Chem. Soc.*, **122**(13), 3015–3020, (2000).

48. J. M. Seminario, A. G. Zacarias, and P. A. Derosa, Analysis of a dinitro-based molecular device, *The J. Chem. Phys.*, **116**(4), 1671–1683, (2002).

49. J. Han and V. H. Crespi, Asymmetry in negative differential resistance driven by electron–electron interactions in two-site molecular devices, *Appl. Phys. Lett.*, **79**(17), 2829–2831, 2001.

50. S. Lakshmi and S. K. Pati, Current-voltage characteristics in donor-acceptor systems: Implications of a spatially varying electric field, *Phys. Rev. B*, **72**(19), 193410 (2005).

51. S. Dutta, S. Lakshmi, and S. K. Pati, Comparative study of electron conduction in azulene and naphthalene, *Bull. of Materials Sci.*, **31**(3), 353–358 (2008).

52. R. G. Pearson, Absolute electronegativity and hardness: Application to inorganic chemistry, *Inorganic Chem.*, **27**(4), 734–740 (1988).

53. R.S. Payal and S. Balasubramanian, Dynamic atomic force microscopy for ionic liquids: Massless model shows the way, *ChemPhysChem*, **13**(13), 3085-3086 (2012).

54. M.L. Perrin, E. Burzuri and H. van der Zant, Single-molecule transistors, *Chem. Soc. Rev.*, **44**, 902–919 (2015).

55. Na Xin, X. Li, C. Jia, Y. Gong, M. Li, S. Wang, G. Zhang, J. Yang and X. Guo, Tuning charge transport in aromatic-ring single-molecule junctions via ionic liquid gating, *Angewandte Chemie International Edition*, **57**(43), 14026–14031 (2018).

56. A. Bergvall, K. Berland, P. Hyldgaard, S. Kubatkin and T. Loefwander, Graphene nanogap for gate-tunable quantum-coherent single-molecule electronics, *Phys. Rev. B*, **84**, 15541 (2011).

57. X. Li, B. Xu, X. Xiao, X. Yang, L. Zang and N. Tao, Controlling charge transport in single molecules using electrochemical gate, *Faraday Discussions*, **131**, 111–120 (2006).

58. T. Welton, Room-temperature ionic liquids. Solvents for synthesis and catalysis, *Chem. Rev.*, **99**(8), 2071–2084 (1999).

59. B.G. Trewyn, C.M. Whitman and V.S.-Y. Lin, Morphological control of room-temperature ionic liquid templated mesoporous silica nanoparticles for controlled release of antibacterial agents, *Nano Lett.*, **4**(11), 2139–2143 (2004).

60. M.J. Earle, K.R. Seddon, C.J. Adams and G. Roberts, Friedel–Crafts reactions in room temperature ionic liquids, *Chem. Commun.*, 2097–2098 (1998).

61. S. Pandey, Analytical applications of room-temperature ionic liquids: A review of recent efforts, *Analytica Chimica Acta*, **556**(1), 38–45 (2006).

62. T. Y. Zhao and T. Bostrom, Application of ionic liquids in solar cells and batteries: A review, *Current Organic Chem.*, **19**(6), 556–566 (2015).

63. M. Matsumoto, Y. Saito, C. Park, T. Fukushima and T. Aida, Ultrahigh-throughput exfoliation of graphite into pristine 'single-layer' graphene using microwaves and molecularly engineered ionic liquids, *Nature Chem.*, **7**, 730–736 (2015).

64. A. Mondal, S. Das and S. Balasubramanian, Recent advances in modeling green solvents, *Current Opinion in Green and Sustainable Chem.*, **5**, 37–43 (2017).

65. A. Mondal and S. Balasubramanian, Quantitative prediction of physical properties of imidazolium based room temperature ionic liquids through determination of condensed phase site charges: A refined force field, *J. Phys. Chem. B*, **118**(12), 3409–3422 (2014).

66. A. Mondal and S. Balasubramanian, Refined all-atom potential for imidazolium-based room temperature ionic liquids: Acetate, dicyanamide, and thiocyanate anions, *J. Phys. Chem. B*, **119**(34), 11041–11051 (2015).

67. R.S. Payal and S. Balasubramanian, Orientational ordering of ionic liquids near a charged mica surface, *ChemPhysChem*, **13**(7), 1764–1771 (2012).

68. M. Matsumoto, D.S. Shimizu, R. Sotoike, M. Watanabe, Y. Iwasa, Y. Itoh, and T. Aida, Exceptionally high electric double layer capacitances of oligomeric ionic liquids, *J. American Chem. Soc.*, **139**(45), 16072–16075 (2017).

69. A.M.O'Mahony, D.S. Silvester, L. Aldous, C. Hardacre and R.G. Compton, Effect of water on the electrochemical window and potential limits of room-temperature ionic liquids, *J. Chem. and Engin. Data*, **53**(12), 2884–2891 (2008).

Chapter 19

Predictive Models of Multi-scale Behavior of Materials: Mechanistic versus Machine Learning Schemes

Raagya Arora and Umesh V. Waghmare[*]

Theoretical Sciences Unit and School of Advanced Materials Jawaharlal Nehru Centre for Advanced Scientific Research, Jakkur, Bangalore 560064, Karnataka, India
*waghmare@jncasr.ac.in

A model of a system is an approximate representation of its real structure and behavior. Computer simulations of a material mimic its model numerically, and their accuracy in prediction of material-specific behavior is singularly determined by how realistic the model is. The construction of a mechanistic model of a material involves identification of the degrees of freedom relevant to the property of interest. The model captures effective interactions between these by integrating out the rest, a step known as *coarse-graining* which filters out the features of short length and time scales. Reducing number of degrees of freedom while maintaining close similarity to material's real behavior is the challenge and science of modeling. *First-principles* density functional theory provides the fundamental input at the shortest scales in hierarchical modeling to capture of phonons, phase transitions, dislocations and continuum description of materials. When separation between relevant and irrelevant variables is not straightforward or there are several coupled groups of relevant variables, machine learning can be useful in developing predictive models. While machine learning needs big data and result in uninterpretable models, machine learning combined with dimensional analysis and physical laws seems promising in developing insightful models of complex material behavior from relatively *small* data.

1. Introduction

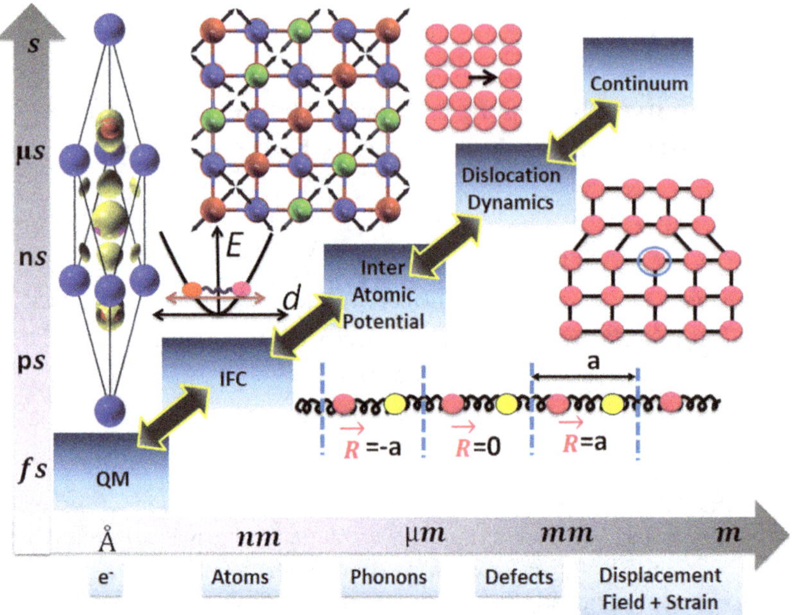

Fig. 1. Multi-scale structure, degrees of freedom and governing mechanics of a crystalline material.

At the fundamental level, the structure of a material is specified by spatio-temporal arrangement of electrons and nuclei or atoms, and changes in this structure in response to external fields determine its properties. The structure and properties of a material are thus intimately related to each other. The spatial structure of a material varies over several decades of scales (see Figure 1), from electronic length-scales of sub-angstrom features, bond-lengths of angstrom scales to dislocations involving micron length-scales. Similarly, the temporal structure of a material involves electronic time-scales at femto seconds, vibrations at pico-seconds and microstructural dynamics at time scales longer than nano-seconds. These multi-scale structural features interestingly govern the behavior of a material relevant to most technological applications. Predictive models of materials' behavior necessarily involve a framework for consistently capturing the dependencies between structural processes occurring at varied length and time-scales, which are governed by quantum mechanics at the smallest scales, dislocation dynamics at meso-scale and continuum mechanics at macroscopic scales (see Figure 1). Such modeling of materials has traditionally been based on the principles of symmetries and statistical mechanics, and is recently impacted by advances in machine learning particularly when the mechanistic schemes are not quite effective. The goal of this chapter is to introduce fundamental ideas in these modeling schemes, with illustrations of their applications to structural phase transitions and functional properties of materials.

2. Principles of Modeling and Simulation

A model of a system is an approximate representation of its real structure and behavior. For example, an architectural model gives a reasonably clear picture of a building, or a museum model of a wind-mill gives an idea of how it works. In the present context, we focus on computable models of a material, which necessarily are approximate mathematical description of its structure and behavior. They are typically derived to capture the material behavior at certain length and time-scales and involve approximations of ignoring the material's features at other scales or effectively including their effects. Such a model is *computable* because it is based on a finite number of *degrees of freedom (dof)* or variables or entities, while a real material in principle has infinite of these. Reduction in the number of degrees of freedom in representing a system with a model while maintaining close similarity to its real behavior is the challenge and art of modeling. Thus, most models involve simplification and attempt to capture the essence of a material system at given scales.

A computable model is a *parametrized* mathematical function of material features in which the numerical parameters vary from one material to another. For example, specific heat (C_v) of a metal as a function of temperature T is $C_v = aT + bT^2$, where a and b are the parameters, whose numerical values vary from one metal to another. These parameters may be derived from fundamental physics (*"first-*

principles" or "*ab initio*") or by fitting to empirical data. Thus, such a model is material-specific, and prediction of such model parameters is central to the prediction or design of new materials.

Transferability of a model is defined by its accuracy in capturing the behavior of a material in conditions outside the domain used in its derivation. Sometimes, the parametrized form of a model itself may *not* be transferable from one material to another. *Interpretability* of a model derives from its form, giving insight into the physical mechanism governing the property. For example, linear dependence of specific heat on T essentially highlights the presence of almost free electrons near the Fermi level of a material, which dominate the specific heat at low temperatures. We note that there is also *universality* in behavior of rather different models, particularly in the conditions where the model's behavior is scale-invariant, such as near a critical point of phase transitions. While this is important and central to understanding of phase transitions in materials, our focus here is on development of material-specific models, motivated by the dream of computational discovery of new materials.

2.1. *Mechanistic modeling*

In the mechanistic scheme of modeling of materials, a central object is the Hamiltonian H or the energy as a function of degrees of freedom, which governs their dynamics. At T=0 K, the system takes the ground state of H, corresponding to minimization of the energy in classical analysis. At T>0 K, thermodynamic behavior of the material as a collection of large number of degrees of freedom can be determined within the framework of statistical mechanics. Here, we will use the partition function within classical statistical mechanics to illustrate the modeling principles. Construction of a model of a material (see Figure 2) involves identification of a set of dofs or variables u's which are relevant to the property of interest. We label the rest of the variables v's as the irrelevant variables. Such separation of degrees of freedom is often based on (a) *scales* governing a physical phenomenon of interest and (b) the nature and strength of their coupling and interactions. A model

Hamiltonian can be derived from the full Hamiltonian of a system $H(u, v)$, by integrating out the irrelevant degrees of freedom using

$$e^{\frac{-H_{model}(u)}{k_B T}} = \int dv\, e^{\frac{-H(u,v)}{k_B T}}. \tag{1}$$

Here, u and v denote the groups of relevant variables and irrelevant variables respectively. This procedure leads to a simple form of the model Hamiltonian (also known as effective Hamiltonian), when the coupling between u and v is weak and simple. For example, if the dependence of H on v is quadratic, and v's couple quadratically with u:

$$H_{model}(u) \approx \underset{v}{\text{Min}}\, H(u, v) \tag{2}$$

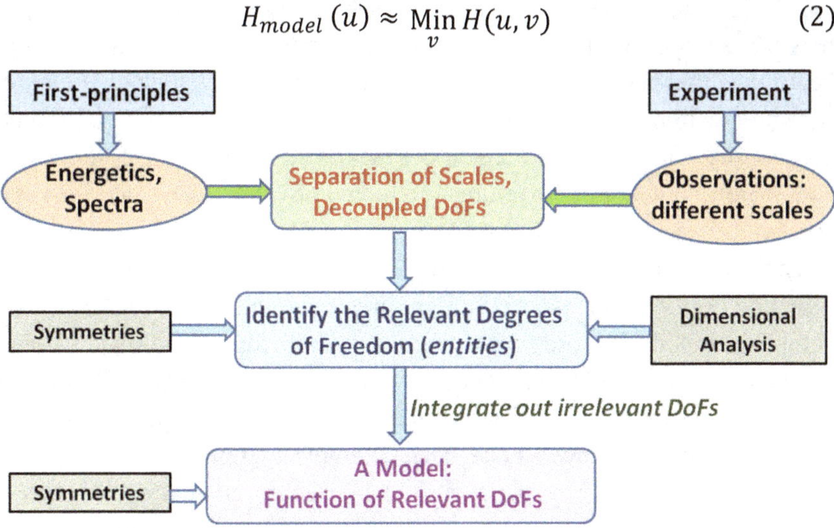

Fig. 2. Mechanistic construction of a model of a material.

While Equation (2) gives an approximate way to carry out the integral in Equation (1), it gives a general and very powerful approach to development of models, as minimization can be carried out readily using various efficient algorithms. It means that the irrelevant degrees of freedom relax to their lowest energy for a given configuration of u's. Secondly, this approach is physically very intuitive and insightful: if some of the relevant variables are unidentified and left in the group of v's, the resulting H_{model} picks up unusual features, for example, multivalued-ness. Equation (1) represents ideas from the theory of renormalization group and gives effective interactions between u's as renormalized by v's, and Equation (2) achieves it approximately. Symmetry properties of the relevant degrees of freedom and of the system constrain the functional form of the model Hamiltonian: H_{model} is invariant under all symmetry operations in the group of symmetries of the system. The nature of interactions between u's is determined by their symmetry properties and physical dimensions (see Figure 2).

2.2. *Models using machine learning*

It is not always possible to start with the fundamental Hamiltonian and determine the behavior of a material using the mechanistic modeling scheme (section 2.1). For example, many phenomena in materials, such as ductility, fracture and dielectric breakdown field involve processes and structural features spanning many length and time scales. In the context of unconventional superconductivity in oxides, it is often not possible to determine even the electronic quantum ground state (in Equation 1). At the same time, a huge amount of work has been completed on all of these problems, due to their fundamental appeal and technological importance. Thus, there exists a large amount of information, or *big data*, on various features and aspects of these materials, and there even have been attempts to find patterns in these datasets and predict new materials, like quasicrystals.[1] Such work had been driven by physical and chemical intuition into the large materials data available.

Such data-based approach has recently undergone a revolution due to significant advances in machine learning techniques and computing resources. Significant efforts since 2011 have resulted in the development of open databases of materials properties determined with first-principles calculations[2] In this scheme of modeling, the goal is to model the relationship between property of a material with several of its features (see Figure 3), whose functional form is unknown. For example, while the unique dependence of many-electron quantum ground state energy on electron density is guaranteed by density functional theory,[3] its functional form is unknown. Machine learning scheme of modeling is quite general and powerful, and has the potential to capture seemingly unexpected relationships between materials properties and their features (see Figure 3), varying from composition, structural features, experimental microscopy images to even synthetic conditions.

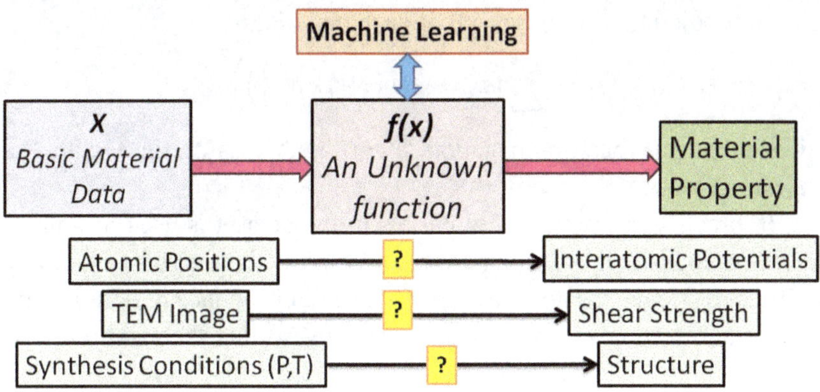

Fig. 3. Machine learning, a unified framework for modelling.

While there are many techniques of machine learning being employed in materials science, we highlight two techniques belonging to two rather distinct categories: (a) deep learning, artificial neural networks (ANN), and (b) feature selection algorithms, like Bootstrapped Projected Gradient Descent (BoPGD). While ANNs aim to capture the complex relationship in the data available using *all* the features, the BoPGD aims to identify *finger-print descriptors*, and build a model as a function of these. Thus, BoPGD is closer in spirit to the mechanistic modelling scheme, fingerprint descriptors being analogous to the relevant degrees of freedom.

Artificial neural networks are simplified numerical models of biological neurons. A fundamental unit of an ANN is the perceptron (see Figure 4). The ith perceptron receives inputs d_j's, which are weighted by factors W_{ij}'s, and its output y_i is given by a sigmoid function $\int$ (e.g. smoothened unit step function) of the cumulative inputs (weighted sum),

$$f_i = \int \left(\Sigma_j W_{ij} d_j \right), \qquad (3)$$

in way a similar to how electrical signals transmit in a biological neuron. A deep-learning neural network (see inset of Figure 4) is made of several layers, each consisting of several perceptrons, and the weighting parameters W_{ij}'s define the model represented by such an ANN. These parameters are determined from the large dataset available on the property (modeled with the ANN as y=M(W_{ij},d$_j$)) by minimizing the error in prediction of property (p) of each material μ:

$$E_{RR}(W_i) = \sum_{\mu} \left(p_\mu - M\left(\left\{ W_{ij}, d_j^\mu \right\}\right) \right)^2 \tag{4}$$

using the error back-propagation algorithm,[4] a procedure called as *training* of the ANN.

In practice, a fraction of materials in the dataset is used in training, i.e. determination of W_{ij} parameters, and the rest is used in validation of the trained model through *testing* the accuracy of its prediction. Such training and testing sets are picked randomly so that the resulting ANN model is robust, and insensitive to the choice of sub-datasets. It is quite

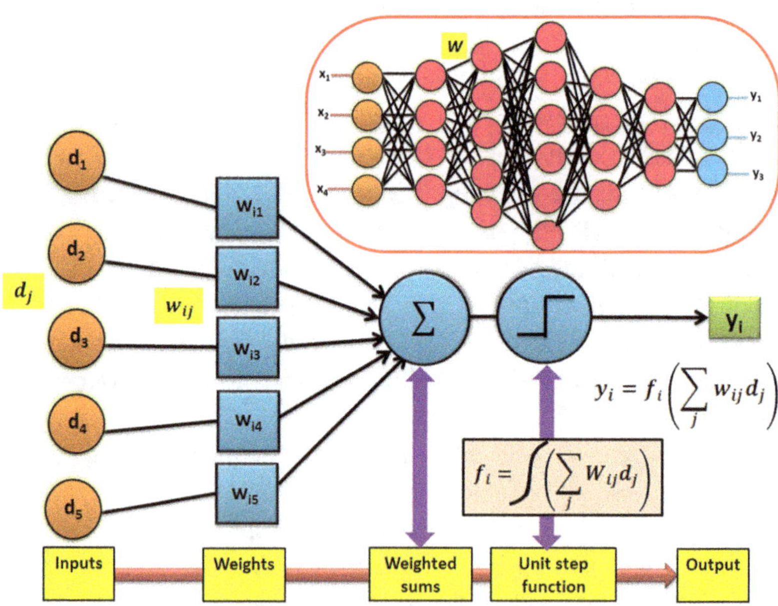

Fig. 4. A schematic illustration of a perceptron, a unit of an artificial neural network and deep learning neural network (inset), consisting of several layers of perceptrons.

evident that the resulting ANN model has a rather complex functional form based on highly nonlinear sigmoid function, nested as many times as the number of layers in the ANN. It is often not straight-forward to identify which of the input variables, d_j's, are relevant or important to the property being modeled by the ANN. Indirect attempts are often made by perturbing the values of these variables and assessing the sensitivity of the outcome y to judge the relative relevance of these variables.

BoPGD scheme,[5] on the other hand, involves determination of finger-print descriptors, a minimal set of variables d_j's, which are relevant to the property of interest, a process called as the feature selection. First of all, the notion of descriptors is totally generalized in this scheme,[6] and each variable d_j is called as a descriptor. This is based on the idea that each variable has a physical meaning and is a descriptor (measure or pointer to) of some property of a material. After the preprocessing of the data (d_j and p_j) for each material μ, which involves elimination of outliers and normalization of the data to a uniform scale, an expanded set of descriptors is generated by considering various *nonlinear* combination of the primary variables d_j's called the composite descriptors. This step is called as engineering of descriptors.[6] An essentially linear model of this expanded set of descriptor variables is given by:

$$M(\{d_j, W_j\}) = \sum_j^N W_j d_j,\qquad(5)$$

The model parameters W_i's are determined by minimizing the error function (given by Equation 4), demanding that a minimum number of W_i's should be nonzero:

$$W_i^{model}: \ \underset{w_i}{\mathrm{Min}}\ E_{RR}(w_i) \ \text{with min} \ \langle\langle W \rangle\rangle_i .\qquad(6)$$

Fingerprint descriptors, defined by the minimal set of d_i's for which W_i's are nonzero, facilitate obtaining insights into the mechanism governing the material property being modeled. However, a domain expert is typically needed to interpret the model and derive physical or chemical insights, as the composite descriptors can be quite complicated.

2.3. *Simulation of material behavior*

Simulation of a material on computers involves the study of the dynamical evolution or response of its model subjected to conditions corresponding to real experiment, by solving the equations of dynamics of the dofs in the model. A simulation tries to mimic the model of a real material, and hence the accuracy of its results depends most importantly on the model's efficacy in representing the reality. The mechanistic modeling scheme results in a model that can be readily used in such simulations, to determine the thermodynamic behavior of a material. While machine learning model of a general property (e.g. dielectric constant) of a material is not useful in simulating its thermodynamic behavior, the machine learning scheme can surely be used in modeling the interatomic interactions and exploring the thermodynamic stability of the phases of a material.[7]

A state of a material is represented by a configuration defined by a vector Γ_i in the phase space spanned by dofs (entities) in the model. For

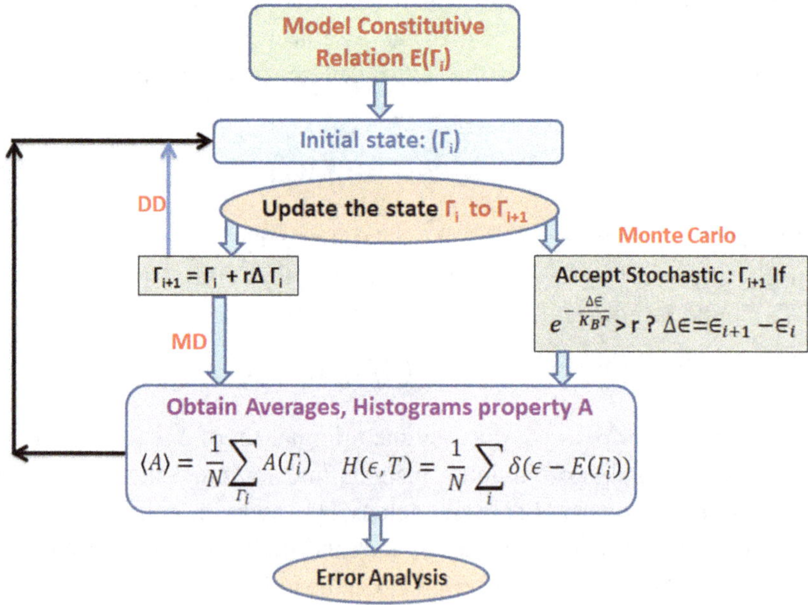

Fig. 5. A flowchart of generic simulation of a model. Here, DD and MD mean dislocation dynamics and molecular dynamics (U V Waghmare, Physics News (2015)).

example, Γ_i in an atomistic simulation is defined by positions and velocities of atoms in the finite system being simulated. In a simulation of a Heisenberg model, a configuration is defined by spins located at the positions of magnetic atoms (typically a lattice). In meso-scale simulation of a material subjected to mechanical stress field, a configuration may be defined by lines of dislocations discretized into line-segments.

A simulation (see Figure 5) of a material system typically starts with an initial configuration, Γ_1, and updates it to generate new configuration iteratively. In molecular dynamics simulations, such an update is based on integration of Newton's laws of motion, while it is carried out stochastically with Metropolis algorithm for importance sampling in a Monte Carlo simulation. At the end of the simulation, thermodynamic properties of the material are obtained as averages of the properties over the configurations generated during the simulation. In a dislocation dynamics (DD) simulation, one essentially monitors the response of model system manifesting as redistribution of dislocation lines arising from applied stress fields, evolving to a new microstructure. We note that one requires access to force acting on each entity (or dot) in molecular dynamics or dislocation dynamics simulations while evaluation of energy for each configuration is adequate in a Monte Carlo simulation. In sections 3 and 4, we will present the mechanistic scheme of hierarchical modeling that bridge description at different scales.

3.　The Total Energy Function

We now introduce here the concept of total energy function, which provides a fundamental framework for understanding the structure-property relationship of a material. It also forms the basis of first-principles simulations of molecules and materials, which have been key to emergence of the area of computational materials science since the 1980's.

Most technologically important properties of a material arise from the interactions between atomic nuclei and electrons, and their quantum dynamics. Thus, a material can be modeled as a soup of nuclei and electrons (see Figure 6(a)). At low temperatures, the structure of a material is determined by low energy states of this model. Atomic positions define the atomic structure of a material and are the relevant degrees of freedom in determination of its structure. Thus, we would like to eliminate the electronic degrees of freedom (Equation 2), noting that electrons are notably lighter and faster than nuclei, and assuming that they redistribute themselves to minimize the energy when atomic positions change. This is called as the Born Oppenheimer or adiabatic approximation, in which electrons follow the atomic motion by remaining in their instantaneous ground state. Electronic motion needs to be treated within quantum framework (indicated as waves in Figure 6(a)) and atomic motion can be treated classically in simulations of many properties at high temperature. Thus, we express their energy in terms of atomic positions R (relevant dofs), and quantum wavefunctions (Ψ) of electrons (irrelevant dofs), and obtain the model for dynamics of atoms using Equation 2:

$$E_{total}^{(R)} = \overset{Min_{QM}}{\Psi} \ \mathrm{E}\,[\Psi, \mathrm{R}]\,. \tag{7}$$

Here, minimization of E with respect to Ψ needs solution of an interacting many body quantum ground state problem, which is very hard to solve even on the most powerful computer. However, a breakthrough in tackling this was the density functional theory (DFT) of Kohn and Hohenberg[3] and subsequent development over the next two decades.[8] This has opened up the area of *first-principles* simulations of materials, which have been proven to be very powerful tools in (a) understanding experiments, (b) complementing experiments by accessing information that is hard to measure in laboratories, and (c) prediction of new materials and structure and even new functionalities in existing materials. A practical scheme of solving Equation (7) within DFT involves mapping the many-body problem onto an effective single electron problem, and gives access to electronic structure, often called as the *band structure* of a crystalline material.

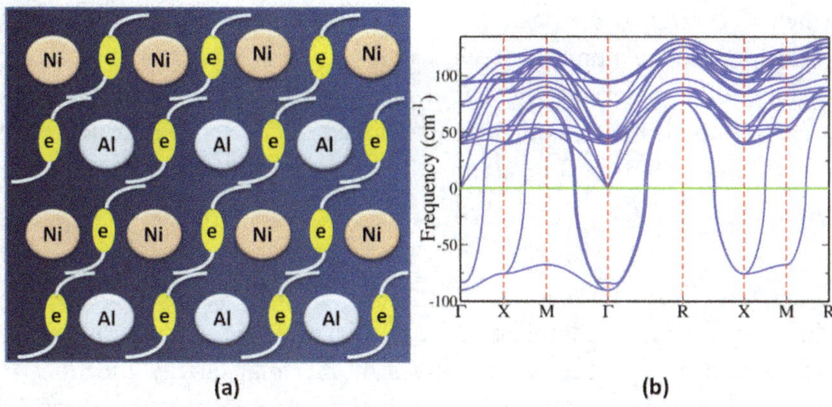

Fig. 6. (a) A schematic of a material modelled as a soup of electrons (treated as waves) and nuclei (treated as classical particles), and (b) dynamical structure of $Sn_{0.75}Ge_{0.25}Te$.

The total energy function $E_{total}(R)$ is the fundamental objects relevant to most techniques of atomistic simulations of materials. Minimization of the total energy with respect to atomic positions facilitates prediction of the atomic structure of a material essentially from scratch, with knowledge of just atomic numbers (chemical constituents of a material) and an initial guess for the structure. External fields like electric, magnetic, stress and strain fields can be readily included in this framework, and derivatives of the total energy with respect to these fields give measurable conjugate properties of a material. For example, electric enthalpy of a material in the presence of electric field ς_α can be obtained by minimization of the total energy with respect to atomic positions (using Equation 2):

$$H\left(\varsigma_\alpha\right) = \operatorname*{Min}_{R} E_{tot}\left(\varsigma_\alpha, R\right).\tag{8}$$

The first derivative of H with respect to electric field gives the electric polarization of the material. Moreover, the linear compliance that gives response to electric field, the dielectric susceptibility, is the second derivative of the H with respect to electric field:

$$\epsilon_{\alpha\beta} = \delta_{\alpha\beta} - \frac{4\pi}{\Omega}\frac{\partial^2 H}{\partial\varsigma_\alpha\partial\varsigma_\beta},\tag{9}$$

which is central to application of a material in capacitors, field effect transistors, and a number of electronic devices. Similarly, the elastic modulus of a material, which is a measure of its stiffness relevant to structural applications, is obtained as mixed of the second derivative of total energy with respect to strain:

$$C_{\alpha\beta\gamma\delta} = \frac{1}{V} \frac{\partial^2 E_{tot}}{\partial \epsilon_{\alpha\beta} \partial \epsilon_{\gamma\delta}} \,. \qquad (10)$$

Many other linear compliances (e.g. piezoelectric response) can be obtained as mixed second derivatives of the total energy function. In first-principles density functional theoretical framework, these can be obtained efficiently using DFT linear response. Many open-source[9] and commercial[10] softwares are available and have been well-optimized to carry out such simulations, and have now become commonplace in research in materials science.

Before closing this section, we highlight that the hessian of total energy function, the *3Nx3N* matrix of second derivatives with respect to displacements N atoms in three Cartesian directions, is the interatomic force constant (IFC in Figure 1) matrix that can also be obtained within the DFT-linear response framework. Eigen-spectrum of a closely related dynamical matrix gives the frequencies of normal modes or vibrations or phonons. If some of these eigenvalues are negative, *i.e.* the frequencies are imaginary, the corresponding eigenmodes constitute the lattice instabilities of material (see Figure 6(b)). It means that the system is not a local minimum of energy and is unstable with respect to atomic displacements given by the eigenvector of the unstable mode. Frequencies of the long wavelength modes (q $\rightarrow$ 0) in phonon spectrum at Γ (Figure 6(b)) are measurable with IR and Raman spectroscopies, and their eigenvectors facilitate the interpretation of experimental vibrational spectra, where access to the eigenmodes is not possible. As we will see in the next Section, phonon spectra form the basis of modeling thermodynamic properties of a material.

4. Thermodynamics of Materials

While a local minimum of total energy function (Equation 7) with respect to atomic positions gives its stable structure at T=0 K, temperature dependent effects on structure and properties are often significant, and need to be captured with further modeling. As the temperature is increased, atoms in a material vibrate with respect to their average positions, and phonons (structural excitations) are excited. While the harmonic terms (IFC matrices) in the Taylor expansion of the total energy function in terms of atomic displacements of a material are used in obtaining its phonon spectrum, anharmonic interactions among phonons are responsible for important phenomena like lattice expansion, thermal conductivity and structural phase transitions. While ab initio molecular dynamics and Monte Carlo simulations can in principle be used to determine such thermodynamic behavior, their computational cost can be prohibitively high. We note that computational times of first-principles calculations of total energy and its hessian scale as $O(N^3)$ and $O(N^4)$ respectively, N being number of atoms in the periodic unit cell.

4.1. *Quasi-harmonic modeling of thermodynamics of materials*

When a material does *not* undergo qualitative changes in its structure as a function of temperature, and the relevant anharmonic interactions among phonons are weak, it suffices to model phonon frequencies as a function of cell volume or lattice parameters of a crystal. It amounts to including the third order anharmonic interactions between optical and acoustic phonons in the analysis of thermodynamic properties. Temperature dependent relative stability of two structures of a material is determined by their free energies, which are not directly accessible to MD or MC simulations. More importantly, MD and MC simulations involve classical statistical mechanics of phonons, which often breaks down at low temperatures, as phonons need to be treated with Bose-Einstein statistics. To this end, a practical method to obtain free energies relies on the phonon spectra ($\omega_{q\mu}$) obtained from the IFCs described in Section 3:

$$F^{QH}(\mathrm{R}) = E_{tot}(\mathrm{R}) + \frac{k_B T}{N} \sum_{q,\mu=1}^{N} log\left[2sinh\frac{\hbar\omega_{q\mu}}{2k_B T}\right]. \qquad (11)$$

Such calculations of free energies have been quite successful in analysis of thermodynamic stability of stacking faults in semiconductors like SiC[11] and metals like Ti[12]. They also highlight the importance of soft (low frequency) phonons to the stability of a material.

4.2. *Modeling structural transitions: Strong anharmonicity*

Many crystalline materials undergo structural phase transitions as a function of temperature, which are fundamental to the emergence of their functional properties of technological importance. For example, $PbTiO_3$ is a perovskite oxide having cubic structure at high temperatures, undergoes a transition to a non-centrosymmetric ferroelectric phase as the temperature is lowered, acquiring (a) piezoelectric property that makes it functionally useful in electromechanical sensors and actuators and (b) switchable spontaneous polarization that makes it useful in non-volatile memory devices. Both properties are exhibited only by the low temperature phases lacking inversion symmetry. Similarly NiTi undergoes a martensitic phase transformation as a function of temperature that equips it with the functional property of shape memory, which is useful in a number of biomedical and aerospace applications.

At a structural phase transition, there is a sharp change in some of the structural parameters at the transition temperature, as a result of highly anharmonic phonon-phonon interactions. In fact, the high temperature structure of such a material is unstable at T=0 K, as reflected in unstable modes in its phonon spectrum (see for example, Figure 6(b)). Imaginary frequencies of such modes cannot be used meaningfully in the quasi-harmonic model of free energy (Equation 11), necessitating the use of a more rigorous statistical mechanical analysis of anharmonic interactions, possibly with MD or MC simulations. [13]

In such a modeling scheme[13] (see Figure 7), one uses the high temperature (high symmetry) structure as the reference structure and analyzes its total energy landscape expressed as a Taylor expansion in atomic displacements describing its structural distortions. For example, in the context of ferroelectric transition in $PbTiO_3$, its cubic structure is used as the reference structure. All its structural instabilities are evident in phonon dispersion of the reference cubic structure. The first step in modeling here involves identification of the relevant degrees of freedom, which are clearly given by the symmetry invariant subspace constituted of modes belonging to the bands containing lattice instabilities. Higher energy (frequency) phonons of the full Hamiltonian are then integrated out readily within a harmonic approximation, amounting to coarse-graining in time-domain: the shortest time-scale in the resulting model are longer than the ones in the full Hamiltonian. The resulting H_{model} is expressed conveniently using a localized basis (phonon Wannier functions) to span the subspace of relevant phonons, and its functional form is essentially the symmetry invariant Taylor expansion containing anharmonic terms up to 6th or 8th. All parameters in this model are derived from first-principles calculations. Analysis of this lattice model Hamiltonian using MD or MC simulations allows prediction of behavior of the material near its structural phase transition, and epitaxial strain-temperature phase diagrams.[13] The latter are useful in guiding experimentalists in the development of epitaxial films with desired structural transitions and associated functional properties.

The second step in hierarchical modeling here involves coarse-graining in real-space, leading to model of free energy as a functional of field of collective dof, which is the order parameter of the structural phase transition. In the context of ferroelectric transition, it is polarization. To derive such free energy function, one requires two constructs: (a) a polostat, which allows MD simulation of the model Hamiltonian maintaining its average polarization at a given value P (similar to the thermostat), and (b) thermodynamic integration, in which the changes in free energy with evolution of the system from the reference cubic structure $P=0$ to the ferroelectric state are summed up:

$$F(P) = F(P = 0) + \int_0^P \frac{\partial F}{\partial P'} dP' . \tag{12}$$

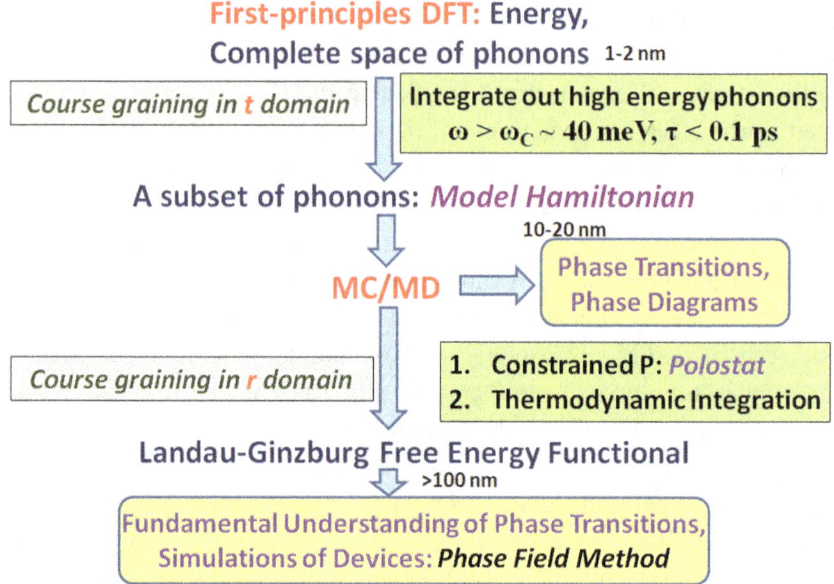

Fig. 7. Hierarchical modeling and simulations of structural phase transitions (Ref. 13).

The resulting free-energy as a function of order parameter P, is the Ginzburg–Landau model of structural phase transition, which allows finite element simulation of a ferroelectric material at much longer length-scales, for example when it is used in devices. We note that this methodology is somewhat similar to the phase field method often used in metallurgy to model and simulate the microstructure of a material. Such modeling has demonstrated how strain-phonon coupling is responsible for ferroelectric ordering and domain structure in perovskite oxides.[13]

5. Modeling Disorder and Defects in Solids

Defects and disorder are invariably present in most materials and have important consequences to their behavior. Implicit in most of the models and simulations described above was the use of periodic boundary conditions, in which a periodic unit or a box containing atoms is repeated in space. Even disordered states of a material are modeled with a periodic box, size of which restricts the length-scales associated with disorder. We briefly summarize modeling tools used in studies of (a) chemical substitutional lattice site occupancy disorder and (b) structural disorder with no inherent lattice structure like in amorphous solids.

5.1. *Chemical substitutional disorder*

Many technologically important materials are solid solutions or substitutional alloys. For example, one of the most useful ferroelectrics is $PbTi_{1-x}Zr_xO_3$ (PZT), which is a solid solution. Most materials used in structural elements are alloys, e.g. Ni–Al alloys for aerospace applications or Mg, Ti alloys for automobile applications. Common to these material systems is the disorder or randomness in occupation of lattice sites with two types of atoms. Point defects (e.g. oxygen vacancies in ZnO) associated with vacancies can also be treated as a substitutional chemical disorder. It is essential to include effects of such a disorder on the material's properties. There are two techniques commonly used in modeling such chemical disorder: (a) Cluster Expansion (CE) technique, and (b) Site Occupancy Disorder (SOD) technique.

In the CE technique, the identity of an atom occupying a lattice site is represented with Ising-like spin. For example, $\sigma_i=1$ when the atom at i^{th} site is Ti, and $\sigma_i=-1$, when it is Zr in PZT. Thus, any chemical configuration of PZT can be represented with Ising spins at B-site of its crystal lattice. First-principles DFT energy landscape is modeled with a generalized Ising model, in which the model Hamiltonian is the sum of the cluster terms in Ising spins, such as 2-body $\sigma_i\,\sigma_j$, 3-body $\sigma_i\,\sigma_j\,\sigma_k$, ...[14] Coefficients of these terms are fit to the relaxed DFT energies of select chemically ordered configurations of the solid solution or alloy. The resulting model is simulated with the Monte Carlo method to assess the tendency of chemical (dis)ordering in the material. Indeed, the CE can be used to model other properties of a material as well: CE can be used to model band-gaps, dielectric constant.

In the SOD technique,[15] a relatively small periodic cell is used to consider *all* chemically ordered states. However, symmetry of the crystalline lattice is used to reduce this large set to a subset (much smaller) of symmetry inequivalent configurations. Total energies and other properties of interest are determined for each of these symmetry-inequivalent states. Statistical mechanical analysis uses these configurations weighted by the number of symmetry equivalent states of each. The advantage of SOD technique over the CE method is that one can obtain electronic and other properties directly and exactly, but the disadvantage is that one can simulate only systems of small size and substitutional concentrations for which the number of symmetry-inequivalent states is small. Such analysis has been used effectively in cationic and anionic substitution in oxides,[16] showing how cation and anion substitutions tune conduction and valence bands respectively.

5.2. *Periodic models of amorphous structures*

Models of *structurally disordered* states of a material simulated within periodic boundary conditions can be obtained using MD simulations of its liquid state and supercooling it down to a temperature below its freezing point, and relaxing it to a nearby local minimum of energy. Similarly, it can be obtained starting from its crystalline state, but with lattice expanded by a suitable amount to achieve the density of its amorphous state. From the phonon dispersion of such an expanded lattice, its lattice instabilities are determined. Volume preserving relaxation of its structural supercell distorted with random linear combinations of a number of these instabilities yields a structurally disordered state. We note that an accurate representation of the structural disorder needs consideration of many realizations of such model disordered states.

5.3. *Discrete dislocation dynamics*

Dislocations are topological line defects in a crystal, which are very common in ductile materials, particularly metals. They are the carriers of plastic deformation, which involve permanent change in the size or shape of a material in response to applied stress. Dislocations involve structural distortions and strain fields over micron length scales, which are much too large for DFT calculations at present. While atomistic simulations of dislocations have been reported, they are typically for relatively simple solids due to the limited availability of accurate interatomic potentials. Interactions between dislocations are mediated by acoustic phonons and can be derived by integrating out the acoustic phonons in the spirit of Equation (2). With access to the force-field or interaction potential between dislocations, their response to external loading is simulated by dynamically evolving their structure represented with discretized dislocation line segments. Such dislocation dynamics[17] leads to the prediction of micro-structure of a material.

6. Phenomenological Models

6.1. *Ginzburg–Landau free energy*

Atomistic simulation of a material used in a device, e.g. a ferroelectric in a sensor, is not feasible due to forbiddingly large number of degrees of freedom involved. In such case, a continuum Ginzburg-Landau like model, gives free energy expressed as a symmetry invariant polynomial function in the order parameter(s) with high-temperature phase as the reference:

$$G(P) = G_o + \frac{A}{2}(T - T_C)P^2 + \frac{B}{4}P^4 + \frac{C}{6}P^6, \qquad (13)$$

with coefficients fit to experiment and some information obtained from first-principles. Such a model can be used in a finite element simulation.[18] In the problems, where descriptions of a material at different scales is required concurrently, one may use a quasi-continuum method.[19]

6.2. *Ductility versus Brittleness of a Material*

As discussed in Section 5.2, plastic deformation of a material involves processes at various length-scales. At the tip of a micro-crack possibly present in a material, there is competition between two processes (see Figure 8): (a) propagation of the crack further through creation of surfaces, measured with the surface energy γ_s, and (b) blunting of the crack tip by nucleation of dislocations, as measured by unstable stacking fault energy γ_{us}.[20] If the latter wins, the material is ductile, while it is brittle if the former process is relatively more favorable. One can define a ductility parameter[21]:

$$D = \frac{\gamma_s}{\gamma_{us}} \tag{14}$$

If D is small, the material is likely to be brittle and vice versa. While this is an oversimplified picture of complicated ductile versus brittle behavior of a material, it is a useful descriptor in computational design of materials with improved ductility and toughness. Both energy parameters γ_s and γ_{us} can be readily determined from DFT calculations and have been used in predicting improved ductility of $MoSi_2$ with substitution of V.[21]

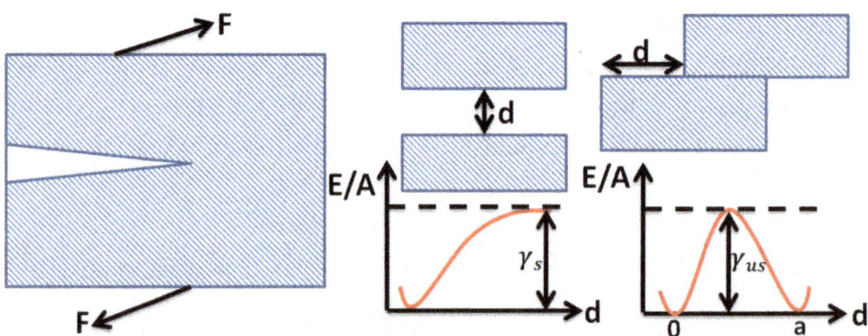

Fig. 8. Competing processes of cleavage and slip at the crack tip that result in creation of surfaces or slip of crystal planes.

7. Machine Learning Constrained by Dimensional Analysis and Scaling Laws

Some of the criticisms on the predictive models of material properties obtained with machine learning are (a) ANN models are complex and not physically interpretable, giving no insights, (b) the models are not transferrable in predicting properties of materials of different kinds, and (c) they require big data while the databases of materials are typically small. Recently, an attempt[5] has been made to address these by constraining a feature selection algorithm, BoPGD, with (a) dimensional analysis through the use of Buckingham Pi theorem to determine dimensionless descriptors, and (b) empirical relationships between the descriptors. Learning from data and physical laws, it claims to give predictive models that are simple, transferable and physically interpretable, and can be derived from relatively *small* data. It was demonstrated in deriving a model for dielectric breakdown field (electric analogue of fracture stress) of semiconductors and insulators:

$$F_b = A\frac{E_g}{d^{3.5}} + B, \tag{15}$$

where A and B are the parameters fit to the data available on dielectric breakdown field F_b, bandgap E_g and nearest neighbor distance d of about 90 compounds. This form was shown to apply to new set of compounds with greater accuracy than the one derived earlier using LASSO.[22]

8. Summary

A computer simulation of a material aims to mimic its model numerically, and its accuracy in predicting the material-specific behavior is singularly determined by how realistic the model is. To this end, mechanistic models of a material focus on the degrees of freedom relevant to the problem of interest, effectively integrating out the irrelevant degrees of freedom. This is known as coarse-graining, which results in filtering our high frequency or short-wavelength features in time and spatial domains respectively. Such mechanistic modeling provides fundamental insights into mechanisms of seemingly complex behavior of materials, which depend on structure and processes occurring at many scales. First-principles density functional theoretical framework stands at the bottom of the hierarchical modeling of materials, and can be a powerful starting point for developing models for phonons, phase transitions, defects, dislocations and even continuum of materials as used in devices. When the separation between relevant and irrelevant variables is not straightforward or there are too many coupled groups of relevant variables, machine learning can be quite powerful in developing predictive models. While the traditional machine learning schemes need big data and result in models that are too complex and uninterpretable, recent ideas of machine learning constrained by dimensional analysis and scaling laws seem promising in developing insightful models of complex behavior based on relatively *small* data.

Acknowledgement

UVW acknowledges support from a J.C Bose National fellowship of the department of Science and Technology, Government of India, Sheikh Saqr fellowship and funding from a DST project to support India-Korea Virtual Network Centre in Computational Materials Science.

References and Citations

1. Rabe, K. M., Kortan, A. R., Phillips, J. C., & Villars, P., Quantum diagrams and the prediction of new ternary quasicrystals, *Phys. Rev. B*, **43**(7), 6280 (1991).

2. Nosengo, N. Can artificial intelligence create the next wonder material?, *Nature News*, **533**(7601), 22 (2016).

3. a) Hohenberg, P., & Kohn, W., Inhomogeneous electron gas, *Phys. Rev.*, **136** (3B), B864 (1964).
   b) Kohn, W., & Sham, L. J., Self-consistent equations including exchange and correlation effects, *Phys. Rev.* **140**(4A), A1133 (1965).

4. a) Bhadeshia, H. K. D. H., Neural networks in materials science, *ISIJ Int.,* **39**(10), 966 (1999).
   b) Bhadeshia, H. K. D. H., Neural networks and information in materials science, *The ASA Data Sci. J.*, **1**(5), 296 (2009).

5. Kumar, N., Rajagopalan, P., Pankajakshan, P., Bhattacharyya, A., Sanyal, S., Balachandran, J., & Waghmare, U.V., Machine learning constrained with dimensional analysis and scaling laws: simple, transferable, and interpretable models of materials from small dataset, *Chem. Mater.*, **31**(2), 314 (2018).

6. Pankajakshan, P., Sanyal, S., de Noord, O. E., Bhattacharya, I., Bhattacharyya, A., & Waghmare, U. V., Machine learning and statistical analysis for materials science: stability and transferability of fingerprint descriptors and chemical insights, *Chem. Mater.*, **29**(10), 4190 (2017).

7. Behler, J., Martoňák, R., Donadio, D., & Parrinello, M., Metadynamics simulations of the high-pressure phases of silicon employing a high-dimensional neural network potential, *Phys. Rev. Lett.*, **100**(18), 185501 (2008).

8. Payne, M. C., Teter, M. P., Allan, D. C., Arias, T. A., & Joannopoulos, A. J., Iterative minimization techniques for ab initio total-energy calculations: molecular dynamics and conjugate gradients., *Rev. Mod. Phys.*, **64**(4), 1045 (1992).

9. a) https://www.quantum-espresso.org/, b) https://www.abinit.org/, c) https://departments.icmab.es/leem/siesta/, d) https://www.dftb.org/

10. a) https://www.vasp.at/, b) http://susi.theochem.tuwien.ac.at/

11. Thomas, T., Pandey, D., & Waghmare, U. V., Soft modes at the stacking faults in SiC crystals: First-principles calculations, *Phys. Rev. B,* **77** (12), 121203 (2008).

12. Bhogra, M., Ramamurty, U., & Waghmare, U. V., Smaller is plastic: polymorphic structures and mechanism of deformation in nanoscale HCP metals, *Nano Lett.,* **15** (6), 3697 (2015).

13. Waghmare, U. V., First-principles theory, coarse-grained models, and simulations of ferroelectrics, *Acc. Chem. Res.*, **47** (11), 3242 (2014)

14. Sanchez, J. M., Cluster expansion and the configurational theory of alloys, *Phys. Rev. B*, **81**(22), 224202 (2010).

15. R Grau-Crespo, Waghmare, U. V., Ed. B Rai, *Molecular Modeling for the Design of Novel Performance Chemicals and Materials*, 303, CRC Press (2012)

16. Manjunath, K., Prasad, S., Servottam, S., Waghmare, U. V., & Rao, C. N. R., Hg_2NF, analogue of HgO, *European J. Inorganic Chem.*, **2019**(19), 2398 (2019).

17. a) LeSar, R., Simulations of dislocation structure and response, *Annu. Rev. Condens. Matter Phys.*, **5**(1), 375 (2014)., (b) Richard LeSar, *Introduction to Computational Materials Science: Fundamentals to Applications*, Cambridge University Press (2013)., (c) Allen, M., & Tildesley, D., *Computer Simulation of Liquids,* Oxford Science (1990).

18. Tadmor, E. B., Waghmare, U. V., Smith, G. S., & Kaxiras, E., Polarization switching in $PbTiO_3$: an ab initio finite element simulation, *Acta Materialia,* **50**(11), 2989 (2002).

19. a) Tadmor, E. B., Ortiz, M., & Phillips, R., Quasicontinuum analysis of defects in solids, *Philos. Mag. A,* **73**(6), 1529 (1996), b) http://qcmethod.org/

20. Rice, J. R., Dislocation nucleation from a crack tip: an analysis based on the Peierls concept, *J. Mech. Phys. Solids*, **40**(2), 239 (1992).

21. Waghmare, U. V., Kaxiras, E., & Duesbery, M. S., Modeling brittle and ductile behavior of solids from first- principles calculations, *Physica Status Solidi* (b), **217**(1), 545 (2000).

22. Tibshirani, R., Regression shrinkage and selection via the lasso, *J. Royal Statistical Society: Series B (Methodological)*, **58**(1), 267 (1996).

Chapter 20

Computational Materials Design Using DFT Databases and Descriptors

Shobhana Narasimhan[*]

*Theoretical Sciences Unit and School of Advanced Materials,
Jawaharlal Nehru Centre for Advanced Scientific Research, Jakkur,
Bangalore 560064, Karnataka, India*
[*]*shobhana@jncasr.ac.in*

The search for novel materials for targeted applications can be greatly aided and speeded up by following a program of rational materials design, making use of ab initio density functional theory computations. In this review, some recent efforts to use high throughput computation to carry out such a program are described. As the chemical space to be searched can be very large, the process can be considerably hastened by formulating 'descriptors'. Some landmark efforts, from the literature, to develop descriptors for three-dimensional crystal structure prediction, are summarized. Finally, a recent study in which descriptors were developed for the structure of two-dimensional host-guest self-assembled monolayers on a substrate, is recapitulated.

1. Introduction

To date, the periodic table consists of 118 elements, of which 92 are naturally occurring, and the remaining have been artifically synthesized. Out of these elements, one can form compounds; it is estimated that about 3000 such materials exist in nature as minerals. These materials exhibit a large diversity in their physical and chemical properties. It is this diversity that makes their use in technology key to human progress – so much so that this progress is usually labelled by the 'new' material whose adoption by humans serves as an indicator of a significant advance in civilization: the Stone Age, the Bronze Age, the Copper Age, the Iron Age, . . . Particularly noteworthy among these age-defining materials is bronze: a man-made alloy of copper and tin, that is not found in nature, and possesses superior and desirable properties compared to its constituent elements.

By combining the elements to create novel materials that are not found in nature, one might hope to considerably 'improve' upon a desired target property. For example, one might hope to find a material that is harder than diamond, or a better electrical conductor than silver, or a more efficient and/or cheaper catalyst for a given chemical reaction. How does one go about searching for such a material? Until recently, the 'best' material for a given purpose was found by accident, or by trial and error, or by brute force: systematically searching through a very large number of candidate systems. For example, the catalyst for the Haber-Bosch process for synthesizing ammonia from nitrogen and hydrogen was obtained after searching through about 4000 possible candidates, Of these, it was determined that osmium was the most efficient catalyst; however, for reasons of economic viability, iron was instead chosen as the commercially used catalyst. Similarly, Thomas Edison reportedly tested about 6000 materials before determining that carbonized bamboo was the optimal material to make light bulb filaments.

Even with modern combinatorial methods in the laboratory, and high throughput computing, the brute force method becomes impractical because of the vast chemical space that has to be explored. For example, it has been estimated that there are about 10^{11} stable ternary compounds, and about 10^{18} stable quarternary compounds.[1] An exhaustive search through such an enormous space is clearly impossible, be it experimental or computational. There is therefore a great need for some informed guesses, or predictions, about where in this vast chemical space one might be likely to find materials that are most likely to possess the desired physical or chemical properties. Such predictions can be considerably aided by the formulation of 'descriptors'.

A descriptor is some combination of the microscopic parameters that characterize a material, that correlates well with an observable macroscopic property of that material. To be useful in practice, a descriptor should be quick to evaluate: typically, much faster than either carrying out the experimental measurement of that property, or of calculating that property by performing (for example) an *ab initio* density functional theory calculation.

In order to estimate the desired property, one may need just one descriptor: e.g., in an organic compound, if the property P is the energy required to break a carbon-carbon bond, then the descriptor D could be the number of shared electrons in that bond (single bond, double bond or triple bond). Or one may need two descriptors – the most familiar example of this is the periodic table, where the rows and the columns of the periodic table function as two descriptors $\{\mathcal{D}_r, \mathcal{D}_c\}$ and the property $\mathcal{P}$ could be one of a number of different things, e.g., whether the elemental material is a gas, liquid or solid at ambient conditions; or whether it is an insulator, semiconductor or metal. The example of the periodic table serves to illustrate one important feature of descriptors: that properties cluster in descriptor space. It also, famously, serves to illustrate the predictive power of descriptors: when Mendeleev first formulated the periodic table, several elements, e.g., Ge and Ga, were yet to be discovered, yet their properties could be predicted with a high degree of accuracy because of the missing 'holes' in descriptor space.

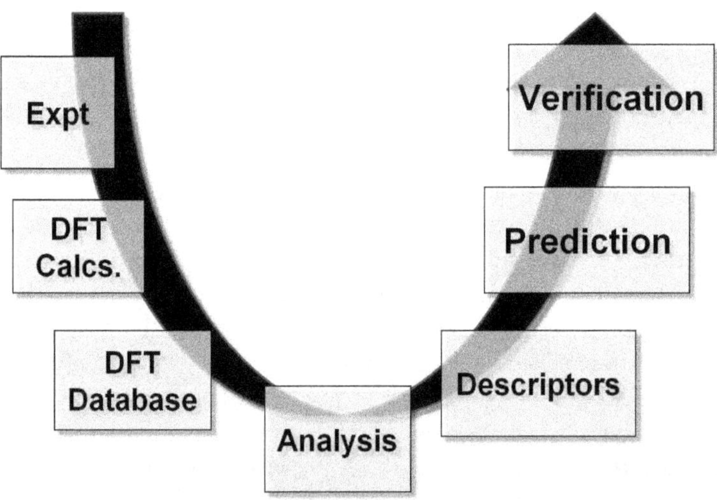

Fig. 1.   Steps that are typically followed to develop a descriptor or descriptors to aid materials discovery.   See the text for further explanation of the various steps in the process.

Figure 1 shows a schematic representation of the steps one might typically follow in order to develop descriptors. One first assembles experimentally available information on the class of materials to be studied, especially about possible crystal structures. Then, one performs DFT calculations to assemble a database of first principles results. This data then has to be analyzed and mined – to do this, one uses some combination of physical/chemical intuition and numerical techniques such as regression and (increasingly, nowadays) machine learning on a training set that might constitute all of the data contained in the database, or some subset of it.

Finally, the descriptors have to be validated, by checking the predictions made using the descriptors, against further first principles calculations.

2. High Throughput Computing and DFT Databases

The first stage in the formulation of descriptors is the assembling of a database containing information among which one will search for patterns, thereby finally identifying the descriptor. This information could be gained either from experiments, or theory, or a combination of both. Typically, when one is interested in, ultimately, having the ability to predict the properties of compounds, such databases contain information both about the constituent elements and about the properties of the compounds built up out of these elements, and one is interested in finding correlations between elemental features and the properties of the compounds.

Examples of elemental features include atomic number, atomic mass, electron affinity, ionization potential, electronegativity and atomic size. We note that some of these could possibly be characterized in multiple ways – e.g., to measure atomic size, one could use the ionic, covalent, metallic or van der Waals radius. Examples of material properties which some combination of these elemental features may possibly correlate with include crystal structure, band gap (and thus metallic/semiconductor/insulator nature), heat of formation, electronic conductivity, thermal conductivity, thermoelectric coefficient and catalytic activity.

With the advent of fast computers and accurate methods for calculating materials properties, increasingly, such databases are being assembled using high-throughput computing. In such an approach, automated workflows are used to compute the properties of a large number ($\sim$ thousands) of materials using *ab initio* density functional theory (DFT) calculations. One can measure the success of such high-throughput endeavors in two ways: (i) their ability to reproduce experimental data, where available, and (ii) their ability to successfully predict the existence and properties of novel materials that do not form a part of the original database. To date, experience has shown that DFT has a very high success rate in the former goal, and an increasingly encouraging success rate in the latter goal.

However, before proceeding to assemble a database of results using DFT calculations, and then use it to formulate descriptors, one has to be aware of both the strengths as well as the limitations of the DFT approach. The formal basis of DFT (founded on the Hohenberg-Kohn theorems[2] and the Kohn-Sham formulation[3]) guarantees that the total energy and charge density in the ground state, and therefore properties that directly depend on these, should be accurately computed in DFT calculations. Thus, DFT is very good at correctly reproducing and/or predicting the crystal structure and stability of materials; however, electronic properties such as the band gap are not guaranteed to be correct. In fact, it is known that standard DFT typically underestimates band gaps by 30% – 100%.[4] DFT does usually succeed in qualitatively predicting whether a material will be a metal or an insulator/semiconductor – though even here, there are some well known failures. For example, if exchange-correlation interactions are described using the generalized gradient approximation (GGA),[5] germanium is incorrectly predicted to be a semi-metal. Such errors can be corrected by using so-called 'hybrid' functionals[6–8] or performing GW calculations,[9,10] though the price one pàys for the increased accuracy is a considerable increase in computational cost.

An example of a project whereby one can gauge the accuracy of DFT calculations is presented by the study of Curtarolo *et al.*, who performed a high-throughput (HT) search for stable intermetallics, considering binary compounds formed out of 30 different *d*-electron metals, i.e., 435 possible pairs of metals.[11] In 224 (51.5%) of the cases, both experiment and HT showed that the two metals mix (i.e., form stable intermetallics – see the green dots in the lower triangle in Figure 2), in 117 (26.9%) of the cases, both experiment and HT showed that the two metals are immiscible (see the gray dots in the lower triangle in Figure 2); i.e., agreement was obtained for 341 (78.4%) of the 435 pairs. For the relatively small number of remaining cases, there was either disagreement between experiment and HT (red and blue triangles in Fig. 2) or a lack of available experimental data. The authors also report that the experimental structure was reproduced for 96.7% of the cases. We will return to discussion of the results of this study further below.

In an example where high throughput computing was used to perform a targeted search for materials for a desired application, Castelli *et al.* considered all possible cubic perovskites of the form ABO_3, where A and B are 51 different non-radioactive metals, to look for catalysts for photoelectrochemical water splitting, in order to generate hydrogen.[12] A suitable candidate should first of all be stable, i.e., it should have a heat of formation (using water and hydrogen gas as references) that is negative (the authors relaxed this criterion to allow the heat of formation to be less than 0.2 eV/atom, so as to also consider slightly metastable compounds). In addition, to be efficient at splitting water, its band gap should be 1.5 to 3.0 eV

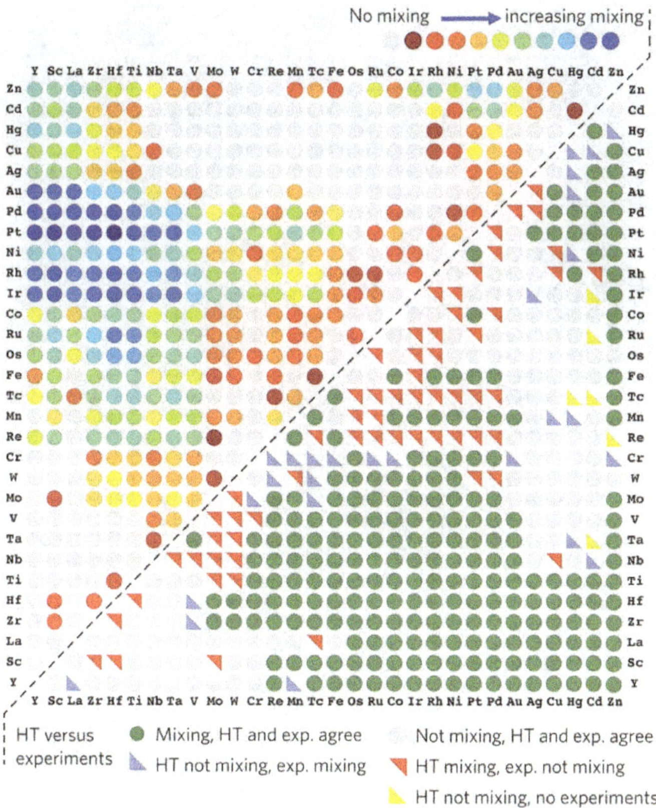

Fig. 2. Results from a high-throughput DFT study on binary intermetallics, taken from Ref. 11. Top left triangle: ordering tendency of the mixtures, as defined in the main text, for elements ordered by Pettifor's chemical scale. Grey circles indicate no ordering, whereas darker blue circles indicate increasing capability to form stable compounds. Bottom right triangle: comparison of high-throughput (HT) versus experimental results. Green and grey circles denote agreement between calculation and experimental data on the existence (green) or absence (grey) of ordered compounds. Purple (red) triangles indicate disagreement of HT predictions of compound absence (existence) versus experimental existence (absence). Yellow triangles indicate that data is unavailable for comparison.

(we have already noted above that standard DFT tends to severely underestimate band gaps. The authors report that they have largely overcome this problem by making use of the GLLB-SC functional[13] when computing band gaps). These limits on band gap were decided upon based on the requirement for overlap with the solar spectrum, and the need to be above the water-splitting threshold. Upon imposing these criteria on formation

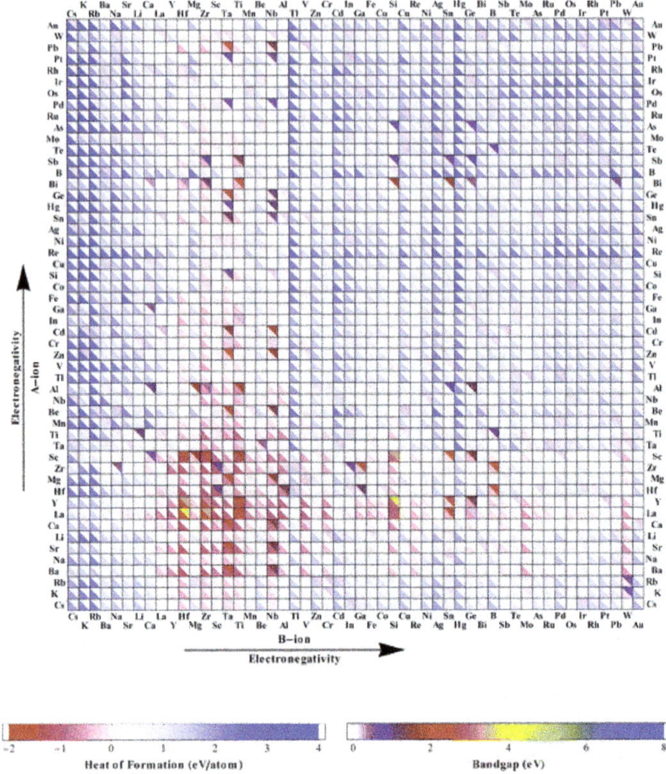

Fig. 3.   Results from a high-throughput DFT study on binary metal perovskite oxides, taken from Ref. 12. In each small square, the color of the lower left triangle shows the formation energy, while the color of the upper right triangle indicates the band gap. For formation energies, a red color indicates stability, while blue indicates instability. For band gap, a red color indicates a favorable color in the range from 1.5 – 3.0 eV.

energies and band gaps, the authors were left with only 43 out of the 2704 compounds screened by them. A third criterion was then imposed – that the band gaps should not just have a suitable magnitude, but should be appropriately aligned with respect to the hydrogen evolution potentials. The pool of candidate materials was then narrowed down to 10. A similar search was also performed for oxynitrides, resulting in 5 candidates. Of these, 4 were compounds that had already been previously known to split water, however one ($MgTaO_2N$) was a newly identified candidate that could now be considered for experimental synthesis and further exploration.

Yet another recent example of a targeted application of high throughput searches is the work of Mounet *et al.*, who wished to identify new two-dimensional (2D) materials, starting from DFT calculations on three-dimensional (3D) bulk materials.[14] They started from 108,423 unique experimentally known 3D compounds contained in various crystallographic databases, and performed DFT calculations, incorporating van der Waals interactions, on them. The structures were then analyzed, searching for chemically bonded subunits that are held together by weak van der Waals interactions, with a criterion based on comparing interatomic distances to van der Waals radii. In this way, they identified 5,619 materials as 'layered'. For a subset of 3,210 of these, the exfoliation energy was computed using DFT. One measure of the strength of the van der Waals interactions (and hence the ease of exfoliation) is the difference in binding energy when DFT calculations were performed with and without the inclusion of van der Waals interactions (see Figure 4). If the binding energy fell below a certain threshold (that depended on the van der Waals functional used) the parent compound was defined to be 'easily exfoliable'. In this way, they identified 1,036 easily exfoliable and 789 'potentially exfoliable' compounds. Of these, 18 lead to unary (elemental) monolayers, including all known elemental monolayers, such as graphene and phosphorene. The results of this study vastly expand the space of potential 2D compounds. We note that further calculations have to be performed (and were, in many cases, performed by the authors) to check the mechanical and dynamical stability of such possible 2D materials.

As of now, several large databases containing either experimental or computational data, or a combination of the two, have been accumulated, and can be searched and mined. Two of the largest structural databases are the Inorganic Crystal Structure Database (ICSD)[15] and the crystallographic open database (COD),[16] which contain hundreds of thousands of entries. In the USA, the Materials Genome Initiative has been established to systematically assemble computational data. Among the notable DFT initiatives and repositories of data are the Materials Project,[17] the Materials Cloud,[18] NOMAD,[19] AFLOWLIB,[20] and OQMD.[21]

3. Examples of Descriptors

The examples listed above show that computer power has now made exhaustive searches possible. However, such an approach can also be – well, exhausting – and, as mentioned already above, becomes computationally

S. Narasimhan

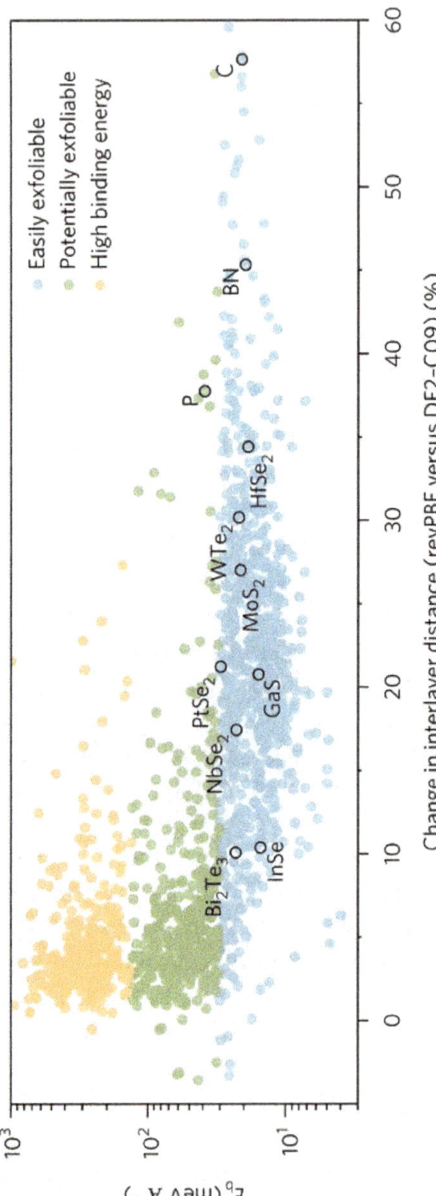

Fig. 4. DFT results for the binding energy E_b versus the change in the interlayer distance, for 1,535 3D layered compounds, when the calculations are performed using the revPBE (no van der Waals interactions) and the DF2-C09 functionals (van der Waals interactions incorporated). Materials classified as easily exfoliable are colored blue. Points corresponding to well-known 2D materials are labeled. Taken from Ref. 14.

unfeasible when one wants to vary three or more constituent elements systematically. This is where one can be considerably aided by formulating descriptors.

In other words, one would like to arrange the data in such a way that one sees patterns and trends in the data. In Figure 3 above, an attempt has been made to facilitate this by arranging the metals (both rows and columns) in order of their electronegativity. If, for example, either the band gap or the heat of formation were sensitive to a quantity like the sum or difference of the electronegativities of the two metals A and B, one should see a trend in the colors in Figure 3, and a clustering of colors. This doesn't quite happen – suggesting that electronegativities may not (at least by themselves) be a particularly good descriptor in this case.

Let us now consider Figure 2 above; earlier we had discussed the lower triangle, which indicates the amount of agreement between experiment and HT result. We turn our attention now to the upper triangle. Here, each intersection of a row and column is colored according to a measure of the formation energy of the intermetallic, which in turn is a measure of the ordering tendency of the mixture. To a certain extent, one sees a clustering of the colors – though this clustering is not perfect, i.e., the variation in the formation energy as one progresses along a row or column is not quire monotonic or even smooth. In this case, the metals have again been placed according to a possible descriptor – viz., 'Pettifor's chemical scale', which we will now go on to discuss.

3.1. *Descriptors for structural prediction of 3D compounds*

Pettifor originally suggested a chemical scale for the metals in the periodic table, as a descriptor for the structure of binary intermetallics.[22] He assigned a number χ to each metal [see Figure 5(a)], so that when a two dimensional structure map was drawn in (χ_A, χ_B) space, one found a good separation into various domains according to the structure of the binary intermetallic AB [see Figure 5(b)]. In his paper, Pettifor mentions that the values were reverse-engineered, to give *a posteriori* good separation in the two-dimensional structure map, with certain constraints placed on how χ varied as one progressed through the periodic table. Further, for the elements Be to F, χ was set equal to the value of the electronegativity.

Pettifor's achievement is quite remarkable, given that structure prediction is one of the hardest problems in materials science. An example of a challenging problem in this class is the structure of the octet compounds

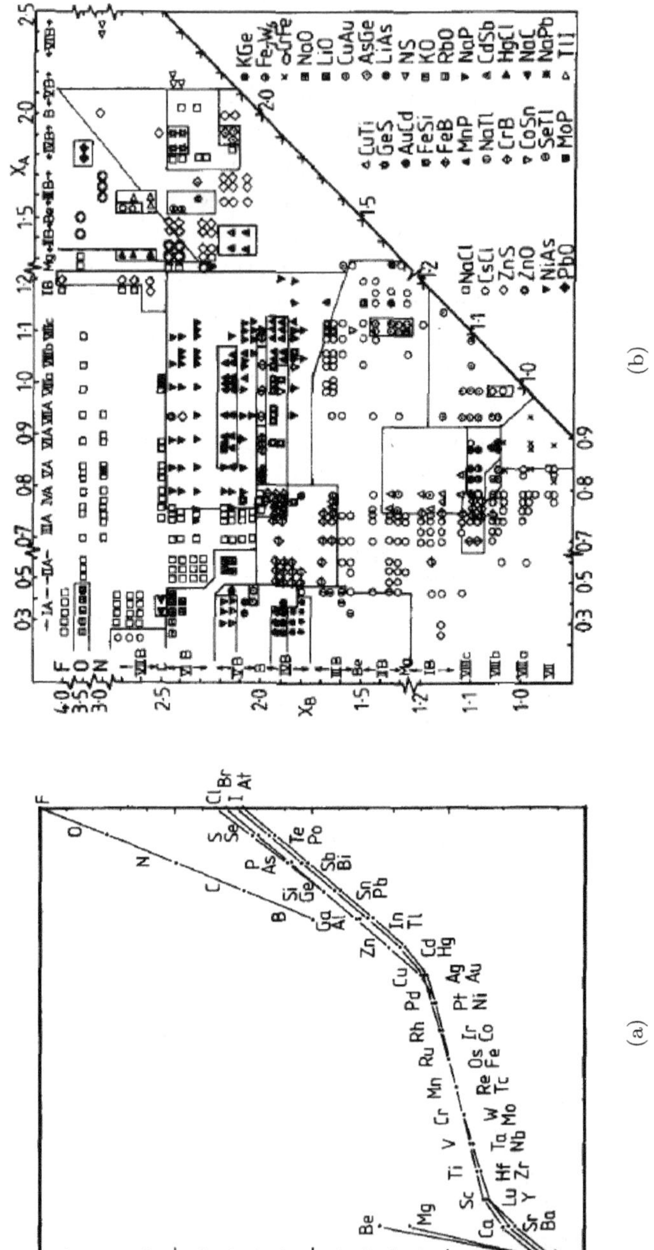

Fig. 5. Pettifor's chemical scale. (a) The values of the chemical scale χ for selected elements in the periodic table. (b) Use of the chemical scale results in compounds AB with different structures clustering in a two-dimensional structure map where the two axes are χ_A and χ_B. Figures taken from Ref. 22.

(binary compounds formed of two elements whose valency adds up to 8); these compounds can form, e.g., in the rock salt, zinc blende or wurtzite structures, and it is famously hard to predict which of these structures will be lowest in energy for a particular compound, from a knowledge of properties of the two individual elements alone. Pettifor's chemical scale also results in a separation in (χ_A, χ_B) space of octet compounds into clusters consisting of different structures.

Previous attempts had also been made to identify descriptors for this problem of the structures of octet compounds. For example, Phillips and Van Vechten showed that a separation into different structural groups in descriptor space could be achieved by making use of the pair of descriptors (C, E_h), where C and E_h are the average heteropolar and homopolar energy gaps, respectively.[23,24] Note that C and E_h are not straightforward to compute, requiring various assumptions to be made.[23] C and E_h depend on the dielectric constant and nearest-neighbor distance, which are (relatively) expensive to compute.

Such early attempts to formulate descriptors are characterized by the use of chemical intuition, and various models, and it is often hard to understand what procedure was followed to come up with the descriptor, and to reproduce it to deal with other properties or other classes of systems. However, this is no longer true: in recent years, the approach to developing descriptors has undergone a paradigm shift, with the use of systematic numerical methods, such as regression analysis and machine learning.

Such an approach was applied recently, to this particular problem of the structural prediction of the octet compounds, by Ghiringhelli *et al.*[25] For 82 binary materials AB, DFT calculations were performed on AB in the rock salt, wurtzite and zinc blende structures, optimizing the lattice constant and obtaining the total energy in the ground state. Scientific insight does continue to play a role, in that it was used to define 23 parameters that defined a 'feature space' out of which a descriptor could be built up. For each constituent element (A or B), these were the ionization potential (IP), the electron affinity (EA), the energies of the highest occupied and lowest unoccupied Kohn-Sham levels, and the (peak) radii of the s, p and d orbitals, as well as data on the equilibrium distance, binding energy and HOMO-LUMO gaps of AA, BB and AB dimers. These were then combined, taking sums, differences, products, ratios and powers, so as to find the best descriptors, using a machine learning algorithm.

The best one-dimensional descriptor found by them was

$$D_1 = (\mathrm{IP}(B) - \mathrm{EA}(B))/r_p(A)^2, \tag{1}$$

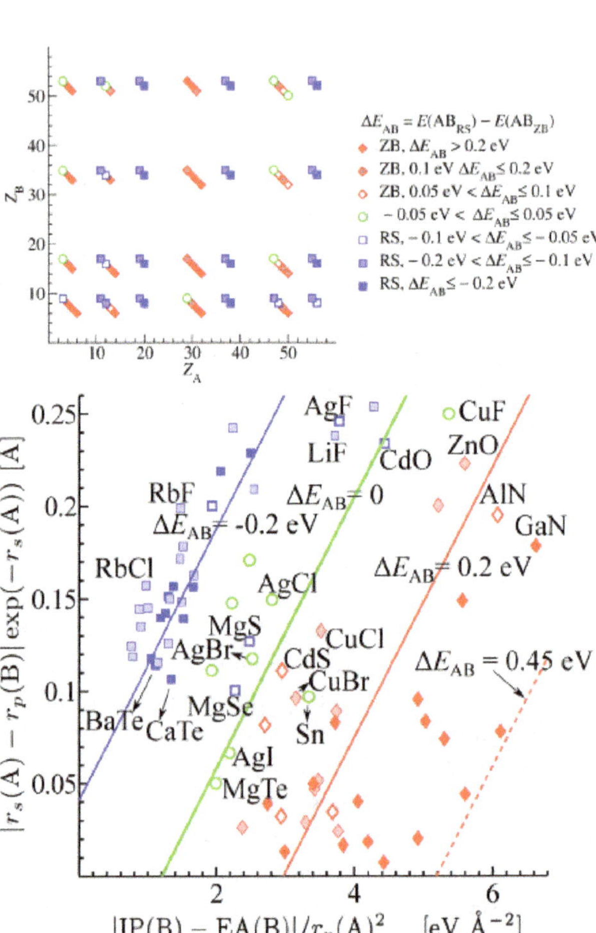

Fig. 6. Results for descriptors for the structures of octet compounds AB, taken from Ref. 25. In the top panel, the atomic numbers Z_A and Z_B are chosen as the two descriptors, resulting in a lack of separation in descriptor space. In the bottom panel, the two descriptors D_1 and D_2 identified by a machine learning algorithm are used to define the axes. Colors and symbols indicate the calculated energy differences between the rock salt and zinc blende structures, as obtained from a DFT calculation. Note that in the bottom panel, a clear clustering of symbols and colors is seen, suggesting that descriptors have been identified successfully.

and the best two-dimensional descriptor was (D_1, D_2), where

$$D_2 = |r_s(A) - r_p(B)|/\exp[r_s(A)], \tag{2}$$

where r_s and r_p are the peak radii of the relevant s and p orbitals, respectively. Figure 6 shows a plot in two-dimensional descriptor space of the 82 different materials. Not only does one get a separation in this space between those materials that prefer the rock salt structure (blue and green symbols) and those that prefer the zinc blende structure (red symbols), one also sees that their position in this space is an indicator of the relative energy difference between the rock salt and zinc blende structures. This is one of the advantages of this study over previous attempts at this problem. Another advantage is that a systematic attempt has been made to obtain the 'best' descriptor for each dimensionality. However, one pays a cost for this: one can see that the descriptors obtained are rather complex in form, and non-intuitive. They do not lend themselves easily to physical interpretation, though they are useful. Another feature of the pair of descriptors is that we do not have one axis corresponding to the properties of the element A and another corresponding to the element B, both D_1 and D_2 involve both A and B. Moreover, they do so in a non-symmetric way, which may seem counter-intuitive.

3.2. *Descriptors for prediction of structures of 2D self assembled monolayers*

Self assembled structures consist of networks of organic molecules bound together by weak non-covalent interactions, such as hydrogen bonds and van der Waals interactions. They have attracted much attention in recent years because of the possibility of using bottom-up self assembly to construct devices with nanoscale features that are hard to engineer using top-down machining techniques. Particular interest has been devoted to the possibility of constructing molecular motors by using the process of self assembly.

Self assembled structures can be one-dimensional, two-dimensional or three-dimensional. A great deal of the literature on self-assembly is devoted to two-dimensional (2D) architectures, comprised of a single self assembled monolayer (SAM) of organic molecules, deposited on a substrate. The substrate frequently consists of a metal such as gold, or can also be a material such as graphite. It has been suggested that such systems can be used to construct next-generation devices such as molecular electronic circuits and solar cells, or can be used as coatings.

It is of interest to know whether the monolayer will be periodic (crystalline) or random (glassy). Also, in the case of periodic structures, it is of interest to know whether the (host) structure is characterized by the presence of pores, which can accommodate other molecules (guests). Interestingly, in some cases, the presence of the guest molecules can also alter the delicate energy balance between various types of non-covalent interactions that determines the structure of the host assembly, and thereby trigger a structural phase transition.

In order to have the ability to construct made-to-measure self-assembled architectures, one would like to know, beforehand, which molecules to pick so as to achieve a desired geometry. For this purpose, one would be greatly aided if one could formulate descriptors that would permit one to predict the structure of the host assembly or host-guest assembly, using a knowledge only of the structure or formula of the isolated host or guest molecules. This problem was attempted by the present author, in collaboration with S. Ghosh, together with the experimental group of P. Zalake and K.G. Thomas.[26]

Note that for such a program to be carried out effectively, it is necessary to first construct a DFT database with reliable data on structures and energetics. This problem is challenging because of the difficulty in accurately capturing weak non-covalent interactions in DFT calculations. While the energetics of hydrogen bonding is generally accepted to be captured reasonably reliably, it is well-known that a shortcoming of standard DFT calculations is the inability to treat van der Waals interactions accurately. However, this problem has, to a large extent, been overcome in recent years, with the introduction of new van der Waals functionals, as well as semi-empirical treatments of van der Waals interactions within a DFT framework.[27–29]

In this joint experimental and theoretical study,[26] a 'training set' was used, consisting of three host molecules and five guest molecules (see Figure 7). Note that it is quite unusual to use such an approach on such a small training set; this is very far from constituting 'big data'! However, as will be shown further below, the approach proves remarkably (and indeed, surprisingly) successful even when applied to such a small data set. The three host molecules (labeled PE4A, PE4B and PE3A) are somewhat similar in structure, being different carboxylic acid derivatives of phenyleneethynylene. However, they also differ in important ways. PE3A has a shorter central backbone than PE4A and PE4B; it also has only two side alkoxy chains, whereas the other two molecules have four side chains. With the length L defined as the distance between the centers of the phenyl rings at the two ends of the central backbone, $L = 2.35$ nm for PE4A and PE4B, whereas $L = 1.39$ nm for PE3A. Further, PE4A and PE3A have four COOH terminal groups (two at each end of the molecule), whereas PE4B has only two COOH terminal groups (two at one end of the molecule). Note that the labeling convention used for the host molecules is PEnX, where n is the number of phenyl rings in the central backbone, and X gives further structural information about the terminal groups.

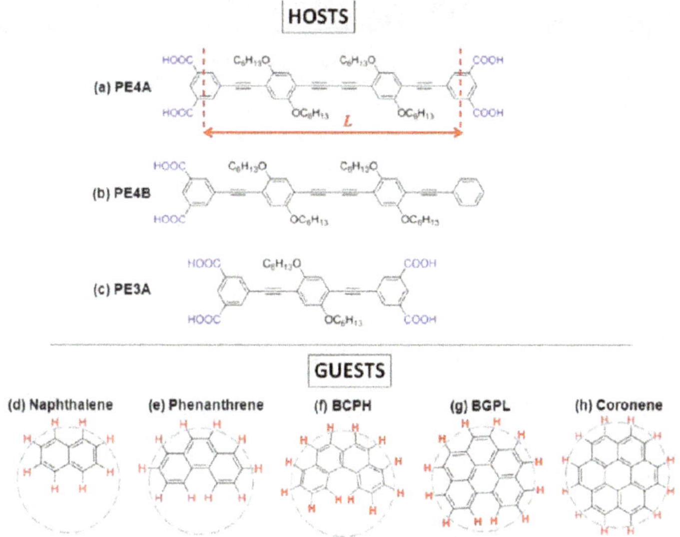

Fig. 7. Structures of the host and guest molecules used as the training set; figure taken from Ref. 26. The host molecules are labeled (a) PE4A, (b) PE4B and (c) PE3A. The red arrow indicates the length L of the central backnone. The guest molecules are (d) naphthalene (e) phenanthrene, (f) benzo-c-phenanthrene, (g) benzo-ghi-perylene, and (h) coronene. The dashed circles are drawn so as to pass through the largest number of H atoms on the periphery of the guest molecules, as described in the text.

In principle, each host molecule can self-assemble in two types of periodic patterns, each with a different symmetry, one of which is called the linear pattern (LIN), and the other a hexagonal pattern (HEX) (see Figure 8). The HEX is characterized by the presence of large hexagonal cavities, whereas the LIN features much smaller pentagonal cavities; the former pattern is significantly more porous than the latter. Guest molecules can

494   S. Narasimhan

be accommodated within these cavities, further below it is discussed which molecules can possibly be guests for these patterns.

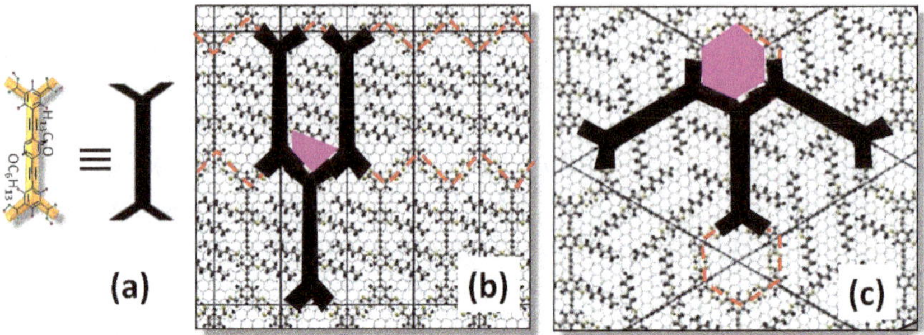

Fig. 8.   Possible self-assembled architectures of the host molecules, shown as an example for PE4A. (a) Schematic representation of the host molecule, (b) the LIN pattern, (c) the HEX pattern. Dashed red lines indicate the hydrogen bonds between terminal COOH groups, and the magenta pentagon and hexagon indicate cavities that can accommodate guest molecules.

In reality, for each host molecule (and later, when considering also guest molecules, for each host+guest combination) only one of the patterns, the HEX or the LIN, will be lowest in energy, and will therefore (in principle) be seen in experiments. However, in DFT calculations, one can perform calculations on both patterns, and calculate not only which pattern is lowest in energy, but also the energy difference between the two patterns.

The small structural differences between these three host molecules result in their forming three different types of self-assembled patterns when deposited on graphite. Scanning Tunneling Microscopy (STM) experiments show that PE4A forms the HEX pattern, whereas PE4B forms the LIN pattern. As for PE3A, it forms a random glasslike pattern, though there are small domains where locally the LIN pattern is seen (see Figure 9). These observations can be explained by DFT calculations, which show that for PE4A, the HEX pattern is lower in energy than the LIN by 1.07 meV/Å^2, whereas for PE4B, the LIN is lower in energy than the HEX by 0.96 meV/Å^2. For PE3A, the calculations show that the LIN is lower than the HEX by the relatively small amount of 0.26 meV/Å^2; since this energy difference is small, there is a competition between these two patterns, resulting in the formation of a glassy pattern.[30]

The guest molecules, naphthalene, phenanthrene, benzo-c-phenanthrene (BCPH), benzo-ghi-perylene (BGPL) and coronene are shown in Fig-

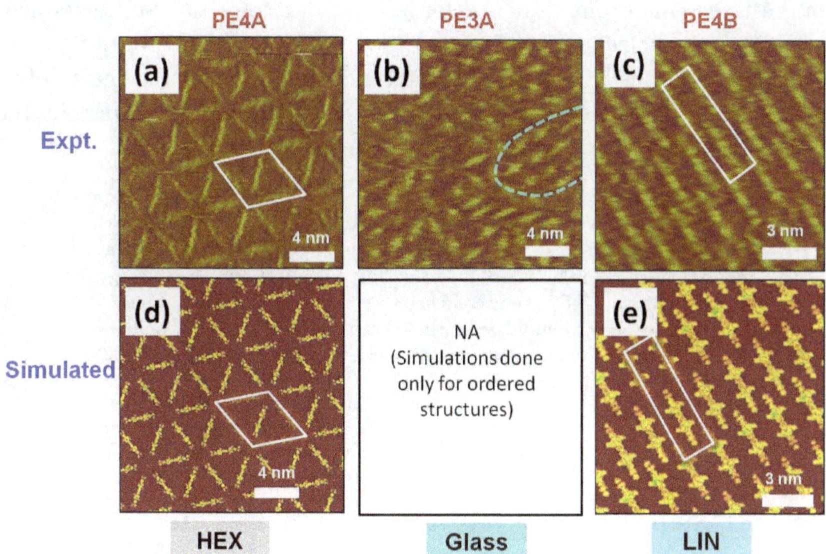

Fig. 9.   Experimental and simulated STM images of patterns formed by host molecules, figure modified from Ref. 26. The top and bottom panels show experimental and theoretically simulated STM images, respectively. PE4A, PE3A and PE4B form HEX, glassy and LIN patterns, respectively. The rectangles and rhombi indicate the boundaries of a unit cell for the pattern. The dashed green curve in panel (b) encloses a region where, locally, the LIN pattern is observed.

ure 7(d)–(h). From these figures, it is evident that the first four molecules can be viewed as radial fragments of coronene. The guest molecules have H atoms on their periphery, that can form hydrogen bonds with the atoms on the interior of the cavities in the HEX and LIN patterns of the host molecules.

Experiments show that upon co-depositing host and guest molecules, when the host molecule is PE4A, the observed pattern is the HEX always, regardless of which guest molecule is used. This is confirmed by DFT calculations, which show that the HEX is lower than the LIN for all PE4A+guest combinations (see Table 1). In the case of PE4B, when the guest is naphthalene or phenanthrene, the pattern remains LIN, as it was for the bare host; however for the remaining guest molecules there is a structural transition to the HEX pattern. This can be understood partly on steric grounds, as the larger guest molecules cannot fit into the small pentagonal cavities of the LIN pattern; they can only fit into the larger hexagonal cavities of the HEX pattern. The results when the host is PE3A are particularly interesting, where one sees a disorder-to-order transition on introducing certain guest molecules. The energy lowering on making several hydrogen bonds between the host and the guest is sufficient to lower the energy of the HEX pattern relative to that of the LIN, so that the ordered HEX pattern becomes favored.

Table 1. Favored patterns for host+guest combinations, from experiment and DFT calculations. The labels HEX/LIN/Glass indicate which kind of pattern is observed experimentally for the particular host+guest combination. The numbers give the difference in Gibbs free energy, as computed using DFT, between the HEX and LIN patterns, in meV/Å^2. A negative/positive number implies that the HEX/LIN is favored. Note that theory and experiment always agree. Values taken from Ref. 26.

↓Host/Guest→	None	Naph	Phen	BCPH	BGPL	Coro
PE4A	HEX −1.07	HEX −0.96	HEX −0.94	HEX −1.86	HEX −2.07	HEX −2.34
PE3A	Glass 0.26	Glass 0.17	HEX −0.07	HEX −0.08	HEX −0.09	HEX −1.12
PE4B	LIN 0.94	LIN 0.82	LIN 0.72	HEX −0.11	HEX −0.38	HEX −0.63

To obtain the numbers contained in Table 1, one computes from DFT, the change in Gibbs free energy upon adsorption of the self-assembled monolayer on the substrate, this is given by

$$\Delta G = (1/A)(E_{\text{host+guest}/\text{G}} - N_{\text{host}}\mu_{\text{host}} - N_{\text{guest}}\mu_{\text{guest}} - N_G\mu_G), \qquad (3)$$

where $E_{\text{host+guest}/\text{G}}$ is the total energy from DFT of the host+guest system on graphene; μ_{host}, μ_{guest} and μ_G are the chemical potentials of the host molecules, the guest molecules and graphene, respectively; and N_{host}, N_{guest} and N_G are the corresponding number of molecules or atoms of the respective species in the unit cell for the host+guest system on graphene. The numbers in Table 1 are obtained by calculating the difference in ΔG for the HEX and LIN patterns for each combination of host+guest. This then comprises the database of DFT values which constitutes the training set which can be used to find descriptors.

One first looks for a host descriptor. For this, one makes use of the features in which the three host molecules differ: the length of the molecule L, the number of terminal COOH groups N_{COOH}, and the number of alkoxy side chains N_{alkoxy}. Accordingly, one looks for a host descriptor of the form:

$$\eta = (N_{\text{COOH}} \times L^{\alpha})/[1 + (N_{\text{alkoxy}})^{\beta}]. \qquad (4)$$

This form for η is motivated partly by chemical intuition. The values for α and β are obtained by doing a linear regression, so that the best correlation is obtained between the values of $(\Delta G)_{\text{HEX}} - (\Delta G)_{\text{LIN}}$ and the descriptor η. This results in obtaining $\alpha = 1.0$ and $\beta = 1/8 = 0.125$ [see Figure 10(a)].

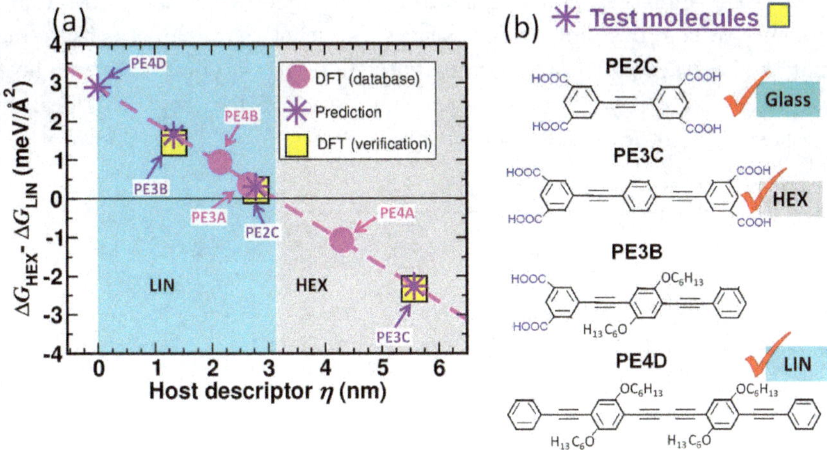

Fig. 10. Determination and validation of host descriptor η. (a) Differences in Gibbs free energy between the competing HEX and LIN structures correlate linearly with η. The magenta dots show the results from DFT for the 3 host molecules in the training sets. The indigo stars show the predictions, using the descriptor η, for this difference in free energy, for the test molecules shown in (b). The yellow squares show the results from DFT calculations on the test molecules. (b) Structures of the test molecules used to verify the host descriptor.

To validate the host descriptor, one considers test molecules, depicted in Figure 10(b). While these are broadly similar to the host molecules depicted in Figure 7, they differ in important particulars – e.g., molecule PE2C is much shorter, molecule PE3C has no side chains, and molecule PE4D has no terminal COOH groups. It is a matter of a few seconds to compute the value of η for each of the four test molecules, and accordingly, by looking at the position on the magenta dashed line in Figure 10(a), one can make a prediction about what pattern the test molecule will form. If the point [see the indigo stars in Figure 10(a)] falls well in the cyan region, a LIN pattern should be favored, if it falls well in the gray region, a HEX pattern should be favored, and if it falls near the cyan/gray boundary, a glassy pattern should form. For three of the molecules (PE2C, PE3C and PE4D), these predictions match with experimental data; for the fourth molecule (PE4B)

498   *S. Narasimhan*

experimental data is not yet available. One can however do even better than such qualitative matching – subsequent DFT calculations (yellow squares in Figure 10) show that the descriptor η can not only predict the qualitative pattern, but also the magnitude of the difference in free energy between the HEX and LIN patterns (see the close agreement between the yellow squares and indigo stars in Figure 10). This suggests that Eq. 4 results in an effective host descriptor. This success is rather remarkable, considering that the host descriptor was formulated using a data set of only three host molecules. Note that, as was the case for Eqs. 1 and 2 above, this form of the host descriptor does not lend itself easily to physical interpretation.

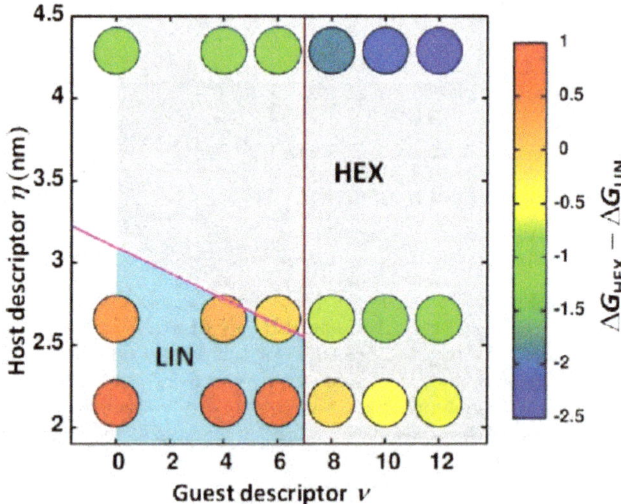

Fig. 11.   Structural map in 2D descriptor space taken from Ref. 26. The symbols indicate the 15 host+guest pairs, as well as the three host-only systems. The colors of the circles indicate the value of the difference in ΔG in the HEX and LIN patterns, as computed from DFT, according to the color bar on the right of the figure. The cyan and gray regions indicate regions where the LIN and HEX patterns are favored, respectively.

To construct a guest descriptor, one first considers the polygon formed by the peripheral H atoms on each guest molecule. One draws a circle through this polygon, such that it passes through a maximal number of vertices of this polygon (see Figure 7). This maximal number of vertices then constitutes the guest descriptor ν. One can demonstrate that the difference in ΔG between the HEX and LIN patterns correlates well with ν.[26]

Now, one can examine how differences in ΔG behave in the two-dimensional descriptor space (η, ν). This is shown in Figure 11, where the colors of the dots represent $(\Delta G)_{\text{HEX}} - (\Delta G)_{\text{LIN}}$, as computed from DFT, and the background color represents whether the HEX or LIN is favored. One sees that one now obtains a structural phase diagram, with a clustering in descriptor space. The colors of the dots vary smoothly from red in the lower left corner (LIN strongly favored) to blue in the upper right corner (HEX strongly favored). This phase diagram confirms yet again that effective host and guest descriptors have been found. Now, for any other pair of host and guest molecules (belonging, broadly, to this class of molecules), one need only compute the corresponding values (η, ν) to find out the pattern that the host+guest combination would favor, when deposited on graphite or graphene.

4. Conclusions

We have seen above that the process of discovering novel materials for specific targeted applications can be carried out using computational materials design within the framework of ab initio density functional theory. The brute force approach towards this is to carry out high throughput computational searches, using automated workflows. Worldwide, several large databases of computational data (supplemented, in some cases, with experimental results) are available, generally accessible to researchers through open source protocols.

The chemical space in which one looks for new materials is vast, and one would like guidance regarding which smaller region, in this vast space, one should turn one's attention to, i.e., one would like a procedure that can identify promising candidates that merit further detailed calculations using first principles techniques, and/or experimental investigations. To identify promising candidates in this way, one makes use of descriptors. Such descriptors can be developed using a combination of intuition, physical arguments and numerical data mining. Their use can save considerable time and effort.

We have reviewed descriptors for the particular problem of structure prediction – either for 3D crystals (especially the octet compounds) or of 2D self-assembled monolayers. We have seen that the descriptors that emerge, while effective, appear non-intuitive. It is possible that future work may help us understand the structure of the descriptors better. One can also identify descriptors for other properties. For example, Nørskov and

Hammer identified the 'd-band center' as a descriptor for the catalytic efficacy of transition metals.[31] Other possible descriptors for catalytic efficacy include the generalized coordination number,[32] the effective coordination number,[33] and the dual center descriptor.[34] Similarly, descriptors have been found recently for charge transfer and binding in 2D material on metal systems, and for the efficacy of aliovalent doping in oxides.[35,36] As this approach gains popularity, it is envisaged that several more descriptors will be developed, for a larger number of properties.

References

1. O. Isayev, C. Oses, C. Toher, E. Gossett, S. Curtarolo, and A. Tropsha, Universal fragment descriptors for predicting properties of inorganic crystals, *Nature Commun.* **8**, 15679 (2017).
2. P. Hohenberg and W. Kohn, Inhomogeneous electron gas, *Phys. Rev.* **136**, B864–B871 (1964).
3. W. Kohn and L.-J. Sham, Self-consistent equations including exchange and correlation effects, *Phys. Rev.* **140**, A1133–A1138 (1965).
4. A. Morales-Garcia, R. Vallero and F. Illas, An empirical, yet practical way to predict the band gap in solids by using density functional band structure calculations, *J. Phys. Chem. C* **121**, 18862 (2017).
5. J.P. Perdew, K. Burke and M. Ernzerhof, Generalized gradient approximation made simple, *Phys. Rev. Lett.* **77**, 3865–3868 (1996).
6. J. Heyd, G.E. Scuseria and M. Ernzerhof, Hybrid functionals based on a screened Coulomb potential, *J. Chem. Phys.* **118**, 8207 (2003).
7. A.D. Becke, Density-functional exchange-energy approximation with correct asymptotic behavior, *Phys. Rev. A* **38**, 3093–3100 (1988).
8. C. Lee, W. Yang and R.G. Parr, Development of the Colle-Salvetticorrelation-energy formula into a functional of the electron density, *Phys. Rev. B* **37**, 785–789 (1988).
9. L. Hedin, New method for calculating the one-particle green's function with application to the electron-gas problem, *Phys. Rev.* **139**, A796–A823 (1965).
10. M.S. Hybertsen and S.G. Louie, Electronic correlation in semiconductors and insulators: Band gaps and quasiparticle energies, *Phys. Rev. B* **34**, 5390 (1984).
11. S. Curtarolo, G.L.W. Hart, M.B. Nardelli, N. Mingo, S. Sanvito, and O. Levy, The high-throughput highway to computational materials design, *Nature Materials* **12**, 191–201 (2013).
12. I.E. Castelli, T. Olsen, S. Datta, D.D. Landis, S. Dahl, K.S. Thygesen, and K.W. Jacobsen, Computational screening of perovskite metal oxides for optimal solar light capture, *Energy and Environ. Sci.* **5**, 5814–5819 (2012).
13. O. Gritsenko, R. van Leeuwen, E. van Lenthe and E.J. Baerends, *Phys. Rev. A* **51**, 1944 (1995).
14. N. Mounet, M. Gibertini, P. Schwaller, D. Campi, A. Merkys, A. Marrazzo,

T. Sohier, I.E. Castelli, A. Cepellotti, G. Pizzi and N. Marzari, *Nature Nanotechnol.* **13**, 246–252 (2018).

15. www2.fiz-karlsruhe.de/icsd_home.html

16. http://crystallography.net

17. https://materialsproject.org/; A. Jain, S.P. Ong, G. Hautier, W. Chen, W.D. Richards, S. Dacek, S. Cholia, D. Gunter, D. Skinner, G. Ceder and K. Persson, Commentary: The Materials Project: A materials genome approach to accelerating materials innovation, *APL Materials* **1**, 011002 (2013).

18. https://materialscloud.org/; G. Pizzi, A. Cepellotti, R. Sabatini, and N. Marzari, AiiDA: automated interactive infrastructure and database for computational science, *Comput. Mater. Sci.* **111**, 218–230 (2016).

19. https://nomad-coe.eu/; C. Draxl and M. Scheffler, NOMAD: The FAIR concept for big data-driven materials science, *MRS Bulletin* **43**, 676–682 (2018).

20. aflowlib.org; S. Curtarolo, W. Setyawan, S. Wang, J. Xue, K. Yang, R.H. Taylor, L.J. Nelson, G.L.W. Hart, S. Sanvito, M.B. Nardelli, N. Mingo and O. Levy, AFLOWLIB.ORG: A distributed materials properties repository from high-throughput ab initio calculations, *Comput. Mater. Sci.* **58**, 227–235 (2012).

21. oqmd.org; S. Kirklin, J.E. Saal, B. Meredig, A. Thompson, J.W. Doak, M. Aykol, S. Rühl and C. Wolverton, The open quantum materials database (OQMD); assessing the accurace of DFT formation energies, *Comput. Mater.* **1**, 15010 (2015).

22. D.G. Pettifor, A chemical scale for crystal structure maps, *Solid State Commun.* **51**, 31–34 (1984).

23. J.C. Phillips, Ionicity of the chemical bond in crystals, *Rev. Mod. Phys.* **42**, 317–356 (1970).

24. J.A. van Vechten, Quantum theory of electronegativity in covalent systems: I. electronic dielectric constant, *Phys. Rev.* **182**, 891–905 (1969).

25. L.M. Ghiringhelli, J. Vybiral, S.V. Levchenko, C. Draxl and M. Scheffler, Big data of materials science: critical role of the descriptor, *Phys. Rev. Lett.* **114**, 105503 (2015).

26. P. Zalake, S. Ghosh, S. Narasimhan and K.G. Thomas, Descriptor-based rational design of two-dimensional self-assembled nanoarchitectures stabilized by hydrogen bonds, *Chem. Mater.* **29**7170–7182 (2017).

27. M. Dion, H. Rydberg, E. Schröder, D.C. Langreth, and B.I. Lundqvist, Van der Waals density functional for general geometries, *Phys. Rev. Lett.* **92**, 26401 (2004).

28. A. Tkatchenko, R.A. DiStasio Jr., R. Car and M. Scheffler, Accurate and efficient method for many body van der waals, *Phys. Rev. Lett.* **108**, 236402 (2012).

29. S. Grimme, Semiempirical GGA-type density functional constructed with a long-range dispersion correction, *J. Comput. Chem.* **27**, 1787–1799 (2006).

30. H. Zhou, H. Dang, J.H. Yi, A. Nanci, A. Rochefort, J.D. Wuest, Frustrated 2D molecular crystallization, *J. Am. Chem. Soc.* **129**, 13774–13775 (2007).

31. B. Hammer and J. Nørskov, Electronic factors determining the reactivity of metal surfaces, *Surf. Sci.* **343**, 211 (1995).

32. F. Calle-Vallejo, J.I. Martinez, J.M. Garcia-Lastra, P. Sautet and D. Loffreda, Fast prediction of adsorption properties for platinum nanocatalysts with generalized coordination numbers, *Angew. Chem. Int. Ed.* **53**, 8316 (2014).

33. P. Ghosh, R. Pushpa, S. de Gironcoli and S. Narasimhan, Effective coordination number: a simple indicator of activation energies for NO dissociation on Rh(100) surfaces, *Phys. Rev. B* **80**, 233406 (2009).

34. N. Mammen, S. de Gironcoli and S. Narasimhan, Substrate doping: a strategy for increasing the reactivity of gold nanoclusters by tuning *sp* bands, *J. Chem. Phys.* **143**, 144307 (2015).

35. D. Sharma, G. Gautam and S. Narasimhan, A simple descriptor for binding and charge transfer at blue phosphorene–metal interfaces, *Appl. Surf. Sci.* **492**, 16–22 (2019).

36. S. Ghosh, N. Mammen and S. Narasimhan, A descriptor for the efficacy of aliovalent doping of oxides an its application for the charging of supported Au clusters, *J. Phys. Chem. C* (2019).

Chapter 21

Mechanical Behaviour of Glasses and Amorphous Materials

Anshul D. S. Parmar* and Srikanth Sastry[†]

Jawaharlal Nehru Centre for Advanced Scientific Research, Jakkur, Bangalore 560064, Karnataka, India
[†]*sastry@jncasr.ac.in*

A wide range of materials can exist in microscopically disordered solid forms, referred to as amorphous solids or glasses. Such materials – oxide glasses and metallic glasses, to polymer glasses, and soft solids such as colloidal glasses, emulsions and granular packings – are useful as structural materials in a variety of contexts. Their deformation and flow behaviour is relevant for many others. Apart from fundamental questions associated with the formation of these solids, comprehending their mechanical behaviour is thus of interest, and of significance for their use as materials. In particular, the nature of plasticity and yielding behaviour in amorphous solids has been actively investigated. Different amorphous solids exhibit behaviour that is apparently diverse and qualitatively different from those of crystalline materials. A goal of recent investigations has been to comprehend the unifying characteristics of amorphous plasticity and to understand the apparent differences among them. We summarise some of the recent progress in this direction. We focus on insights obtained from computer simulation studies, and in particular those employing oscillatory shear deformation of model glasses.

1. Introduction

Amorphous materials are ubiquitous in nature, and arise as among the common solid forms of a wide range of substances. The most familiar are various forms of silica glass, but chalcogenide glasses and amorphous metallic alloys or metallic glasses[1] are other common examples of what are some times termed *hard glasses*. At the other end of the spectrum are a variety of

---

*Current Address: Laboratoire Charles Coulomb, UMR 5221 CNRS-Université de Montpellier, Montpellier 34095, France. anshul-deep-singh.parmar@umontpellier.fr
[†]Corresponding author.

soft materials, which superficially form a very distinct class, but share some fundamental characteristics with hard glasses.[2,3] Some examples are dense colloidal assemblies, gels, emulsions, pastes and foams. These materials are typically studied for their flow or rheological properties, but many of them are observed to possess a *yield stress* below which they exhibit solid-like response and above which they exhibit viscous flow. Other common materials that may deserve mention are polymer glasses and granular materials. Many biological and geophysical formations on relevant scales are well described as amorphous solids, and their analysis is therefore informed by the understanding of amorphous solids (statistics of earthquakes is a familiar and oft-quoted example).

A common feature of these substances is the lack of microscopic order, unlike crystalline materials. In terms of their mechanical and flow behaviour, all these materials exhibit solid-like behaviour, in that upon the application of small external stresses, they respond elastically, with deformations being proportional to the applied stresses, and reversible, so that they return to the original state when the external stresses are removed. At larger applied stresses, they display *plastic* deformation, which remain when the applied stresses are removed. The point of onset of plasticity, the yield point, is thus an important material characteristic, which precedes various forms of failure of these materials. This broad and simplified description of mechanical behaviour is superficially the same as that for crystalline materials, and thus one must clarify what distinguishes amorphous materials from crystalline solids. A key difference is the translational order present in crystals, and the consequent possibility to have well defined structural defects. In particular, the plastic response of crystals is described in terms of the presence and movement of dislocation defects. In amorphous solids, there is no translational order, and thus no well defined notion of defects that are instrumental in plastic deformation. A related key distinction is that since amorphous solids are typically obtained as preparation protocol dependent, out of equilibrium, materials, the amorphous state is not unique for a given substance. The preparation method plays a key role in their observed behaviour, and a modification of the structure is often a key element in the response to applied stresses (see Figure 1).[2,3,11–15]

Studies of mechanical behaviour of amorphous solids focus on these key distinctions from the crystalline state, but it may be noted that the description of crystalline solids on large scales shares some of these features.[16] A second important general consideration pertains to the notion of solidity of either crystalline or amorphous materials. On general grounds, it has been

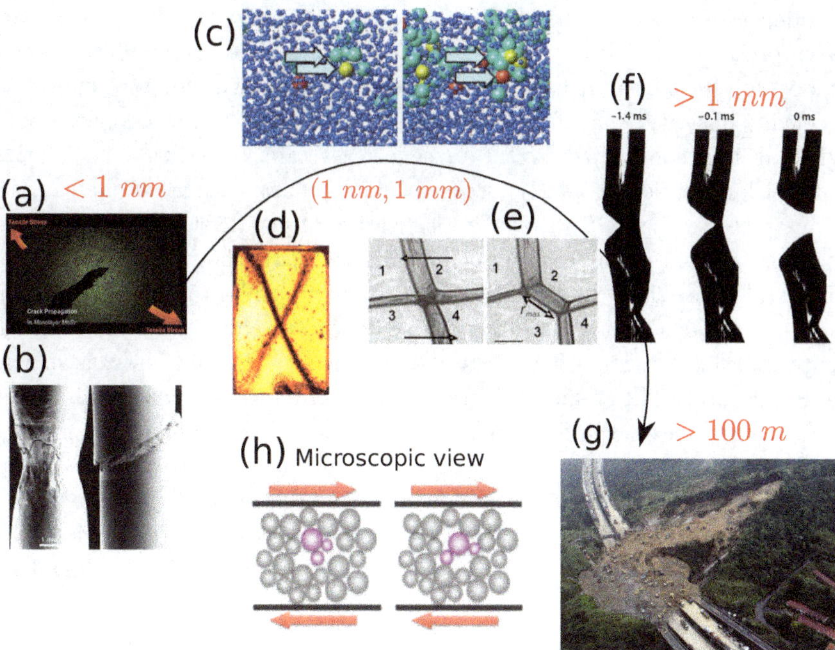

Fig. 1.   Examples of amorphous solids and deformations.   (a) Image of a crack with atomic resolution in a planar MoS_2 sample.[4] (b) Bulk metallic glass pillars under uniaxial stress. The yield behaviour of a composite glass reinforced with dendrites becomes more ductile.[5] (c) The plastic events in a slowly sheared colloidal system is represented in terms of a change in the neighbours of colloids. The color map and the size scale represent the number of nearest neighbors lost during a plastic event.[6] (d) The strain localisation in a granular pack of micrometer size glass beads under biaxial compression.[7] (e) Localized rearrangements in a bubble raft. The interface between neighboring bubbles exhibits a $T1$ event, triggered by applied strain.[8] (f) The stretching of a polymer melt. At sufficiently high strain rates the polymer melt behaves in a way similar to a solid.[9] (g) Landslide hit highway in Malaysia (source: China Press, The Straits Times). (h) The mechanical response to applied deformation is illustrated to involve local rearrangements.[10]

argued that a solid at finite temperature, under finite external stresses, will always deform and flow at finite shear rates and the distinction between solids with a finite shear modulus and viscoelastic fluids must be made with some care.[17,18] Thus, many attempts to understand yielding as a well defined transition focus on the limit of zero temperature, as discussed below. On the other hand, an understanding of behaviour in limiting cases forms the basis for describing finite temperature and shear rate behaviour.[19]

Figure 2 shows an illustration of stress-strain curves of amorphous solids, displaying some typical behaviours observed. In some materials, the dependence of stress on the strain is gradual, and smoothly approaches an asymptotic steady flow value. In other cases, one observes a stress overshoot, followed by a drop, before approaching the steady flow value. It is typically observed that in the cases where one observes a stress overshoot, yielding is accompanied by the presence of strain localisation, *i. e.*, the spatial localisation of plastic strain within narrow regions of the sample (illustrated in Figure 3(a)). Such localisation of plastic strain is, for example, observed in metallic glasses, and leads to brittle failure (Figure 3(b)). The stress overshoot and the flow stress also depend on the rate at which strain is applied. The yielding and flow behaviour can also be represented in terms of steady state flow stresses as a function of strain rates, as shown in Figure 2(c). The steady flow stresses typically increase with strain rates but in some cases, the flow curves are non-monotonic (illustrated in Figure 2(c)). The non-monotonic flow curves are observed to correspond to the presence of *shear bands* within which the bulk of the flow becomes localised (illustrated in Figure 3).[20] The presence of strain localisation and shear banding represent a key feature of yielding behaviour that requires explanation.

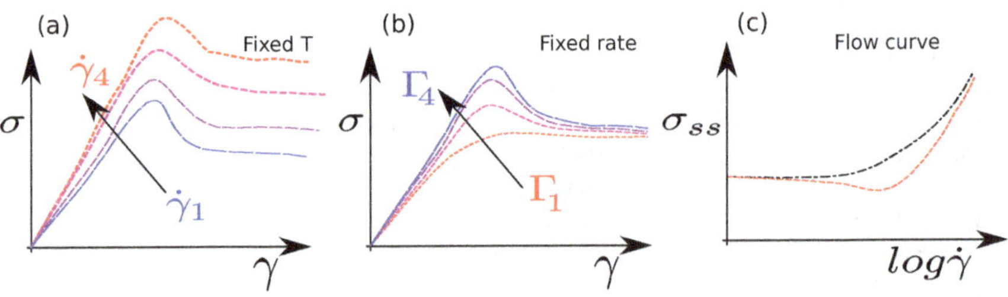

Fig. 2. An illustrative image showing stress strain curves with and without overshoots, and flow curves. (a) A sketch of the stress-strain curves for a glass sheared at different strain rates. Larger rates ($\dot{\gamma}_1 < \dot{\gamma}_2 < \dot{\gamma}_3 < \dot{\gamma}_4 < ...$) lead to larger yield stresses and larger steady state stresses.[14] (b) A sketch of stress-strain curves for glasses prepared at different cooling rates. The lower cooling rates ($\Gamma_1 > \Gamma_2 > \Gamma_3 > \Gamma_4...$) result in better annealed glasses. More annealed glasses show larger maximum stresses. (c) Steady-state flow curves, i.e., dependence of the steady-state shear stress σ_{ss} on the shear rate $\dot{\gamma}$ for a yield stress fluid. Homogeneous flow corresponds to a monotonic flow curve, of the Herschel-Bulkley form.[21] A regime with a negative slope in the flow curve is generally associated with instability and the presence of shear bands.[20]

Thus, some of the key questions to address in discussing the mechanical behaviour of amorphous solids pertains to the nature of elementary processes of plasticity that occur in amorphous solids, the nature of interactions among them, the nature of the yielding transition, the depen-

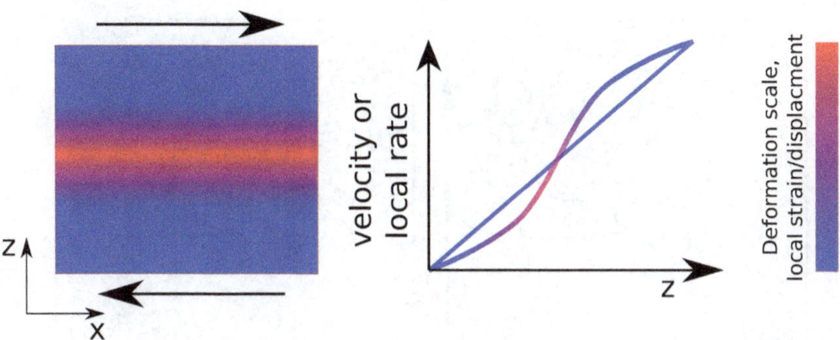

Fig. 3. Illustrative images illustrating (left) strain localization and (right) shear banding under flow. (Left) The blue region corresponds to lower strain and the red regime corresponds to localization of strain. (Right) The velocity profile in a shear banding system, compared with uniform flow.

dence of the mechanical response on the preparation history and structural properties of the amorphous solids, the nature of structural change during deformation, and the presence or otherwise of strain localisation and shear banding. We address these questions in this review, providing a summary of progress in answering them. Only some selected themes are summarised, and many omitted, to permit an adequate discussion of the topics chosen. Hence, the overview presented should be treated more as a perspective than a comprehensive review. In Section 2, we provide a general background and overview of theoretical and computational approaches to develop a description of mechanical behaviour and yielding in amorphous solids. We do not discuss relevant experimental results specifically, but these are well summarised in various reviews we mention.[1-3] Section 3 discusses selected recent results, that address the nature of the yielding transition, the influence of sample preparation and annealing, and shear banding. Section 4 contains a summary of current status and outlook for further investigations.

2. Background and Previous Work

The nature of rearrangements leading to plastic deformation have been investigated over several years, as a step in developing a description of plasticity in amorphous solids. Early work by Argon and co-workers[11] attempted to understand such rearrangements by considering disordered bubble rafts as tractable model systems for glasses. These investigations indicated that under shear deformation, plastic rearrangements involved the participation

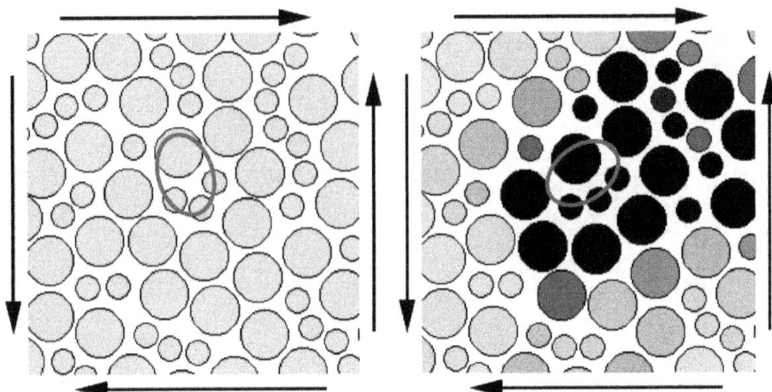

Fig. 4. An illustration of plastic rearrangements at the particle scale: Particle positions before and after a shear transformation event. The darker area represents the region that deviated from the affine deformation field. During this event, the indicated cluster of one large and three small particles changes orientation.[22]

of a small number of bubbles. These localised rearrangements have been termed *shear transformations*, and have been subsequently observed in numerous computer simulations and experiments.[12,22–25] A schematic view, obtained from computer simulations of a two dimensional glass model[22] is shown in Figure 4. Although one may envisage circumstances where plastic deformations may not be so localised, and will not occur as isolated events (see Ref.[3] for a recent review and detailed discussion), the idea of localised shear transformations or shear transformation zones (STZ) underlies many lines of modeling plasticity. Unlike dislocations in crystals, such STZs do not have a well defined structural characterization, nor does one *a priori* have the ability to identify locations where plastic events will occur. Considerable effort has been directed therefore at arriving at ways of identifying the locations of plastic events, based on an analysis of local structure, energies, local yield stress, soft modes etc.,[26–30] including the deployment of machine learning approaches.[31,32] However, the possibility of predicting with high reliability the locations of plastic events remains a subject of ongoing exploration.

Based on the idea that STZs exist, may be created or destroyed, and can undergo transitions between different states in response to applied shear, Falk and Langer[14,22] proposed a description of plasticity termed STZ theory. In this description, the population of STZs is described to be governed by an effective or mechanical temperature χ. Based on a nonequilibrium thermodynamic formulation of the STZs,[14,34] expressions for the evolu-

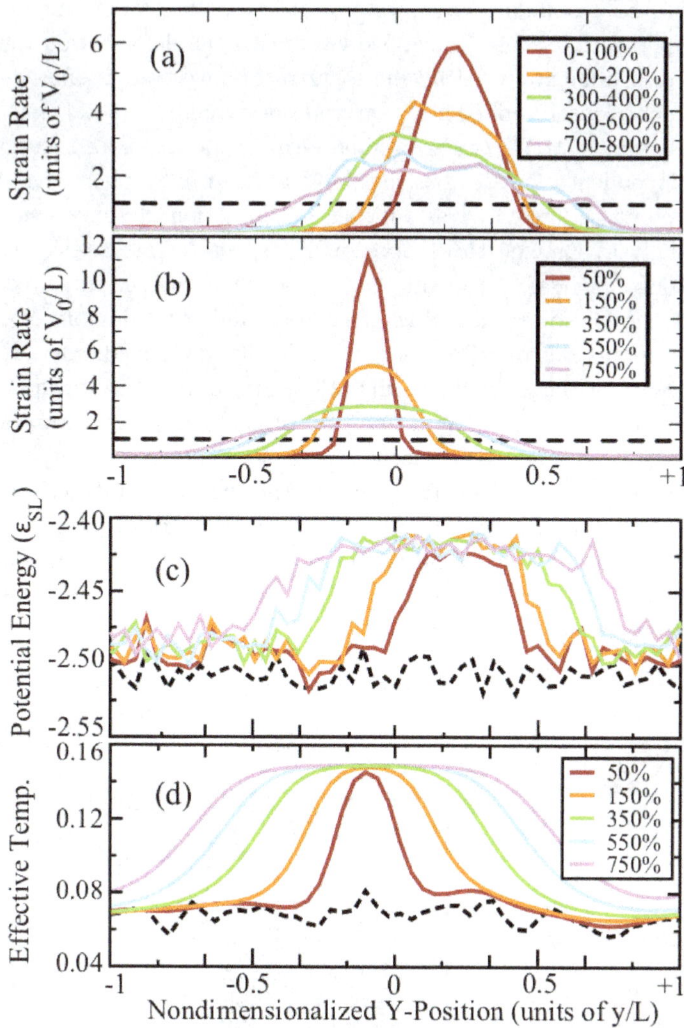

Fig. 5. Comparison of simulation results and STZ theory. (a) Simulated position (Y) dependent strain rates, as functions of the scaled position for various strains, when the system is subjected to a global strain at a fixed rate. The dark dashed line is the imposed strain rate. (b) Corresponding estimates of the strain rates from the STZ theory. (c) Simulated potential energy per atom as a function of position for various strains. (d) The STZ prediction of effective temperature as a function of position. The dark dashed lines in panels (c) and (d) show the initial values for the potential energy and effective temperature.[33]

tion of the STZ population, the rate of plastic deformation, and the effective temperature itself χ (which evolves during the deformation towards a steady state value governed by the shear rate) have been proposed. Yielding as a function of applied stress arises as a dynamical instability beyond a critical applied stress. In the original formulation, the STZ theory has no spatial dependence, but it does so in an extended form where the mechanical temperature is treated as varying as a function of space and time.[33] These analyses manage to capture many aspects of mechanical response in amorphous solids. An example is shown in Figure 5, which compares computer simulation results of the formation and spreading of shear bands with the predictions of STZ theory.[26,33] In the mentioned work, the local average potential energies from simulations are compared with the effective temperatures. Although the notion of a mechanical temperature and the comparison with energies are both reasonable and well motivated, there is so far no concrete prescription for *measuring* this temperature. The other key notion, of an STZ population, with distinct states of occupation, also requires a concrete prescription to directly access it which is currently missing, and has indeed motivated some of the research to identify locations in a glass that are involved in plasticity.[26–29]

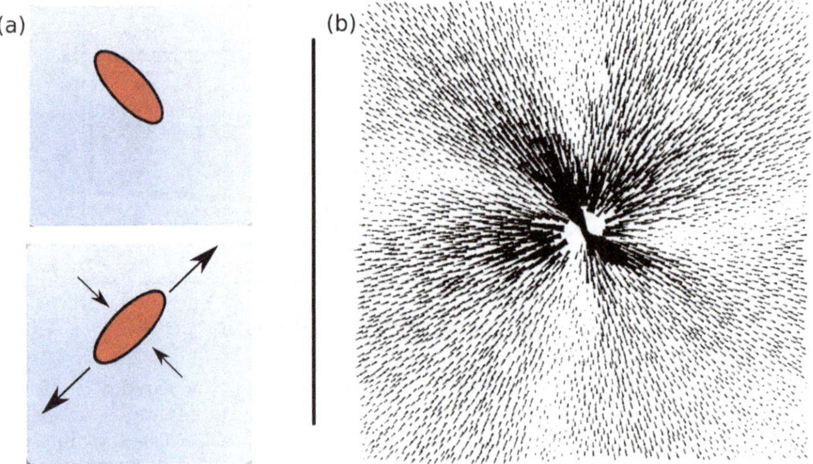

Fig. 6. (a) Schematic representation of an Eshelby inclusion. (b) The tangential non-affine displacement field during a plastic event from computer simulations.[23]

A more elaborate understanding of plastic events is provided by considering the nature of the perturbation they cause to the surrounding solid.

Considering an idealisation of a point shear transformation, under other simplifying assumptions, Picard *et al.*[35] derived an expression for the displacement field (correspondingly the strain field, and assuming linear elasticity, the stress field) in two dimensions, which exhibits a *quadrupolar* form:

$$\sigma_{xy}(\mathbf{r}) \propto \int d\mathbf{r}' G(\mathbf{r} - \mathbf{r}') \epsilon_{xy}^{pl}(\mathbf{r}')$$

where ϵ_{xy}^{pl} is the plastic shear strain at some point, which causes an incremental stress σ_{xy} elsewhere in the system, with the propagator G given by

$$G(r, \theta) \propto \frac{cos(4\theta)}{r^2}.$$

This result carries over to three dimensions, with the propagator varying as $1/r^3$. The analysis by Picard *et al.* is related to earlier work by Eshelby,[36] on the elastic field caused by an *inclusion* in an elastic medium that undergoes a change of form, illustrated schematically in Figure 6(a). Signatures of non-affine displacements arising from a localised plastic event have been investigated in a number of numerical and experimental studies.[12,22–25] An early example is shown in Figure 6(b). These analyses reveal that the effect of a localised plastic event is long ranged, bearing a characteristic anisotropic form, which has been found to be a valid description for amorphous solids.

Such propagation of stresses constitutes a mechanism for long ranged and anisotropic interactions between past and prospective regions of plasticity, and has consequences for theoretical descriptions and modeling. On the one hand, some investigators have argued that the long range nature of the interactions between plastic events implies that mean field descriptions should have a high degree of validity for real systems.[37] On the other hand, a class of models, termed elastoplastic models, have been developed taking cognizance of the explicit nature of stress propagation.[3] We describe the general characteristics of elastoplastic models and indicate some developments based on their study first.

Elastoplastic models represent an amorphous solid as being composed of cells, each of which may be thought of as representing a subvolume of the size of a plastic event. Each cell has a stress value, and a local yield stress value that is drawn from a distribution that forms part of the description of the model. A binary variable represents whether a plastic event is triggered at the cell or not. The stress at all sites increases monotonically with

time with the increase of strain (typically at a fixed strain rate). A plastic
event is triggered when the stress at a given site exceeds the local yield
stress at that site. The stress then is propagated to other sites according to
the propagator discussed earlier (or simpler ones) at a specified rate, and
it relaxes at the site of the plastic events. Many variations of this basic
strategy can be considered, including varying rates at which a plastic event
is triggered, inclusion of thermal activation etc. The expected variation of
stress at a given site is shown in Figure 7. In addition to the fully explicit
models of the kind described, various works have also considered mean field
versions, such as the Hebraud-Lequeux model. This model describes the
time evolution of the distribution of local stresses, dependent on a parameter α that controls stress diffusion, in proportion to the number of sites
that display plastic activity.[38] One obtains a transition from Newtonian
fluid behaviour at low shear rates to yield stress fluid behaviour obeying
the Herschel-Bulkley constitutive law, as α is varied. Elastoplastic models
have also been employed to study *avalanches*, correlated plastic events that
occur in tandem, typically in the steady flow regime.[39,40] Avalanches in the
steady state are relevant if one is interested in the dynamics of plasticity in
that regime[41] (for example, strain localisation, which we discuss in the next
section), and has been of interest in the context of intermittent dynamics
in earthquakes and other driven systems.

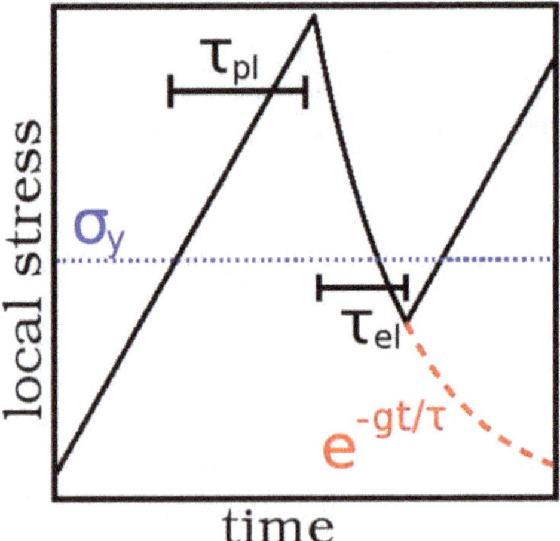

Fig. 7.   Schematic representation of the variation of local stress in an elastoplastic model.
For stresses larger than the threshold σ_Y, a site yields with a probability $1/\tau_{pl}$ (plastic
rate). The site is active for a duration given by the restructuring time τ_{el}.[42]

The nature of avalanches approaching yielding have also been addressed by many studies, motivated by analogies with related systems. We mention two such cases, which will be referred to in discussion of recent results in the next section. Dahmen and co-workers[37] have proposed a model for avalanches and yielding, motivated by ideas emerging from the study of the movement of interfaces in the presence of pinning sites, that predicts avalanches with a power law distribution with a cut off, with a universal power law exponent of $-3/2$, and a mean avalanche size that diverges as the yielding point is approached. While a variety of results have been compiled to show the validity of this description,[43,44] recent results described in the next section raise questions on some aspects of such a description of yielding. Another model that has been used as a reference is the random field Ising model (RFIM), in which, in addition to the external field H, a random field is applied at each site. The strength of the random field R is a key control parameter, and in the H-R plane, the zero temperature phase diagram of this system exhibits a critical point at a critical disorder R_c, and at lower R, a spinodal limit defines the equivalent of the yielding point. Analysis in Ref. 45 indicates that at the transition point, the susceptibility (associated with avalanche size) does not diverge, unlike the mean field case. The implications of this picture to yielding in amorphous solids are discussed in the next section.

Another perspective from which plasticity and yielding in glasses and amorphous solids has been analysed is to focus on the energy landscape properties of the deformed glasses. Such analysis has typically been conducted in the athermal quasi-static limit (AQS), in which the solids are studied in the zero temperature limit, subject to quasistatic strain, such that the glasses always reside in energy minima.[46–54] Plastic rearrangements arise as a result of the loss of stability of the energy minimum that the system resides in, which constitutes a saddle-node bifurcation.[48,49] Such analyses have been extensively conducted through computer simulations of atomic glasses and some results will be discussed in the next section.

3. Recent Investigations

A key question that has been addressed by many workers is the nature of the avalanches and how they may behave as the yielding transition is approached. We begin by a discussion of work in this regard, based on an energy landscape perspective. Procaccia and co-workers[48,55] have analysed the scaling of the stress drop $\Delta\sigma$ and energy drops ΔU during steady state avalanches, observing that they both scale with N, with $\Delta\sigma \sim N^{\beta}$, $\Delta U \sim N^{\alpha}$, with $\alpha - \beta = 1$, and it has been found that for both two and three dimensions, $\alpha = 1/3$, and $\beta = -2/3$. On the other hand, it was found that $\beta \approx -0.62$ for the first plastic event in a freshly prepared glass (which is not universal but appears to depend on the glass preparation), and is related to the behaviour at small values of the distribution of distances to the next plastic event, $\Delta\gamma$, which is generally expected to vary as $P(\Delta\gamma) \sim \Delta\gamma^{\eta}$. Analysing the first plastic event of a freshly quenched glass, Karmakar *et al.*[48] argued (and showed) that the full distribution agrees well with the Weibull distribution, leading to a relationship $\beta = -1/(1+\eta)$. Hentschel *et al.*[56] considered the statistics of intervals to yield in finite strain windows below the yielding point, and made the interesting observation that β values, from $\beta = -0.62$ at $\gamma = 0$ to $\beta = -2/3$ for strain γ above the yield strain, vary non-monotonically, approaching lower values in between (Figure 8(a)). Based on the analysis of a Fokker-Planck equation for the evolution of the Hessian eigenvalues, they argued that for finite strain, $\eta = 0$, and if the relation $\beta = -1/(1+\eta)$ remains valid in this regime, one would expect $\beta = -1$, which would correspond to having avalanches of unit size all the way to the yielding transition.

A similar analysis, on the basis of elasto-plastic models and scaling analysis was performed by Jin *et al.*,[57,58] who concluded that for a given value of η (θ in their notation), the mean avalanche size would scale as $< S > \sim N^{\theta/(1+\theta)}$. In this case, exponent values are greater than zero and vary continuously with stress (Figure 8(b)), as in Ref. 56. The non-zero values of the observed exponent are argued to imply system-spanning avalanches for all stresses (strains) below yielding. Similar observations are made by Ozawa *et al.*[59] (discussed further below) albeit with the variation that with annealing of the glasses, the exponent θ approaches zero in the pre-yield regime. In attempting to find a reconciliation, Lerner *et al.*[54] have suggested that system sizes may matter strongly in accessing the relevant regime of the distributions of distances to local yield, infeasible at the present.

We next discuss briefly some developments in discussing the nature of the yielding transition, before returning to the question of avalanches. A number of recent studies have addressed the nature of the yielding transition, with an emerging consensus that one must understand it as a spinodal limit. Based on a mean field infinite dimensional description of the glass transition for hard spheres, yielding has been described as a spinodal

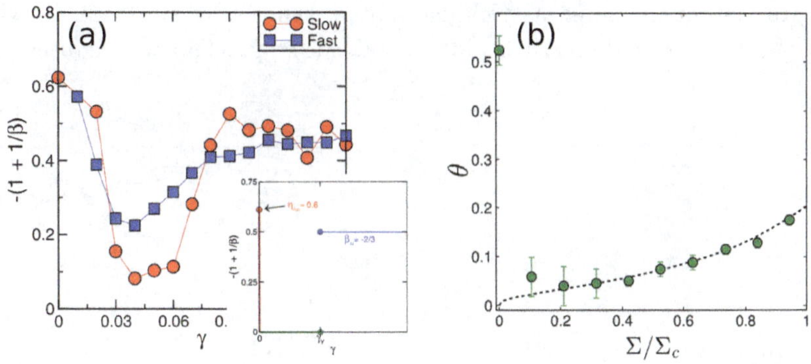

Fig. 8.   (a) The exponent $\eta = -(1 + \frac{1}{\beta})$ for different strain windows from an AQS simulation of a two-dimensional atomic glass for fast and slowly quenched samples. The inset shows the theoretical expectation based on the analysis in Ref. 56. (b) The exponent $\theta\,(\eta)$ for an elastoplastic model exhibits similar behaviour and has been interpreted to suggest criticality of avalanches at all strains below yielding.[58]

point,[60,61] and subsequent simulation studies have worked out some of the details of these calculations for finite dimensional systems.[62]   Jaiswal *et al*[52] explored the idea that yielding was associated with a delocalisation of glass configurations in configuration space. They considered an ensemble of initially nearby configurations, and computed the distribution of overlaps between them as a function of the imposed strain. This distribution is initially peaked at large overlaps, but becomes bimodal as the strain is increased. Yielding is identified as a *first order* transition, associated with the low overlap peak of the distribution becoming larger in weight (Figure 9). This formulation is familiar in replica theories of the glass transition, and in Ref. 53 and 63, a replica analysis of this problem is carried out, with a description of yielding as a spinodal instability, and is associated with the formation of shear bands. Application of random first order transition theory (RFOT) to the strength of glasses and shear banding has also been discussed by Wolynes and co-workers.[64,65] To understand the origins of shear bands, Dasgupta *et al*.[50,51] considered the energetics of arrangements of Eshelby-like inclusions in two dimensions, and showed that an organization of a large density of aligned inclusions along a line/plane become lower energy solutions than an inclusion-free solid, discontinuously at a yielding density. Such a calculation is informative, but does not provide a full analysis of under what conditions a shear banding instability may be observed in an amorphous solid. Shear bands form in some sheared glasses but not all. The evidence for the presence of avalanches that grow in size, or are system spanning, as yielding is approached is correspondingly varied. Some investigations that address and clarify these points are discussed below.

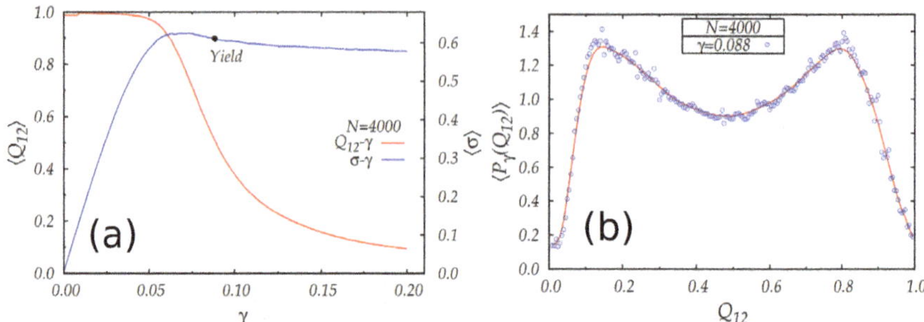

Fig. 9.   (a) Averaged stress *vs.* strain shown along with the overlap function $< Q_{12} >$ for the same strain γ Ref. 52. (b) The averaged probability distribution function $P_\gamma(Q_{12})$ at $\gamma_y = 0.088$. At the "yield" strain the distribution has two peaks with equal weights.

Many investigations of mechanical response have been conducted employing cyclic or oscillatory deformation of glasses, either under the AQS protocol, or at finite rates and/or temperatures.[47,66–73] Under cyclic deformation, with the strain amplitude as the control parameter, it is found that the glasses evolve in properties, but in distinct ways above and below a threshold amplitude that is identified as the yielding amplitude.[66,70,73] Glasses anneal (*i.e.*, reach lower energy states) with repeated cycles below the yielding strain, with the degree of annealing being the highest at the yielding transition (Figure 10). Glasses with different degrees of annealing behave very differently with uniform shear (poorly annealed glasses show a monotonic increase in stress *vs.* strain, whereas better annealed glasses show a stress overshoot), but considering the maximum stress as a function of the strain amplitude γ_{max}, a very different picture emerges. The stress strain curves show a discontinuous jump at the yielding amplitude. The nature of avalanches below and above yielding are qualitatively different. Avalanches below yielding are finite, and do not grow upon approach to the yielding point, or with system size, whereas avalanches above yielding are system spanning, and show the $N^{1/3}$ scaling previously described. Thus, the conclusion from this work is that yielding occurs as a discontinuous transition, and avalanche sizes below yielding show no indication of growth upon approaching the yielding point, nor with system size. The nature of the yielding transition in cyclic shear is strongly influenced by the annealing of the glasses that is observed. Cyclic deformation has been employed, in addition to studies of yielding, in understanding transitions from reversible to irreversible behaviour,[74,75] and memory formation,[76] which we do not discuss here, but details of which may be found in Ref.[76]

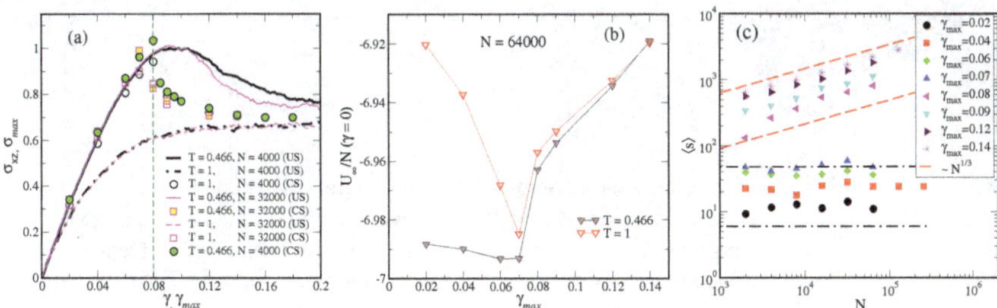

Fig. 10. (a) Averaged stress-strain curves for uniform shear and cyclic shear for a 3D model glass (Kob-Andersen BMLJ), instantly quenched from the liquid at two different temperatures, $T = 1.00$ and $T = 0.466$. Maximum stress σ_{max} versus γ_{max} is shown for cyclic shear. The vertical line at $g_{max} = 0.08$ indicates the sharp yielding transition observed. (b) Asymptotic energy per particle at $\gamma = 0$ plotted against the corresponding strain amplitude. Energies decrease with γ_{max} until the yield strain amplitude is reached ($\gamma_{max} \leq \gamma_y$), after which they increase with γ_{max}. (c) Mean avalanche size versus system size N shows no significant system size dependence for $\gamma_{max} \leq \gamma_y$ but a clear $N^{1/3}$ dependence above. The discontinuous change in behaviour marks the yielding transition, seen at $\gamma_{max} = 0.08$. Ref. 70.

For amplitudes larger than the yielding amplitude, the cyclically sheared glasses exhibit strain localisation, that is stable in characteristics for a given amplitude (the width, for example) but some interesting features are observed (Figure 11) – The energies of the particles within the shear band become the highest possible (*top of the landscape*, as discussed in the context of inherent structures (energy minima) in glass forming liquids[77–79]) while the energies outside the shear band continue to decrease with increase of strain amplitude, and display logarithmic relaxation in cycle number as also seen in liquid simulations[71,73] and for homogeneous glasses below the yielding amplitude. These observations are relevant for our discussion of shear banding and strain localisation below.

The implications of annealing on the nature of yielding has been considered by Ozawa *et al.*[59], who considered glasses that were annealed to different degrees, obtained by equilibrating a model liquid over a wide range of temperatures, and quenching them rapidly to zero temperature. They find that for low enough initial temperatures (or degree of annealing), the glasses exhibit yielding associated with a sudden drop in stress. Such sud-

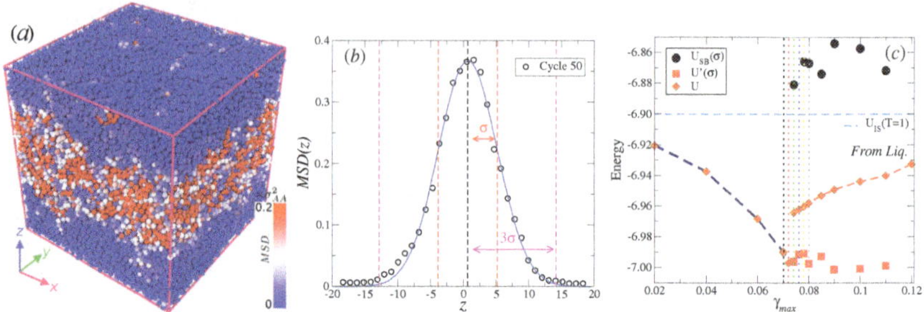

Fig. 11.   (a) A sample configuration of particles from the steady state, for a cyclially sheared glass, at strain amplitude $\gamma_{max} = 0.09$ above the yield strain. The color map is based on the displacement of particles between two consecutive cycles at zero strain. Highly mobile particles (particles with squared displacement $> 0.2\sigma_{AA}^2$) are colored in red, whereas particles in blue move considerably less. (b) Mean squared displacement between two cycles is shown as a function of the coordinate z in the shear direction, along with a Gaussian fit. (c) The mean energy changes discontinuously across the yielding transition. The subsystem outside the shear band is annealed for amplitudes even beyond the yield amplitude, whereas particles within the shear band have higher energies.[73]

den stress jumps are associated with strong strain localisation. But as the degree of annealing reduces, a more gradual transition with and without stress overshoots is found (Figure 12). These authors interpret their results in terms of the behaviour seen in the random field Ising model (providing evidence for such a comparison), and argue that the spectrum of yielding behaviours observed can be rationalised in terms of the zero temperature phase diagram of the RFIM, with a random critical point separating brittle and ductile yielding behaviours.

Ozawa *et al.*[59] also analyse a mean field elastoplastic model, paying particular attention to the initial distribution of stresses that captures the degree of annealing. In such a calculation also, the trend observed with annealing above is recaptured. A very similar investigation is also found in Ref. 80 who draw similar conclusions. It is interesting that the mean field model in[37] with *weakening* produces similar results. A somewhat intriguing feature of all these analyses is that they capture, at the mean field level, behaviour that is supposed to be associated with strong spatial correlations (strain localisation), and is also expected to arise from a specific form of the stress redistribution.

The presence of shear bands in flow or strain localisation have been studied extensively in the past, both experimentally and in computer simulations, in soft glassy materials and hard glasses, as described in Ref. 2,3 and

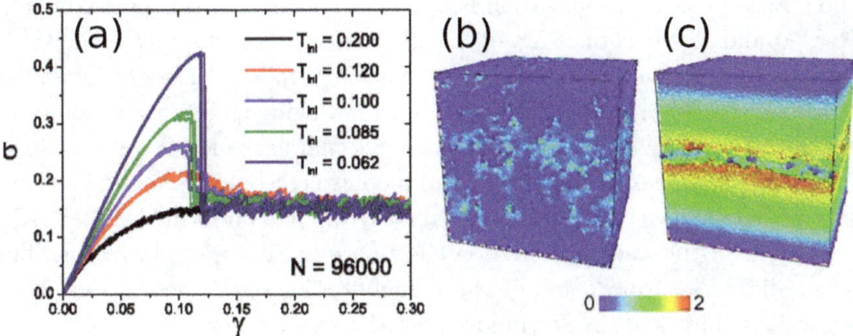

Fig. 12. (a) The stress-strain curves for samples prepared at several temperatures (T_{ini}), which determine their degree of annealing. The role of annealing in the response is evident, with larger discontinuous jumps with higher annealing, below a critical T_{ini}, above which the stress-strain curves are continuous. (b) & (c) Snapshots of the system at the yield point for cases with continuous (b) and discontinuous (c) stress-strain curves. The color code is based on non-affine displacements between the initial configuration and the configuration at the yield strain.[59]

81–88. In the context of hard glasses, strain localisation has been observed and discussed in the context of yielding, as already mentioned several times. Apart from the work mentioned above, there have been some observations, starting from,[26] that a correlation exists with the annealing of glasses and strain localisation during yield. However, a physical understanding of the connection is only now beginning to emerge. On the other hand, one may also inquire as to whether the strain localisation observed at yielding will persist or the shear band will grow and engulf the system as the strain increases. In the rheological context of mainly soft glasses, the persistence or otherwise of shear bands, and dependence on shear rates and other parameters, have been analysed by a variety of investigations.[20,42,87–93] The physical picture of the theoretical analyses invoke some form of competition between rates of *aging*, which is treated as a spontaneous process, and *rejuvenation*, which is often associated with shearing. Following Coussot and Ovarlez,[20] Martens *et al.*[42] analysed a related mechanism, where the relaxation time taken by an activated region to relax stresses plays a key role (refer to Figure 7). When this time scale τ_{el} gets long enough, permanent shear bands arise. In a mean field calculation, this can be seen in non-monotonic flow curves (Figure 13) and in a simulation, in the form of localisation of shear. In the latter case, shear banding is observed only when an Eshelby-type stress propagator is used but not when a short range propagator is used. This latter observation seems to support the idea that the nature of stress propagation is of key importance, as in the analysis in Ref. 50 and 51. But once again, mean field or other abstract models[42,91] appear to capture shear banding even without explicit consideration of the nature of stress propagation in space. Further, aging and rejuvenation rates are not available when one considers AQS dynamics, and where the effects of annealing appear in the rheological models is also not clear. In the context of cyclic shear discussed above, even though performed under AQS conditions, aging and rejuvenation effects are present and play a role, but a key difference from some of the model studies mentioned is that both aging and rejuvenation are induced by the shear deformation itself, heterogeneously in space. It would be interesting to have theoretical analyses that will encompass these scenarios.

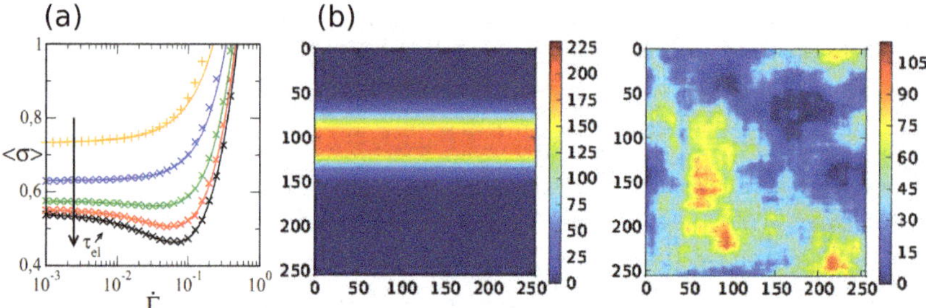

Fig. 13. (a) Flow curves from a mean field calculation, and (b) shear band formation or otherwise in an elasto-plastic model simulation employing an Eshelby-type vs short range propagator.[42]

4. Outlook

The discussion in the previous two sections clearly indicates that much progress has been made in recent times in answering some of the key questions related to yielding in amorphous solids, concerning the nature of elementary processes, their interactions, the nature of the yielding transition, the dependence of the mechanical response on the preparation history and structural properties of the amorphous solids, and the presence or otherwise of strain localisation and shear banding during yielding and flow. There is a convergence of different approaches that at the outset are not very related (e.g., elasto-plastic models vs. glass transition theory), and increasingly, a convergence of what the nature of the yielding transition is. The degree of annealing of glasses has been seen as a key factor, and for well annealed glasses, a sudden, discontinuous yielding transition without precursors appears to be the commonly observed behaviour. Annealing is expected to change the structure, but whether one can structurally identify locations where plastic deformation events can take place is an interesting subject of ongoing investigation. A discontinuous yielding transition resulting from a spinodal instability have been arrived at in a variety of ways. But the nature of the spinodal (whether critical, not, and what kind of divergences may be expected, if any) will continue to be investigated in the near future, but the questions are sharper than they were before. There isn't yet fully clarity or consensus on the question of the nature of avalanches before yielding, in particular whether they would be system spanning or not. The origins, nature and role of aging and rejuvenation effects is another area where open questions remain to be answered. As described, cyclic shear offers an interesting avenue to investigate key aspects of yielding but current modeling approaches are limited in this regard. For example, elastoplastic models are rule based *automata*, which do not easily permit the notion of reversing the strain. Energy or free energy based formulations, such as those of Jagla[89,90,94] are very interesting in this context to explore as starting points. At finite shear rates and finite temperatures, the location and nature of yielding can change substantially (e.g. Ref. 73). Systematically revisiting those features will be important after the athermal limit is well understood.

After this survey of theoretical approaches, it may well be appropriate to end by returning to practical matters, and ask if these ideas will in anyway "help make better glass", paraphrasing P. W. Anderson.[95] Indeed, a key feature of the behaviour of well annealed glasses is brittle failure, which is not a desirable feature, while other properties of such annealed glasses are. Are there ideas for how such failure can be avoided? Based on the understanding of the nature of the yielding instability, various investigators have proposed the inclusion of pinning centers and micro-alloying as a way of curtailing run away instabilities[96–102] and thereby make annealed glasses more resistant to brittle failure, and more ductile. More such insights may arise from the development of a unified view of yielding and flow of amorphous solids in the coming years.

References

1. C. A. Schuh, T. C. Hufnagel, and U. Ramamurty, Mechanical behavior of amorphous alloys, *Acta Materialia.* **55**(12), 4067–4109 (2007).

2. D. Bonn, M. M. Denn, L. Berthier, T. Divoux, and S. Manneville, Yield stress materials in soft condensed matter, *Rev. Mod. Phys.* **89**(3), 035005 (2017).

3. A. Nicolas, E. E. Ferrero, K. Martens, and J.-L. Barrat, Deformation and flow of amorphous solids: A review of mesoscale elastoplastic models, *Rev. Mod. Phys.* **90**(4), 045006 (2018).

4. T. H. Ly, J. Zhao, M. O. Cichocka, L.-J. Li, and Y. H. Lee, Dynamical observations on the crack tip zone and stress corrosion of two-dimensional mos 2, *Nature Commun.* **8**, 14116 (2017).

5. D. C. Hofmann, J.-Y. Suh, A. Wiest, G. Duan, M.-L. Lind, M. D. Demetriou, and W. L. Johnson, Designing metallic glass matrix composites with high toughness and tensile ductility, *Nature.* **451**(7182), 1085 (2008).

6. P. Schall, D. A. Weitz, and F. Spaepen, Structural rearrangements that govern flow in colloidal glasses, *Science.* **318**(5858), 1895–1899 (2007).

7. A. Le Bouil, A. Amon, S. McNamara, and J. Crassous, Emergence of co-operativity in plasticity of soft glassy materials, *Phys. Rev. Lett.* **112**(24), 246001 (2014).

8. A.-L. Biance, S. Cohen-Addad, and R. Höhler, Topological transition dynamics in a strained bubble cluster, *Soft Matter.* **5**(23), 4672–4679 (2009).

9. Q. Huang, N. J. Alvarez, A. Shabbir, and O. Hassager, Multiple cracks propagate simultaneously in polymer liquids in tension, *Phys. Rev. Lett.* **117**(8), 087801 (2016).

10. L. Bocquet, A. Colin, and A. Ajdari, Kinetic theory of plastic flow in soft glassy materials, *Phys. Rev. Lett.* **103**(3), 036001 (2009).

11. A. Argon, Plastic deformation in metallic glasses, *Acta Metallurgica.* **27**(1), 47–58 (1979).

12. C. A. Schuh, T. C. Hufnagel, and U. Ramamurty, Mechanical behavior of amorphous alloys, *Acta Materialia.* **55**(12), 4067–4109 (2007).

13. P. Schall and M. van Hecke, Shear bands in matter with granularity, *Ann. Rev. Fluid Mech.* **42** (2010).

14. M. L. Falk and J. S. Langer, Deformation and failure of amorphous, solidlike materials, *Annu. Rev. Condens. Matter Phys.* **2**(1), 353–373 (2011).

15. J. L. Barrat and A. Lemaître. Heterogeneities in amorphous systems under shear. In eds. L. Berthier, G. Biroli, J. P. Bouchaud, L. Cipelletti, and W. v. Saarloos, *Dynamical Heterogeneities in Glasses, Colloids, and Granular Media*, chapter 8, pp. 264–297. Oxford Science Publications, Oxford (2011).

16. J. P. Sethna, M. K. Bierbaum, K. A. Dahmen, C. P. Goodrich, J. R. Greer, L. X. Hayden, J. P. Kent-Dobias, E. D. Lee, D. B. Liarte, X. Ni, et al., Deformation of crystals: Connections with statistical physics, *Ann. Rev. Mater. Res.* **47**, 217–246 (2017).

17. F. Sausset, G. Biroli, and J. Kurchan, Do solids flow?, *J. Statist. Phys.* **140**(4), 718–727 (2010).

18. H. Yoshino and M. Mézard, Emergence of rigidity at the structural glass transition: a first-principles computation, *Phys. Rev. Lett.* **105**(1), 015504 (2010).

19. P. Nath, S. Ganguly, J. Horbach, P. Sollich, S. Karmakar, and S. Sengupta, On the existence of thermodynamically stable rigid solids, *Proc. Nat. Acad. Sci.* **115**(19), E4322–E4329 (2018).

20. P. Coussot and G. Ovarlez, Physical origin of shear-banding in jammed systems, *The European Phys. J. E.* **33**(3), 183–188 (2010).

21. W. H. Herschel and R. Bulkley, Konsistenzmessungen von gummibenzollösungen, *Colloid & Polymer Science.* **39**(4), 291–300 (1926).

22. M. L. Falk and J. S. Langer, Dynamics of viscoplastic deformation in amorphous solids, *Phys. Rev. E.* **57**(6), 7192–7205 (1998).

23. C. E. Maloney and A. Lemaître, Amorphous systems in athermal, quasistatic shear, *Phys. Rev. E.* **74**(1), 16118–16118 (2006).

24. V. Chikkadi and P. Schall, Nonaffine measures of particle displacements in sheared colloidal glasses, *Phys. Rev. E.* **85**(3), 031402 (2012).

25. K. Jensen, D. A. Weitz, and F. Spaepen, Local shear transformations in deformed and quiescent hard-sphere colloidal glasses, *Phys. Rev. E.* **90**(4), 042305 (2014).

26. Y. Shi, M. B. Katz, H. Li, and M. L. Falk, Evaluation of the disorder temperature and free-volume formalisms via simulations of shear banding in amorphous solids, *Phys. Rev. Lett.* **98**(18), 185505 (2007).

27. J. Ding, S. Patinet, M. L. Falk, Y. Cheng, and E. Ma, Soft spots and their structural signature in a metallic glass, *Proc. Nat. Acad. Sci.* **111**(39), 14052–14056 (2014).

28. M. L. Manning and A. J. Liu, Vibrational modes identify soft spots in a sheared disordered packing, *Phys. Rev. Lett.* **107**(10), 108302 (2011).

29. S. Patinet, D. Vandembroucq, and M. L. Falk, Connecting local yield stresses with plastic activity in amorphous solids, *Phys. Rev. Lett.* **117**(4), 045501 (2016).

30. B. Tyukodi, S. Patinet, S. Roux, and D. Vandembroucq, From depinning transition to plastic yielding of amorphous media: A soft-modes perspective, *Phys. Rev. E.* **93**(6), 63005 (2016).

31. E. D. Cubuk, S. S. Schoenholz, J. M. Rieser, B. D. Malone, J. Rottler, D. J. Durian, E. Kaxiras, and A. J. Liu, Identifying structural flow defects in disordered solids using machine-learning methods, *Phys. Rev. Lett.* **114** (10), 108001 (2015).

32. E. D. Cubuk, S. S. Schoenholz, E. Kaxiras, and A. J. Liu, Structural properties of defects in glassy liquids, *The J. Phys. Chem. B.* **120**(26), 6139–6146 (2016).

33. M. L. Manning, J. S. Langer, and J. Carlson, Strain localization in a shear transformation zone model for amorphous solids, *Phys. Rev. E.* **76** (5),056106 (2007).

34. J. Ashwin, E. Bouchbinder, and I. Procaccia, Cooling-rate dependence of the shear modulus of amorphous solids, *Phys. Rev. E.* **87**(4), 042310 (2013).

35. G. Picard, A. Ajdari, F. Lequeux, and L. Bocquet, Elastic consequences of a single plastic event: A step towards the microscopic modeling of the flow of yield stress fluids, *European Phys. J. E.* **15**(4), 371–381 (2004).

36. J. D. Eshelby, The determination of the elastic field of an ellipsoidal inclusion, and related problems, *Proc. Royal Soc. London. Series A. Mathematical and Physical Sciences.* **241**(1226), 376–396 (1957).

37. K. A. Dahmen, Y. Ben-Zion, and J. T. Uhl, Micromechanical model for deformation in solids with universal predictions for stress-strain curves and slip avalanches, *Phys. Rev. Lett.* **102**(17), 175501 (2009).

38. P. Hébraud and F. Lequeux, Mode-coupling theory for the pasty rheology of soft glassy materials, *Phys. Rev. Lett.* **81**(14), 2934–2937 (1998).

39. M. Talamali, V. Petäjä, D. Vandembroucq, and S. Roux, Strain localization and anisotropic correlations in a mesoscopic model of amorphous plasticity, *Comptes Rendus Mecanique.* **340**(4-5), 275–288 (2012).

40. C. Liu, E. E. Ferrero, F. Puosi, J.-L. Barrat, and K. Martens, Driving rate dependence of avalanche statistics and shapes at the yielding transition, *Phys. Rev. Lett.* **116**(6), 65501 (2016).

41. K. M. Salerno and M. O. Robbins, Effect of inertia on sheared disordered solids: Critical scaling of avalanches in two and three dimensions, *Phys. Rev. E.* **88**(6), 062206 (2013).

42. K. Martens, L. Bocquet, and J.-L. Barrat, Spontaneous formation of permanent shear bands in a mesoscopic model of flowing disordered matter, *Soft Matter.* **8**(15), 4197–4205 (2012).

43. K. A. Dahmen, Y. Ben-Zion, and J. T. Uhl, A simple analytic theory for the statistics of avalanches in sheared granular materials, *Nature Phys.* **7** (7), 554 (2011).

44. J. T. Uhl, S. Pathak, D. Schorlemmer, X. Liu, R. Swindeman, B. A. Brinkman, M. LeBlanc, G. Tsekenis, N. Friedman, R. Behringer, et al., Universal quake statistics: from compressed nanocrystals to earthquakes, *Scient. Rep.* **5**, 16493 (2015).

45. S. K. Nandi, G. Biroli, and G. Tarjus, Spinodals with disorder: From avalanches in random magnets to glassy dynamics, *Phys. Rev. Lett.* **116** (14), 145701 (2016).

46. D. L. Malandro and D. J. Lacks, Relationships of shear-induced changes in the potential energy landscape to the mechanical properties of ductile glasses, *The J. Chem. Phys.* **110**(9), 4593–4601 (1999).

47. D. J. Lacks and M. J. Osborne, Energy landscape picture of overaging and rejuvenation in a sheared glass, *Phys. Rev. Lett.* **93**(25), 255501 (2004).

48. S. Karmakar, E. Lerner, and I. Procaccia, Statistical physics of the yielding transition in amorphous solids, *Phys. Rev. E.* **82**(5), 55103 (2010).

49. H. Hentschel, S. Karmakar, E. Lerner, and I. Procaccia, Do athermal amorphous solids exist? *Phys. Rev. E.* **83**(6), 061101 (2011).

50. R. Dasgupta, H. G. E. Hentschel, and I. Procaccia, Microscopic mechanism of shear bands in amorphous solids, *Phys. Rev. Lett.* **109**(25), 255502 (2012).

51. R. Dasgupta, O. Gendelman, P. Mishra, I. Procaccia, and C. A. Shor, Shear localization in three-dimensional amorphous solids, *Phys. Rev. E.* **88**(3), 032401 (2013).

52. P. K. Jaiswal, I. Procaccia, C. Rainone, and M. Singh, Mechanical yield in amorphous solids: A first-order phase transition, *Phys. Rev. Lett.* **116**(8),

85501–85501 (2016).

53. G. Parisi, I. Procaccia, C. Rainone, and M. Singh, Shear bands as manifestation of a criticality in yielding amorphous solids, *Proc. Nat. Acad. Sci.* **114**(22), 5577–5582 (2017).

54. E. Lerner, I. Procaccia, C. Rainone, and M. Singh, Protocol dependence of plasticity in ultrastable amorphous solids, *Phys. Rev. E.* **98**(6), 063001 (2018).

55. E. Lerner and I. Procaccia, Locality and nonlocality in elastoplastic responses of amorphous solids, *Phys. Rev. E.* **79**(6), 066109 (2009).

56. H. Hentschel, P. K. Jaiswal, I. Procaccia, and S. Sastry, Stochastic approach to plasticity and yield in amorphous solids, *Phys. Rev. E.* **92**(6), 062302 (2015).

57. J. Lin, T. Gueudré, A. Rosso, and M. Wyart, Criticality in the approach to failure in amorphous solids, *Phys. Rev. Lett.* **115**(16), 168001 (2015).

58. J. Lin and M. Wyart, Mean-field description of plastic flow in amorphous solids, *Phys. Rev. X.* **6**(1), 011005 (2016).

59. M. Ozawa, L. Berthier, G. Biroli, A. Rosso, and G. Tarjus, Random critical point separates brittle and ductile yielding transitions in amorphous materials, *Proc. Nat. Acad. Sci.* **115**(26), 6656–6661 (2018).

60. C. Rainone, P. Urbani, H. Yoshino, and F. Zamponi, Following the evolution of hard sphere glasses in infinite dimensions under external perturbations: Compression and shear strain, *Phys. Rev. Lett.* **114**(1), 015701 (2015).

61. P. Urbani and F. Zamponi, Shear yielding and shear jamming of dense hard sphere glasses, *Phys. Rev. Lett.* **118**(3), 38001 (2017).

62. Y. Jin, P. Urbani, F. Zamponi, and H. Yoshino, A stability-reversibility map unifies elasticity, plasticity, yielding and jamming in hard sphere glasses, *Sci. Adv.* **4**(12) (2018).

63. I. Procaccia, C. Rainone, and M. Singh, Mechanical failure in amorphous solids: Scale-free spinodal criticality, *Phys. Rev. E.* **96**(3), 032907 (2017).

64. A. Wisitsorasak and P. G. Wolynes, On the strength of glasses, *Proc. Nat. Acad. Sci.* **109**, 16068–16072 (2012).

65. A. Wisitsorasak and P. G. Wolynes, Dynamical theory of shear bands in structural glasses, *Proc. Nat. Acad. Sci.* **114**(6), 1287–1292 (2017).

66. D. Fiocco, G. Foffi, and S. Sastry, Oscillatory athermal quasistatic deformation of a model glass, *Phys. Rev. E.* **88**, 020301 (2013).

67. I. Regev, T. Lookman, and C. Reichhardt, Onset of irreversibility and chaos in amorphous solids under periodic shear, *Phys. Rev. E.* **88**(6), 062401 (2013).

68. N. V. Priezjev, Heterogeneous relaxation dynamics in amorphous materials under cyclic loading, *Phys. Rev. E.* **87**(5), 052302 (2013).

69. I. Regev, J. Weber, C. Reichhardt, K. A. Dahmen, and T. Lookman, Reversibility and criticality in amorphous solids, *Nature Commun.* **6**(1), 8805–8805 (2015).

70. P. Leishangthem, A. D. S. Parmar, and S. Sastry, The yielding transition in amorphous solids under oscillatory shear deformation., *Nature Commun.* **8**, 14653 (2017).

71. P. Das, A. D. Parmar, and S. Sastry, Annealing glasses by cyclic shear deformation, *arXiv preprint arXiv:1805.12476* (2018).

72. N. V. Priezjev, The yielding transition in periodically sheared binary glasses at finite temperature, *Comput. Mater. Sci.* **150**, 162–168 (2018).

73. A. D. Parmar, S. Kumar, and S. Sastry, Strain localization above the yielding point in cyclically deformed glasses, *Phys. Rev. X.* **9**(2), 021018 (2019).

74. D. J. Pine, J. P. Gollub, J. F. Brady, and A. M. Leshansky, Chaos and threshold for irreversibility in sheared suspensions, *Nature.* **438**(7070), 997 (2005).

75. L. Corte, P. M. Chaikin, J. P. Gollub, and D. J. Pine, Random organization in periodically driven systems, *Nature Phys.* **4**(5), 420 (2008).

76. N. C. Keim, J. D. Paulsen, Z. Zeravcic, S. Sastry, and S. R. Nagel, Memory formation in matter, *Rev. Mod. Phys.* **91**, 035002 (2019).

77. S. Sastry, P. G. Debenedetti, and F. H. Stillinger, Signatures of distinct dynamical regimes in the energy landscape of a glass-forming liquid, *Nature.* **393**(6685), 554 (1998).

78. A. Angell, Thermodynamics: liquid landscape, *Nature.* **393**(6685), 521 (1998).

79. M. Utz, P. G. Debenedetti, and F. H. Stillinger, Atomistic simulation of aging and rejuvenation in glasses, *Phys. Rev. Lett.* **84**(7), 1471 (2000).

80. M. Popović, T. W. de Geus, and M. Wyart, Elastoplastic description of sudden failure in athermal amorphous materials during quasistatic loading, *Phys. Rev. E.* **98**(6), 040901 (2018).

81. T. Divoux, D. Tamarii, C. Barentin, and S. Manneville, Transient shear banding in a simple yield stress fluid, *Phys. Rev. Lett.* **104**(20), 208301 (2010).

82. R. L. Moorcroft, M. E. Cates, and S. M. Fielding, Age-dependent transient shear banding in soft glasses, *Phys. Rev. Lett.* **106**(5), 055502 (2011).

83. R. L. Moorcroft and S. M. Fielding, Criteria for shear banding in time-dependent flows of complex fluids, *Phys. Rev. Lett.* **110**(8), 086001 (2013).

84. S. M. Fielding, Shear banding in soft glassy materials, *Rep. Prog. in Phys.* **77**(10), 102601 (2014).

85. T. Divoux, M. A. Fardin, S. Manneville, and S. Lerouge, Shear banding of complex fluids, *Ann. Rev. Fluid Mech.* **48**, 81–103 (2016).

86. P. Chaudhuri and J. Horbach, Onset of flow in a confined colloidal glass under an imposed shear stress, *Phys. Rev. E.* **88**(4), 040301 (2013).

87. G. P. Shrivastav, P. Chaudhuri, and J. Horbach, Yielding of glass under shear: A directed percolation transition precedes shear-band formation, *Phys. Rev. E.* **94**(4), 042605 (2016).

88. V. V. Vasisht, G. Roberts, and E. Del Gado, Emergence and persistence of flow inhomogeneities in the yielding and fluidization of dense soft solids, *arXiv preprint arXiv:1709.08717* (2017).

89. E. Jagla, Strain localization driven by structural relaxation in sheared amorphous solids, *Phys. Rev. E.* **76**(4), 046119 (2007).

90. E. Jagla, Shear band dynamics from a mesoscopic modeling of plasticity, *J. Statis. Mech.: Theory and Experiment.* **2010**(12), P12025 (2010).

91. R. Radhakrishnan and S. M. Fielding, Shear banding of soft glassy materials in large amplitude oscillatory shear, *Phys. Rev. Lett.* **117**(18) (2016).
92. D. D. Alix-Williams and M. L. Falk, Shear band broadening in simulated glasses, *Phys. Rev. E.* **98**(5), 053002 (2018).
93. V. V. Vasisht, M. L. Goff, K. Martens, and J.-L. Barrat, Permanent shear localization in dense disordered materials due to microscopic inertia, *arXiv preprint arXiv:1812.03948* (2018).
94. E. Jagla, Different universality classes at the yielding transition of amorphous systems, *Phys. Rev. E.* **96**(2), 023006 (2017).
95. D. L. Anderson, Through the glass lightly, *Science.* **267**(5204), 1618–1618 (1995).
96. H. G. E. Hentschel, M. Moshe, I. Procaccia, and K. Samwer, Microalloying and the mechanical properties of amorphous solids, *Philos. Mag.* **96**(14), 1399–1419 (2016).
97. G. R. Garrett, M. D. Demetriou, J. Chen, and W. L. Johnson, Effect of microalloying on the toughness of metallic glasses, *Appl. Phys. Lett.* **101**(24), 241913 (2012).
98. P. Tsai, K. Hsu, J. Ke, H. Lin, J. Jang, and J. Huang, Microalloying effect of si on mechanical properties of ti based bulk metallic glass, *Mater. Technol.* **30**(sup3), 161–165 (2015).
99. S. González, Role of minor additions on metallic glasses and composites, *J. Mater. Res.* **31**(1), 76–87 (2016).
100. L. Jiang, J. Li, P. Cheng, G. Liu, R. Wang, B. Chen, J. Zhang, J. Sun, M. Yang, and G. Yang, Microalloying ultrafine grained al alloys with enhanced ductility, *Scientific Rep.* **4**, 3605 (2014).
101. B. P. Bhowmik, P. Chaudhuri, and S. Karmakar, Effect of pinning on the yielding transition of amorphous solids, *arXiv preprint arXiv:1808.09723* (2018).
102. B. P. Bhowmik, S. Karmakar, I. Procaccia, and C. Rainone, Particle pinning suppresses spinodal criticality in the shear banding instability, *arXiv preprint arXiv:1903.03020* (2019).

Index

9 789811 211324